D0875493

Methods in Enzymology

Volume 79
INTERFERONS
Part B

METHODS IN ENZYMOLOGY

EDITORS-IN-CHIEF

Sidney P. Colowick   Nathan O. Kaplan

Methods in Enzymology

Volume 79

Interferons

Part B

EDITED BY

Sidney Pestka
ROCHE INSTITUTE OF MOLECULAR BIOLOGY
NUTLEY, NEW JERSEY

1981

ACADEMIC PRESS

A Subsidiary of Harcourt Brace Jovanovich, Publishers

New York   London   Toronto   Sydney   San Francisco

ACADEMIC PRESS, INC.
111 Fifth Avenue, New York, New York 10003

United Kingdom Edition published by
ACADEMIC PRESS, INC. (LONDON) LTD.
24/28 Oval Road, London NW1 7DX

Library of Congress Cataloging in Publication Data
Main entry under title:

Interferons.

(Methods in enzymology; v. 78-79)
Includes bibliographical references and indexes.
1. Interferon. I. Pestka, Sidney. II. Series.
[DNLM: 1. Interferon--Analysis. W1 ME9615K v. 79
etc. / QW 800 I619]
QP601.M49 vol. 78-79   599.02'95   81-17545
ISBN 0-12-181979-5 (v. 79)   AACR2

PRINTED IN THE UNITED STATES OF AMERICA

81 82 83 84   9 8 7 6 5 4 3 2 1

*To Robert, Sharon, and Steven
in pursuit of excellence*

Table of Contents

Section I. Techniques for Chromatography and Analysis of Interferons

Section II. Isolation and Translation of Interferon Messenger RNA

Section III. Methods to Study Mode of Interferon Action

A. General Methodology and Considerations

B. Phosphorylation and cAMP in Interferon Action

D. Effect of Interferon on Protein Synthesis and Other Cellular Processes

Section IV. Methods to Study Interferon Activity at the Cellular Level

Section V. Effects of Interferon on the Cell Membrane, Cell Surface, and Cytoskeleton

Section VI. Relationships of Interferon to the Immune System

Section VII. Somatic Cell Genetics of Interferons

Section VIII. Preparation and Assay of Antibodies against Interferons

Section IX. Cloning of the Interferon Genes and Complementary DNAs

Contributors to Volume 79

Article numbers are in parentheses following the names of contributors.
Affiliations listed are current.

CORRADO BAGLIONI (31, 32), *Department of Biological Sciences, State University of New York at Albany, Albany, New York 12222*

DAVID BARNES (46), *Department of Biological Sciences, University of Pittsburgh, Pittsburgh, Pennsylvania 15260*

KURT BERG (58), *Institute for Medical Microbiology, Bartholin Building, University of Aarhus, DK-8000 Aarhus C, Denmark*

SHELBY L. BERGER (9), *Laboratory of Pathophysiology, National Cancer Institute, National Institutes of Health, Bethesda, Maryland 20205*

CONNIE S. BIRKENMEIER (9), *The Jackson Laboratory, Bar Harbor, Maine 04609*

PAUL H. BLACK (41), *Department of Microbiology, Boston University School of Medicine, and The Hubert H. Humphrey Cancer Research Center, Boston University, Boston, Massachusetts 02118*

J. EDWIN BLALOCK (43), *Department of Microbiology, University of Texas Medical Branch, Galveston, Texas 77550*

ROGER J. BOOTH (60), *Department of Medicine, University of Auckland School of Medicine, Private Bag, Auckland, New Zealand*

LARRY BRINK (4, 5), *Department of Physiological Chemistry and Pharmacology, Roche Institute of Molecular Biology, Nutley, New Jersey 07110*

R. E. BROWN (24, 25, 26, 27), *Imperial Cancer Research Fund Laboratories, London WC2A 3PX, England*

DEREK C. BURKE (69), *Department of Biological Sciences, University of Warwick, Coventry CV4 7AL, England*

P. J. CAYLEY (27), *Imperial Cancer Research Fund Laboratories, London WC2A 3PX, England*

MICHAEL A. CHIRIGOS (49), *Laboratory of Chemical Pharmacology, Division of Cancer Treatment, National Cancer Institute, National Institutes of Health, Bethesda, Maryland 20205*

JEAN CONTENT (18), *Institut Pasteur du Brabant, B-1040 Brussels, Belgium*

ALAN F. COWMAN (83), *The Walter and Eliza Hall Institute of Medical Research, Post Office Royal Melbourne Hospital, Victoria 3050, Australia*

BARBARA DALTON (70, 74), *Department of Microbiology, The Medical College of Pennsylvania, Philadelphia, Pennsylvania 19129*

ERIK DE CLERCQ (18), *Rega Institute for Medical Research, Minderbroedersstraat 10, B-3000 Leuven, Belgium*

EDWARD DE MAEYER (15, 59), *Institut Curie-Biologie, Université de Paris-Sud, 91405 Orsay, France*

LUK DE WIT (18), *Institut Pasteur du Brabant, B-1040 Brussels, Belgium*

CARL W. DIEFFENBACH (16), *Division of Biophysics, The Johns Hopkins University, School of Hygiene and Public Health, Baltimore, Maryland 21205*

PAUL DOETSCH (33), *Department of Biochemistry, Temple University School of Medicine, Philadelphia, Pennsylvania 19140*

DAVID A. EPSTEIN (23), *Laboratory of Experimental Pathology, National Institute of Arthritis, Diabetes and Digestive and Kidney Diseases, National Institutes of Health, Bethesda, Maryland 20205*

MARIAN EVINGER (45), *Columbia University Medical Center, 630 West 168 Street, New York, New York 10032*

ERNESTO FALCOFF (62), *INSERM U196, Section de Biologie, Institut Curie, 75005 Paris, France*

REBECCA FALCOFF (34, 36), *Fondation Curie-Institut du Radium, 26 Rue d'Ulm, Paris 5, France*

W. R. FLEISCHMANN, JR. (52, 53), *Department of Microbiology, University of Texas Medical Branch, Galveston, Texas 77550*

W. A. FLEMING* (63), *Department of Microbiology and Immunobiology, The Queen's University of Belfast, Grosvenor Road, Belfast, BT12 6BN, Antrium, Northern Ireland*

WOLF H. FRIDMAN (62), *Laboratoire d'Immunologie Clinique, Section Medicale, Institut Curie, 25 rue d'Ulm, 75005 Paris, France, and Laboratoire d'Immunologie Cellulaire, IRSC, 94800 Villejuif, France*

ROBERT M. FRIEDMAN (21, 23, 29, 38, 42, 54, 55), *Laboratory of Experimental Pathology, National Institute of Arthritis, Diabetes and Digestive and Kidney Diseases, National Institutes of Health, Bethesda, Maryland 20205*

C. S. GILBERT (26), *Imperial Cancer Research Fund Laboratories, London WC2A 3PX, England*

R. R. GOLGHER (26), *Departamento de Microbiologia, Instituto de Ciencias Biologicas, Universidade Federal de Minas Gerais, Caixa Postal 2486, 30,000 Belo Horizonte MG, Brasil*

JAMES J. GREENE (16), *Division of Biophysics, The Johns Hopkins University, School of Hygiene and Public Health, Baltimore, Maryland 21205*

MITCHELL GROSS (81), *Roche Institute of Molecular Biology, Nutley, New Jersey 07110*

JORDAN U. GUTTERMAN (64), *Developmental Therapeutics, The University of Texas M.D. Anderson Hospital and Tumor Institute, Houston, Texas 77030*

EDWARD A. HAVELL (72, 73), *Trudeau Institute, Inc., Saranac Lake, New York 12983*

RONALD B. HERBERMAN (57), *Laboratory of Immunodiagnosis, National Cancer Insti-*

tute, National Institutes of Health, Bethesda, Maryland 20205

IVER HERON (58), *Institute for Medical Microbiology, Bartholin Building, University of Aarhus, D-8000 Aarhus C, Denmark*

MICHAEL J. M. HITCHCOCK (9), *Experimental Therapeutics, Bristol Laboratories, Syracuse, New York 13201*

DONNA S. HOBBS (76), *Roche Institute of Molecular Biology, Nutley, New Jersey 07110*

YUN-TE HOU (11), *Department of Antivirals, Institute of Virology, Chinese Academy of Medical Sciences, 100 Ying Xing Jie, Xuan Wu Qu, Peking (Beijing) 100052, China*

A. G. HOVANESSIAN (24, 35), *Unité d'Oncologie Virale, Institut Pasteur, 25 rue du Dr. Roux, Paris 15, France*

JIRO IMAI (30), *Laboratory of Chemistry, National Institute of Arthritis, Diabetes and Digestive and Kidney Diseases, National Institutes of Health, Bethesda, Maryland 20205*

ANNA D. INGLOT (75), *Polish Academy of Sciences, Ludwik Hirszfeld Institute of Immunology and Experimental Therapy, Ul Czerska 12, 53-114 Wroclaw, Poland*

HOWARD M. JOHNSON (61), *Department of Microbiology, University of Texas Medical Branch, Galveston, Texas 77550*

MARGARET I. JOHNSTON (18, 29), *Laboratory of Chemistry, National Institute of Arthritis, Diabetes and Digestive and Kidney Diseases, National Institutes of Health, Bethesda, Maryland 20205*

WOLFGANG K. JOKLIK (39, 40), *Department of Microbiology and Immunology, Duke University Medical Center, Durham, North Carolina 27710*

NATHAN O. KAPLAN (50), *Department of Chemistry and the Cancer Center, University of California, San Diego, La Jolla, California 92093*

DAVID J. KEMP (83), *The Walter and Eliza Hall Institute of Medical Research, Post Office Royal Melbourne Hospital, Victoria 3050, Australia*

* Deceased.

IAN M. KERR (24, 25, 26, 27, 28), *Imperial Cancer Research Fund Laboratories, London WC2A 3PX, England*

M. KILLEN (63), *Department of Microbiology and Immunobiology, The Queen's University of Belfast, Grosvenor Road, Belfast, BT12 6BN, Antrium, Northern Ireland*

A. KIMCHI (20), *Department of Virology, Weizmann Institute of Science, Rehovot 76100, Israel*

M. KNIGHT (24, 28), *Imperial Cancer Research Fund Laboratories, London WC2A 3PX, England*

HSIANG-FU KUNG (1), *Department of Molecular Genetics, Hoffmann-La Roche Inc., Nutley, New Jersey 07110*

T. KUWATA (68), *Department of Microbiology, School of Medicine, Chiba University, Chiba 280, Japan*

P. M. LAD (21), *Laboratory of Nutrition and Endocrinology, National Institute of Arthritis, Diabetes and Digestive and Kidney Diseases, National Institutes of Health, Bethesda, Maryland 20205*

FRANK R. LANDSBERGER (56), *The Rockefeller University, New York, New York 10021*

E. LEE (42), *Department of Microbiology, Georgetown University Medical and Dental Schools, Washington D.C. 20007*

E. J. LEFKOWITZ (53), *Department of Microbiology, University of Texas Medical Branch, Galveston, Texas 77550*

P. LENGYEL (19), *Department of Molecular Biophysics and Biochemistry, Yale University, New Haven, Connecticut 06511*

WARREN P. LEVY (6), *Unigene Laboratories, Inc., Nutley, New Jersey 07110*

JOHN A. LEWIS (34, 36), *Department of Anatomy and Cell Biology, Downstate Medical Center, SUNY, Brooklyn, New York 11203*

PIN-FANG LIN (67), *Department of Biology, Yale University, New Haven, Connecticut 06511*

CHIH-PING LIU (13), *Department of Genetics, University of Wisconsin, Madison, Wisconsin 53706*

RUSSELL MCCANDLISS (8, 10, 78, 82), *Roche Institute of Molecular Biology, Nutley, New Jersey 07110*

T. A. MCNEILL (63), *Department of Microbiology and Immunobiology, The Queen's University of Belfast, Grosvenor Road, Belfast, BT12 6BN, Antrium, Northern Ireland*

SHUICHIRO MAEDA (79, 80, 81), *Roche Institute of Molecular Biology, Nutley, New Jersey 07110*

R. K. MAHESHWARI (21, 38, 54), *Howard University Cancer Center, Washington D.C. 20060, and Laboratory of Experimental Pathology, National Institute of Arthritis, Diabetes and Digestive and Kidney Diseases, National Institutes of Health, Bethesda, Maryland 20205*

JOHN MARBROOK (60), *Department of Cell Biology, University of Auckland, Private Bag, Auckland, New Zealand*

PATRICIA A. MARONEY (32), *Department of Biological Sciences, State University of New York at Albany, Albany, New York 12222*

E. M. MARTIN (24, 35), *School of Biological Sciences, Flinders University, South Australia 5042, Australia*

HIDEO MASUI (46, 50), *Department of Biology, University of California, San Diego, La Jolla, California 92093*

G. MERLIN (20), *Department of Virology, Weizmann Institute of Science, Rehovot 76100, Israel*

DAVID L. MILLER (1), *Department of Molecular Biology, NYS Institute for Basic Research in Mental Retardation, 1050 Forest Hill Road, Staten Island, New York 10314*

MICHAEL A. MINKS (31), *Friedrich Miescher-Institut, Postfach 273, CH-4002 Basel, Switzerland*

KAORU MIYAZAKI (46), *Division of Enzymology, Institute for Protein Research,*

Osaka University, Yamada-Kami, Suita-shi, Osaka 565, Japan

JOSEPH D. MOSCA (33), *Department of Biochemistry, Temple University School of Medicine, Philadelphia, Pennsylvania 19140*

JOHN MOSCHERA (2), *Biopolymer Research Department, Hofmann-La Roche Inc., Nutley, New Jersey 07110*

JAMES S. MURPHY (48), *The Rockefeller University, New York, New York 10021*

YASUITI NAGANO (44), *National Sagamihara Hospital, Sagamihara 228, Japan*

TIMOTHY W. NILSEN (32), *Department of Biological Sciences, State University of New York at Albany, Albany, New York 12222*

KENNETH OLDEN (38), *Howard University Cancer Center, Washington,D.C. 20060*

JOHN R. ORTALDO (57), *Laboratory of Immunodiagnosis, National Cancer Institute, National Institutes of Health, Bethesda, Maryland 20205*

KURT PAUCKER* (70, 74), *Department of Microbiology, The Medical College of Pennsylvania, Philadelphia, Pennsylvania 19129*

SIDNEY PESTKA (1, 8, 10, 45, 76, 77, 81, 82), *Roche Institute of Molecular Biology, Nutley, New Jersey 07110*

LAWRENCE M. PFEFFER (48, 56), *The Rockefeller University, New York, New York 10021*

PAULA M. PITHA (51), *The Johns Hopkins Oncology Center, Biochemical Virology Laboratory, Baltimore, Maryland 21205*

SUSAN B. REBER (9), *Cytogenetics Laboratory, State University of New York Upstate, Syracuse, New York 13210*

NANCY L. REICHENBACH (33), *Department of Biochemistry, Temple University School of Medicine, Philadelphia, Pennsylvania 19140*

D. M. REISINGER (35), *School of Biological Sciences, Flinders University, South Australia 5042, Australia*

M. REVEL (20), *Department of Virology, Weizmann Institute of Science, Rehovot 76100, Israel*

W. K. ROBERTS (24, 26), *Department of Microbiology, University of Colorado Medical Center B175, Denver, Colorado 80262*

MENACHEM RUBINSTEIN (3), *Department of Virology, Weizmann Institute of Science, Rehovot 76100, Israel*

FRANK H. RUDDLE (13, 65, 66, 67), *Departments of Biology and Human Genetics, Yale University, New Haven, Connecticut 06511*

HIROSHI SAITO (44), *National Sagamihara Hospital, Sagamihara 228, Japan*

CHARLES E. SAMUEL (22, 37), *Department of Biological Sciences, Section of Biochemistry–Molecular Biology, University of California, Santa Barbara, Santa Barbara, California 93106*

GORDON SATO (46), *Department of Biology, University of California, San Diego, La Jolla, California 92093*

YOSUKE SAWADA (33), *Department of Biochemistry, Temple University School of Medicine, Philadelphia, Pennsylvania 19140*

A. SCHATTNER (20), *Department of Virology, Weizmann Institute of Science, Rehovot 76100, Israel*

A. SCHMIDT (20), *Department of Virology, Weizmann Institute of Science, Rehovot 76100, Israel*

EDWARD W. SCHRODER (41), *Department of Microbiology, Boston University School of Medicine, Boston, Massachusetts 02118*

L. A. SCHWARZ (52), *Department of Microbiology, University of Texas Medical Branch, Galveston, Texas 77550*

PRAVINKUMAR B. SEHGAL (17), *The Rockefeller University, New York, New York 10021*

JOHN E. SHIVELY (7), *Department of Immunology, City of Hope Research Institute, Duarte, California 91010*

* Deceased.

L. SHULMAN (20), Department of Biology, Yale University, New Haven, Connecticut 06511

GENE P. SIEGAL (9), Division of Surgical Pathology, Department of Laboratory Medicine and Pathology, University of Minnesota Medical School, Minneapolis, Minnesota 55455

R. H. SILVERMAN (26, 28), Imperial Cancer Research Fund Laboratories, London WC2A 3PX, England

DORIS L. SLATE (13, 65, 66, 67), Pfizer Central Research, Groton, Connecticut 06340

SHERYL SLIMMER (50), Department of Chemistry and the Cancer Center, University of California, San Diego, La Jolla, California 92093

ALAN SLOMA (8, 10, 78, 82), Roche Institute of Molecular Biology, Nutley, New Jersey 07110

MARK E. SMITH (71), Laboratory of Chemical Biology, National Institute of Arthritis, Diabetes and Digestive and Kidney Diseases, National Institutes of Health, Bethesda, Maryland 20205

GARY SPITZER (64), Developmental Therapeutics, The University of Texas M.D. Anderson Hospital and Tumor Institute, Houston, Texas 77030

T. SREELVALSAN (42), Department of Microbiology, Georgetown University Medical and Dental Schools, Washington D.C. 20007

DENNIS W. STACEY (12), Roche Institute of Molecular Biology, Nutley, New Jersey 07110

THEOPHIL STAEHELIN (1, 76), Pharma Research Division, F. Hoffmann-La Roche & Co., Ltd., CH-4002 Basel, Switzerland

CHRISTIAN STÄHLI (76), Pharma Research Division, F. Hoffman-La Roche & Co., Ltd., CH-4002 Basel, Switzerland

G. R. STARK (25), Department of Biochemistry, Stanford University Medical Center, Stanford, California 94305

STANLEY STEIN (2, 4, 5), Department of Molecular Genetics, Hoffmann-La Roche Inc., Nutley, New Jersey 07110

ROBERT J. SUHADOLNIK (33), Department of Biochemistry, Temple University School of Medicine, Philadelphia, Pennsylvania 19140

HIDEO TAKESHIMA (78), Roche Institute of Molecular Biology, Nutley, New Jersey 07110

IGOR TAMM (48, 56), The Rockefeller University, New York, New York 10021

M. N. THANG (14), Equipé de Recherche 238 du Centre National de la Recherche Scientifique and Unité 245 de l'Institut National de la Santé et de la Recherche Médicale, Institut de Biologie Physico-Chimique, 13 rue Pierre et Marie Curie, 75005 Paris, France

TUOMO TIMONEN (57), Department of Pathology, University of Helsinki, SF-00290 Helsinki 29, Finland

PAUL F. TORRENCE (23, 29, 30), Laboratory of Chemistry, National Institute of Arthritis, Diabetes and Digestive and Kidney Diseases, National Institutes of Health, Bethesda, Maryland 20205

MICHAEL G. TOVEY (47), Institut de Recherches Scientifiques sur le Cancer, Laboratory of Viral Oncology, BP 8, 94800 Villejuif, Cedex, France

PAUL O. P. TS'O (16), Division of Biophysics, The Johns Hopkins University, School of Hygiene and Public Health, Baltimore, Maryland 21205

JÜRGEN VAN DER BOSCH (46), Department of Biology, University of California, San Diego, La Jolla, California 92093

DHARMVIR S. VERMA (64), Developmental Therapeutics, The University of Texas M.D. Anderson Hospital and Tumor Institute, Houston, Texas 77030

I. K. VIJAY (38), Department of Dairy Science, University of Maryland, College Park, Maryland 20742

DONALD M. WALLACE (9), Laboratory of Pathophysiology, National Cancer Institute, National Institutes of Health, Bethesda, Maryland 20205

D. WALLACH (20), Department of Virology, Weizmann Institute of Science, Rehovot 76100, Israel

EUGENIA WANG (56), *The Rockefeller University, New York, New York 10021*

B. R. G. WILLIAMS (26, 35), *Infectious Diseases Research Institute, Hospital for Sick Children, University of Toronto, Toronto, Ontario M5G 1X8, Canada*

D. WOLF (20), *Department of Virology, Weizmann Institute of Science, Rehovot 76100, Israel*

D. H. WRESCHNER (26, 28), *Department of Microbiology, Tel Aviv University, Ramat Aviv, Tel Aviv, Israel*

JOSEPH M. WU (33), *Department of Biochemistry, New York Medical College, Valhalla, New York 10595*

CHONG-TAI YANG (11), *Department of Antivirals, Institute of Virology, Chinese Academy of Medical Sciences, 100 Ying Xing Jie, Xuan Wu Gu, Peking (Beijing) 100052, China*

Preface

This second volume on interferon reflects the rapid progress in this field. Although a single volume was originally conceived, it required two volumes to include all the contributions. The methodology in this field continues to evolve quickly. To assist in the dissemination of these new procedures, many authors sent in contributions that had not yet appeared elsewhere. I am grateful to and thank all the authors for their excellent contributions.

The staff of Academic Press has provided superb support in bringing these volumes to fruition. Special thanks are owed to Sophie Cuber, whose dedication, organization, and extraordinary eye for detail provided outstanding editorial assistance; and to Robert Pestka, for his thoroughness and intensive effort in preparing the comprehensive subject index of each volume.

During the preparation of these volumes, my family provided the nurture and humor to sustain this effort. Joan has accepted and borne many of my responsibilities while providing the milieu for this and additional work to proceed efficiently. Robert, Sharon, and Steven have provided many joys and much good humor vital to me.

SIDNEY PESTKA

METHODS IN ENZYMOLOGY

EDITED BY

Sidney P. Colowick and Nathan O. Kaplan

VANDERBILT UNIVERSITY
SCHOOL OF MEDICINE
NASHVILLE, TENNESSEE

DEPARTMENT OF CHEMISTRY
UNIVERSITY OF CALIFORNIA
AT SAN DIEGO
LA JOLLA, CALIFORNIA

METHODS IN ENZYMOLOGY

EDITORS-IN-CHIEF

Sidney P. Colowick Nathan O. Kaplan

Introduction

Since the discovery of interferon by Isaacs and Lindenman (1,2) and similar findings by Nagano and Kojima (3), an enormous literature about interferon has developed. A large number of reviews on interferon are available in journals, compendia, and monographs (see refs. 4–21 for a sampling). Volumes 78 and 79 summarize the methodology underlying this progress in interferon research.

Advances in interferon research have provided the foundations for purification and cloning of genes. In Volume 78, the amino acid sequences

```
                                                      ↓
AAT CGT AAA GAA GGA CAT CTC ATA TAA ATA GGC CAT ACC CAT GGA GAA AGG ACA TTC TAA CTG CAA CCT
                            ─────────── -100
                                                                           S1
                                                                          Met Thr Asn Lys
TTC GAA GCC TTT GCT CTG GCA CAA CAG GTA GTA GGC GAC ACT GTT CGT GTT GTC AAC ATG ACC AAC AAG

                                                                  1
Cys Leu Leu Gln Ile Ala Leu Leu Leu Cys Phe Phe Thr Thr Ala Leu Ser Met Ser Tyr Asn Leu Leu
TGT CTC CTC CAA ATT GCT CTC CTG TTG TGC TTC TTC ACT ACA GCT CTT TCC ATG AGC TAC AAC TTG CTT

Gly Phe Leu Gln Arg Ser Ser Asn Phe Gln Cys Gln Lys Leu Leu Trp Gln Leu Asn Gly Arg Leu Glu
GGA TTC CTA CAA AGA AGC AGC AAT TTT CAG TGT CAG AAG CTC CTG TGG CAA TTG AAT GGG AGG CTT GAA
                            100

Tyr Cys Leu Lys Asp Arg Met Asn Phe Asp Ile Pro Glu Glu Ile Lys Gln Leu Gln Gln Phe Gln Lys
TAC TGC CTC AAG GAC AGG ATG AAC TTT GAC ATC CCT GAG GAG ATT AAG CAG CTG CAG CAG TTC CAG AAG
                                                                  200

Glu Asp Ala Ala Leu Thr Ile Tyr Glu Met Leu Gln Asn Ile Phe Ala Ile Phe Arg Gln Asp Ser Ser
GAG GAC GCC GCA TTG ACC ATC TAT GAG ATG CTC CAG AAC ATC TTT GCT ATT TTC AGA CAA GAT TCA TCT

Ser Thr Gly Trp Asn Glu Thr Ile Val Glu Asn Leu Leu Ala Asn Val Tyr His Gln Ile Asn His Leu
AGC ACT GGC TGG AAT GAG ACT ATT GTT GAG AAC CTC CTG GCT AAT GTC TAT CAT CAG ATA AAC CAT CTG
            300

Lys Thr Val Leu Glu Glu Lys Leu Glu Lys Glu Asp Phe Thr Arg Gly Lys Leu Met Ser Ser Leu His
AAG ACA GTC CTG GAA GAA AAA CTG GAG AAA GAA GAT TTC ACC AGG GGA AAA CTC ATG AGC AGT CTG CAC
                                                  400

Leu Lys Arg Tyr Tyr Gly Arg Ile Leu His Tyr Leu Lys Ala Lys Glu Tyr Ser His Cys Ala Trp Thr
CTG AAA AGA TAT TAT GGG AGG ATT CTG CAT TAC CTG AAG GCC AAG GAG TAC AGT CAC TGT GCC TGG ACC

                                                                              166
Ile Val Arg Val Glu Ile Leu Arg Asn Phe Tyr Phe Ile Asn Arg Leu Thr Gly Tyr Leu Arg Asn END
ATA GTC AGA GTG GAA ATC CTA AGG AAC TTT TAC TTC ATT AAC AGA CTT ACA GGT TAC CTC CGA AAC TGA
    500

AGA TCT CCT AGC CTG TGC CTC TGG GAC TGG ACA ATT GCT TCA AGC ATT CTT CAA CCA GCA GAT GCT GTT
                                                  600

TAA GTG ACT GAT GGC TAA TGT ACT GCA TAT GAA AGG ACA CTA GAA GAT TTT GAA ATT TTT ATT AAA TTA
                                                                              700
                                                                     ↓
TGA GTT ATT TTT ATT TAT TTA AAT TTT ATT TTG GAA AAT AAA TTA TTT TTG GTG CAA AAG TCA ACA TGG
                                              ───────

CAG TTT TAA TTT CGA TTT GAT TTA TAT AAC CAT CCA TAT TAT AA
                            800
```

FIG. 1. The nucleotide sequence of the gene for human fibroblast interferon. The amino acid sequence of fibroblast interferon predicted from this DNA sequence is shown. (Taken from references 22 and 23.)

```
AAA GCA AAA ACA GAC ATA GAA AGT AAA ACT AGG CAT TTA GAA AAT GGA AAT TAG TAT GTT CAC TAT TTA AGA
                                                                       -100

CCT ATG CAC AGA GCA AAG TCT CCA GAA AAC CTA GAG CCA CTG GTT CAA GTT ACC CAC CTC AGG TAG CCT AGT

         S1
                    Met Ala Arg Ser Phe Ser Leu Leu Met Val Val Leu Val Leu Ser Tyr Lys Ser
GAT ATT TGC AAA ATC CCA ATG GCC CGG TCC TTT TCT TTA CTG ATG GTC GTG CTG GTA CTC AGC TAC AAA TCC

                 1
Ile Cys Ser Leu Gly Cys Asp Leu Pro Gln Thr His Ser Leu Arg Asn Arg Arg Ala Leu Ile Leu Leu Ala
ATC TGC TCT CTG GGC TGT GAT CTG CCT CAG ACC CAC AGC CTG CGT AAT AGG AGG GCC TTG ATA CTC CTG GCA
                                                              100

Gln Met Gly Arg Ile Ser Pro Phe Ser Cys Leu Lys Asp Arg His Glu Phe Arg Phe Pro Glu Glu Glu Phe
CAA ATG GGA AGA ATC TCT CCT TTC TCC TGC TTG AAG GAC AGA CAT GAA TTC AGA TTC CCA GAG GAG GAG TTT

Asp Gly His Gln Phe Gln Lys Thr Gln Ala Ile Ser Val Leu His Glu Met Ile Gln Gln Thr Phe Asn Leu
GAT GGC CAC CAG TTC CAG AAG ACT CAA GCC ATC TCT GTC CTC CAT GAG ATG ATC CAG CAG ACC TTC AAT CTC
200

Phe Ser Thr Glu Asp Ser Ser Ala Ala Trp Glu Gln Ser Leu Leu Glu Lys Phe Ser Thr Glu Leu Tyr Gln
TTC AGC ACA GAG GAC TCA TCT GCT GCT TGG GAA CAG AGC CTC CTA GAA AAA TTT TCC ACT GAA CTT TAC CAG
                                               300

Gln Leu Asn Asp Leu Glu Ala Cys Val Ile Gln Glu Val Gly Val Glu Glu Thr Pro Leu Met Asn Glu Asp
CAA CTG AAT GAC CTG GAA GCA TGT GTG ATA CAG GAG GTT GGG GTG GAA GAG ACT CCC CTG ATG AAT GAG GAC
                                                                      400

Phe Ile Leu Ala Val Arg Lys Tyr Phe Gln Arg Ile Thr Leu Tyr Leu Met Glu Lys Lys Tyr Ser Pro Cys
TTC ATC CTG GCT GTG AGG AAA TAC TTC CAA AGA ATC ACT CTT TAT CTA ATG GAG AAG AAA TAC AGC CCT TGT

Ala Trp Glu Val Val Arg Ala Glu Ile Met Arg Ser Phe Ser Phe Ser Thr Asn Leu Gln Lys Arg Leu Arg
GCC TGG GAG GTT GTC AGA GCA GAA ATC ATG AGA TCC TTC TCT TTT TCA ACA AAC TTG CAA AAA AGA TTA AGG
         500
Arg Lys Asp END
AGG AAG GAT TGA AAA CTG GTT CAT CAT GGA AAT GAT TCT CAT TGA CTA ATG CAT CAT CTC ACA CTT TCA TGA
                                                              600

GTT CTT CCA TTT CAA AGA CTC ACT TCT ATA ACC ACC ACA AGT TGA ATC AAA ATT TCC AAA TGT TTT CAG GAG
                                                              700

TGT TAA GAA GCA TCG TGT TTA CCT GTG CAG GCA CTA GTC CTT TAC AGA TGA CCA
```

FIG. 2. The nucleotide sequence of a gene for human leukocyte interferon. The amino acid sequence of leukocyte interferon predicted from this DNA sequence is shown. (Taken from references 22 and 23.)

of human fibroblast interferon and human leukocyte interferon as predicted from the DNA sequences are given. The sequences of human fibroblast (Fig. 1) and leukocyte (Fig. 2) interferon genes (22,23) provide explicit examples of how rapid progress has been (see this volume [78] for additional references). The absence of intervening sequences in these and other human interferon genes is striking (22–25). The amino acid sequences of the interferons as deduced from the DNA sequences of these genes are shown in these figures as well. On determination of the primary structures of several human leukocyte interferons isolated from buffy coat cells, it was surprising, however, to discover that their amino acid sequences (Fig. 3) were ten amino acids shorter than predicted at the carboxy terminal end (26). These results emphasize the need for integrating both DNA and protein sequence information. Advances in methodology of purification and amino acid sequence determination of proteins de-

1 CYS ASP LEU PRO GLN THR HIS SER LEU GLY(10) SER ARG ARG THR LEU MET LEU LEU ALA GLN(20)

21 MET ARG LYS ILE SER LEU PHE SER CYS LEU(30) LYS ASP ARG HIS ASP PHE GLY PHE PRO GLN(40)

41 GLU GLU PHE GLY ASN GLN PHE GLN LYS ALA(50) GLU THR ILE PRO VAL LEU HIS GLU MET ILE(60)

61 GLN GLN ILE PHE ASN LEU PHE SER THR LYS(70) ASP SER SER ALA ALA TRP ASP GLU THR LEU(80)

81 LEU ASP LYS PHE TYR THR GLU LEU TYR GLN(90) GLN LEU ASN ASP LEU GLU ALA CYS VAL ILE(100)

101 GLN GLY VAL GLY VAL THR GLU THR PRO LEU(110) MET LYS GLU ASP SER ILE LEU ALA VAL ARG(120)

121 LYS TYR PHE GLN ARG ILE THR LEU TYR LEU(130) LYS GLU LYS LYS TYR SER PRO CYS ALA TRP(140)

141 GLU VAL VAL ARG ALA GLU ILE MET ARG SER(150) PHE SER LEU SER THR(155)

FIG. 3. The amino acid sequence of leukocyte interferons α_2 and β_1 from sequence analysis of the proteins. The solid lines under the sequences represent the tryptic peptides obtained. (Taken from reference 26.)

scribed in these volumes made these achievements possible with 100–200 μg of the proteins.

We have entered a new era of interferon research. The new stones to be put in place will indeed make the structure resilient. Understanding its physiological significance will provide another elegant tribute to the structure of life.

SIDNEY PESTKA

REFERENCES

1. A. Isaacs and J. Lindenmann, *Proc. R. Soc. London Ser. B* **147,** 258 (1957).
2. A. Isaacs, J. Lindenmann, and R. C. Valentine, *Proc. R. Soc. London Ser. B* **147,** 268 (1957).
3. Y. Nagano and Y. Kojima, *C. R. Seances Soc. Biol. Ses Fil.* **152,** 1627 (1958).
4. S. Baron and F. Dianzani (eds.), *Tex. Rep. Biol. Med.* **35,** 1 (1977).
5. M. H. Ng and J. Vilček, *Adv. Protein Chem.* **26,** 173 (1972).

6. C. Colby and M. J. Morgan, *Annu. Rev. Microbiol.* **25,** 333 (1971).

7. E. De Clercq and T. C. Merigan, *Annu. Rev. Med.* **21,** 17 (1970).

8. M. Ho and J. A. Armstrong, *Annu. Rev. Microbiol.* **29,** 131 (1975).

9. J. Vilček, "Interferon." Springer-Verlag, Berlin and New York, 1969.

10. N. B. Finter, "Interferons and Interferon Inducers," 2nd ed. North-Holland Publ., Amsterdam, 1973.

11. D. A. Stringfellow, "Interferon and Interferon Inducers. Clinical Applications." Dekker, New York, 1980.

12. R. B. Pollard and T. C. Merigan, *Pharmacol. Ther.* (Part A) **2,** 783 (1978).

13. P. F. Torrence and E. De Clercq, *Pharmacol. Ther.* (Part A) **2,** 1 (1977).

14. R. M. Friedman, *Pharmacol. Ther.* (Part A) **2,** 425 (1978).

15. D. L. Slate and F. H. Ruddle, *Pharmacol. Ther.* (Part A) **4,** 221 (1979).

16. K. G. Mogensen and K. Cantell, *Pharmacol. Ther.* (Part A) **1,** 369 (1977).

17. W. A. Carter, S. S. Leong, and J. S. Horoszewicz, *in* "Antiviral Mechanisms and the Control of Neoplasia" (P. Chandra, ed.), p. 663. Plenum, New York, 1979.

18. W. E. Stewart II, "The Interferon System." Springer-Verlag, Berlin and New York, 1979.

19. S. Pestka, *in* "Dimensions in Health Research—Search for the Medicines of Tomorrow" (H. Weissbach, ed.), p. 29. Academic Press, New York, 1978.

20. S. Pestka, M. Evinger, R. McCandliss, A. Sloma, and M. Rubinstein, *in* "Polypeptide Hormones" (R. F. Beers, Jr. and E. G. Bassett, eds.), p. 33. Raven, New York, 1980.

21. S. Pestka, S. Maeda, D. S. Hobbs, W. P. Levy, R. McCandliss, S. Stein, J. A. Moschera, and T. Staehelin, *in* "Cellular Responses to Molecular Modulators" (W. A. Scott, R. Werner, and J. Schultz, eds.), Miami Winter Symposia, Vol. 18. Academic Press, New York, 1981, in press.

22. S. Maeda, R. McCandliss, T.-R. Chiang, L. Costello, W. P. Levy, N. T. Chang, and S. Pestka, *in* "Developmental Biology Using Purified Genes" (D. Brown and C. F. Fox, eds.), ICN–UCLA Symposia on Molecular and Cellular Biology, Vol. XXIII. Academic Press, New York, 1981, in press.

23. S. Pestka, S. Maeda, D. S. Hobbs, T.-R. C. Chiang, L. L. Costello, E. Rehberg, W. P. Levy, N. T. Chang, N. R. Wainwright, J. B. Hiscott, R. McCandliss, S. Stein, J. A. Moschera, and T. Staehelin, *in* "Proceedings of the 'Third Cleveland Symposium on Macromolecules: Recombinant DNA'" (A. G. Walton, ed.). Elsevier, Amsterdam, 1981, in press.

24. M. Streuli, S. Nagata, and C. Weissmann, *Science* **209,** 1343 (1980).

25. R. M. Lawn, J. Adelman, T. J. Dull, M. Gross, D. Goeddel, and A. Ullrich, *Science* **212,** 1159 (1981).

26. W. P. Levy, M. Rubinstein, J. Shively, U. Del Valle, C.-Y. Lai, J. Moschera, L. Brink, L. Gerber, S. Stein, and S. Pestka, *Proc. Natl. Acad. Sci. U.S.A.,* in press (1981).

Methods in Enzymology

Volume 79

INTERFERONS

Part B

Section I

Techniques for Chromatography and Analysis of Interferons

[1] The Crystallization of Recombinant Human Leukocyte Interferon A

By DAVID L. MILLER, HSIANG-FU KUNG, THEOPHIL STAEHELIN, and SIDNEY PESTKA

Crystallization of a substance satisfies one of the classical criteria for homogeneity. Crystallization may also provide a useful purification step. Furthermore, when large, ordered single crystals can be obtained, the molecule's tertiary structure may be determined by X-ray crystallography.

Numerous techniques have been developed for the crystallization of proteins[1]; however, no generalized procedure has been discovered, and many proteins remain uncrystallized. The most widely used approach involves the addition to the protein solution of a crystallizing agent, which is commonly a salt, such as ammonium sulfate or ammonium citrate, or an organic solvent, such as ethanol or 2-methyl-2,4-pentanediol.

One of the most versatile crystallizing agents is polyethylene glycol (PEG), which combines some of the characteristics of the salts and the organic solvents.[2,3] Recombinant human leukocyte interferon A (IFLrA) crystallizes readily from PEG, occasionally in large single crystals.

Materials

Recombinant human leukocyte interferon A (IFLrA), isolated as previously described[4,5]
Polyethylene glycol 4000
N-2-Hydroxyethylpiperazine-*N'*-2-ethane sulfonic acid (HEPES)
Sodium azide
Dimethyldichlorosilane
Nine-well glass spot plate

Procedure

IFLrA (4 ml, 0.2 mg/ml) was dialyzed overnight at 4° against 0.01 M HEPES, adjusted to pH 7.1 with NH_4OH. The solution was concentrated fivefold by centrifugal evaporation (Speed-Vac, Virtis), redialyzed against

[1] A. McPherson, Jr., *in* "Methods of Biochemical Analysis" (D. Glick, ed.), Vol. 23, p. 249. Academic Press, New York, 1976.
[2] K. B. Ward, B. C. Wishner, E. E. Lattman, and W. E. Love, *J. Mol. Biol.* **98**, 161 (1975).
[3] A. McPherson, Jr., *J. Biol. Chem.* **251**, 6300 (1976).
[4] T. Staehelin, D. S. Hobbs, H.-F. Kung, C.-Y. Lai, and S. Pestka, *J. Biol. Chem.*, in press (1981).
[5] T. Staehelin, D. S. Hobbs, H.-F. Kung, and S. Pestka, this series, Vol. 78 [72].

METHODS IN ENZYMOLOGY, VOL. 79

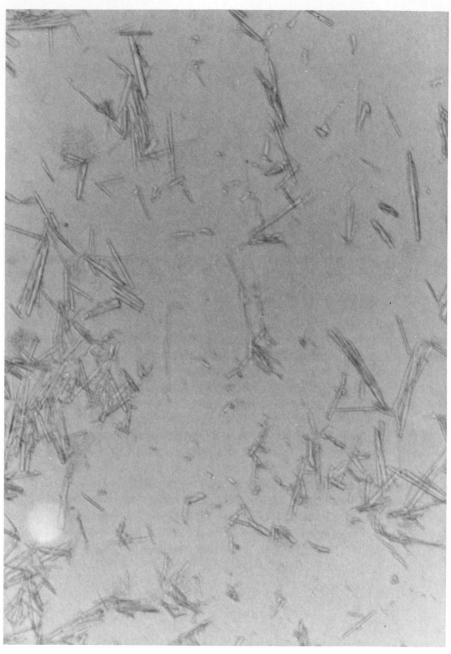

Fig. 1. Needle crystals of IFLrA. The crystals had an approximate diameter of 0.01 mm and ranged in length from 0.05 to 0.15 mm.

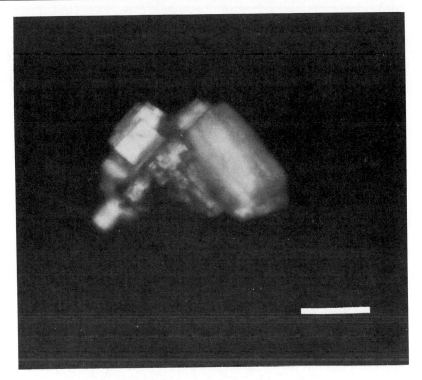

Fig. 2. Complex crystal of IFLrA. Bar = 0.1 mm.

0.01 M HEPES (pH 7.1), and further concentrated to 5 mg/ml. The final concentration was determined by ultraviolet spectrophotometry.

The spot plate was siliconized prior to use by dipping it in 5% (v/v) dimethyldichlorosilane in carbon tetrachloride, and then baking it at 180°. It was washed with water and rebaked. In each of four wells, 20 μl of the interferon solution was placed. A PEG solution (200 mg/ml) containing NaN$_3$ (0.5 mg/ml) was added to each well to give final concentrations of 20, 40, 60, and 100 mg/ml. The droplets were immediately mixed using a micropipettor, and the spot plate was placed above a solution of 100 mg/ml PEG contained in a crystallizing dish. The dish was sealed and kept at 4°. After 1–3 days, crystals appeared in each of the wells (Fig. 1). After several more days, larger crystals (Fig. 2) appeared in some of the droplets.

The crystals were separated from the liquid phases by centrifugation, washed with 0.05 M HEPES (pH 7.1) containing 10% PEG, and redissolved in 0.05 M HEPES, pH 7.1. Bioassays of the solutions prepared from the 10% PEG mixture revealed that the crystals contained interferon ac-

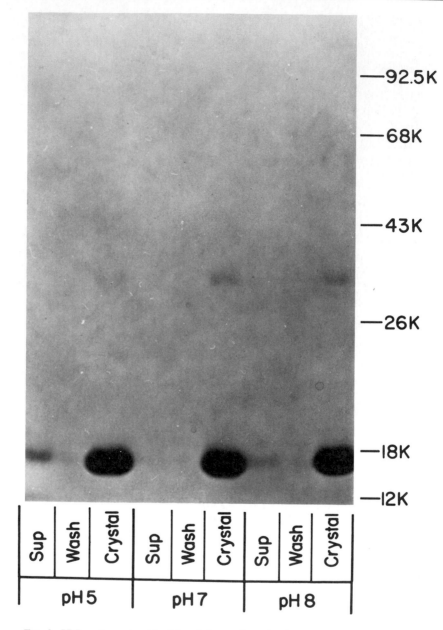

FIG. 3. SDS–polyacrylamide slab gel electrophoresis of crystals of IFLrA and supernatant after crystallization. SDS–polyacrylamide gel electrophoresis was performed as described previously.[5] The molecular weight standards used were purchased from Bethesda Research Laboratories.

Recovery of Interferon in Crystals[a]

Crystallization condition	Crystals		Supernatant	
	Antiviral activity (units/ml)	Percentage recovery	Antiviral activity (units/ml)	Percentage recovery
pH 5	4.0×10^8	(94)	2.5×10^7	(6)
pH 7	6.0×10^8	(99)	4.5×10^6	(1)
pH 8	8.0×10^8	(99)	4.5×10^6	(1)

[a] The crystals were washed in 0.05 M HEPES (pH 7.1) containing 10% PEG and redissolved in 0.05 M HEPES (pH 7.1). The total antiviral activity recovered in the crystals and in the supernatant was assayed as described.[6] Values in parentheses represent the percentage recovered in the various fractions.

tivity (see the table). Most of the interferon activity ($>90\%$) was recovered in the crystals (table). The crystallized protein was indistinguishable from the uncrystallized IFLrA by gel electrophoresis (Fig. 3), which confirms that the crystalline product is uncleaved and intact.

It is not necessary to crystallize interferon A from concentrated solutions. Microcrystals appear in solutions of 0.2 mg/ml interferon, 10% PEG. This method thus may be useful for concentrating dilute solutions of interferon. With the preparation of appropriate crystals of IFLrA and other interferons, it will be possible to initiate X-ray crystallography to elucidate its tertiary structure.

[6] S. Rubinstein, P. C. Familletti, and S. Pestka, *J. Virol.* **37**, 755 (1981).

[2] High-Performance Liquid Chromatography and Picomole-Level Detection of Peptides and Proteins

By Stanley Stein and John Moschera

The instrumentation and methodology described below have been used for the isolation and analysis of human leukocyte and fibroblast interferons, as presented in Vol. 78 of this series.

Fluorescamine Column Monitor

The reagent fluorescamine is dissolved in a nonhydroxylic solvent, such as acetone, and is added to a solution of a primary amine at pH 9.[1]

[1] S. Udenfriend, S. Stein, P. Böhlen, and W. Dairman, *Science* **178**, 871 (1972).

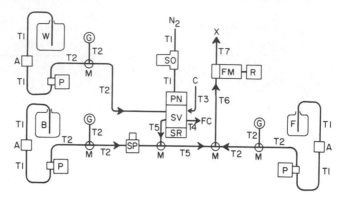

FIG. 1. Schematic diagram of the column monitoring system. The numbers in parentheses refer to items listed in column 1 of the table. W, water; F, fluorescamine in acetone; B, borate buffer; A, air trap (12); P, pump (13); G, pressure gauge (3a); M, tee fitting (22); SV, sampling valve (18); PN, pneumatic actuator (20); SR, spring return (21); C, column; FC, fraction collector; SP, septum injector (27, 28); FM, fluorometer (1); R, recorder (2); X, waste; SO, solenoid air valve (25); T1, 0.063 in. i.d., 0.125 in. o.d. (15d); T2, 0.031 in. i.d., 0.063 in. o.d. (15c); T3, 0.012 in. i.d., 0.063 in. o.d. (15a), ca. 15 in. length; T5, 0.023 in. i.d., 0.063 in. o.d. (15b), 2 in. length; T-6, 0.023 in. i.d., 0.063 in. o.d. (15b), 6 ft. length; T7, 0.012 in. i.d., 0.063 in. o.d. (15a), 6 ft. length (to give 40–60 psi back pressure); N_2, ca. 80 psi.

The fluorogenic reaction is complete within a fraction of a minute. Since both the reagent and its hydrolysis by-products are nonfluorescent, the fluorescence may be measured without further manipulations. The reactive groups in peptides are the α-amino group at the amino terminus and the ϵ-amino group of lysine. Peptides having neither a lysine residue nor a free α-amino group (e.g., prolyl, pyroglutamyl, or N-acetyl) will go undetected.

A schematic diagram of the column monitor is shown in Fig. 1.[2] Its mode of operation is as follows: Column effluent passes through a stream-sampling valve and then to a fraction collector. The sampling valve may be either of two types—one having an internal chamber (<10 μl), the other having external loops (>10 μl) (Fig. 2). At regular intervals (usually 10 sec), this valve injects aliquots of the column effluent into a stream of water. The aliquots are transported to a mixing tee, where 0.3 M boric acid adjusted to pH 9.3 with lithium hydroxide is added and then to another tee where 0.02% fluorescamine in acetone is added. The mixture then flows through a filter fluorometer and then to waste. Fluorescence is plotted with time on a chart recorder. The flow rates of the water, borate, and fluorescamine pumps are 15, 30, and 15 ml/hr, respectively. The lengths and diameters of the segments of connecting Teflon tubing

[2] P. Böhlen, S. Stein, J. Stone, and S. Udenfriend, *Anal. Biochem.* **67**, 438 (1975).

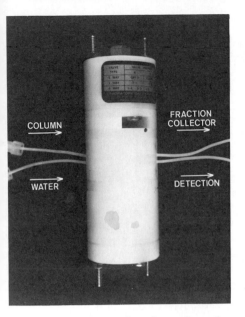
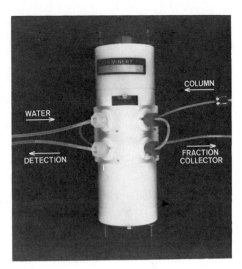

FIG. 2. Two styles of sampling valves are shown. Both valves are driven at regular intervals alternately between two positions by a pneumatic actuator and a spring return. The valve in the left panel (Cat. No. 0414-3080) has a slider with three holes of equal volume. These sliders are interchangeable and are available with chambers of 2, 5, 10, and 20 μl. Two of the chambers are in the flow paths at any given time (the center and top or the center and bottom chambers). The valve on the right (Cat. No. 0414-2270) has two loops of equal volume. The loops may be readily changed, the minimum sampling volume being about 10 μl. The flow paths of the column effluent and the water alternate between the two loops.

are given in the legend to Fig. 1. After the instrument has been constructed, the pumps should be set first with water and acetone to avoid possible precipitation in the tubing due to incorrect flow rates. High purity (amine-free) reagents should be used. All components described herein are listed in the table along with the suppliers.

The chromatography system is depicted in Fig. 3. There is little advantage in using an expensive HPLC pump. Instead, the same pulsatile pump is used for both chromatography and detection. Samples may be applied to the column with a high-pressure injection valve or drawn through the pump. A four-way valve (Fig. 3) is used in the latter method. Sample is drawn into the tubing with the syringe to displace the air. After switching the valve, the sample will be sucked through the pump onto the column. The valve is then returned to its original position, avoiding entry of air into the pump.

Gradient elution is a necessity, and a system employing a proportion-

LIST OF COMPONENTS FOR DETECTION AND CHROMATOGRAPHY SYSTEMS

Component	Quantity	Supplier	Catalog No.
1. Fluorometer	1	Gilson Medical Electronics, Box 27, Middleton, WI 53562	Spectra/Glo
2. Recorder	1	Linear Instruments, 17282 Eastman Ave., Irvine, CA 92714	585
3. Pressure Gauge			
a. 0–200 psi	3	Navtec Industries, 1095 Route 110, Farmingdale, NY 11735	6530-200
b. 0–4000 psi	1		6530-4000
4. Gauge adapter	4	R. S. Crum, 1167 Globe Ave., Mountainside, NJ 07092	SS-100-7-4
5. Tee fitting, steel	1	R. S. Crum	SS-100-3
6. N_2 regulator adapter	1	R. S. Crum	B-200-7-4
7. Sample injection valve	1	Rheodyne, Inc., 2809 Tenth St., Berkeley, CA 94710	7010
8. Column inlet filter	1	Rheodyne Inc.	7302
9. Spinbar	1	Bel-Art Products, Pequannock, NJ 07440	F-37119
10. N_2 regulator	1	Matheson, East Rutherford, NJ 07073	9-580
11. Chrontrol	2	Lindburg Enterprises, 4888 Ronson Court, San Diego, CA 92111	CT-4
12. Air trap	3	Beckman Instrument Co., 1117 California Ave., Palo Alta, CA 94304	326187
13. Pump	4	Laboratory Data Control, P. O. Box 10235, Riviera Beach, FL 33404	196-0066-01
14. Luer adapter		Laboratory Data Control	
15. Teflon tubing	2	Laboratory Data Control	0414-5020

	Qty	Supplier	Part No.
a. small bore	4		T063012
b. medium bore	2		T063023
c. large bore	2		T063031
d. large	2		T125063
16. Steel tubing (0.063 in. o.d. × 0.02 in. i.d.)	5 ft.		600011
17. Tube end fittings		Laboratory Data Control	
a. small	3	Laboratory Data Control	0414-3600
b. large	1		0414-3680
18. Four-way valve or sample injection valve	3	Laboratory Data Control	0414-2270
		Laboratory Data Control	0414-3080
19. Three-way valve	1	Laboratory Data Control	0414-2250
20. Pneumatic actuator	2	Laboratory Data Control	0414-2350
21. Spring return	2	Laboratory Data Control	0414-2370
22. Tee fitting, Teflon	5	Laboratory Data Control	0414-3950
23. Flanging tool	1	Laboratory Data Control	0414-3750
24. Coupling	1	Laboratory Data Control	0414-5000
25. Solenoid air valve	2	Laboratory Data Control	0414-3410
26. Glass connector	3	Laboratory Data Control	0414-3940
27. Silicone septum	1	Laboratory Data Control	0414-5200
28. Tee fitting, Teflon	1	Altex Scientific Inc., 1780 Fourth St., Berkeley, CA 94710	200-22

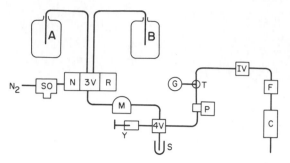

FIG. 3. High-performance liquid chromatography (HPLC) system. The gradient is formed with a pneumatically driven proportioning valve. Samples may be injected onto the column or drawn through the pump. The numbers in parentheses refer to items listed in column 1 of the table. A, initial buffer; B, final buffer; SO, solenoid air valve (25); N, pneumatic actuator (20); R, spring return (21); 3V, three-way valve (19); M, mixing chamber (26); 4V, four-way valve (18); Y, syringe; S, sample; N_2, nitrogen pressure (ca. 80 psi); P, pump (13); T, tee connector (5); G, pressure gauge (3b, 4); IV, high-pressure injection valve (7); F, in-line filter (8); C, column. Stainless steel tubing is used at the high-pressure side of the pump.

ing valve is needed with a single pump. There are several alternatives, ranging from the expensive LKB (Rockville, Maryland) system to an inexpensive one illustrated in Fig. 3 and listed in the table.[3] In this system, a microprocessor-timer controls an electric solenoid air valve. In the "on" position, the air drives a pneumatically operated three-way valve to allow buffer B to be pumped. In the "off" position, the valve springs back to allow buffer A to be pumped. A multistep program is written, each step comprising a repeated off–on combination. For example, 27 sec off/3 sec on gives 90% buffer A/10% buffer B. A mixing chamber between the three-way valve and the pump smoothes out the steps into a continuous gradient. This same type of microprocessor may be used to operate the sampling valve of the column monitor or to shut off the pumps and recorder automatically. A more sophisticated microprocessor for controlling the entire instrument has been introduced by Eldex Labs (Menlo Park, CA).

Bypass Assay of Protein Concentration

A tee fitting (Fig. 4) is used for manual injection of sample into the detection system (Fig. 1). This technique is useful for determining accurately the protein concentration (versus a proper calibration standard) in individual samples in the nanogram or microgram range. One arm of the tee is fitted with a silicone septum (cut to half-thickness and trimmed) held

[3] R. V. Lewis, *Anal. Biochem.* **98**, 142 (1979).

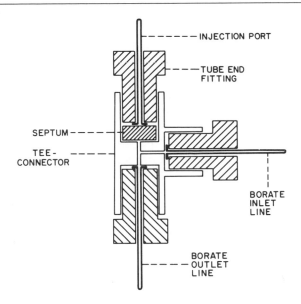

FIG. 4. Cut-away view of the septum bypass injector.

in place with a tubing end-fitting. Teflon tubing is used as a guide for the Hamilton syringe. The design of this simple injector minimizes dead space; this results in sharper peaks.

Fluorescence is linear with amount of protein injected over more than two orders of magnitude (Fig. 5). The technique is applicable at the low picomole level for amino acids and peptides or the low nanogram level for proteins having a normal lysine content. Samples may be hydrolyzed with acid or alkali prior to the bypass assay in order to increase sensitivity or reveal blocked peptides.

Chromatography

Whenever possible, volatile solvents and buffers are employed. After chromatography, peptides and proteins are readily recoverable by evaporation. Furthermore, distillation over ninhydrin will produce amino acid- and peptide-free eluents. Pyridine, acetic acid, formic acid, and n-propanol are most commonly used. However, inorganic buffer salts may also be used.

Samples should be free of particulate matter to protect the pump and the HPLC column. Generally, a preliminary purification step, such as gel filtration, affinity chromatography, or acid precipitation is used prior to HPLC. Large volumes (up to liters) can be applied to the HPLC column

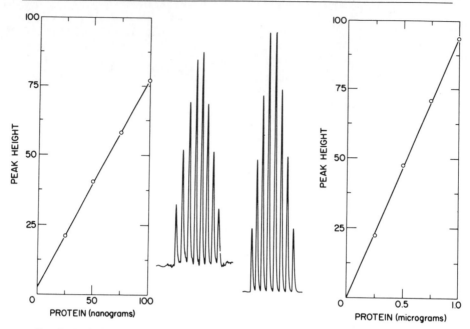

FIG. 5. Analysis of protein with the bypass injector. Varying volumes (2.5, 5.0, 7.5, and 10 μl) of bovine serum albumin (10 ng/μl on left and 100 ng/μl on right) were injected at 2-min intervals. Average peak height versus amount of protein was plotted. The values are uncorrected for blank values.

(in a medium that will not elute the peptide of interest) without compromising efficiency. A standard 25 × 0.46 cm HPLC column has a loading capacity in the range of 5–50 mg of protein. Recovery can become a problem at the low microgram level, although some experiments with ^{125}I-labeled proteins and peptides have given quantitative recoveries at the nanogram level. At low peptide concentrations, it is advisable to use only polypropylene tubes and pipette tips. Disposable tubes should be used and gloves should be worn for the sake of cleanliness.

Reverse-phase chromatography has been used most extensively in this laboratory. Small peptides (<2000 daltons) are generally resolved on a C_{18} column, whereas large peptides and small proteins are run on a C_8 column. For large proteins (>50,000 daltons), it is necessary to use a large-pore silica (20–50 nm) as the matrix, rather than the standard 6–10 nm silica.[4] The cyanopropyl and alkylphenyl reverse-phase supports have been found to behave in a similar fashion to the C_8 support. However, a reverse-phase diphenyl support[4] has shown unique properties.

[4] R. V. Lewis, A. Fallon, S. Stein, K. D. Gibson, and S. Udenfriend, *Anal. Biochem.* **104**, 153 (1980).

Reverse-phase chromatography typically is carried out with a gradient of n-propanol within the limits of 0 to 60%.[5] The pH is held constant within the range of 3 to 5.5 with pyridine and formic or acetic acid. High ionic strength (>0.3 M pyridine) is used. Since a small change in the propanol concentration can markedly affect retention, shallow gradients (e.g., 0.3% n-propanol increment per column volume) achieve the best separation. A flow rate of 20 ml/hr with a 25×0.46 cm column is generally used in the final step of a purification. Going above 40 ml/hr can result in considerable peak broadening for proteins.[6] However, large volumes of sample may be rapidly pumped onto the column, but then the flow rate should be decreased prior to starting the gradient. It is usually possible to achieve difficult purifications by sequentially using a combination of different reverse-phase columns or n-propanol gradients at different pH levels.

For rechromatography, it is better to dilute out the propanol and pump a large volume onto the column, rather than lyophilize and redissolve. Hydrophobic proteins and peptides might stick to the walls of the polypropylene tubes unless a relatively high propanol concentration is present (up to one-half of that required for elution from the reverse-phase column). It is advisable to periodically check the cleanliness of the column by running a "blank" gradient (no sample injected). In some cases this is done both before and after the actual sample is run.

Ion-exchange HPLC is also applicable to peptides and proteins, especially in cases where the presence of organic solvent would be detrimental to biological activity. Silica-based ion-exchange supports from Separation Industries (Orange, NJ) have been successfully used by the authors. Enzyme separations are achieved under elution conditions similar to those used with cellulose-based ion-exchange supports, but with much better resolution.[7,8]

It should be noted that HPLC columns can also be monitored by ultraviolet absorption, which is simpler with regard to the instrumentation involved. However, unless short wavelengths (approximately 210 nm) are used, the sensitivity and general applicability to peptides does not approach that of fluorescamine detection. The choice of eluents is severely limited to those that have a low extinction coefficient at this wavelength.

A variety of peptides and proteins have been isolated and analyzed by

[5] M. Rubinstein, S. Stein, L. D. Gerber, and S. Udenfriend, *Proc. Natl. Acad. Sci. U.S.A.* 74, 3052 (1977).
[6] B. N. Jones, R. V. Lewis, S. Paabo, K. Kojima, S. Kimura, and S. Stein, *J. Liq. Chromatogr.* 3, 1373 (1980).
[7] S. H. Chang, R. Noel, and F. E. Regnier, *Anal. Chem.* 48, 1839 (1976).
[8] S. H. Chang, K. M. Goodling, and F. E. Regnier, *J. Chromatogr.* 125, 103 (1976).

the methodology described in this chapter. They include rat β-endorphin (3500 daltons),[9] rat β-lipotropin (11,000 daltons),[5] camel pro-opiocortin (33,000 daltons),[10] bovine enkephalin-containing peptides and proteins (700–24,000 daltons),[11-13] and human interferons (18,000 daltons).[14,15] The HPLC of large proteins, such as collagen chains and native collagen, has been accomplished.[4] This methodology is not complete without the ability to carry out amino acid analysis at the picomole level (see this volume [4]).

[9] M. Rubinstein, S. Stein, and S. Udenfriend, *Proc. Natl. Acad. Sci. U.S.A.* **74**, 4969 (1977).
[10] S. Kimura, R. V. Lewis, L. D. Gerber, L. Brink, M. Rubinstein, S. Stein, and S. Udenfriend, *Proc. Natl. Acad. Sci. U.S.A.* **76**, 1756 (1979).
[11] A. S. Stern, R. V. Lewis, A. Kimura, J. Rossier, L. D. Gerber, L. Brink, S. Stein, and S. Udenfriend, *Proc. Natl. Acad. Sci. U.S.A.* **76**, 6680 (1979).
[12] S. Kimura, R. V. Lewis, A. S. Stern, J. Rossier, S. Stein, and S. Udenfriend, *Proc. Natl. Acad. Sci. U.S.A.* **77**, 1681 (1980).
[13] R. V. Lewis, A. S. Stern, S. Kimura, S. Stein, and S. Udenfriend, *Science* (in press).
[14] M. Rubinstein, S. Rubinstein, P. C. Familletti, R. S. Miller, A. A. Waldman, and S. Pestka, *Proc. Natl. Acad. Sci. U.S.A.* **76**, 640 (1979).
[15] S. Stein, C. Kenny, H.-J. Friesen, J. Shively, U. Del Valle, and S. Pestka, *Proc. Natl. Acad. Sci. U.S.A.* **77**, 5716 (1980).

[3] High-Performance Liquid Chromatography of Interferon Tryptic Peptides at the Subnanomole Level

By MENACHEM RUBINSTEIN

Principle

Peptides, and more frequently proteins, have been analyzed by treatment with proteolytic enzymes such as trypsin or chymotrypsin followed by fractionation of the peptide mixture by chromatography, electrophoresis, or a combination of both. Two-dimensional paper or thin-layer chromatography, or electrophoresis in one direction followed by chromatography in a perpendicular direction, has been used frequently to generate a "tryptic fingerprint."[1] As this name suggests, the method is used for identification of a specific protein, for comparing proteins, and sometimes for detection of characteristic proteins, as in certain genetic disorders, such as sickle cell anemia.[2]

[1] V. M. Ingram, *Biochim. Biophys. Acta* **28**, 539 (1958).
[2] C. Baglioni, *Biochim. Biophys. Acta* **48**, 392 (1961).

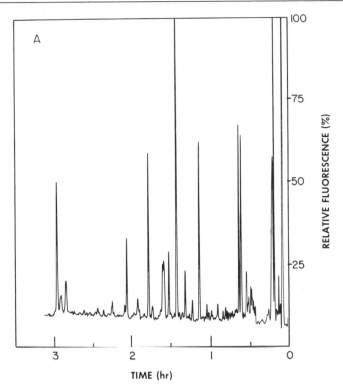

FIG. 1. High-performance liquid chromatography of tryptic peptides of human leukocyte interferon species α_1, β_3, and γ_2 (A–C, respectively). Chromatography was performed on an Ultrasphere octyl column as described under Procedure. From Rubinstein *et al.*[5]

The use of column chromatography, particularly high-performance liquid chromatography (HPLC) for peptide maps, offers several advantages over paper or thin-layer chromatography. The high resolving power of reverse-phase HPLC permits the analysis of highly complicated peptide mixtures in relatively short times and with excellent reproducibility; the column itself is stable, i.e., there is no change in elution position of marker peptides even after repeated use over a period of several months; the recoveries depend on the peptide, but in many cases are better than 90%. Column chromatography can be automated, it is easier to quantify by measuring peak heights rather than the size of spots, and the resolved peptides are collected in tubes ready for further analysis without the extraction step required in thin-layer or paper chromatography.

Most fingerprinting methods apply ninhydrin for visualization of the resolved peptides. In the procedure described here, the fluorescamine-monitoring system[3] is used for detection and quantitation of the column

[3] S. Stein and J. Moschera, this volume [2].

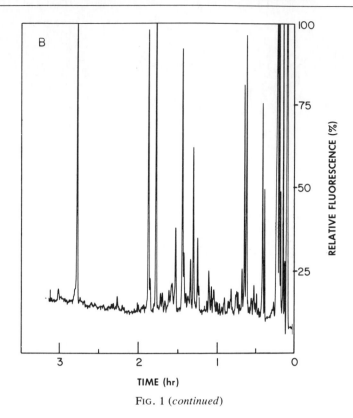

FIG. 1 (*continued*)

eluate. This system offers high sensitivity and the ability to recover most of the peptide in its underivatized form. Reverse-phase chromatography resolves peptides according to their relative hydrophobicity. The more hydrophobic the side chains on a peptide, the longer it will be retained on the column. Since larger peptides carry more hydrophobic groups, they will be retained longer. However, addition of one or more hydrophilic amino acids to a given peptide decreases its elution time in spite of the increased molecular weight.

The eluents that have been used most widely in reverse-phase chromatography are acetonitrile–water or methanol–water mixtures. Their low viscosity permits high flow rates, and their transparency to ultraviolet (UV) light allows monitoring of the peptides by their absorbance in the UV range. In this procedure, *n*-propanol is used for chromatography of larger polypeptides for a number of reasons. Larger peptides require higher concentrations of an organic solvent for elution. However, many polypeptides are not very soluble in solutions containing a large proportion of such organic solvents. Lower concentrations of the more hydrophobic *n*-propanol function as well as do high concentrations

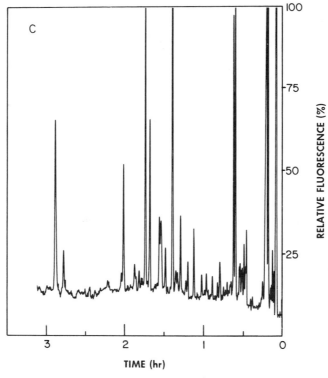

FIG. 1 (*continued*)

of acetonitrile in eluting large polypeptides without the risk of precipitation. The only disadvantage of *n*-propanol is its high viscosity, which does not permit high flow rates. This is not a real limitation because large polypeptides diffuse relatively slowly and require low flow rates for maximal resolution.

The combination of high resolution offered by HPLC with the high sensitivity offered by fluorescamine detection provides clear peptide fingerprints even from large proteins that are available at subnanomole quantities.

Materials and Methods

All the species of human leukocyte interferon were obtained in a pure state by a procedure based on HPLC.[4] Columns of octyl and octadecyl silica (Ultrasphere octyl and Ultrasphere octadecyl prepacked columns, 4.6 × 250 mm, 5 μm particle size, Altex-Beckmann, Berkeley, California) were used. Pyridine, formic acid, *n*-propanol, and acetone were HPLC-

---

[4] M. Rubinstein and S. Pestka, this series, Vol. 78 [67].

grade and were distilled over ninhydrin. All column buffers contained thiodiglycol (0.01%) as an antioxidant. Deionized water was further purified by a system consisting of an activated charcoal filter, a mixed-bed ion exchanger, and a 0.45 μm membrane filter. Additional information on the fluorescamine detector is given in this volume.[3]

Procedure

Samples of interferon (250 pmol) were digested with tosylphenyl-alanine chloromethylketone-treated trypsin (0.16 μg) for 16 hr at 37° in 50 μl of 0.1 N NaHCO$_3$. Then 2-mercaptoethanol (5 μl) was added, and the sample was incubated for 30 min at 37°. After incubation, the sample was brought to pH 3 by the addition of pyridine and formic acid to a final concentration of 0.14 M and 0.5 M, respectively, and applied to the Ultrasphere-octyl column. The column was eluted at a flow rate of 0.5 ml/min with a linear gradient of n-propanol 0 to 40% (v/v) in 0.03 M pyridine/0.1 M formic acid (pH 3) for 3 hr. The column effluent was monitored with the automatic fluorescamine system. Typical results are shown in Fig. 1.[5]

[5] M. Rubinstein, W. P. Levy, J. A. Moschera, C.-Y. Lai, R. D. Hershberg, R. T. Bartlett, and S. Pestka, *Arch. Biochem. Biophys.,* in press (1981).

[4] Amino Acid Analysis of Proteins and Peptides at the Picomole Level: The Fluorescamine Amino Acid Analyzer

By STANLEY STEIN and LARRY BRINK

As in the case of interferon, only microgram amounts of purified proteins may be available for structural analysis. The ability to carry out amino acid analysis at the picomole level is therefore critical. The most modern commercial amino acid analyzers, which use the colorimetric reagent ninhydrin, require that the levels of each amino acid be at least 100 pmol, usually closer to 1 nmol. Fluorometric detection, on the other hand, is applicable at the low picomole level. Although various approaches using fluorogenic reagents are possible, this chapter describes an instrument that uses the reagent fluorescamine for post-column reaction.[1-3] Preparation of the reagents required for analysis at the picomole level is also presented.

[1] S. Stein, P. Böhlen, J. Stone, W. Dairman, and S. Udenfriend, *Arch. Biochem. Biophys.* 155, 202 (1973).
[2] A. M. Felix and G. Terkelsen, *Arch Biochem. Biophys.* 157, 177 (1973).
[3] A. M. Felix and G. Terkelsen, *Anal. Biochem.* 56, 610 (1973).

METHODS IN ENZYMOLOGY, VOL. 79

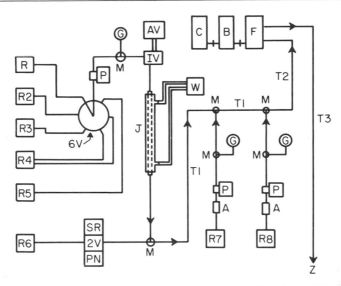

FIG. 1. Schematic diagram of the fluorescamine amino acid analyzer. The numbers in parentheses refer to numbers in column 1 of the table. R1–R5 are reservoirs (7) pressurized at about ca. 7 psi and containing column buffers A, B, C, D, and NaOH, respectively; R5 is a polypropylene bottle with the same fittings as R1–R4; R6 contains N-chlorosuccinimide and is fabricated of heavy-wall polystyrene; R7, borate buffer; R8, fluorescamine; A, air trap (17); B, integrator (3); C, recorder (2); F, fluorometer (1); G, pressure gauge (12); J, jacketed column (5, 6); M, mixing tee (14); P, pump (4); PN, pneumatic activator (10); SR, spring return (11); 2V, two-way valve (9); 6V, six-way valve (8); AV, automatic sample injection valve (15); IV, sample injection valve (16); W, water circulator (13); Z, waste; T1, 0.012 in. i.d., 0.063 in. o.d., 6 in. length; T2, 0.023 in. i.d., 0.063 in. o.d., 10 ft. length; T3, 0.023 in. i.d., 0.063 in. length, 20 ft. length.

Instrumentation

A schematic diagram of the amino acid analyzer is given in Fig. 1. Four eluents are selected in sequence using a rotary (six-way) valve. The eluents are kept under argon pressure (approximately 7 psi). The selected eluent is pumped through an automatic sample injection valve onto a jacketed column containing sulfonated polystyrene resin. The column effluent passes sequentially through three tee fittings. At the first tee a solution of N-chlorosuccinimide may be added by gas pressure pumping. The flow of this reagent is controlled by a two-way valve, which is opened only during the elution of proline (or hydroxyproline) and closed at all other times. The column effluent then mixes with borate buffer and fluorescamine at the next two tee fittings, passes through the flow cell of the filter fluorometer, and goes to waste. The fluorescence pattern is plotted on a dual-pen chart recorder with an integrator to measure the

LIST OF COMPONENTS FOR AMINO ACID ANALYZER[a]

Component	Quantity	Supplier	Catalog No.
1. Fluorometer	1	Varian Instrument Division, Palo Alto, CA	Fluorochrom
2. Recorder	1	Linear Instruments, Irvine, CA 92714	585
3. Integrator	1	Spectra Physics, Santa Clara, CA	System I
4. Pumps	3	Laboratory Data Control, Riviera Beach, FL	396-31
5. Column	1	Glenco Scientific, Houston, TX	Series 3202
6. Resin	5 g	Durrum Chemical Corp., Sunnyvale, CA	DC-4A
7. Pressurized reservoirs	5	Omnifit Inc., Cedarhurst, NY	—
8. Six-way valve	1	Altex Scientific Inc., Berkeley, CA	202-01
9. Two-way valve	1	Altex Scientific Inc.	201-50
10. Pneumatic actuator	1	Altex Scientific Inc.	201-57
11. Spring return	1	Altex Scientific Inc.	201-58
12. Pressure gauges		Enerpac, Butler, WI	
a. 0–4000 psi	1		—
b. 0–200 psi	2		—
13. Water circulator	1	Brinkmann Instrument, Westbury, NY	Lauda
14. Mixing tees	6	Valco Instrument Co., Houston, TX	—
15. Automatic injection valve	1	Valco Instrument Co.	AH-CST-16-HPax-HC
16. Injection valve	1	Rheodyne, Inc., Cotati, CA	7010
17. Bubble traps	2	Beckman Instrument Co., Palo Alto, CA	326187

[a] Teflon tubing, fittings, and other small components are listed in the table of Chapter [2] of this volume.

areas of the peaks. A listing of the components and the suppliers is given in the table. A suitable electronic controller for the automated buffer changes and the automated sample injection is available from Eldex Labs (Menlo Park, CA).

Reagents

Sodium citrate for amino acid analysis (E. Merck, catalog No. 6430); sodium chloride (Suprapur, E. Merck, catalog No. 6406), and sodium

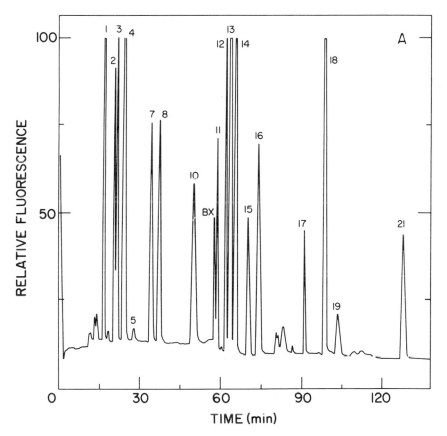

FIG. 2. Amino acid analysis of a hydrolyzate of human leukocyte interferon. This chromatogram represents about 0.3 μg of protein. 1, Aspartic acid; 2, threonine; 3, serine; 4, glutamic acid; 5, cysteine; 6, proline; 7, glycine; 8, alanine; 10, valine; 11, methionine; 12, isoleucine; 13, leucine; 14, norleucine (internal standard); 15, tyrosine; 16, phenylalanine; 17, histidine; 18, lysine; 19, ammonia; 21, arginine. Cysteine is determined in panel A, whereas proline is determined in panel B by the addition of N-chlorosuccinimide, which destroys the cysteine peak.

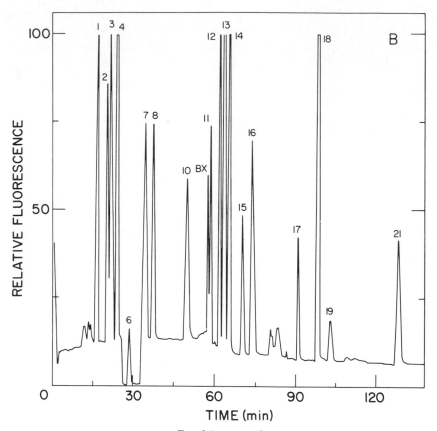

FIG. 2 (*continued*)

hydroxide (Suprapur, MCB, catalog No. SX589) were obtained from MCB Manufacturing Chemists, Inc. (Cincinnati, Ohio). EDTA (catalog No. 6354) was from Eastman Kodak Co. (Rochester, New York).

Acetone and 2-propanol were distilled over ninhydrin. Constant-boiling hydrochloric acid was prepared over sodium dichromate. House-distilled water was further purified through mixed-bed deionizer and activated charcoal cartridges (Hydro Service and Supplies, Durham, North Carolina). Thiodiglycol (25% v/v), lithium hydroxide, and pHix preservative were from Pierce Chemical Co. (Rockford, L.I., New York).

The column buffers were prepared as follows to a final volume of 1 liter with water. Buffer A: 17.2 g of sodium citrate (0.175 N Na$^+$), 25 ml of 2-propanol, 5 drops of pHix, adjusted to pH 3.15 with hydrochloric acid. Buffer B: 17.2 g of sodium citrate, 5 drops of pHix, adjusted to pH 3.45. Buffer C: 19.6 g of sodium citrate (0.20 N Na$^+$), 0.5 ml of thiodiglycol, 5 drops of pHix, adjusted to pH 4.10. Buffer D: 44.1 g of sodium citrate, 32.1 of g sodium chloride ($1.0 N$ Na$^+$), 0.5 ml of thiodiglycol, 5 drops of pHix, about pH 7.9. Wash: 7 g of sodium hydroxide ($0.2 N$).

The sample diluent was prepared by adjusting buffer A to pH 2.2 with hydrochloric acid. The detection system buffer is $0.2 M$ boric acid adjusted to pH 9.6 with a saturated solution of lithium hydroxide. Fluorescamine is dissolved in acetone (0.2 g/liter). N-Chlorosuccinimide (Eastman, Rochester, New York) is dissolved in water ($5 \times 10^{-4} M$).

Chromatography

The column is preheated to 60° and then packed with a slurry of the resin in $0.2 N$ NaOH. The resin is added and packed in several steps, care being taken to avoid air bubbles, to a height of 33 cm (10% higher than the resin bed will be at the end of packing). The column is washed at 15 ml/hr with $0.2 N$ NaOH for 1.5 hr and then for 1 hr with buffer A. The complete set of buffers is cycled through the column before use.

Chromatography is carried out at 18 ml/hr at 57° with the following program: A, 15 min; B, 17 min; C, 24 min; D, 41 min; NaOH, 5 min; A, 18 min. Borate buffer and fluorescamine solution are pumped at 35 and 18 ml/hr, respectively. When used, the N-chlorosuccinimide is delivered at 6 ml/min by adjusting the pressure on the reservoir.

Two chromatograms depicting hydrolyzates representing 17 pmol (0.3 μg) of human leukocyte interferon are given in Fig. 2. In Fig. 2A proline is not being determined, whereas in Fig. 2B proline is determined at the expense of cysteine. The N-chlorosuccinimide used for proline detection deaminates α-amino acids, such as cysteine, to nonfluorogenic products.

[5] Amino Acid Analysis of Protein Bands on Stained Polyacrylamide Gels

By STANLEY STEIN and LARRY BRINK

Amino acid analysis can be carried out on proteins that are present as bands on stained polyacrylamide gels.[1] After electrophoresis, the gel is stained with Coomassie Blue and destained extensively with 7% acetic acid–10% methanol in the usual manner. The bands of interest (and blank regions) are excised with a razor blade and briefly rinsed with water. The

[1] S. Stein, C. H. Chang, P. Böhlen, K. Imai, and S. Udenfriend, *Anal. Biochem.* **60,** 272 (1974).

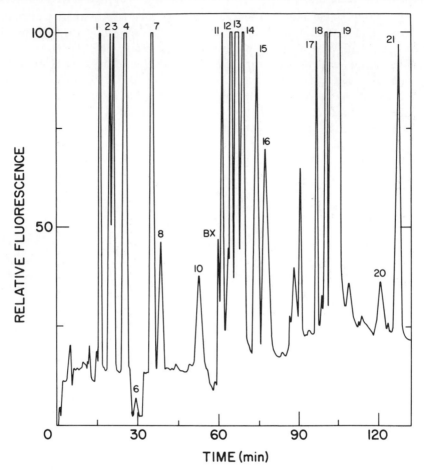

FIG. 1. Amino acid analysis of a hydrolyzate of human fibroblast interferon present as a stained band on a gel. This chromatogram represents about 230 ng of protein. 1, Aspartic acid; 2, threonine; 3, serine; 4, glutamic acid; 6, proline; 7, glycine; 8, alanine; 10, valine; 11, methionine; 12, isoleucine; 13, leucine; 14, norleucine (internal standard); 15, tyrosine; 16, phenylalamine; 17, histidine; 18, lysine; 19, ammonia; 20, tryptophan area; 21, arginine.

slices are placed in hydrolysis tubes, dried *in vacuo,* and then hydrolyzed *in vacuo* in 200 μl of constant-boiling hydrochloric acid, containing 1% thioglycolic acid, at 110° for 24 hr. After hydrolysis, the hydrolyzate is removed from the gel residue, dried *in vacuo,* dissolved in the pH 2.2 diluent, and run on the fluorescamine amino acid analyzer (see this volume [4]).

A chromatogram of a hydrolyzate of human fibroblast interferon is presented in Fig. 1. The amount of hydrolyzate injected onto the column was about 12 pmol (230 ng) of interferon. Even though the response per mole for ammonia is close to one-thousandth that of a typical amino acid, the ammonia peak is the largest one in the chromatogram. Analysis of the basic amino acids would not be possible if ninhydrin or o-phthalaldehyde were used for detection, owing to the tremendous ammonia peak that would ensue.

With this procedure it is not necessary to extract the protein from the gel in order to carry out amino acid analysis. Extraction from the gel typically gives poor recovery of protein and introduces amino acid contamination.

This procedure offers a unique opportunity to characterize a protein. In many instances gel electrophoresis is the simplest method for separating a protein from a mixture, and it is applicable at the low microgram level. In our studies on purified human fibroblast interferon, a minor contaminant of about 40,000 daltons was detected on gels. Amino acid analysis revealed it to be a dimer of interferon by its identity in composition with the 20,000-dalton band.[2]

[2] H. J. Friesen, S. Stein, M. Evinger, P. C. Familletti, J. Moschera, J. Meienhofer, J. Shively, and S. Pestka, *Arch. Biochem. Biophys.* **206**, 432 (1981).

[6] Manual Edman Sequencing Techniques for Proteins and Peptides at the Nanomole Level

By WARREN P. LEVY

A major obstacle in determining the primary sequence of proteins has been the availability of sufficient amounts of pure material for sequencing. Standard techniques for automatic or manual sequencing that use the Edman reaction[1] require milligram amounts of protein. Purification protocols for many proteins yield only microgram amounts of homogeneous material, necessitating the development of new methods for sequence analysis.

The manual dansyl-Edman technique has been applied successfully to peptides[2,3] and intact proteins[4] at the nanomole level. However, this technique requires the removal of an aliquot of the peptide at each cycle for

[1] P. Edman, *Acta Chem. Scand.* **4**, 283 (1950).
[2] C. J. Bruton and B. S. Hartley, *J. Mol. Biol.* **52**, 165 (1970).
[3] W. R. Gray and J. F. Smith, *Anal. Biochem.* **33**, 36 (1970).
[4] A. M. Weiner, T. Platt, and K. Weber, *J. Biol. Chem.* **247**, 3242 (1972).

hydrolysis and subsequent analysis, increasing the repetitive losses normally encountered in Edman sequencing. The tremendous sensitivity of the method, which has been the major reason for its widespread use, has been approached by recent improvements in the analysis of subnanomolar quantities of phenylthiohydantoin (PTH) amino acids.[5-7]

Automatic spinning cup sequenators have been successfully remodeled for microsequence analysis after extraordinary investments of time, manpower, and resources.[8-10] Although these automatic methods are clearly required for long peptides or intact proteins, a simple manual technique would be beneficial in the analysis of small peptides or of the first few residues of a large peptide. We report here a manual microsequencing strategy that has been used to determine the primary sequence of nanomolar and subnanomolar quantities of small peptides (nine amino acids or less) produced by tryptic digestions of human leukocyte interferon.[11] An excellent theoretical analysis of the Edman chemistry and typical manual sequencing procedures has already been published[12] and will not be repeated here.

Materials

Polypropylene microcentrifuge tubes (1.5 ml; Walter Sarstedt, Inc., Princeton, New Jersey) are used exclusively throughout this procedure to minimize losses of peptides due to adsorption. Ultraviolet-absorbing impurities are removed from the tubes by washing with acid. Tubes are soaked in 2 N HCl for at least 4 hr, rinsed in ultrapure water, and air-dried. The polypropylene tubes are preferable to acid-washed glass vials, which have been shown to adsorb large quantities of peptides and to inhibit the extraction and washing steps by selectively adsorbing the emulsified droplets produced during vigorous mixing. All manipulations of liquids are done with polypropylene micropipette tips (Fisher Scientific Co., Springfield, New Jersey).

Phenylisothiocyanate (PITC), trifluoroacetic acid (TFA), and N,N-dimethyl-N-allylamine (DMAA) are Sequanal grade from Pierce Chemical

[5] C. L. Zimmerman, E. Apella, and J. J. Pisano, *Anal. Biochem.* **77,** 569 (1977).
[6] M. Abrahamsson, K. Gröningsson, and S. Castensson, *J. Chromatogr.* **154,** 313 (1978).
[7] A. J. Bhown, J. E. Mole, A. Weissinger, and J. C. Bennett, *J. Chromatogr.* **148,** 532 (1978).
[8] B. Wittmann-Liebold, *Hoppe Seyler's Z. Physiol. Chem.* **354,** 1415 (1973).
[9] M. W. Hunkapiller and L. E. Hood, *Biochemistry* **17,** 2124 (1978).
[10] M. W. Hunkapiller and L. E. Hood, *Science* **207,** 523 (1980).
[11] W. P. Levy, M. Rubinstein, J. Shively, U. Del Valle, C.-Y. Lai, J. Moschera, L. Brink, L. Gerber, S. Stein, and S. Pestka, in preparation.
[12] G. E. Tarr, this series, Vol. 47, p. 335.

Co., Rockford, Illinois. Heptane, ethyl acetate, and acetonitrile are from Burdick and Jackson Laboratories, Inc., Muskegon, Michigan. Pyridine is distilled over ninhydrin. Water is purified by glass distillation followed by filtration through charcoal and mixed-bed deionizer cartridges (Hydro Service and Supplies, Inc., Durham, North Carolina). It has been our experience that the use of ultrapure solvents is the most important prerequisite for reproducible results.

Phenylthiohydantoin amino acids are identified by high-performance liquid chromatography (HPLC) with a system (Altex Scientific, Berkeley, California) consisting of an Ultrasphere ODS column (5 μm), Model 420 microprocessor, two Model 110-A pumps, and a Model 153 UV detector. Samples are dried under vacuum with a Speed Vac concentrator (Model SVC100, Savant Instruments, Inc., Hicksville, New York). All centrifugation steps are for 1 min in an Eppendorf microcentrifuge (Model 5412, Brinkmann Instruments, Inc., Westbury, New York).

Procedures

Sample Preparation. The tryptic peptide (1–10 nmol) is transferred to an acid-washed 1.5-ml polypropylene microcentrifuge tube. Norleucine (25–50 nmol) is added as a carrier and internal standard to monitor the efficiency of the reactions. The sample is dried under vacuum, and the tube is flushed with argon gas. Nitrogen gas can be used also, but argon is heavier than air and will remain in the tube when the cap is removed. Peptides have been stored for at least 2 weeks under these conditions with no observable loss in coupling efficiency.

PITC Coupling. The peptide is suspended in 40 μl of DMAA buffer (15 ml of pyridine, 1.18 ml of DMAA, 10 ml of H_2O, pH adjusted to 9.5 with TFA). This buffer provides a higher yield of coupled peptides than others that have been examined. Triethylamine and trimethylamine give inferior yields, presumably owing to the presence of decomposition products that accumulate upon storage and may inhibit the reaction. The vial is flushed with argon gas, and immediately thereafter 3 μl of PITC are added. The reaction proceeds for 30 min at 50°.

Excess reagent is removed by washing twice with 150 μl of heptane–ethyl acetate (10 : 1) and once with 200 μl of heptane–ethyl acetate (2 : 1). Vigorous mixing facilitates removal of the excess reagent and by-products of the reaction. Centrifugation is required for adequate separation of the two phases. The organic phase is discarded after each wash, and the aqueous phase is dried under vacuum after washing is complete.

Cleavage. Trifluoroacetic acid is used for cleavage. Anhydrous HCl has been suggested as a better reagent,[12] but we have found TFA to be adequate for most peptides. Twenty microliters of TFA are added to the

RESOLUTION OF
PHENYLTHIOHYDANTOIN (PTH)
AMINO ACIDS BY
HIGH-PERFORMANCE LIQUID
CHROMATOGRAPHY[a]

PTH amino acid	Elution time (min)
Asp	3.8
Asn	6.5
Glu	7.0
Ser	7.4
Gln	8.0
Thr	8.5
His	9.0
Gly	9.4
Arg	12.7
Ala	13.0
Tyr	15.1
Met	20.2
Val	20.5
Pro	20.7
Trp	23.4
Phe	25.5
Ile	26.3
Lys	26.6
Leu	27.6

[a] The gradient buffers are 10 mM sodium acetate, pH 4.5 (buffer A) and 5 mM sodium acetate, pH 4.5, 50% acetonitrile (buffer B). The PTH amino acids are eluted with a linear gradient of 40 to 75% B for 12.5 min followed by isocratic elution at 75% B for 17.5 min. The column temperature is maintained at 62°. These conditions provide a maximum sensitivity of approximately 20 pmol.

vial, and the sample is incubated at 50° for 20 min. The sample is then dried under vacuum.

Extraction. Forty microliters of 30% pyridine are added to dissolve the sample, and the anilinothiazolinone (ATZ) amino acid is separated from the residual peptide by extracting three times with 150-μl portions of benzene–ethyl acetate (1 : 2). Centrifugation is required after mixing to provide adequate phase separation.

The organic layers from each extraction are pooled and dried under vacuum. The aqueous layer containing the peptide is supplemented with 25–50 nmol of norleucine and dried under vacuum. After flushing with argon gas, a second cycle of Edman degradation can be performed as described above.

Conversion. The dried organic phase containing the ATZ amino acid is suspended in 40 μl of 1 N HCl. Conversion of the ATZ amino acid to the PTH derivative of the amino acid is complete after 10 min at 80°.

The PTH amino acid is removed from the aqueous phase by three extractions with 50-μl portions of ethyl acetate. The ethyl acetate layers are pooled, dried under vacuum, and suspended in 30 μl of acetonitrile.

Identification of PTH Amino Acid. The PTH amino acids are identified with an HPLC system described under Materials. The individual PTH amino acids are resolved within 30 min under our experimental conditions (see the table).

Discussion

The microsequencing strategy described here has been used to determine the complete primary sequence of approximately 1-nmol amounts of two pentapeptides produced by tryptic digestion of human leukocyte interferon, and partial sequence data on a number of larger peptides.[11] This methodology is generally applicable for microsequencing of nanomole and subnanomole amounts of peptides.

Acknowledgments

The author would like to thank S. Stein and S. Udenfriend for use of their facilities, J. Moschera for the HPLC system, L. Gerber and L. Brink for expert technical assistance, S. Kimura, M. Rubinstein, and R. Lewis for helpful suggestions, and S. Pestka for support of this research.

[7] Sequence Determinations of Proteins and Peptides at the Nanomole and Subnanomole Level with a Modified Spinning Cup Sequenator

By JOHN E. SHIVELY

The most widely used approach to determine extended amino acid sequences of proteins or peptides is the Edman degradation scheme,[1] which was published in automated form in 1967 by Edman and Begg.[2] The

[1] P. Edman, *Acta Chem. Scand.* **4**, 283 (1950).
[2] P. Edman and G. Begg, *Eur. J. Biochem.* **1**, 80 (1967).

automated instrument is routinely used to sequence proteins in about the 100–400 nmol range and, depending on the protein, gives the NH_2-terminal sequence of 10–40 amino acids. Since increasing the sensitivity of sequence analysis is a major goal of protein chemistry, attention has focused on improving in the instrument performance and in the detection of the resulting phenylthiohydantoin (PTH) amino acids. The following advances are now available.

1. Substantial improvements in instrument performance, and reagent and solvent purification methods, were described in 1973 by Wittman-Liebold.[3]
2. The ability to separate and detect all of the common PTH-amino acid derivatives in the subnanomole range was made possible by the use of reverse-phase, high-performance liquid chromatography.[4]
3. The description of a device that permits the automatic conversion of ATZ to PTH derivatives of amino acids was made in 1976 by Wittman-Liebold et al.[5]
4. The use of polybrene as a carrier in the spinning cup prevents solvent washout of small peptides[6] or low nanomole to subnanomole amounts of proteins.[7]

The first description of the combined use of these improvements appeared in a 1978 report by Hunkapiller and Hood.[7] The purpose of this chapter is to describe the improvements and to give practical criteria that critically assess the instrument performance. The chemistry and operational description of an automated sequencer are thoroughly discussed by Niall[8] in this series. One of the applications of the instrument described here is the sequence determination of human leukocyte interferon, which is currently available in the nanomole range. Representative results from these studies are discussed.

Abbreviations and Proprietary Names. Phenylisothiocyanate (PITC), phenylthiocarbamyl (PTC), phenylthiohydantoin (PTH), anilinothiazolonine (ATZ), HFBA (heptafluorobutyric acid), high-performance liquid chromatography (HPLC), methylthiohydantoin (MTH), N,N,N^1N^1-tetrakis-(2-hydroxypropyl)ethylenediamine (Quadrol), polytetrafluoro-

---

[3] B. Wittman-Liebold, *Hoppe-Seyler's Z. Physiol. Chem.* **354**, 1415 (1973). The miniature diaphragm valves are described by Graffunder *et al.* in U.S. Patent No. 4,168,724 owned by Max-Planck-Gesellschaft, Federal Republic of Germany.

[4] C. L. Zimmerman, E. Apella, and J. J. Pisano, *Anal. Biochem.* **77**, 569 (1977).

[5] B. Wittman-Liebold, H. Graffunder, and H. Kohls, *Anal. Biochem.* **75**, 621 (1976).

[6] D. A. Klapper, C. E. Wilde, and J. D. Capra, *Anal. Biochem.* **85**, 126 (1978).

[7] M. W. Hunkapiller and L. E. Hood, *Biochemistry* **17**, 2124 (1978).

[8] H. D. Niall, this series, Vol. 27, p. 942.

ethylene (PTFE or Teflon), polytrifluorochloroethylene (Kel-F), trifluoroacetic acid (TFA), dithiothreitol (DTT).

Instrument Design

The modifications described here were performed on a Beckman 890C
sequenator but in principle can be applied to other instruments by allowing for changes from one model to another. The following components
should be removed from the sequenator: (a) entire low-vacuum system
including all lines, manifolds, and connections; (b) high-vacuum line and
valve; (c) reagent, solvent, and waste diaphragm valves (5 items); (d)
three-way valves and manifold that operates diaphragm valves; (e) reagents and solvents and their connections except for vent and pressurized
lines: (f) cup assembly including cover plate for drive housing (save spinning cup and cup drive); (g) fraction collector; (h) refrigeration unit.

The replacement components that follow are described by systems and
require machine and glass shop fabrication work. Swage and pipe fittings
required to make necessary connections in the systems described are
omitted from this discussion for the sake of brevity and simplicity. The
reader is advised to plan carefully for these needs. In addition, it is necessary to have wiring diagrams for the instrument and an understanding of
their operating principles. Teflon tubing is connected to Teflon plugs or
inserts via "pull through" fittings; i.e., heat and stretch the tubing to a
smaller inner diameter (i.d.), insert into predrilled hole which has smaller
i.d. than the nominal outer diameter of the tubing, pull the tubing through
until it is tightly sealed, and trim off excess.

High-Vacuum System

The drive housing is connected to a high-capacity (400 liters/min or
greater) pump capable of achieving a vacuum of at least 1 μm via a liquid
nitrogen trap and two electrically activated valves. One convenient and
reliable source of the vacuum pump, electrically actuated valves, and
stainless steel tubing, together with "quick fit" connectors, is Leybold-
Heraeus.[9] The trap and Dewar flask should be made of stainless steel. The
trap has a 2–3-liter capacity and is equipped with "quick fit" flanges. The
valves are stainless steel, right angle, have an 0.5 inch i.d. (KF 16), and
are connected together. The plunger of one of the valves should be drilled
through to create a 0.1 mm leak for restricted vacuum.[10] Thus, when the
undrilled valve is actuated, a restricted vacuum function is obtained; and
when both valves are actuated, a hard vacuum is obtained. The valves are

---

[9] Leybold Heraeus, Trivac, Model D16A, Leybold Heraeus, Monroeville, Pennsylvania.
[10] M. W. Hunkapiller, personal communication.

wired to relays that deliver voltage to the valves when the appropriate program signal is received. The valves should be connected to the drive housing via a one-piece adapter that is machined to fit the place of the "Bimba" valve, projects through the floor of the housing, and terminates in a "quick fit" flange. Alternatively, connection is made to a new drive housing as described below.

Drive Housing

The drive housing is modified to accept a more reliable cup assembly drive. There are two choices in drive-housing modifications: one is to alter the existing housing; the other is to machine a new one. The existing housing can be modified as follows.

1. Weld on the high-vacuum port adapter.
2. Machine a flat top plate (stainless steel), which replaces the old slanted cup assembly cover plate.[10] The new plate is mounted in the housing with the three existing screws, is sealed with the existing O-ring, and has a center hole for the spinning cup. Three 2.75-inch metal posts (0.5-inch diameter) are mounted on the plate. The assembly of the plate and posts, together with the Teflon plug, is shown in Fig. 1.
3. The N_2 cell solenoid and connection remain the same, but should be leak-tested at high vacuum.
4. The N_2 flush solenoid is left in place. The line to the cup is replaced with Teflon tubing, which is connected to the solenoid with Swage fittings and to the Teflon plug in the cup with a pull-through fitting.
5. The old restricted vacuum inlet is capped.
6. The cell vacuum sensor should be connected to the housing via a valve (manual or electrically actuated). All new fittings should be thoroughly leak-tested at high vacuum.

Although the above modifications of the drive housing are adequate for microsequencing, it is highly recommended to machine a new drive housing of single-piece construction with "quick-fit" flanges welded in place of the pipe-threaded fittings in the old design. Single-piece construction eliminates the O-ring seal at the metal cup in the bottom of the drive housing. This particular seal is the single most common source of vacuum leaks in the drive housing. The design of the single-piece unit should otherwise follow closely the dimensions of the Beckman drive housing. In addition, it is recommended that O-ring grooves be machined larger to fit the commercially available Teflon encapsulated O-rings.[11] Encapsulated O-rings eliminate background introduced by leaching of elastomers into the cup by exposure to organic solvent vapors.

[11] CAPFE O-rings, Ace Glass Inc., Vineland, New Jersey.

FIG. 1. Photograph showing reaction cup assembly adapted from a design by M. W. Hunkapiller and L. E. Hood. A bell jar is mounted on drive-housing adapter plate (A) that has recessed O-ring. Teflon plug (B) that has a recessed O-ring (C) fits on top of bell jar; vacuum and pressure seal is made by pressing top plate (D) against bell jar (E). Three support posts (F) are mounted on drive-housing adapter plate.

Cup Assembly

The cup assembly is shown in Fig. 1. The design principle[10] is to create a seal on the bell jar by pressing a top metal plate against the Teflon plug, which seals to the bell jar via an O-ring and in turn is pressed against the top plate of the drive housing, which has an O-ring in it. This method of creating a seal against vacuum and pressure is superior to the design in the unmodified instrument. The dimensions of the plug and bell jar are not critical, but should fit easily into the space between the metal posts on the drive housing plate. The O-rings should be Teflon encapsulated. The plug has three Teflon lines that are sealed via pull-through fittings. One line is connected to the N_2 flush solenoid, the second is the scoop line, and the third is the common reagent and solvent delivery line. The scoop line is fitted into a glass elbow for support and should be cut on an angle as described in the Beckman sequencer manual. When the plug is removed from the assembly, care is taken to tilt it out of the groove so as not to damage the scoop line. The delivery line should be trimmed to almost touch the knob at the bottom of the spinning cup when the plug is in place. This ensures a gentle, uniform delivery to the cup and prevents disruption of the film by turbulence or splashing.

Solvent–Reagent Delivery System

A major problem with the diaphragm valves in the Beckman sequencer is the tendency for the diaphragm to crack and leak. These valves are

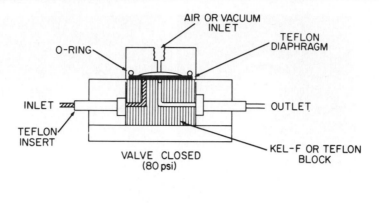

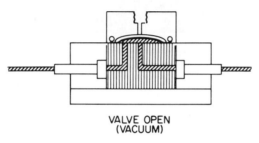

FIG. 2. Schematic of Wittman–Liebold miniature diaphragm valve.[3] Air pressure presses a Teflon diaphragm against either a Kel-F or Teflon block to provide a liquid seal on an inlet line, which in turn is held against the Kel-F or Teflon block by a metal fitting (not shown). When vacuum is supplied to the upper block via a three-way electrically actuated valve, the Teflon diaphragm is displaced upward and allows liquid to flow from the inlet to the outlet side. The outlet side is connected to a center running zigzag line (more clearly shown in Fig. 3), which is open on the top of the block. The Teflon diaphragm forms the top seal on this line. The whole assembly is held together by screws (not shown) that apply a seal on the upper portion of the Teflon diaphragm by an O-ring. The O-ring does not come in contact with liquid.

replaced with the miniature valves designed by Wittman-Liebold.[3] The operation of a single valve is shown in Fig. 2. A Kel-F or Teflon block of rough dimension 1.5 × 3.0 cm (length of block varies depending on the number of valves) is drilled for inlet and outlet lines that are sealed with a Teflon diaphragm (3 × 3 cm) that is pressed against the block by air pressure (120 psi). When vacuum is applied to the diaphragm, it bows upward and allows liquid to flow from inlet to outlet. Air or vacuum is supplied to the valve via a three-way solenoid. The three-way solenoid operation is similar to that in the unmodified instrument. It is necessary to machine two manifolds for their operation. One manifold supplies air to all three-way valves and is connected to a compressed air tank. The other

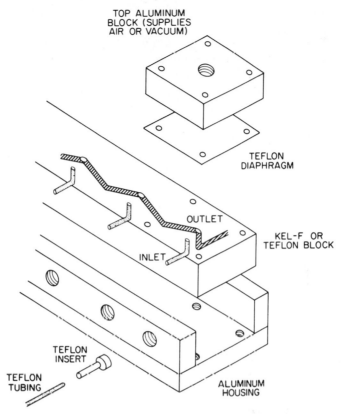

FIG. 3. Schematic of Wittman–Liebold cell delivery valve[3] that illustrates how a series of inlets for reagents and solvents exit through a single outlet. See Fig. 1 for details of the operation of a single portion of this unit.

manifold supplies vacuum to the three-way valves and is connected to a low-capacity vacuum pump.[12] A key design feature of the diaphragm valves is that there are many inlet lines to the valves, but only a single outlet line. Figure 3 illustrates how the system works. When a valve is actuated, liquid flows from the inlet line to a common zigzag line that exits (in this case) to the cup. The delivery system for the cup consists of inlet lines for three solvents, three reagents, and N_2. The N_2 line at the far end of the block acts as a "blowout" to displace liquid from the block to the cup. In this system no liquids or vapors are in contact with O-rings. The Teflon lines are sealed to Teflon inserts by "pull through" fittings. The Teflon inserts are sealed against the diaphragm block by a metal adapter

[12] Sargent-Welch Direct Drive, Model 8805, Sargent-Welch Scientific Co., Skokie, Illinois.

(not shown in Fig. 3) that fits over the insert and threads into the aluminum housing.

A second series of diaphragm valves, which are contiguous with the cup delivery system, supplies reagent, solvent, and N_2 blowout to the automatic conversion flask. A third series of diaphragm valves controls vacuum, N_2, and waste to the conversion flask, and delivery from the cup to the conversion flask. A flow diagram illustrating the functions and relationships among the diaphragm valves, cup, and conversion flask is shown in Fig. 4. Liquid flow from the scoop to the diaphragm valve may be directed to either a waste bottle (waste$_1$) or to the conversion flask. Delivery from the conversion flask to the fraction collector is achieved by pressurizing the converion flask via N_2 blowout (line on far right side of conversion flask), which forces liquid up through the center tube in the

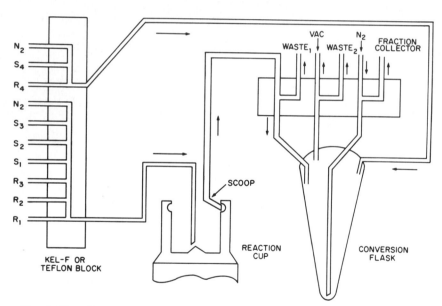

FIG. 4. Flow diagram of sequencer operation. Valves on left deliver reagent and solvents to reaction cup (Beckman spinning cup) or conversion flask. When valves are activated (see Figs. 1 and 2) liquid flows through single exit line to either cup or flask and is entirely displaced by a stream of N_2 from the blowout valve (N_2). Delivery from the cup to the flask is made through the scoop line (same as in the Beckman design) via a three-way valve of similar design to the reagent–solvent delivery valves. This valve determines whether delivery is to the flask or to waste. In addition, the conversion flask is supplied with three-way valves that supply the following functions: vacuum or an exit to waste, N_2 or delivery to the fraction collector. Delivery to the fraction collector is made possible by pressurizing the flask with N_2 via the blowout function. The conversion flask design and operation is the same as described by Wittman-Liebold et al.[5]

TABLE I
LIST OF REAGENTS AND SOLVENTS[a]

R_1, 5% PITC in n-heptane
R_2, 0.33 M Quadrol buffer[b] (pH 9.0) in n-propanol–water (3 : 4, v/v)
R_3, HFBA
R_4, TFA–water (25%)
S_1, benzene
S_2, ethyl acetate–0.1% acetic acid
S_3, 1-chlorobutane–0.001% DTT
S_4, Acetonitrile–0.001% DTT

---

[a] Reagent and solvent compositions and purifications are similar to those described by Hunkapiller and Hood.[7]

[b] Quadrol (purchased from Wyandotte Chemicals Corp.) is purified by removal of volatile components in a molecular still and gives a negative Tollen's test for aldehydes. It is warmed and dissolved in n-propanol–water (3 : 4, v/v) to give a 1 M solution that is adjusted to pH 9.0 with TFA. The stock solution is diluted to 0.33 M in 25% n-propanol–water (v/v).

conversion flask that runs to the bottom of the flask. The conversion flask operation is essentially identical to that described by Wittman-Liebold *et al.*[5] The composition of the reagents and solvents is listed in Table I. The conversion flask and the diaphragm valves required for its operation (see Fig. 4) are mounted in a portion of the space previously occupied by the fraction collector in the unmodified instrument. A hot water circulator[13] is mounted beneath the conversion flask in the space formerly occupied by the compressor. Necessary tubing is run through the floor of what was previously the fraction collector chamber.

The reagent and solvent bottles are replaced with bottles with ground-glass joints (standard taper 24/40) that accept appropriately machined Teflon plugs. The vent, pressurize, and delivery tubes are "pulled through" the plug to create seals without the use of O-rings. The plugs are held in place on the ground-glass joints with screw clamps. This system eliminates the 5 O-rings used per bottle in the unmodified sequenator.

Fraction Collector

A fraction collector[14] is mounted in an enclosed Lucite box adjacent to the conversion flask in what was previously the fraction collector chamber. An inert atmosphere is supplied to the enclosed box via a needle

---

[13] Lauda hot water circulating bath, Model T-1, Lauda, West Germany.
[14] Gilson microfractionator, slave unit, Gilson Medical Electronics Inc., Middleton, Wisconsin.

valve connected to the N_2 system and the box. The top of the box is removable in order to remove and replace tubes in the fraction collector. Tubing connecting the appropriate diaphragm valve (see Fig. 4) to the fraction collector is run through a sealed grommet mounted in the side of the Lucite box. The fraction collector is advanced by a signal from an auxiliary relay actuated by the "fraction collector advance" signal from the programmer. The length of the signal pulse in milliseconds can be controlled by an RC circuit if necessary. The power cable and cable to the relay are run through grommets in the Lucite box.

Since the fractions collected are the stable PTH derivatives, it is not considered necessary to refrigerate them as is done for the less stable ATZ derivatives in the unmodified sequenator. However, it is advisable to store them in an oxygen-free atmosphere as much as possible. Experience in this laboratory suggests that either nitrogen or argon, as used by Hunkapiller and Hood,[7] is adequate.

Wiring and Programmer Changes

In order to effect the changes described above, it is necessary to break certain existing interlocks on the Beckman sequenator and to add new functions. The interlocks that prevent delivery of R_4 or S_4 when vacuum functions are engaged are broken. The new functions can be picked up from unused pins on the programmer and wired to existing switches (or functions) that are not otherwise used in the modified sequenator. All wiring changes require the use and understanding of the circuit diagrams supplied by Beckman with their instrument. Since the operation of a given function is simply the activation of a given solenoid, a good strategy would be to replace the outdated relay logic system with transitor–transitor logic (TTL) or its equivalent. Unfortunately a detailed description of the required wiring changes is beyond the scope of this chapter, and only a general guide will be given here.

Conversion Flask Controls. The former five-way control for the fraction collector functions (VAC, DRY, N_2, OFF, AUTO) is used for operation of the conversion flask; i.e., VAC activates the three-way valve for vacuum to the conversion flask, N_2 actuates the valve for N_2, and DRY actuates N_2 *and* vacuum. The former control labeled "FRAC. COLL." is relabeled "CONV. FLASK" and actuates the solenoid, which controls delivery from the scoop to the conversion flask. Delivery from the scoop to waste ($WASTE_1$) is controlled by the previous "WASTE" function and is relabeled "$WASTE_1$." The waste function on the conversion flask ($WASTE_2$) is controlled by the addition of a new function mounted directly above $WASTE_1$ on the control panel. This new function is picked up at pin 41 on the programmer.

Blowout. The new blowout functions that deliver N_2 to either the cup or the conversion flask are achieved by adding one new function that is interlocked to either WASTE$_1$ (cup) or WASTE$_2$ (conversion flask) to give two functions. The new function labeled "BLOW OUT" is mounted between HIGH VAC and WASTE$_2$ and is actuated by pin 39 on the programmer. The combination of BLOW OUT and WASTE$_1$ delivers N_2 to the cell (plus WASTE$_1$), and BLOW OUT and WASTE$_2$ delivers N_2 to the conversion flask (plus WASTE$_2$).

Fraction Collector. The wiring for the fraction collector advance function has been discussed earlier. The new function for delivery to the fraction collector is mounted on the control panel next to WASTE$_2$ and is labeled "COLLECTOR." This function is picked up from pin 42 on the programmer.

Vacuum. The "RESTRICTED" vacuum function is used to actuate the undrilled vacuum valve discussed earlier. The "HIGH" vacuum function activates both valves. The "LOW" vacuum function is either not used or can be rewired to actuate a solenoid that opens to the cup vacuum sensor. If this function is used, it should be relabeled "CELL SENSOR" and is now under programmer control. This feature is useful if one wishes to monitor cup vacuum only at certain times during a sequencer run.

Sequencer Program

The programs published by Hunkapiller and Hood[7] illustrate the operation of the modified sequencer. Experience over the past 2 years in this laboratory has prompted us to modify the Hunkapiller and Hood program. Table II presents the key features of our program. Delivery of R_3 (HFBA) and R_2 (Quadrol) is followed by repeated solvent precipitation steps in order to assure complete removal of R_3 or R_2 from the manifold and delivery lines. This strategy, although time consuming, has increased repetitive yields from 94 to 97% in some cases. One critical aspect of microsequencing is the extended solvent rinses that decrease both background and repetitive yields. It is necessary to balance these effects by comparing results obtained with decreased solvent rinses. It should be noted that delivery times will vary significantly among different instruments. Also it is necessary to increase cell pressure over that recommended by Beckman (from 40 to 60 mm Hg) in order to assure delivery from the scoop to the conversion flask.

Identification of PTH Amino Acids by HPLC

An important aspect of the sequencing strategy in the low nanomole to subnanomole range is the analysis of the PTH amino acids by reverse-

TABLE II

PROGRAM FOR MICROSEQUENCER[a]

Step No.	Cell	Cup speed	Time (sec)	Conversion flask
1	Cell pressurize	L	4	
2	Cell dry	L	10	
3	R_3 delivery	L	44	
4	Blow out/waste	L	50	
5	Cleavage	L	150	CF vac
6	Cell dry	L	100	
7	Restricted vac/N_2	L	100	
8	High vac/N_2	L	40	
9	Cell pressurize	L	4	
10	Cell dry	L	10	
11	S_3 delivery	L	7	
12	Blow out/waste	L	60	
13	Restricted vac/N_2	L	120	
14	Cell pressurize	L	4	
15	S_3 delivery	L	5	
16	Blow out/waste	L	60	
17	Restricted vac/N_2	L	120	
18	Cell pressurize	H	4	
19	S_3 delivery	H	2	
20	Blow out/waste	H	60	
21	S_3 delivery to CF (4 ml)	H	100	CF dry
22	Blow out/CF	H	100	
23	Cell dry	H	100	
24	Restricted vac/N_2	H	100	
25	High vac	L	100	
26	Cell pressurize	L	4	
27	Cell dry	L	10	
28	R_3 delivery	L	44	
29	Blow out/waste	L	60	
30	Cleavage	L	100	CF vac
31	Cleavage	L	10	CF dry
32	Cleavage	L	20	CF dry
33	Cleavage	L	2	CF dry/R4 delivery
34	Cleavage	L	60	Blow out/N_2/waste
35	Restricted vac/N_2	L	100	Conversion
36	High vac/N_2	L	40	Conversion
37	Cell pressurize	L	4	Conversion
38	Cell dry	L	10	Conversion
39	S_3 delivery	L	7	Conversion
40	Blow out/waste	L	60	Conversion
41	Restricted vac/N_2	L	120	Conversion
42	Cell pressurize	L	4	Conversion
43	S_3 delivery	L	5	Conversion
44	Blow out/waste	L	60	Conversion

TABLE II (*continued*)

Step No.	Cell	Cup speed	Time (sec)	Conversion flask
45	Restricted vac/N$_2$	L	120	Conversion
46	Cell pressurize	H	4	Conversion
47	S$_3$ delivery	H	2	Conversion
48	Blow out/waste	H	60	Conversion
49	S$_3$ delivery/waste (4 ml)	H	100	Conversion
50	Blow out/waste	H	80	Conversion
51	Cell dry	H	100	Conversion
52	Restricted vac/N$_2$	H	100	Conversion
53	High vac/N$_2$	H	100	Conversion
54	High vac	H	200	Conversion
55	Cell pressurize	H	4	Conversion
56	Cell dry	H	10	Conversion
57	R$_1$ delivery	H	2	Conversion
58	Blow out/waste	H	50	Conversion
59	R$_2$ delivery	H	33	Conversion
60	Blow out/waste	H	50	Conversion
61	Coupling	L	10	Conversion
62	Coupling	H	10	Conversion
63	Coupling	L	10	Conversion
64	Coupling	H	500	Conversion
65	R$_2$ delivery	H	7	Conversion
66	Blow out/waste	H	50	Conversion
67	Coupling	L	10	Conversion
68	Coupling	H	10	Conversion
69	Coupling	L	10	Conversion
70	Coupling	H	100	CF dry
71	Coupling	H	100	CF vac/N$_2$
72	Coupling	H	200	CF vac
73	Coupling	H	20	CF dry
74	Coupling	H	50	S4 delivery (2 ml)
75	Coupling	H	30	Blow out/N$_2$/waste
76	Coupling	H	40	Blow out/FC
77	Coupling	H	50	S$_4$ delivery (2 ml)
78	Coupling	H	30	Blow out/N$_2$/waste
79	Cell dry	H	200	Blow out/FC
80	Restricted vac/N$_2$	H	200	Blow out/N$_2$/waste
81	High vac/N$_2$	H	200	CF dry
82	Cell pressurize	L	4	(step FC)
83	Cell dry	L	10	
84	S$_1$ delivery	L	5	
85	Blow out/waste	L	50	
86	Restricted vac/N$_2$	L	100	
87	Cell pressurize	L	4	
88	S$_1$ delivery	L	3	
89	Blow out/waste	L	50	

(*continued*)

TABLE II (*continued*)

Step No.	Cell	Cup speed	Time (sec)	Conversion flask
90	Restricted vac/N$_2$	L	100	
91	Cell pressurize	H	4	
92	S$_1$ delivery	H	1	
93	Blow out/waste	H	80	
94	S$_1$ delivery (15 ml)	H	300	
95	S$_2$ delivery (30 ml)	H	500	
96	Blow out/waste	H	80	
97	Cell dry	H	100	CF dry
98	Restricted vac/N$_2$	H	200	CF dry
99	High vac	H	200	CF dry

[a] Dry is equivalent to N$_2$/waste; CF is conversion flask. Vent and pressurize steps are not shown. Volumes delivered for selected steps are shown in parentheses; all other deliveries are 1 mm below undercut.

phase HPLC. The sample work-up in this laboratory is as follows. An internal standard (1–2 nmol of MTH-valine) is added to each fraction collected. The sample is transferred to a 12-ml conical tube with ground-glass joints (standard taper 14/20) that fit on a Buchler flask evaporator,[15] and evaporated to dryness under vacuum. The sample is dissolved in 50 μl of acetonitrile and transferred to a 0.4-ml polyethylene micro test tube[16]; the conical tube is rinsed with an additional 25 μl of acetonitrile, which is added to the previous rinse. All glassware is acid cleaned, and polypropylene plastic tips that are used on micropipettes are prerinsed with acetonitrile. The amount of sample injected onto the HPLC is determined by the nanomoles of sample sequenced (e.g., if 1 nmol of protein is sequenced, inject between 10 and 25% of the sample), and can be monitored by the size of the internal standard peak. The Waters Associates chromatography system employed in this work consists of two Model 6000A pumps, a Model UK6 injector, a Model 440 dual-channel detector (254 and 313 nm), and a Model 660 gradient programmer. Gradient elution and integration are controlled by a Spectra Physics SP 4000 integration system. A Hewlett–Packard Model 7128A two-pen chart recorder is connected to the recorded outputs of the SP 4000 data integration modules. Chromatography is performed on a DuPont Zorbax ODS column. This system is capable of resolving all 20 of the common PTH amino acids and gives a linear calibration from 10 pmol to 100 nmol at 254 nm. This system has the following attractive features, which are essential to sequencing in the subnanomole range: (*a*) sharp peaks; (*b*) reproducible elution times

[15] Evapo-mix, Buchler Instruments, Fort Lee, New Jersey.
[16] Bio-Rad, Richmond, California.

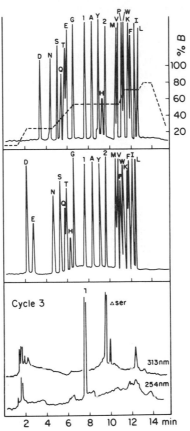

FIG. 5. *Upper panel:* Separation of 0.5–1.0 nmol each of phenylthiohydantoin (PTH)-amino acid standards on a DuPont Zorbax ODS column using gradient elution at 45° and 1.3 ml/min between buffer A (0.02 M sodium phosphate, pH 4.5, 20% MeCN) and buffer B (0.02 M sodium phosphate, pH 3.0, 75% MeCN). Percentage of B is shown on the scale to the right. Recycle time is 25 min. Internal standards included are MTH-valine (1) and PTH-aminoisobutyric acid (2). Not shown are the PTH derivatives of cysteine and arginine. PTH-Cys elutes between P and W; PTH-Arg between Y and M. Full scale is 0.20 absorbance unit. *Middle panel:* Run made under identical conditions except that buffer A has a pH of 5.83. *Lower panel:* Chromatogram from cycle 3 of a 1.0 nmol run of sperm whale myoglobin with detection at 254 nm and 313 nm. Major peak at 254 nm is internal standard (1). Major peak at 313 nm is a "dehydro" derivative of PTH-Ser (Ser). Full scale is 0.05 absorbance unit at 254 nm and 0.025 at 313 nm.

(± 2 sec); (c) short recycle times (25 min); and (d) baseline resolution of nearly all peaks.

Examples of standard runs are shown in Fig. 5. The top and middle panels show the effect of pH change on the separation of the PTH amino acids on the reverse-phase column. The shift in elution times of Asp, Glu,

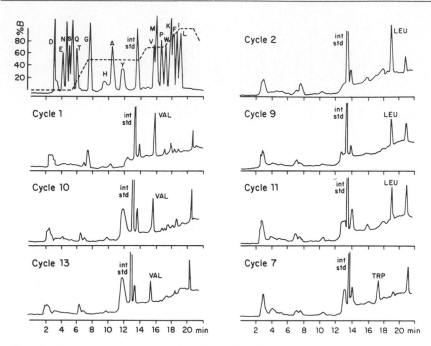

FIG. 6. Chromatograms for representative cycles of a 1.0-nmol sequencer run on sperm whale myoglobin. Upper left panel shows a standard run with buffers similar to that shown in the middle panel of Fig. 4, but a different elution schedule. Phenylthiohydantoin-norvaline was used as an internal standard. Full-scale absorbance is 0.05 unit.

His, and Arg can be used to advantage if there is a question regarding their identification. When 5–10 nmol of protein are sequenced, there are no problems detecting PTH amino acids for 0–30 cycles. On extended or subnanomole runs, Ser and Thr are detected as "dehydro" derivatives, and His and Arg give lower responses than the other amino acids. These runs often necessitate the detection of Ser and Thr at 313 nm[7] and require the injection of larger amounts of cycles suspected to contain Arg or His. An example of the analysis of cycle 3 from a sequencer run on 1.0 nmol of sperm whale myoglobin is shown in the bottom panel of Fig. 5. Little or no PTH-Ser is detected at 254 nm, but a substantial peak (at this attenuation) is detected at 313 nm (MTH-Val does not absorb strongly at 313 nm).

The analysis of several cycles from 1.0 nmol sequencer run on sperm whale myoglobin (96% repetitive yield) is shown in Fig. 6. This run was performed over a year ago in this laboratory and illustrates several types of background peaks, most of which can be eliminated (the gradient elution program is also different from that shown in Fig. 5). The broad polar peak in the beginning of the chromatogram is due to DTT buildup (or

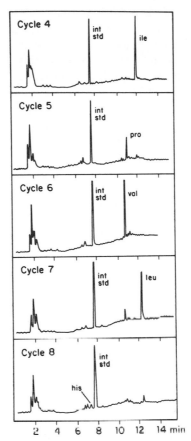

FIG. 7. Chromatograms for representative cycles of a 1.0-nmol sequencer run on a tryptic fragment from human leukocyte interferon. Buffers and gradient elution are identical to center panel on Fig. 4.

breakdown products) in S_3 and can be dramatically reduced by regularly acid-cleaning the S_3 reservoir. The broad peak preceding the internal standard is due to a contaminant often found in commercially available Quadrol and is easily removed by molecular distillation (see Table I). A non UV-absorbing contaminant (also removed by distillation) in Quadrol causes peak splitting of the internal standard (the peak eluting directly after internal standard). Another source of peak splitting in this area of the chromatogram is the presence of Quadrol, which may not be completely removed from the cup when programs with decreased S_1/S_2 rinse volumes are used. The elution of Quadrol at this portion of the chromatogram has been confirmed by collection of the eluent and analysis by mass spectrometry (unpublished data). The impurity peak eluting at the end of the chromatogram (>20 min) is due to a substance accumulating in S_1 (benzene). We find that benzene does not ordinarily contain impurities, but may pick them up from handling, contact with vacuum grease, or exposure to air. A thorough acid-cleaning of all reagent and solvent bottles is recommended on a routine basis.

Figure 7 depicts several representative chromatograms obtained from the sequence analysis of 1.0 nmol of a tryptic fragment of human leukocyte interferon.[17] The yield of the first residue was 50%, an average repetitive yield was 94%, and 19 out of 20 residues were identified by this analysis. These chromatograms illustrate the low backgrounds and resulting high sensitivity that can be reproducibly obtained with the modified Sequenator described in this chapter.

Acknowledgments

The author is a member of the City of Hope Cancer Research Center, which is supported by National Cancer Institute Grant CA 16434. He gratefully acknowledges the support of Dr. Charles W. Todd, and the technical assistance of William C. Schnute, Jr., Ursino Del Valle, and Mark J. Levy.

[17] Fragment obtained from W. Levy and S. Pestka.

Section II

Isolation and Translation of Interferon Messenger RNA

[8] Isolation and Cell-Free Translation of Human Interferon mRNA from Fibroblasts and Leukocytes

By RUSSELL MCCANDLISS, ALAN SLOMA, and SIDNEY PESTKA

The isolation of interferon messenger RNA (mRNA) is important for studying the expression and regulation of the interferon genes. A number of different methods for the isolation of active mRNA from a wide variety of sources have been described.[1] In each case, the primary concern has been either the rapid separation of the RNA from nucleases or the rapid inactivation of nucleases. Human fibroblast interferon mRNA has been isolated in a number of laboratories using modifications of phenol extraction procedures.[2-6] In each case, the mRNA was isolated and, by use of one of a variety of translation systems, shown to code for biologically active human fibroblast interferon. Human leukocyte interferon mRNA was isolated from a lymphoblastoid cell line also by use of a phenol extraction procedure.[7] The isolation of biologically active interferon mRNA from induced human leukocytes has been difficult because of the high levels of nucleases present in these cells. Phenol extraction procedures have proved to be unreliable for the isolation of leukocyte interferon mRNA, the usual result being extensive degradation of the RNA.

Rapid lysis of the cells and inactivation of the ribonucleases may be achieved using modifications of the method described by Chirgwin *et al.*[8] for the isolation of mRNA from rat pancreas, which contains high levels of ribonuclease. This method involves lysis of the cells in guanidine thiocyanate, a powerful denaturing agent, and 2 mercaptoethanol, a strong reducing agent. By this method, it has been possible to isolate interferon mRNA from induced human leukocytes and to demonstrate its activity by *in vitro* translation.

[1] J. M. Taylor, *Annu. Rev. Biochem.* **48,** 681 (1979).

[2] F. H. Reynolds, Jr., and P. M. Pitha, *Biochem. Biophys. Res. Commun.* **59,** 1023 (1974).

[3] S. Pestka, J. McInnes, E. A. Havell, and J. Vilček, *Proc. Natl. Acad. Sci. U.S.A.* **72,** 3898 (1975).

[4] F. H. Reynolds, Jr., E. Premkumar, and P. M. Pitha, *Proc. Natl. Acad. Sci. U.S.A.* **72,** 4881 (1975).

[5] P. B. Sehgal, B. Dobberstein, and I. Tamm, *Proc. Natl. Acad. Sci. U.S.A.* **74,** 3409 (1977).

[6] R. L. Cavalieri, E. A. Havell, J. Vilček, and S. Pestka, *Proc. Natl. Acad. Sci. U.S.A.* **74,** 4415 (1977).

[7] R. L. Cavalieri, E. A. Havell, J. Vilček, and S. Pestka, *Proc. Natl. Acad. Sci. U.S.A.* **74,** 3287 (1977).

[8] J. M. Chirgwin, A. E. Przybyla, R. J. MacDonald, and W. J. Rutter, *Biochemistry* **18,** 5294 (1979).

METHODS IN ENZYMOLOGY, VOL. 79

General Considerations

Precautions are taken to minimize the possibility of ribonuclease contamination at all stages of the preparations, although in the steps using the strong denaturing reagents, this is not as critical. All glassware is baked in an oven overnight at 121°. Where possible, solutions are autoclaved. When components that cannot be autoclaved are used, the solid materials are dissolved in sterile water or buffers. All nonautoclavable equipment is treated with 0.1% diethylpyrocarbonate (Eastman). RNA solutions are stored in a liquid nitrogen freezer.

Isolation of Total RNA from Induced Human Fibroblasts

The procedure originally developed for the isolation of total RNA from human fibroblasts involves phenol extraction of the RNA in the presence of the ribonuclease inhibitors bentonite, polyvinyl sulfate, sodium dodecyl sulfate (SDS), and 8-hydroxyquinoline.[9] Because human fibroblasts do not have abnormally high levels of ribonuclease, the procedure works well.

Solutions

Solution I: 10 mM sodium acetate (pH 5.0), 1.2 mg/ml of polyvinyl sulfate, 0.3 mg/ml of bentonite, 1 mg/ml of 8-hydroxyquinoline, 0.5% (5 mg/ml) of sodium dodecyl sulfate (SDS)

Solution II: 1 M Tris · HCl, pH 9, 1 M NaCl, 0.01 M EDTA

Preparation of Induced Human Fibroblasts. Human fibroblasts (GM-2504A) are induced by a modification of the superinduction procedure with poly(I) · poly(C) described by Cavalieri *et al.*[6] Growth medium (Eagle's minimal essential medium and 10% fetal calf serum) is removed from roller bottles (850 cm^2, Corning glass). For induction, 50 ml of growth medium containing 50 μg of poly(I) · poly(C) (P-L Biochemicals) and 10 μg of cycloheximide per milliliter is added to each bottle. The fibroblast monolayers are incubated with the inducer for 4 hr at 37°. The induction medium is removed, and the cell monolayers are washed once with 50 ml of phosphate-buffered saline (PBS, 0.14 M NaCl, 3 mM KCl, 1.5 mM KH$_2$PO$_4$, 8 mM Na$_2$HPO$_4$). The PBS is removed, and each bottle is incubated at 37° with 10 ml of a trypsin–EDTA solution (Gibco No. 610-5305) until all the cells are detached. The trypsin solutions containing the cells from each bottle are pooled, and heat-inactivated fetal calf serum is added to a concentration of 10% (v/v). The cells are sedimented at 500 g for 15 min, and the supernatant is discarded. The pellets are resuspended in PBS, placed into a single 50-ml centrifuge tube, and resedimented. The

[9] M. Green, T. Zehavi-Willner, P. N. Graves, J. McInnes, and S. Pestka, *Arch. Biochem. Biophys.* **172,** 74 (1976).

cell pellet is stored frozen in the vapor phase of a liquid nitrogen refrigerator until used. Approximately 0.17 g of cells is obtained from an 850-cm² roller bottle.

Procedure. Ten grams of frozen induced fibroblasts are quickly placed in a mixture of 100 ml of solution I, 100 ml of water-saturated phenol, and 100 ml of chloroform and are homogenized for 2 min in an Omnimixer (DuPont Sorvall). The homogenate is heated to 56°, allowed to cool to room temperature for 15 min, and placed in an ice bath for 15 min. The phases are separated by centrifugation for 20 min at 10,000 rpm at 4° in a Sorvall GSA rotor. The aqueous layer is carefully removed. For maximal recovery of nucleic acids, the phenol layer can be reextracted with 50 ml of solution I. The nucleic acids are precipitated by addition of 0.1 volume of 2.4 M sodium acetate, pH 5.5, and 2.5 volumes of absolute ethanol. After 18 hr at $-20°$, the precipitate is collected by centrifugation at 10,000 rpm for 20 min at $-20°$ in a Sorvall GSA rotor. The pellet, which contains both RNA and DNA, is dissolved in 100 ml of water. To the nucleic acid solution are added 10 ml of a solution containing 0.1 M Tris · HCl, pH 7.2, and 0.05 M MgCl$_2$. Ribonuclease-free deoxyribonuclease (Worthington DPFF) is added to a final concentration of 5 μg/ml. The solution is incubated at 23° for 20 min. Eleven milliliters of solution II (0.1 volume) are added, followed by SDS to a final concentration of 1%. After addition of an equal volume of phenol : chloroform : isoamyl alcohol (50 : 50 : 1; v/v/v), the mixture is shaken vigorously for 10 min at room temperature, and the phases are separated by centrifugation at 10,000 rpm for 20 min at 4° in a Sorvall GSA rotor. The aqueous layer is reextracted once as above. The RNA in the aqueous fraction is precipitated by addition of 0.1 volume of 2.4 M sodium acetate, pH 5.5, and 2.5 volumes of absolute ethanol followed by standing overnight at $-20°$. The total RNA is collected by centrifugation and dissolved in sterile water. Analysis of the RNA at this point by sucrose gradient centrifugation or by gel electrophoresis should show the presence of undegraded 4–5 S, 18 S, and 28 S RNA species.

Isolation of Total RNA from Induced Human Leukocytes

Because of the high levels of ribonuclease present in human leukocytes, the procedure described above proved to be unsatisfactory for the routine isolation of leukocyte RNA. A procedure involving lysis of the cells in guanidine thiocyanate has been adapted and has given excellent results.

Solutions

Solution A: The lysis solution contains 4 M guanidine thiocyanate, 0.1 M Tris · HCl, pH 7.5, and 0.1 M 2-mercaptoethanol. Guanidine thiocyanate obtained from Merck, Fluka, or Baker has been found to be of sufficiently good quality for use. Guanidine thiocyanate (472.6 g) is dissolved with heating in 500 ml of water and 200 ml of Tris buffer stock solution (0.5 M Tris · HCl, pH 7.5). The solution is allowed to cool to room temperature, after which 7.15 ml of 2-mercaptoethanol (Eastman) are added and the volume is brought to 1 liter with sterile water. Particulate matter from the guanidine thiocyanate is removed by filtration through a Nalge disposable filter unit. The solution is stable for at least 1 month at room temperature. Strong precautions should be taken to avoid contact of this reagent with the skin.

Solution B: The wash solution contains 6 M guanidine hydrochloride (Schwarz–Mann Ultrapure), 10 mM Na$_2$EDTA, pH 7.0, and 10 mM dithiothreitol. Because of the high purity of the guanidine hydrochloride used, no filtration of this solution is necessary. It is stable when stored at room temperature. Again, care should be taken to avoid contact of this solution with the skin.

Preparation of Induced Human Leukocytes. Human leukocytes (usually obtained from patients with chronic myelogenous leukemia) are induced to produce interferon with Newcastle disease virus as described in Volume 78 [6]. At 10 hr after addition of the virus to the culture, the cells are harvested by centrifugation (10 min at 1500 g) and washed once with PBS. The cells are centrifuged as above, frozen immediately in dry ice, and stored in a liquid nitrogen freezer.

Lysis of Cells. This procedure is performed in a fume hood. The frozen cell pellet is broken into small pieces and the pieces are placed in solution A (20 ml/g wet cells) in the chamber of an Omnimixer. The Omnimixer is run at full speed for 2 min to lyse the cells. The lysate is then centrifuged for 10 min at 12,000 rpm in a Sorvall GSA rotor to remove any debris present.

Isolation of RNA by Ethanol Precipitation. The supernatant containing the RNA and other cellular components is acidified to pH 5 by addition of 0.04 volume of 1 N acetic acid, and RNA is precipitated by addition of 0.5 volume of absolute ethanol. After mixing, the solution is kept at −20° for at least 2 hr. Leaving the mixture at −20° for much longer periods does not significantly increase the yield of RNA and results in the coprecipitation of larger amounts of protein, which may be difficult to redissolve in the next step. The precipitated RNA is collected by centrifugation for 10 min at 8000 rpm in a Sorvall GSA rotor. The supernatant is removed, and the pellet is dissolved in approximately 0.5 volume (relative to the original volume of the lysate) of solution B. Dissolution may require brief heating to 70°. The RNA is again acidified to pH 5 by addition of 0.04 volume of 1

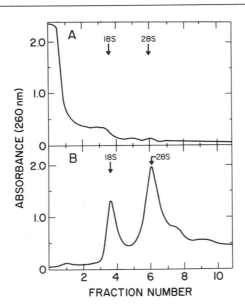

FIG. 1. Sucrose gradient profiles of RNA isolated from human leukocytes by phenol extraction (A) or by the guanidine thiocyanate/2-mercaptoethanol lysis procedure (B). Linear 5 to 20% sucrose gradients in 0.02 M sodium acetate, pH 5.0, are made in SW40 tubes and equilibrated to 4°; 500 μg of total RNA were applied to the gradients, and centrifugation was performed for 20 hr at 30,000 rpm at 4°. Gradients are monitored for RNA by measuring absorbance at 260 nm.[11]

N acetic acid and precipitated by 0.5 volume of ethanol. Precipitation is almost immediate and is quantitative if the solution is cooled in an ice bath and cold ethanol is added. The RNA is collected by centrifugation at 6000 rpm for 10 min in a Sorvall GSA rotor. The procedure with solution B is then repeated once, with a further reduction in volume if possible. The final pellet is dissolved in sterile 0.24 M sodium acetate, pH 5.5, and precipitated by addition of 2.5 volumes of ethanol. The total RNA is best stored as the ethanol precipitate at $-20°$ until further use. The RNA at this stage is undegraded and free of proteins and DNA (DNA is precipitated only by the addition of a higher percentage of ethanol to the guanidine solutions).[10] The sucrose density gradient profiles of total RNA obtained by this procedure and by the phenol extraction procedure are shown in Fig. 1.[11]

Preparation of RNA by Centrifugation through CsCl. A modification of the procedure that is particularly useful for small amounts of cells involves centrifugation of the RNA through a cushion of CsCl so that DNA

[10] R. A. Cox, this series, Vol. 12B, p. 120.
[11] S. Pestka and H. Hintikka, *J. Biol. Chem.* **246**, 7723 (1971).

does not pellet as originally described by Glišin et al. [12] For a preparation from about 1 g of cells, the guanidine thiocyanate lysate is prepared as described above, and 0.1 g of CsCl per milliliter is added. The lysate is layered carefully over a cushion of 6 ml of 5.7 M CsCl in 10 mM EDTA in a tube for the SW27 rotor. The solution is centrifuged for 20 hr at 25,000 rpm at 20°. The glassy pellet at the bottom of the tube consists of RNA. The overlaying solutions are removed by aspiration, and the tubes are drained. The tubes are then cut off near the bottom to prevent contamination of the pellet by anything remaining above on the walls of the tube. The pellets are rinsed once with 70% ethanol, then removed with a flamed forceps into 0.1 M sodium acetate, pH 5.5, 0.01 M 2-mercaptoethanol, and 0.5% Sarkosyl. The RNA is dissolved by heating and agitation with a Pasteur pipette. After the RNA has dissolved, 0.5 volume of water-saturated phenol is added, and the solution is shaken vigorously. One-half volume of chloroform is added, the solution is shaken, and the mixture is centrifuged briefly to separate the phases. The RNA in the aqueous phase is precipitated by addition of 2.5 volumes of ethanol. The total RNA prepared by this method is virtually identical to that obtained as described above and shown in Fig. 1B.

Isolation of mRNA from Total RNA

From this step, the procedures used for both fibroblast and leukocyte interferon mRNAs are identical. It has been shown by several groups that interferon mRNA is polyadenylated.[2,3] Poly(A)-containing RNA is isolated from the total RNA by chromatography on oligo(dT)-cellulose.[9] The RNA solution (30–50 A_{260} units/ml) is adjusted to 0.5 M NaCl by addition of 0.11 volume of 5 M NaCl and then applied to a column containing 5 g of oligo(dT)-cellulose (T-3, Collaborative Research) equilibrated with 10 mM Tris · HCl, pH 7.4, and 0.5 M NaCl at a flow rate of about 20 ml/hr. Nonbound RNA is removed from the column by washing with the same buffer until the optical density is <0.08 A_{260} unit/ml. Poly(A) RNA is then eluted with 10 mM Tris · HCl, pH 7.4. The poly(A) RNA is recycled over the column to decrease contamination by ribosomal and other non-poly(A) RNAs. After the first oligo(dT)-cellulose column, the RNA preparation may be assayed for interferon mRNA activity by injection into Xenopus laevis oocytes as described elsewhere in this volume [10].

Enrichment of Interferon mRNA by Sucrose Gradient Centrifugation

The interferon mRNA may be enriched significantly by separating the RNA on a sucrose gradient. Linear gradients are made by mixing equal volumes of 5% (w/v) and 20% sucrose in 0.02 M sodium acetate, pH 5, in

[12] V. Glišin, R. Crkvenjakov, and C. Byus, *Biochemistry* 13, 2633 (1974).

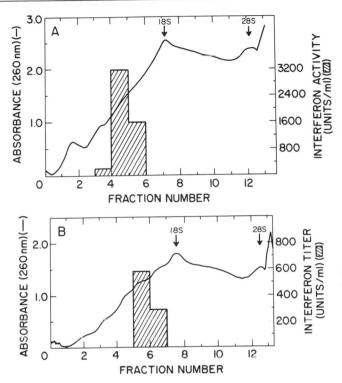

FIG. 2. Sucrose gradient profiles of interferon mRNA from induced human fibroblasts (A) and leukocytes (B). About 500 μg of poly(A) RNA were centrifuged in an SW40 rotor as described in the legend to Fig. 1. Gradients were monitored for absorbance at 260 nm.[11] Fractions of 1.0 ml were collected; 0.1 ml of 2.4 M sodium acetate, pH 5.5, and 2.5 ml of ethanol were added to each 1-ml fraction. After standing overnight at −20°, RNA was sedimented by centrifugation at 10,000 rpm at −20° in an SS34 Sorvall rotor. The RNA pellet was washed once with cold 70% ethanol and then dried in air. RNA is dissolved in sterile water and assayed for interferon mRNA activity (hatched area) by injection into *Xenopus laevis* oocytes (see this volume [10]).

the chambers of a Buchler gradient maker. Usually, gradients are made in Beckman SW40 tubes. After formation, the gradients are equilibrated at 4° for at least 4 hr. Prior to loading, the RNA samples are heated to 80° for 2 min and cooled rapidly in an ice bath to reduce aggregation. The samples are layered over the gradients and are centrifuged for 20 hr at 30,000 rpm at 4° using the SW40 rotor (Beckman). After centrifugation, the gradients are collected with the use of an Isco gradient pump modified so that the flow cell is attached to a Gilford spectrophotometer.[11] The gradients are displaced by pumping 50% sucrose into the bottom of the tube. Fractions are collected, ethanol-precipitated, dissolved in sterile water, and assayed by injection into *Xenopus laevis* oocytes. As shown in Figs. 2A and 2B,

INTERFERON mRNA YIELDS FROM INDUCED FIBROBLASTS AND LEUKOCYTES[a]

	Yield (mg/g cells)		Activity (units/ml)	
	Fibroblast	Leukocyte	Fibroblast	Leukocyte
Total RNA	1.5–2.0	1.0–3.0	ND	ND
Poly(A), I	0.09–0.16	0.1–0.15	100–300	20–80
Poly(A), II	0.04–0.08	0.03–0.24	300–600	50–200
Sucrose gradient (12 S)	0.01–0.02	0.003–0.03	2000–5000	300–2000

[a] The yield of RNA per gram wet weight of cells is given in the table. Interferon activity of mRNA was determined by injection of mRNA into *Xenopus laevis* oocytes at a concentration of 1 mg/ml (see this volume [10]). Interferon titers are expressed in reference units per milliliter. ND not determined.

both human fibroblast and leukocyte interferon mRNAs migrate approximately as 12 S species.

The procedures described yield biologically active interferon mRNAs. Typical yields and interferon titers obtained are given in the table. Either of the procedures described for isolation of total RNA may be used for fibroblast RNA with very similar results. Because of the convenience and reliability of the guanidine thiocyanate lysis procedure, however, we have found it to be preferable. In each case, approximately 10–20 μg of sucrose gradient-purified mRNA can be obtained per gram of cells. Since the leukocytes used are variable with respect to the specific types of cells present from one donor to another, the yields of total RNA from one preparation to another are more variable than those from fibroblasts. The interferon titer produced by the leukocytes varies from one preparation to another, but the interferon titer produced by the cells correlates with the titer obtained from the purified mRNA.

Cell-Free Translation of Interferon mRNA

The interferon mRNA prepared by the procedures described can be translated in many different cell-free systems.[13] However, a most useful and sensitive system is the *Xenopus laevis* oocyte assay described in this volume [10]. It has been translated in an Ehrlich ascites extract,[3] wheat germ extract,[12] and in an mRNA-dependent reticulocyte lysate as described by Pelham and Jackson.[14] The efficiency of the translation in oocytes was about 50- to 1000-fold the efficiency of translation in these cell-

[13] R. L. Cavalieri and S. Pestka, *Tex. Rep. Biol. Med.* 35, 117 (1977).
[14] H. R. B. Pelham and R. J. Jackson, *Eur. J. Biochem.* 67, 247 (1976).

free extracts. The low efficiency may be due to lack of appropriate processing of the product in the cell-free extracts.[15]

Cell-Free Translation of Interferon mRNA in Mouse Ehrlich Ascites Extracts. Interferon mRNA fractions were translated in mouse Ehrlich ascites extracts as described.[3,9] Each 0.05 ml reaction mixture contained the following components: 0.15–0.23 mg of S-30 protein; 70 μg of reticulocyte polysomal salt-wash fraction protein[9]; 0.05–0.1 A_{260} unit of mRNA; 10 mM Tris · HCl (pH 7.5) or 10 mM N-2-hydroxyethylpiperazine-N'-2-ethanesulfonic acid (HEPES) (pH 7.5); 120 mM KCl; 3.5 mM MgCl$_2$; 1 mM ATP; 0.1 mM GTP; 5 mM creatine phosphate; 8 μg of creatine kinase; 2 mM 2-mercaptoethanol; and each of the unlabeled amino acids at a concentration of 40 μM. Incubations were performed for 60 min at 30°. Protein synthesis can be determined by amino acid incorporation into hot trichloroacetic acid-precipitable material with a radioactive amino acid. In addition, the interferon titer of the cell-free incubations stimulated by various mRNA fractions can be assayed directly by a cytopathic effect inhibition assay (see Volume 78 [56]).

[15] S. Pestka, J. McInnes, D. Weiss, E. A. Havell, and J. Vilček, *Ann. N. Y. Acad. Sci.* **284**, 697 (1977).

[9] Preparation of Interferon Messenger RNAs with the Use of Ribonucleoside–Vanadyl Complexes

By Shelby L. Berger, Donald M. Wallace, Gene P. Siegal, Michael J. M. Hitchcock, Connie S. Birkenmeier, and Susan B. Reber

The study of messenger RNA (mRNA) requires dependable methods for obtaining intact, translatable material. Many procedures employ high concentrations of denaturants, such as guanidine hydrochloride, sodium dodecyl sulfate, or sodium lauryl sarcosinate (Sarkosyl), to disrupt cells and inactivate degradative enzymes, particularly ribonucleases.[1] Other techniques make use of exogenous ribonuclease inhibitors, such as heparin or diethyl pyrocarbonate. Both strategies, however, have disadvantages. Strong denaturants permit isolation of intact RNA, but only at the expense of subcellular organization: polysomes are disaggregated, nuclear and cytoplasmic constituents commingle, and interactions between mes-

[1] Reviewed by R. Poulson, *in* "The Ribonucleic Acids" (P. R. Stewart and D. S. Letham, eds.), p. 333. Springer-Verlag, Berlin and New York, 1977.

senger RNAs and their accompanying proteins are eliminated. Ribonuclease inhibitors, on the contrary, are gentler but often inadequate. Nucleases in human peripheral blood lymphocytes are not completely inhibited by either of the substances indicated above.[2] Furthermore, not only is mRNA partially degraded, but also heparin (1–2 mg/ml) and diethyl pyrocarbonate (0.1–0.2%) rupture the nuclei.[2] Diethyl pyrocarbonate also reacts with the adenine moieties of polynucleotides reducing the translational efficiency of affected mRNA molecules.[3] These problems have been overcome, in large part, by the use of ribonucleoside–vanadyl complexes as ribonuclease inhibitors throughout all phases of RNA preparation.[2]

Principle

According to transition state theory, an activated complex or transition state occurs as an intermediate during enzyme catalysis.[4] This intermediate or altered form of the substrate binds very much more tightly to the enzyme than either substrates or products. Dissociation constants for the hypothetical process of separating an enzyme from its activated intermediate may be 10^{-7} the dissociation constant of the enzyme–substrate complex.[5] Hence, stable analogs of transition states are potentially extremely specific and powerful enzyme inhibitors. Ribonucleoside–vanadyl complexes are transition state analogs for activated $2',3'$-cyclic phosphates believed to occur during catalysis by ribonuclease.[6] The phosphate in the transition state is replaced by a $2',3'$-linked oxovanadium ion in the analog. The result is a noncovalent complex similar in geometry to the probable transition state for substrates of ribonuclease. It is interesting to note that K_i for the uridine-vanadyl analog is 1/1200th K_m for uridine $2',3'$-phosphate (cyclic) when pancreatic ribonuclease is used as the enzyme.[6] Ribonucleoside–vanadyl complexes are also excellent inhibitors of the nucleases found in both human peripheral blood lymphocytes[2] and the Namalva lymphoblastoid cell line.[7] These materials are therefore well suited for preventing degradation of interferon mRNAs isolated from leukocytes.

[2] S. L. Berger and C. S. Birkenmeier, *Biochemistry* **18**, 5143 (1979).
[3] R. E. L. Henderson, L. H. Kirkegaard, and N. J. Leonard, *Biochim. Biophys. Acta* **294**, 356 (1973).
[4] L. Pauling, *Am. Sci.* **36**, 51 (1948).
[5] R. Wolfenden, this series, Vol. 46, p. 15.
[6] G. E. Lienhard, I. I. Secemski, K. A. Koehler, and R. N. Lindquist, *Cold Spring Harbor Symp. Quant. Biol.* **36**, 45 (1971).
[7] S. L. Berger, M. J. M. Hitchcock, K. C. Zoon, C. S. Birkenmeier, R. M. Friedman, and E. H. Chang, *J. Biol. Chem.* **255**, 2955 (1980).

Preparation of Ribonucleoside–Vanadyl Complexes[2]

A solution containing 1 mmol each of adenosine, cytidine, guanosine, and uridine in 17 ml of glass-distilled water is prepared. While the ribonucleosides are dissolving in a boiling water bath, 2 M VOSO$_4$ (vanadyl sulfate) is made up in a separate tube and heated in the boiling water bath. One milliliter of hot vanadyl sulfate is added to the dissolved ribonucleosides, and the mixture is flushed for 1–5 min with nitrogen by bubbling the gas through the liquid. The acidic solution is titrated under nitrogen to approximately pH 6.5 with 10 N NaOH and finally to pH 7 with 1 N NaOH while stirring continuously in the boiling water bath. Formation of the complex is indicated by a change in color from bright blue, characteristic of vanadyl sulfate solutions, to green-black, characteristic of ribonucleoside–vanadyl complexes. The solution may become transiently cloudy during the titration, but should be clear, albeit optically dense to visible light, at pH 7. Finally, the solution is diluted to 20 ml with water, resulting in a 200 mM mixture of four ribonucleoside–vanadyl complexes, each at 50 mM. Although ribonucleoside–vanadyl complexes are stable at room temperature, storage at $-20°$ in convenient aliquots is recommended. The complexes can be frozen and thawed with impunity but should be warmed to 37° and mixed vigorously on a vortex mixer before use. The complexes are available commercially from Bethesda Research Laboratories, Rockville, Maryland.

Isolation of Interferon Messenger RNAs

Cells

Lymphocytes can be purified from the blood of unselected normal donors. The methods and the equipment required for obtaining a population highly enriched in small lymphocytes has been detailed previously.[8] The cells are cultured in 10% autologous plasma in stationary cultures at 37° for 1–5 days. The culture conditions, the cellular composition of the cultures, and the yields have been described elsewhere.[8] In some cases lymphocytes have been obtained from plateletphoresis residues, since the residue from a single platelet donor often contains fivefold the lymphocytes of a unit of blood. The cells can be separated from contaminants by applying the methods developed for whole blood indicated above. However, owing to the scarcity of plasma and the high percentage of lymphocytes in the starting material, nylon column filtration and the steps preceding it are omitted. When culturing lymphocytes from residues, pooled AB serum can be used to supplement autologous plasma to maintain a final volume of 10%.

[8] S. L. Berger, this series, Vol. 58, p. 486.

Lymphoblastoid cells of the Namalva line are available informally from a number of laboratories. The cultures described here came originally from G. Klein.[9] They are grown in spinner flasks in volumes of 1–9 liters in buffered RPMI-1640 containing 10% heat-inactivated fetal calf serum and 50 μg/ml of gentamycin.[7,10] Only cultures at saturation density are used for interferon production.

Induction of Interferon

Induction of interferon synthesis is routinely carried out by the addition of 25 hemagglutinin units of Newcastle disease virus, strain B-1 per 10^6 cells.[11] Both normal lymphocytes and lymphoblastoid cells respond well to this treatment by synthesizing leukocyte interferon[12] and a mixture of leukocyte type and fibroblast interferons,[13] respectively. Lymphocytes can also be induced to produce interferon by the addition of 0.2 μg of staphylococcal enterotoxin A per milliliter to the cultures.[14]

Cell Fractionation and Extraction of Cytoplasmic RNA

The procedures specified here service a liter of Namalva cells (1.8 to 2.2×10^6 cells/ml) or actively growing lymphocytes (2.0×10^6 cells/ml). The volumes should be reduced about one-fourth for virus-treated lymphocytes, since these are small, quiescent cells with scanty cytoplasm, similar in their gross metabolic parameters to resting lymphocytes. These methods have been scaled up to handle 30 liters of cells or reduced for 25 ml of cells by maintaining the same proportions except where indicated. They are a modification of published procedures.[15]

Reagents

Low-salt Tris: 20 mM Tris · HCl at pH 7.4, 10 mM NaCl, and 3 mM magnesium acetate
Lysing buffer: low-salt Tris containing 5% sucrose (w/w) and 1.2% Triton N101 (w/w) (Rohm and Haas)
Ace: 50 mM sodium acetate and 10 mM EDTA at pH 5.1
Phenol: only redistilled phenol, stored frozen at $-20°$ in the presence

[9] G. Klein, Stockholm, Sweden.
[10] K. C. Zoon, C. E. Buckler, P. J. Bridgen, and D. Gurari-Rotman, *J. Clin. Microbiol.* **7**, 44 (1978).
[11] K. C. Zoon and F. E. Campball, this series, Vol. 78 [42].
[12] E. A. Havell, B. Berman, C. A. Ogburn, K. Berg, K. Paucker, and J. Vilček, *Proc. Natl. Acad. Sci. U.S.A.* **72**, 2185 (1975).
[13] E. A. Havell, Y. K. Yip, and J. Vilček, *J. Gen. Virol.* **38**, 51 (1978).
[14] M. P. Langford, G. J. Stanton, and H. M. Johnson, *Infect. Immun.* **22**, 62 (1978).
[15] S. L. Berger and C. S. Birkenmeier, *Biochim. Biophys. Acta* **562**, 80 (1979).

of 0.1% 8-hydroxyquinoline, should be used. Before commencing an RNA extraction, the phenol should be thawed at 60–65° and saturated with 0.3 volume of Ace. If the Ace is added to the frozen phenol, thawing is hastened and the process can be carried out equally well at 50°. Once saturated, phenol will remain liquefied at the lower temperatures encountered during the isolation of RNA.

Chloroform : isoamyl alcohol, 94 : 6 (v/v)

Sodium acetate, 2 M

Sodium dodecyl sulfate (SDS), 10%

All buffers are treated with diethyl pyrocarbonate 0.05% (v/v) for 10 min at room temperature, heated to 65–80° for 45 min to destroy unreacted material, and cooled to 4° before use. Since diethyl pyrocarbonate reacts with Tris, some loss of buffer capacity is inevitable.[16] Solvents and detergents do not require treatment.

The two largest variables in the yield of interferon mRNA from leukocytes are the amount actually made by the cells and the time of the harvest. Since the investigator has little or no control over the former, with the exception of mixing competent viruses with healthy cells, the latter should be examined thoroughly. In this laboratory, maximum yields of interferon mRNA from lymphoblastoid cells occurred throughout the interval between 8 and 14 hr after induction; thereafter, yields declined.[7] However, if one wishes to maximize polysomal interferon mRNA, cells should be harvested between 8 and 10 hr after induction.[7] Once the rate of interferon synthesis decreases, usually at 11–12 hr after the addition of virus, polysomes will be virtually devoid of interferon mRNA. Optimal harvest times for virus-treated lymphocytes average approximately 6–10 hr after induction; for staphylococcal enterotoxin A-treated samples, the time of maximum yield is difficult to determine because cells from different donors respond differently to mitogens.

Cells are recovered from the culture fluids by centrifugation at room temperature *before* the time of harvest (the time at which the biological material is subsequently chilled). Thus, depending on the volume of the cultures and the capacity of the centrifuges, approximately 15 min to an hour will elapse during which the leukocytes are gradually cooled from 37° toward room temperature. Immediately afterward, they are brought rapidly to 4°. A detailed protocol follows.

Leukocytes from 1 liter of culture are centrifuged at room temperature for 10 min at 600 g and 1600 g for lymphocytes and lymphoblastoid cells, respectively. The supernatant liquid is removed by aspiration or by careful decantation, and an aliquot is saved in order to determine the interferon titer. The cells are resuspended in approximately 250 ml of ice-cold

[16] S. L. Berger, *Anal. Biochem.* 67, 428 (1975).

Eagle's minimal essential medium or RPMI-1640 at the time of harvest and centrifuged again at 4°. Precise volumes are not critical. Washed cells are dispersed in 12 ml of ice-cold low-salt Tris buffer supplemented with 0.8 ml of 200 mM mixed ribonucleoside–vanadyl complexes and lysed by the addition of 4 ml of ice-cold lysing buffer. After blending rapidly on a vortex mixer, the nuclei are removed by sedimentation at 1200 g for 5 min in the cold. The supernatant fluid, which contains the cytoplasmic components, can be recovered with a pipette. It is essential to remove the supernatant liquid immediately, since prolonged exposure to lysing buffer will gradually rupture the nuclei and release DNA and nuclear RNA into the cytoplasmic preparation. Once cytoplasm has been obtained, it can be used for either the preparation of polysomes[2,7] or the isolation of RNA. For isolating RNA, the cytoplasmic components are transferred to a 250-ml glass bottle fitted with a ground-glass stopper containing 60 ml of Ace buffer and 4 ml of 10% SDS. After adding 80 ml of Ace-saturated phenol at 50°, the mixture is shaken vigorously at room temperature for 5 min. The phenol phase, which initially is yellow owing to dissolved 8-hydroxyquinoline, turns gray-black in the presence of the vanadyl complexes. The aqueous phase[17] is recovered by centrifugation at 4° for 10 min at 4000 g, either in 50-ml plastic tubes or in a 250-ml plastic bottle resistant to phenol. The aqueous sample is then reextracted at room temperature for 5 min with a mixture of 40 ml of saturated phenol and 40 ml of chloroform–isoamyl alcohol in a clean 250-ml stoppered bottle. The phases are separated in plastic bottles[18] by centrifugation at 4° for 10 min and, if the interphase is not clear, extracted by the same methods with 80 ml of chloroform–isoamyl alcohol. The final RNA-rich aqueous layer is transferred to any convenient vessel and treated with 0.05 volumes of 2 M sodium acetate and 2–3 volumes of ice-cold absolute ethanol overnight at −20° to precipitate RNA. The entire procedure should not exceed 90 min from the time of harvest to the ethanol precipitation step.

Isolation of Polyadenylated Molecules

In order to purify interferon mRNA, polyadenylated molecules must be separated from all others. Any standard method can be used.[1] However, in this laboratory, chromatography on columns of oligo(dT)-cellulose rather than poly(U)-Sepharose is preferred because the columns of the former can be regenerated. Neither method results in pure polyadenylated

---

[17] The aqueous phase is less dense and is therefore on top.

[18] There are no plastic bottles available that are truly resistant to chloroform. In practice we use Maxiforce, polypropylene bottles from VWR Scientific Inc. Bottles are exposed to chloroform at 4° for the minimum time necessary.

RNA after one cycle of chromatography; in both cases 25–50% of the interferon mRNA is lost by rechromatographing the sample in order to remove contaminating rRNA.

Buffers and Materials

Binding buffer: 10 mM Tris · HCl at pH 7.4, 1 mM EDTA, 0.5 M NaCl, and 0.5% SDS

Twice-concentrated binding buffer

Wash buffer: 10 mM Tris · HCl at pH 7.4, 1 mM EDTA, 0.1 M NaCl, and 0.5% SDS

Elution buffer: 10 mM Tris · HCl at pH 7.4, 1 mM EDTA, and 0.5% SDS

Oligo(dT)-cellulose: Each fresh sample of 250 mg should be washed before use with 10 ml of binding buffer, 2 ml of *Escherichia coli* tRNA at 1 mg/ml, 10 ml of elution buffer, 3 ml of 0.1 N NaOH, and binding buffer until the pH of the eluate returns to 7.4. The washed resin is stored at 4° and warmed to room temperature before use. After fractionating a sample, the resin is regenerated by the sequential application of binding buffer, NaOH, and binding buffer as described above; the tRNA and the elution buffer are omitted. When 10 samples have been processed, the oligo(dT)-cellulose is discarded.

A sample is prepared for chromatography by dissolving the RNA in water at a concentration of 6 mg/ml based on the absorbance at 260 nm,[19] denaturing it at 65° for 10 min, cooling it rapidly on ice, and diluting it with an equal volume of twice-concentrated binding buffer. The sample is applied at room temperature at a maximum velocity of 3 ml/min to a column 0.6 cm in diameter containing 250 mg of oligo(dT)-cellulose. The column is developed sequentially with 20 ml of binding buffer and 10 ml of wash buffer, and finally poly(A)-containing molecules are removed in 6 ml of elution buffer. The mRNA sample is supplemented with 0.05 volume of 2 M sodium acetate and 3 volumes of ethanol and precipitated at −55° for 15 min. The precipitate is recovered by centrifugation at 10,000 g for 20 min and redissolved in water to determine the absorbance at 260 nm, hence the yield. For purposes of translation mRNA is reprecipitated and redissolved at a higher concentration (0.1–5 mg/ml). We have also used the method of Wallace *et al.*[20] for isolating polyadenylated molecules.

The specific interferon activity of the mRNA may be determined by translating a fixed aliquot in oocytes of *Xenopus laevis* and determining the

---

[19] Phenol must be removed by reprecipitation in ethanol or by extraction with ether in order to obtain an accurate value.

[20] D. M. Wallace, R. Jagus, C. R. Benzie, and J. E. Kay, *Biochem. J.* **184,** 277 (1979).

INTERFERON MESSENGER RNAs (mRNAs) FROM LYMPHOID CELLS[a]

		Yield		
Cell	Inducer	Cytoplasmic RNA (pg/cell)	Polyadenylated mRNA ($10^2 \times$ pg/cell)	Interferon mRNA activity (units/μg mRNA)
Lymphocyte	NDV	1	1	20[b]
Lymphocyte	SEA	2	3	0.3[c]
Namalva	NDV	3	4	80[d]

[a] RNA was isolated from the cytoplasm of NDV-induced cells at 10–12 hr after induction and from SEA-induced lymphocytes at 48 hr after mitogen treatment. Messenger RNAs were obtained by one cycle of chromatography on oligo(dT)-cellulose columns. Each mRNA sample was assayed in triplicate by microinjecting three groups of 10 frog oocytes. The specific activity of interferon mRNA was obtained by averaging the results from a minimum of three completely independent samples. The data are expressed in reference units of interferon produced per microgram of unfractionated mRNA translated in the oocyte. (Although higher specific activities are achievable by purifying the mRNA, such procedures are beyond the scope of this chapter.) The RNA concentrations were determined using $E^{1\%}_{260\ nm} = 230$. NDV, Newcastle disease virus; SEA, staphylococcal enterotoxin A.

[b] >99% Leukocyte interferon; <1% of the interferon remained after treating the sample with anti-leucokyte interferon antibody.

[c] T interferon; the titer was unchanged after treating the sample with either anti-leukocyte interferon antibody or anti-fibroblast interferon antibody.

[d] 80% Leukocyte interferon and 20% fibroblast interferon based on treatment with either anti-leukocyte interferon or anti-fibroblast interferon antibody.

amount of interferon synthesized. Our methods are a refined version of those of Gurdon[21] in which exported proteins, only, are assayed for interferon activity.[7,22] Interferon in the culture fluids bathing the oocytes is best quantified by measuring the inhibition of the cytopathic effect in GM-2504 fibroblasts[23] produced by encephalomyocarditis virus as described by Armstrong.[24] The titers are expressed in units defined by the human leukocyte interferon reference standard.[25]

The table details the yield of RNA and the specific interferon mRNA activity obtained from lymphoblastoid cells and lymphocytes. In lymphocyte cultures the amount of cytoplasmic RNA and polyadenylated mRNA

[21] J. B. Gurdon, C. D. Lane, H. R. Woodland, and G. Marbaix, *Nature (London)* 233, 177 (1971).

[22] A. Colman and J. Morser, *Cell* 17, 517 (1979).

[23] Available from The Human Genetic Mutant Cell Repository, Camden, New Jersey.

[24] J. A. Armstrong, *Appl. Microbiol.* 21, 723 (1971).

[25] Catalog No. 023-901-527, available from The National Institute of Allergy and Infectious Diseases, Bethesda, Maryland.

varies with the inducer. The virus-treated cell, which produces predominantly leukocyte interferon, has less bulk RNA and less polyadenylated mRNA in the cytoplasm than its mitogen-stimulated counterpart. Indeed, the virus-treated lymphocyte, in some respects, is virtually indistinguishable from a resting lymphocyte. Treatment with staphylococcal enterotoxin A, on the contrary, results in major metabolic changes including DNA synthesis; the synthesis of T or immune interferon is only one of a large number of mitogen-induced responses.[26] A comparison of the apparent amount of the two interferon mRNAs recovered from lymphocytes indicates that far more leukocyte interferon than T interferon is produced, using appropriate inducers in each case. However, one must assume that the translational efficiencies in the oocyte of the two mRNAs and the specific antiviral activities of the two types of interferon are comparable in order to draw this conclusion. Neither premise is necessarily correct. Lymphoblastoid cells, as evidenced by the Namalva line, contain more cytoplasmic polyadenylated mRNA per cell and produce more interferon mRNA per microgram of total mRNA than do lymphocytes. The interferon mRNA is approximately one-fifth fibroblast type and four-fifths leukocyte type.[13,27]

Comments

Any comparison of interferon mRNA yields should take into account the inherent errors in the assay methods. Interferon assays, for example, are subject to a twofold uncertainty. When the errors accompanying translation of mRNAs in frog oocytes are considered, the overall uncertainty in the assay of interferon mRNA activity approaches fourfold. Thus, one cannot state with confidence that the specific interferon mRNA activities of virus-treated lymphocytes and lymphoblastoid cells reported in the table are different. An added difficulty comes from the use of columns of oligo(dT)-cellulose or poly(U)-Sepharose to isolate polyadenylated molecules. Samples of unfractionated poly(A)-containing RNA may include amounts of ribosomal RNA in excess of 50% after one cycle of chromatography. The contamination varies considerably even when identical columns, or the same column, are used. The problem can be overcome by rechromatographing the sample, but only with a sacrifice in yield. When one adds to these impediments the need for handling extremely small amounts of material, it is not hard to understand why caution must be exercised in interpreting results and comparing data.

[26] N. R. Ling and J. E. Kay, "Lymphocyte Stimulation," 2nd ed., p. 303. American Elsevier, New York, 1975.
[27] R. L. Cavalieri, E. A. Havell, J. Vilček, and S. Pestka, *Proc. Natl. Acad. Sci. U.S.A.* 74, 3287 (1977).

Ribonucleoside–vanadyl complexes can be used in a wide variety of situations in which a nuclease inhibitor is required. Because they are specifically aimed at ribonucleases, these materials can protect RNA from degradation without affecting other cellular components. They have been used successfully to prepare polysomal interferon mRNA[7] and to prepare and fractionate ribosomes and polysomes.[2] When Namalva cytoplasm was incubated at 4° for 1 hr in the presence of the complexes, no destruction of interferon mRNA was observed.[7] By utilizing ribonucleoside–vanadyl complexes it is no longer necessary to denature proteins or deproteinize solutions early in the isolation of mRNA. Since these substances are easy to prepare and handle, they should be considered whenever RNA extractions are anticipated.

Several months after this paper was submitted for publication, the work was presented at the DNA Recombinant Interferon Cloning Workshop held on September 8–9, 1980 at the National Institutes of Health, Bethesda, Maryland.

[10] Translation of Human Interferon mRNAs in *Xenopus laevis* Oocytes

By ALAN SLOMA, RUSSELL MCCANDLISS, and SIDNEY PESTKA

Gurdon demonstrated that *Xenopus laevis* oocytes can translate injected mRNA.[1] Other studies have shown that oocytes can perform post-translational modifications such as cleavage of polypeptides,[2–4] glycosylation,[5] acetylation,[6] and assembly.[7] It was shown that *Xenopus* oocytes can translate interferon mRNA from poly(I) · poly(C)-induced fibroblasts to synthesize biologically active human interferon.[8–10] Studies have shown a

[1] J. B. Gurdon, C. D. Lane, H. R. Woodland, and G. Marbaix, *Nature (London)* 233, 177 (1971).

[2] B. Mach, C. Faust, and P. Vassalli, *Proc. Natl. Acad. Sci. U.S.A.* 70, 451 (1973).

[3] R. A. Laskey, J. B. Gurdon, and L. V. Crawford, *Proc. Natl. Acad. Sci. U.S.A.* 69, 3665 (1972).

[4] R. L. Jilka, P. C. Familletti, and S. Pestka, *Arch. Biochem. Biophys.* 192, 290 (1979).

[5] R. L. Jilka, R. L. Cavalieri, L. Yaffe, and S. Pestka, *Biochem. Biophys. Res. Commun.* 79, 625 (1977).

[6] A. J. M. Berns, M. Van Kraaikamp, H. Bloemendal, and C. D. Lane, *Proc. Natl. Acad. Sci. U.S.A.* 69, 1606 (1972).

[7] R. H. Stevens and A. R. Williamson, *Nature (London)* 239, 143 (1972).

[8] F. H. Reynolds, Jr., E. Premkumar, and P. M. Pitha, *Proc. Natl. Acad. Sci. U.S.A.* 72, 4881 (1975).

linear relationship between the amount of interferon mRNA injected and of interferon produced.[9,10] Lymphoblastoid,[10] leukocyte,[11] and mouse[12] interferon mRNA can also be translated in the oocyte system.

There are many advantages of using *Xenopus* oocytes over other translational systems to assay interferon mRNA. The most obvious is that the oocytes translate interferon mRNA efficiently. Interferon titers obtained from oocytes were 10- to 500-fold higher than various cell-free systems per microgram of mRNA used. Interferon mRNA has also been translated in heterologous cells,[13-16] but not as efficiently as in oocytes. The interferon titers obtained from heterologous cells can vary greatly. Interferon titers are reproducible in the oocytes and, with moderate care, the procedure is convenient and simple. Another advantage of the oocytes is the small amount of interferon mRNA used in the assay. It is possible to detect interferon produced from 1 ng of total mRNA from induced fibroblasts.[10] At saturation ($\sim$50 ng/oocyte), only a total of 1 μg of total mRNA need be used for injection into 10–20 oocytes.

Materials and Reagents

Preparation of Oocytes. *Xenopus laevis* are obtained from Nasco, Fort Atkinson, Wisconsin (No. LM531LQ). The oocytes obtained from these frogs yield reproducible high titers of interferon. The animal is placed in ice water for 30 min, then sacrificed. The oocytes are removed and immediately placed in 150 ml of oocyte incubation medium.[10]

> Oocyte Incubation Medium
> NaCl, 88 mM
> KCl, 1 mM
> NaHCO$_3$, 2.4 mM
> MgSO$_4$ · 7 H$_2$O, 0.82 mM
> Ca(NO$_3$)$_2$ · 4 H$_2$O, 0.33 mM
> CaCl$_2$ · 6 H$_2$O, 0.41 mM

[9] P. B. Sehgal, B. Dobberstein, and I. Tamm, *Proc. Natl. Acad. Sci. U.S.A.* **74**, 3409 (1977).
[10] R. L. Cavalieri, E. A. Havell, J. Vilček, and S. Pestka, *Proc. Natl. Acad. Sci. U.S.A.* **74**, 3287 (1977).
[11] R. McCandliss, A. Sloma, and S. Pestka, this volume [8].
[12] B. Lebleu, E. Hubert, J. Content, L. De Wit, I. A. Braude, and E. De Clercq, *Biochem. Biophys. Res. Commun.* **82**, 665 (1978).
[13] J. De Maeyer-Guignard, E. De Maeyer, and L. Montagnier, *Proc. Natl. Acad. Sci. U.S.A.* **69**, 1203 (1972).
[14] F. H. Reynolds, Jr. and P. M. Pitha, *Biochem. Biophys. Res. Commun.* **59**, 1023 (1974).
[15] L. H. Kronenberg and T. Friedman, *J. Gen. Virol.* **27**, 225 (1975).
[16] J. J. Greene, C. W. Dieffenbach, and P. O. P. Ts'o, *Nature (London)* **271**, 81 (1978).

Tris base, 7.5 mM
Penicillin G, potassium, 18 units/ml (11 μg/ml)
Streptomycin, 18 μg/ml

The final pH is adjusted to pH 7.6 with HCl and the solution is sterilized by filtration.

Individual oocytes are obtained by gently teasing apart the oocyte sacs with a blunt probe. When the oocytes are separated, they come apart fairly easily. Damaged oocytes should not be used. The oocytes have a distinct animal and vegetal pole. The yolk of the vegetal pole has a light white-green color, and the animal pole is black. Although they vary in size, most oocytes are approximately 1 mm in diameter. The largest oocytes, approximately 1.2–1.5 mm in diameter, tend to leak after injection and should not be used. After isolation, the oocytes are stored at 5°. Oocytes are capable of giving maximum interferon titer after storage for 4–5 days at 5°.

Preparation of RNA. Interferon mRNA isolation and partial purification is described elsewhere in Section II of this volume. The precipitated mRNA is dissolved in sterile H_2O at a final concentration of 0.5–1 mg/ml. The freshly dissolved mRNA is kept on ice until injected. Solutions of mRNA are stored in a liquid nitrogen refrigerator.

Injection Needles. Five-microliter microdispenser tubes (No. 105G, Drummond Scientific) are pulled by a vertical pipette puller (Model 700B, David Kopf Instruments, Tujunga, California). The pulled needles are sealed at the tip. The tips are then broken off with fine scissors under a dissecting microscope to form the needle with a tip of 0.005–0.01 mm in diameter. The microdispenser capillaries consist of uniform bore tubing with 0.27 mm along the cylindrical axis equivalent to 50 nl (5.4 mm equivalent to 1 μl).

Injection of mRNA. Two microliters of interferon mRNA are placed in the bottom of a 60 × 15 mm round petri dish filled with sterile heavy-duty mineral oil. The small aqueous bubble remains at the bottom against the wall of the dish. The injection needle is then clamped to the end of a Brinkman micromanipulator (No. 06-15-00) and attached to tubing connected to a hydraulic pressure system (Fig. 1). Positive pressure fills the needle up to 2 mm from the open tip. The needle is then placed through the mineral oil into the solution of mRNA. Capillary action draws the solution into the tube with the help of negative pressure applied to the system. The sample fills $\frac{1}{2}$–$\frac{3}{4}$ of the injection needle. A small air bubble remains between the RNA solution and the hydraulic fluid (sterile H_2O). A small piece of graph paper with 1-mm divisions is placed on the needle and aligned with the meniscus.

For injection, 10 oocytes are aligned against the edge of a slide taped to

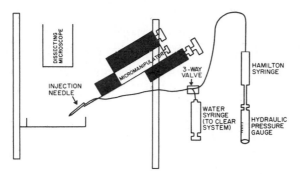

FIG. 1. A schematic outline of injection apparatus. The syringe for clearing the system with H₂O has a capacity of 10 ml. The Hamilton syringe has a capacity of 0.25 ml. The hydraulic pressure is produced by appropriately rotating the screw of the micrometer syringe holder to move the syringe piston linearly. A useful syringe holder is the microinjection unit (No. 1208) from David Kopf Instruments (Tujunga, California). The graduations on the micrometer of this holder provide effective gauge of the hydraulic pressure by accurately noting the linear displacement.

a square (100 × 100 mm) petri dish. The animal or vegetal pole can be injected without affecting the final interferon titer. As the needle enters the oocyte at a 60° angle, positive pressure is applied to the system. Displacement of the meniscus is approximately 0.3 mm for each oocyte injected. This is equivalent to about 50 nl injected per oocyte. The needle is removed from the oocyte, and the remaining 9 are injected in sequence. If the mRNA solution is 1 mg/ml, about 40–50 ng of interferon mRNA are injected per oocyte. It has been previously shown that this concentration is saturating.[10] Immediately after injection, the 10 oocytes are placed in 0.1 ml of fresh oocyte incubation medium and incubated for 18 hr at 23° in a 1.5-ml sterile conical polypropylene tube.

Preparation of Oocyte Extract. After incubation, the oocytes are homogenized manually in the same tube in which they were incubated. This is important because some interferon is secreted into the incubation medium.[17] The extract is then centrifuged for 5 min in an Eppendorf centrifuge. Ninety microliters of supernatant are carefully removed. A lipid layer forms at the top and is toxic to the cells in the interferon assay. Therefore, it is important to remove as little of the lipid as possible if the sample is to be used directly for biological assay of interferon. The 90 μl of supernatant are then centrifuged again for 5 min. The sample is now ready to be assayed for interferon activity as described elsewhere in Volume 78 (see Section IV). The samples are stored at −20° until assayed. Samples appear to be stable at −20° for at least several weeks.

[17] A. Colman and J. Morser, *Cell* 17, 517 (1979).

[11] Translation of Human Fibroblast Interferon mRNA in African Fish Ova (*Tilapiamosaiabica peters* Oocytes)

By HOU YUN-TE and YANG CHONG-TAI

Principle

The induction of interferon in human fetal skin muscle diploid cell cultures by inducers or its superinduction by polyinosinic acid · polycytidylic acid [poly(I) · poly(C)] and cycloheximide results in the activation of the cellular interferon structural gene, leading to the transcription of interferon messenger RNA (mRNA). The total cell poly(A)-RNA, containing interferon mRNA, can be isolated from the induced fibroblasts and has been found to direct the synthesis of human fibroblast interferon in *Tilapiamosaiabica peters* oocytes. These fish ova can be much more easily obtained than those from *Xenopus laevis*.

Preparation of Human Fibroblasts

Human fetal skin muscle diploid cells (HF-7)[1] in their 15th to 35th passages, previously characterized as high producers of fibroblast interferon, are grown in glass bottles in Eagle's minimal essential medium (MEM) with 10% fetal bovine serum, containing 100 μg of penicillin, 100 μg of streptomycin, and 300 μg of glutamine per milliliter. Confluent monolayers with 0.5 to 1.0 $\times$ 10^7 cells/100 cm^2 are formed after 5–7 days of incubation in a 37° incubator. Cultures are used 10–12 days after seeding.

Superinduction Procedure

After removal of the growth medium, MEM containing per milliliter 100–200 μg of poly(I) · poly(C), and 40 μg of cycloheximide (Roth Chemicals, Germany), and adjusted to pH 7.2 is added and incubation at 37° is carried out for 4–7 hr, when the interferon mRNA yield reaches a maximum.[2] To control culture bottles, the same amount of MEM without poly(I) · poly(C) and cycloheximide is added.

Harvesting of Superinduced Human Fibroblasts

After 4–7 hr of incubation at 37°, induced and control cultures are drained and washed four times with Hanks' solution. Each bottle then

[1] Hou Yun-te, *J. Traditional Chinese Med.* **21**, 210 (1980).
[2] Wu Shu-hua, Li Yu-ying, Hou Yun-te, *Acta Acad. Med. Sinicae* **1**, 14 (1979).

METHODS IN ENZYMOLOGY, VOL. 79

receives half the original volume of cooled MEM. The cells are scraped from the glass with a rubber policeman. The cell suspensions are then collected into 250-ml glass centrifuge tubes, chilled on ice, and centrifuged at 4° at 2000 rpm for 20 min.

Preparation of Total RNAs from Superinduced and Noninduced Fibroblasts

Total RNAs are prepared according to the methods described by Pestka et al.,[3] Reynolds et al.,[4] and Green et al.[5] with minor modifications. The final cell pellets are frozen (below −60°) and thawed (at 37°) three times and then homogenized with a homogenizer by adding TSES buffer solution (0.2 M Tris · HCl, 0.05 M NaCl, 0.01 M Na$_2$EDTA, 0.5% sodium dodecyl sulfate, pH 9.0). The homogenates are clarified by centrifugation for 20 min at 1500 rpm. The total nucleic acid is extracted with 0.5 volume of TSES buffer-saturated phenol and 0.5 volume of chloroform–octanol, 24 : 1 (v/v) by vigorous shaking for 30 min at room temperature, followed by centrifugation for 20 min at 2500 rpm at 4°. The aqueous fractions are collected and reextracted three times by the same procedures. The final aqueous fraction is designated as total nucleic acid of induced or control fibroblasts. Potassium acetate (2% w/v) and 2 volumes of cold ethanol are added. The mixture is frozen at −20° for 2 hr and then stirred with a glass rod to remove flocculent DNA. The precipitate is harvested by centrifugation for 20 min at 4000 rpm at 4° and resuspended in a mixture of saline and 2.5 volumes of cold ethanol. After freezing at −20° for 30 min, the precipitate, designated as total RNA, is collected by centrifugation for 30 min at 12,000 rpm at 4°. The resulting total RNA is resuspended and dissolved in saline. To precipitate the RNA, 2 volumes of cold ethanol are added and the tube is stored at −60° before use. The extracted RNA is identified in the spectrophotometer and should conform to the standard: $A_{260}/A_{230} >$ 2.0, $A_{280}/A_{260} < 0.5$.

Oligo(dT)-Cellulose Chromatography

Total RNA at a concentration 75–100 A_{260} units/ml in 0.01 M Tris · HCl buffer containing 0.5 M KCl (pH 7.5) is mixed with oligo(dT)-cellulose (Shanghei Second Reagent Factory), previously equilibrated with the same buffer. After 6 hr at 4°, the mixture is put onto a 1.0 × 4.5 cm column. The

[3] S. Pestka, J. McInnes, E. A. Havell, and J. Vilček, *Proc. Natl. Acad. Sci. U.S.A.* **72**, 3898 (1975).
[4] F. H. Reynolds, E. Premkumar, and P. M. Pitha, *Proc. Natl. Acad. Sci. U.S.A.* **72**, 4881 (1975).
[5] M. Green, P. N. Graves, T. Zehavi-Willner, J. McInnes, and S. Pestka, *Proc. Natl. Acad. Sci. U.S.A.* **72**, 224 (1975).

column is eluted sequentially at a flow rate of 10–15 ml/hr with 100 ml of 0.01 M Tris · HCl buffer containing 0.5 M KCl, pH 7.5 (solution A), at 4° until $A_{260} < 0.03$, then with 100 ml of 0.01 M Tris · HCl, pH 7.5 (solution C) at 20–25° until $A_{260} < 0.03$. Fractions of 4 ml are collected. The absorption at 260 nm of the column fractions is determined. The first peak is characterized as rRNA and tRNA, and the second peak as mRNA.[6] After freezing and drying, the different RNA fractions are stored at −60° for translation experiments in African fish ova.

The oligo(dT)-cellulose column is regenerated by washing with 0.1 M KOH and reequilibrated with 0.01 M Tris · HCl containing 0.5 M KCl, pH 7.5.

The mRNA contents have been found to be more than double in poly(I) · poly(C)-induced human fibroblasts than in noninduced control fibroblasts, indicating the presence of newly produced mRNA in the induced cells.[7]

Translation of Interferon mRNA in Tilapiamosaiabica peters Oocytes

Tilapiamosaiabica peters is a type of African fish, originally from Mozambique in southeast Africa (Fig. 1) and now reared as an edible fish in almost every country of the world. This fish grows very quickly at the optimal temperature of 26–35°. Its oocytes are of comparatively large diameter (1–1.5 mm) and are produced throughout the year. A mature female fish, 15 cm in length, can lay 500 eggs at a time. This fish can be obtained from breeding farms at any season.

For the translation of interferon mRNA, the oocytes are harvested by dissection or by squeezing the cloacal orifice of the fish. The oocytes obtained are suspended in Holtfreter's medium (NaCl, 0.35 g/liter; KCl, 0.005 g/liter; $CaCl_2$, 0.01 g/liter; $NaHCO_3$, 0.02 g/liter) on a glass plate. Four micrograms of mRNA are dissolved in 20 μl of Holtfreter's medium. Handmade micropipettes about 20 μm in diameter and calibrated against a microsyringe are used. Microinjection is carried out under the dissecting microscope with a dosage of 0.2 μg of mRNA per oocyte. In the control group, induced cell rRNA and tRNA, noninduced cell mRNA, and saline are injected instead. Twenty oocytes are used for each group. Injected oocytes are incubated at 25° for 24 hr. After incubation, the oocytes are homogenized in the same Holtfreter's medium that has been used for incubation and then centrifuged at 2000 rpm for 30 min. The supernatant is removed and stored at −60° until it is assayed for biological activity.

[6] M. Green, T. Zehavi-Willner, P. N. Graves, J. McInnes, and S. Pestka, *Arch. Biochem. Biophys.* 172, 74 (1976).

[7] Yang Chong-tai, Wu Shu-hua, Li Yu-ying, Hou Yun-te, *Acta Acad. Med. Sinicae* 2, 1 (1980).

FIG. 1. *Tilapiamosaiabica peters.*

For radiolabeling, 10 μCi of [^{14}C]phenylalanine are dissolved in 20 μl of Holtfreter's medium, mixed with the mRNA preparation, and then injected into the oocytes. The incorporation rate of [^{14}C]phenylalanine into proteins synthesized by the oocytes during translation of mRNA has been found to be significantly increased after injection of induced cell mRNA as compared with oocytes injected with saline.[7]

The translation products in fish ova have been found to contain human interferon by the following criteria:

1. Marked antiviral activity in homologous cell culture is detected with the translation products from mRNA of induced fibroblasts, but not from that of noninduced cells or from the rRNA and tRNA of induced fibroblasts.
2. The antiviral activity is species-specific; it is detected only in human cell cultures, not in chick cell cultures.
3. The antiviral activity can be neutralized by antiserum against interferon prepared from human fibroblasts from which the interferon mRNA has been isolated.
4. The antiviral substance is sensitive to trypsin and insensitive to RNase.
5. Similar results can be obtained with the same mRNA preparation, using heterologous cells as a translation system according to the method described by De Maeyer-Guignard *et al.*[8]

[8] J. De Maeyer-Guignard, E. De Maeyer, and L. Montagnier, *Proc. Natl. Acad. Sci. U.S.A.* **69**, 1203 (1972).

Antiviral activity is measured by cytopathic effect reduction assay with bovine vesicular stomatitis virus in human fibroblast monolayers. All interferon units are corrected to express reference units as defined by titration of a reference human interferon preparation G-023-901-527, supplied by the Reference Reagents Branch of the National Institute of Allergy and Infectious Diseases, NIH.

[12] Microinjection of mRNA and Other Macromolecules into Living Cells

By DENNIS W. STACEY

Techniques for the microinjection of macromolecules into living cells were developed independently by Diacumakos[1] and Graessmann.[2] In both procedures a glass micropipette is loaded with the sample and delicately manipulated until the 1.0-μm tip penetrates into the desired region of a cultured cell, where 1×10^{-11} to 1×10^{-10} ml of sample is expelled under pressure. In the Graessmann technique an inverted microscope is used and microinjection is performed from above the cells. The classical "hanging drop" approach was adapted for the Diacumakos technique, where recipient cells are injected from below. This technique requires a standard microscope and a more complicated micropipette than the Graessmann technique, but it allows observation under the highest available phase-contrast resolution. A modification of the Diacumakos technique that has been used specifically to handle highly concentrated, labile, or extremely small samples will be described here.

Equipment

In the Diacumakos technique of microinjection the recipient cells grow as a monolayer on a coverslip. The coverslip is inverted and supported above a glass microscope stage plate, creating an area that is filled with growth medium in which the micropipette is manipulated. The injections are viewed through the coverslip from above (Fig. 1).

Microscope. The microscope must have a fixed stage (Fig. 2). Phase-contrast optics are highly beneficial, and a long working-distance condenser is needed. Routine injections can be performed with a 40×, flat-field, "high-dry" objective, and oil immersion can be used for more deli-

[1] E. G. Diacumakos, *Methods Cell Biol.* 7, 287 (1973).
[2] A. Graessmann, *Exp. Cell Res.* 60, 373 (1970).

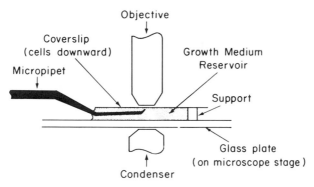

FIG. 1. Diagram of the spatial arrangements of microinjection components in the "hanging drop" approach to microinjection. From Stacey.[4]

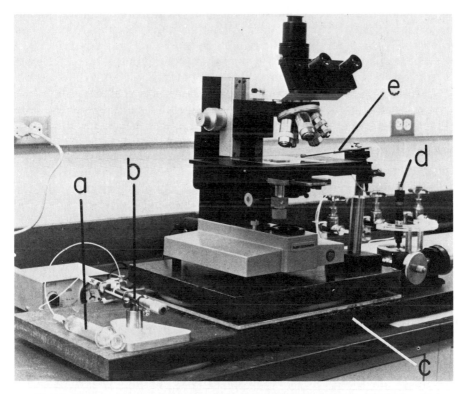

FIG. 2. Photograph of microinjection equipment. The microscope is a Leitz instrument modified by McBain Instrument Co. (Chatsworth, California) so that the nosepiece can be lifted away from the stage during preparation. The stage has been rotated 180° so that adjustments can be made with the left hand. a, glass syringe; b, micrometer-driven syringe; c, shock mount; d, micromanipulator hand-operated control unit; e, instrument holder with a micropipette positioned above the condenser.

cate work. If the immersion oil causes problems after injection, glycerin may be a suitable substitute. A 10× scanning objective is needed to align the micropipette and recipient cells. A Leitz (Rockleigh, New Jersey) object marker is valuable for identification of injected cells. To reduce eye strain it is best to perform injections in a darkened room.

Manipulator. A "joy-stick" type of micromanipulator is preferable for rapid injections. Leitz produces a highly recommended mechanical manipulator whose vertical motion is controlled by a separate knob located near the "joy-stick." This is inconvenient when injections are performed by vertical motions. The pneumatic De Fonbrune series C manipulator (Beaudouin, Retaud, Paris) controls movements in all three dimensions with the "joy-stick" and is very compact, although quite fragile. Unfortunately, adjustments that allow lateral movements in the plate of the coverslip are difficult. This problem is not as important with the more bulky and less precise De Fonbrune series A manipulator, which must have air reservoirs placed in its pneumatic transmission lines for the delicate manipulations required for injecting cultured cells.

The manipulator found most suitable by the author is the hydraulic Narishige Model 103 M (Labtron, Farmingdale, New York). Like the De Fonbrune manipulator, this instrument allows physical separation between the instrument holder and the hand-operated control unit (Fig. 2d). Probe movements are strictly in the plane of the coverslip, and gross adjustments are readily performed.

Microforge. One of the principal advantages of the Graessmann technique is the simplicity of the micropipette used. The required straight pipette can be formed mechanically by a variety of forges primarily designed for electrophysiology, or even by a handmade instrument.[3] A De Fonbrune microforge is required for the Diacumakos technique. This forge must be further modified with a microscope tube that can accommodate two or three objectives to replace the standard dissecting microscope.[1] A long working-distance, high-resolution objective such as the Zeiss (Carl Zeiss, New York City) UD_{40} along with 10× and perhaps 2.5× objectives are used. The advantage of this type of microforge is its flexibility in forming the type of micropipette best suited to the experiment (see below).

Injection System. Two means of creating pressure may be used. One is a glass syringe (Fig. 2a), and the other is a Hamilton (Whittier, California) gastight 1-ml syringe driven by a spring-opposed micrometer (Fig. 2b). These are connected through a high-pressure valve (Chromatronix, Berkeley, California) to the micropipette so that only one operates at a time. Connections are also made with tubing designed for high-pressure chromatography that is flexible enough not to transmit vibrations to the

[3] M. Graessmann and A. Graessmann, *Proc. Natl. Acad. Sci. U.S.A.* **73**, 366 (1976).

micropipette, yet rigid enough that deformation is minimal even at high pressure. The entire system is completely filled with water prior to injection.

The micropipette is secured during injection by a modified Leitz instrument holder. A thread assembly is attached to its base for connection to the high-pressure tubing. The shape of the instrument holder may have to be modified to fit the physical structure of the microscope (Fig. 2e). The instrument holder is supplied with a flexible rubber gasket that must be replaced by Tygon tubing large enough for easy insertion and removal of the 1-mm base of the micropipette. In addition, metal washers must be placed both in front of and behind the tubing. This can be accomplished simply by splitting the metal sleeve supplied with the holder and placing the tubing between the two halves.

Shock Mount. Vibrations of the micropipette must be reduced until almost undetectable. A number of vibration-reducing assemblies are commercially available (for example, Optiquip, Highland Mills, New York). A suitable system can be constructed, however, as follows (Fig. 2c). A heavy stone slab base is supported on top of flexible rubber tubing. On top of the stone slab a $\frac{3}{8}$-in. rubber mat supports a metal plate on which the microscope is placed. If vibrations persist, a third layer of rubber tubing supporting another stone slab may be added. The rubber tubing should be placed in a circular array to stop side-to-side movements. The plate and slabs should be wide and long enough so that the microscope is firmly held in an upright position. Air currents, even those caused by movements nearby, can produce vibrations in the micropipette.

Flame Board. In the Diacumakos technique and in the sample handling procedures described below, capillary tubes must be bent and drawn by hand. An adjustable tiny gas flame is required as described by Diacumakos.[1] An appropriate burner can be constructed of stainless steel tubing or a syringe needle approximately 1 mm in diameter (17-gauge). If a needle is used, it must be cut off square. The flow of gas is regulated by a needle valve, and air currents may be minimized by placing the entire assembly in a covered, three-sided box.

Centrifuge Adapter. Debris may be removed from a sample before injection by placing it in a sealed capillary tube and centrifuging it. In Fig. 3 a Delrin plug is diagrammed which adapts a 1-mm capillary tube to a Sorvall (Newtown, Connecticut) SS34 rotor.[4] In a similar plug constructed for a Beckman SW50.1 rotor (Beckman Instruments, Fullerton, California), a sealed capillary tube remained intact up to 30,000 rpm. Since capillarity prevents reorientation of a sample in a capillary tube, there is no advantage of a swinging-bucket rotor over a fixed-angle rotor.

Capillary Tubes. Several instruments made of 1 mm soft glass capillary

[4] D. W. Stacey, Dissertation, The Rockefeller University, New York City, 1976.

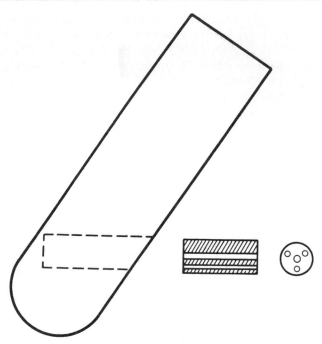

Fig. 3. Specifications for a centrifuge adapter to support three sealed capillary centrifuge tubes (24 mm long, 1 mm in diameter) in a Sorvall SS34 rotor. A Delrin plug is constructed to exactly the size of a typical centrifuge tube ($1\frac{5}{32}$ in. $\times$ $4\frac{3}{8}$ in.). At $2\frac{11}{16}$ in. from the top a $\frac{3}{8}$ in. flat-bottomed hole is drilled $1\frac{3}{16}$ in. at a 56° angle (indicated by the dashed line). An insert $\frac{7}{8}$ in. long and $\frac{3}{8}$ in. in diameter fits snugly into the hole so as to be recessed $\frac{1}{8}$ in. The insert (shown in side and end views) has a 3-48 threaded hole in the middle used for its removal in case a tube breaks. Capillary tubes are inserted into peripheral 0.035–0.040 in. holes (three are shown). From Stacey.[4]

tubing will be described.[5] Approximately 2 hr are needed to construct all the instruments used in 20–30 experiments. All capillary tubes are cleaned in 50% nitric and 50% sulfuric acid, washed, and baked at 300°F overnight. The precursor of the micropipette or *pre-micropipette* (Fig. 4j) is formed by drawing out a 1-mm capillary tube (Fig. 4e) over a small flame. The drawn-out portion should be at least 1.5 cm long and 0.2 mm in diameter, or thick enough to support the capillary tube without bending (Fig. 4f). The drawn-out capillary tube is broken off so that the drawn-out portion is attached to a 3-cm segment of 1-mm capillary tubing (Fig. 4g), the large end of which is fire-polished by placement in a small flame. Two successive (approximately 30%) bends are made just at the beginning of the drawn-out portion (Fig. 4h) and about 1 cm back along the capillary

[5] D. W. Stacey and V. G. Allfrey, *Cell* 9, 725 (1976).

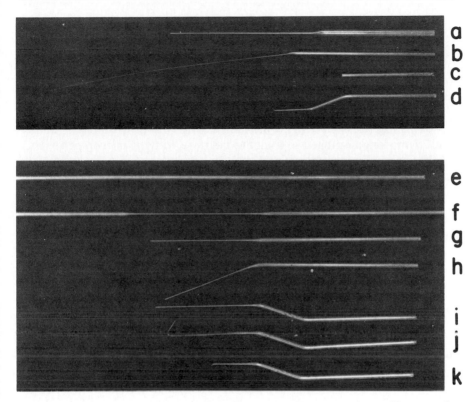

FIG. 4. Photographs of instruments made from capillary tubes: (a) capillary transfer pipette; (b) micropipette filling tube; (c) capillary centrifuge tube; and (d, k) the micropipette. Instruments e through k represent various stages in the formation of the pre-micropipette as described in the text.

tube (Fig. 4i). The base of the capillary tube and drawn-out portion should be parallel and in the same plane. To test this, the base can be rotated rapidly between thumb and forefinger. Finally a small hook must be made in the drawn-out portion (Fig. 4j). After the pre-micropipette is formed, its drawn-out portion is filled with Beckman Desicote by capillary action. Distilled water is then drawn through it, and it is again baked. Two or three minutes are required to make each pre-micropipette.

The *capillary centrifuge tube* described above is formed by sealing one end of a 1-mm glass capillary tube (pretreated with Desicote) over a tiny flame, followed by breaking the tubing at 24 mm (Fig. 4c) with a diamond pencil.

The *micropipette filling tube* is made by drawing out a 1-mm glass capillary tube to its finest dimension, so that it can be inserted from the base of the micropipette through the drawn-out portion up to the orifice (Fig. 4b). This tube is siliconized by drawing Desicote followed by distilled water through it; it is then baked as above. A similar tube, the *capillary transfer pipette,* is made from 2-mm capillary tubing. It is drawn out to produce a segment small enough to fit into the 1-mm capillary centrifuge tube (Fig. 4a). It is also siliconized and baked as above.

Stage Plate. The coverslip with recipient cells attached is supported above a large glass plate ($\frac{1}{8}$ in. in diameter) by two glass supports (2 mm² and 20 mm long) arranged so as not to interfere with the micropipette (Fig. 1). The coverslip must be parallel to the microscope stage. The stage plate can be sterilized with 70% ethanol and ultraviolet irradiation between usages.

Microinjection Procedure

Formation of the Micropipette. This is perhaps the most critical factor in microinjection. The pre-micropipette described above is secured in a Leitz instrument holder and placed in the microforge with the 30° bends in a vertical plane. A 1-g weight is attached to the hooked end of the pre-micropipette, whose drawn-out portion is positioned within the loop of the microforge filament (Fig. 5). As the filament is heated, the drawn-out portion of the micropipette is bent (Fig. 6A,B) and then drawn out (Fig. 6C) to a point (Fig. 6D). The heat of the filament is reduced, and the tip of micropipette is viewed at 40× (Fig. 6E). A small glass bead attached to the extreme end of the filament is brought into contact with the tip of the pre-micropipette (Fig. 6F). If the temperature of the glass bead is correct the tip will adhere, whereupon the filament is raised upward and to the left at a 45° angle. This movement draws the capillary tube to an extremely fine point (Fig. 6G). The filament is finally cooled and used to break off the sharp point to produce a micropipette with orifice dimensions between 0.5 and 1.5 μm in outside diameter. Fire-polishing of the tip is not required.

During the initial steps (Fig. 6A–D) the filament should not touch the glass. As little heat as possible should be used in order to assure thin walls. In the final steps of orifice formation, the heat of the glass bead is critical. A hot bead adheres well to the glass tube but may seal the orifice. A cooler bead may not stick well or else may result in an orifice tapered so abruptly and with walls so thin that the micropipette breaks easily during injection. A good micropipette has a flexible tip and as large an inside to outside diameter ratio as possible. There is no substitute for practice in making micropipettes, followed by determination of their characteristics during microinjection. A fluorescent-labeled protein can be injected for practice. The orientation of the pre-micropipette with respect to a hori-

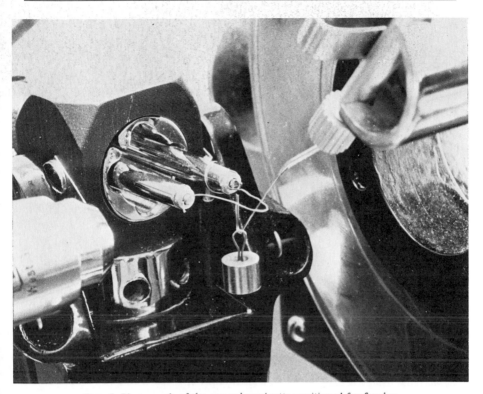

FIG. 5. Photograph of the pre-micropipette positioned for forging.

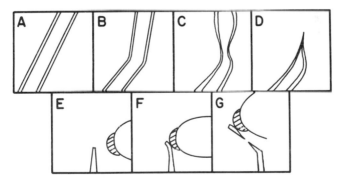

FIG. 6. Illustration of the steps in microforging as described in the text. Panels A–D illustrate the appearance of the drawn-out portion of the pre-micropipette during the initial stages of forging as viewed through a 10× objective. The heated filament, which is not shown here, is removed from the capillary tube a distance equal to approximately twice its diameter. Panels E and F illustrate the final stages of micropipette formation as viewed through a 40× objective.

zontal line can be adjusted on the De Fonbrune microforge. This angle, along with the angle at which the glass bead is drawn away from the initial tip of the pre-micropipette (Fig. 6G), determines the angle at which the orifice approaches the coverslip during microinjection.

Sample Preparation. Injection volumes may average up to 5% of the cell volume. In order to achieve normal cellular concentrations after injection a molecule within the sample must be concentrated at least 20-fold above physiological levels. Many protein solutions can be highly concentrated by dialysis against distilled water or a volatile buffer followed by lyophilization and redissolving in a small volume. For proteins that lose biological activity after lyophilization or that do not redissolve well, chemical means such as precipitation with ammonium sulfate or affinity techniques may be applicable. If a large volume is available, pressure dialysis can be used.

One technique that is effective in concentrating very small volumes involves the use of hollow dialysis fiber (formerly, BioFiber, from Bio-Rad, Richmond, California). Unfortunately these fibers are presently available only as components in a rapid dialysis assembly (Amicon, Lexington, Massachusetts). The sample is drawn into the hollow fiber, which is then sealed at one end while the free end is secured within the solution to be concentrated. Aquacide II (Calbiochem, San Diego, California) or Sephadex (Pharmacia, Uppsala) is placed around the sample-containing fiber for 20 min to 2 hr. The sample is concentrated as it travels down the fiber to replace buffer absorbed by the Aquacide. Extremely high concentrations are obtained at the sealed end of the tube. After concentration the fiber is cleaned and dialyzed for 5–10 min against injection buffer. The small volume of highly concentrated material can be handled as described below. Finally, small volumes can be further concentrated by evaporation.

An RNA solution with $0.2\,M$ sodium chloride or sodium acetate can be concentrated by precipitation with two volumes of ethanol at $-20°$. Added carrier, such as tRNA, will become highly concentrated in the injected sample and may have unwanted biological activity within the recipient cell. With less than 5 μg of RNA, precipitation at $-20°$ should continue for 48 hr. The precipitate can then be placed in a polyallomer ultracentrifuge tube (Beckman) and centrifuged at high speed to ensure pelleting of the precipitate. If excess salt or SDS is present the precipitate may be washed with cold 70% ethanol exactly as above. The sample is redissolved following lyophilization by placing a 2–4-μl drop of injection buffer into the bottom of the tube and thoroughly working the droplet over the entire bottom surface of the centrifuge tube with a 5 μl pipette (Clay Adams, Parsippany, New Jersey). DNA solutions are handled the same as RNA solutions except that freezing of the ethanol precipitate in liquid N_2 or solid CO_2 eliminates the need for long incubation at $-20°$ to promote formation of precipitate. Work with retroviral DNA molecules has shown that DNA molecules up to 10^7 daltons are not sheared during microinjection.

The salt concentration of injected solutions should conform to that within a cell, where potassium ions are generally in excess. Although intracellular salt concentrations vary, the following values are typical: 113 mM K^+, 23 mM Na^+, 3 mM Mg^{2+}, 3 mM PO_4, and 140 mM Cl^-.[6] In practice a 1% KCl solution can be used for injections. Solutions of KCl between 0.1 M and 0.3 M may be suitable along with sucrose solutions from 0.3 M to 1.0 M.

Sample Handling. Several techniques are used for loading samples into the micropipette. Graessmann simply draws the sample through the orifice into the micropipette. It is important to secure the micropipette and view the filling procedure under a microscope. During filling the sample should be kept in a moist atmosphere to delay evaporation. This is a convenient, direct technique that involves little handling of the sample. Its disadvantage is the limitation in amount of sample loaded. Although up to 500 injections can be made from a micropipette filled in this way, it must be remembered that capillary action will fill the orifice region of a micropipette to a certain volume. If sample is expelled beyond this point, culture medium will be drawn back into the micropipette when pressure is released, resulting in dilution of the sample. This process may continue unnoticed by the operator. Fluorescent-labeled protein marker may help determine how many injections can be made. Diacumakos loads sample from the back of the pre-micropipette. This direct technique has no limitation on volume but requires that the sample be loaded prior to forging. If a mistake is made in forging, it is difficult to recover the sample.

The technique of sample handling and micropipette filling found to be most useful to the author requires the formation of a long, whisker-like capillary filling tube (Fig. 4b), which is not required in the above procedures, but the technique has the advantages that (*a*) the sample is handled in a minimum of time following preparation; (*b*) clogging of the orifice is minimized; (*c*) the sample injected is not exposed to the potential of denaturation at an air–water interface or high salt concentration due to evaporation; (*d*) extremely small samples can be handled; and (*e*) loss of sample due to adsorption of glass surfaces is minimized.[5]

Construction of the capillary instruments needed in this technique has been described. After preparation the sample is aspirated into the capil-

Cell Survival. The gross affect of an injection upon the cell can be assessed by injecting fluorescent-labeled hemoglobin and determining

[6] M. Wickson-Ginzburg and A. K. Solomon, *J. Gen. Physiol.* 46, 1303 (1963).

lary transfer pipette and expelled into the bottom of the capillary centrifuge tube. The loaded centrifuge tube is placed into the centrifuge adapter and centrifuged to remove debris (approximately 10 min at 12,000 rpm). After centrifugation the micropipette filling tube is inserted into the capillary centrifuge tube beneath the surface of sample, and approximately 0.2–0.5 μl is aspirated into the tube. The fine tip of the filling tube is then removed from the centrifuge tube, inserted into the base of the preformed micropipette, and forced past the 30° bends into the drawn-out portion up to the orifice. As the orifice is approached, the sample begins to be forced out of the tube under pressure. This continues until 0.1–0.2 μl of sample has been expelled at the orifice in a position for injection, at which time the filling tube is withdrawn from within the micropipette. Unless the orifice is extremely large, more sample is loaded in this way than would be expelled during an injection experiment.

If clogging of the micropipette remains a problem, the pre-micropipette can be filled with distilled water prior to formation of the orifice. The heated filament drives the water away from the site of forging. The micropipette filling tube is inserted into the water-filled micropipette as before, and water behind the orifice is displaced by sample. A small volume of water will remain within the orifice and protect the sample from evaporation, but it must be expelled prior to injections. Since the sample flows past the walls of the filling tubes and micropipette in these micropipette filling techniques, and since the sample in contact with these walls is continually replaced, loss of the sample due to adsorption to glass surface is not a concern. Furthermore, evidence indicates that little adsorption to the siliconized glass occurs. As little as 0.5 μl of sample has been handled by this technique. A sample as large as 2.0 μl has been used several times with the sample stored in the capillary centrifuge tube between usages.

Microinjection. The loaded micropipette is placed in the instrument holder of the injection system and attached to the manipulator so that the tip of the micropipette points upward. The coverslip with recipient cells is placed with the cell downward on the support of the stage plate, and the region between the coverslip and stage plate is filled with growth medium (Fig. 1). To prevent a change in pH, medium with a nonvolatile buffer can be used. The micropipette is positioned immediately above the condenser and near enough to the stage plate that it will not contact the coverslip, which is then also placed above the condenser by adjusting the stage. The 10× objective is used to position the micropipette and cells. The micropipette is brought nearer to the coverslip, and the high-power, phase-contrast objective is moved into position. The micropipette is raised to the cell monolayer, pressure is created within the injection system, and injections begin.

Injection is performed by positioning the orifice of the micropipette beneath the recipient cell and then raising the micropipette until the orifice enters the cell. Pressure within the injection system then forces sample into the cell. Entry of injected solution is immediately apparent. The cytoplasmic granules move away from the site of injection, and the dark texture of the cytoplasm lightens briefly owing to the introduction of a less dense material. Immediately after injection the micropipette is withdrawn from the cell. Injected volumes should be small enough so that no gross deformation of the cell is apparent and so that only local disturbances occur in the cytoplasm. If sample will not enter the cell, it is possible that: (a) the micropipette is clogged; (b) the cell is too thin in the region of contact for the orifice to enter; (c) the micropipette orifice is shaped improperly for injection; or (d) the plasma membrane is blocking the orifice. The orifice can gently be dropped away from the coverslip or raised toward it to solve the latter problem. If the orifice comes near a cell while sample is rapidly being expelled, ruffles may move across the cell and give the superficial appearance that an injection has occurred.

Injections can be made at almost any region of the cell with an extremely fine orifice or with an orifice that is nearly perpendicular to the glass plate. It is convenient to inject near the nucleus where the cell depth is greatest. Nuclear injections should be made with rapid movements. The dark phase-contrast appearance of the nucleus lightens immediately with the introduction of a small injected volume. Nuclear injections should not be large enough visibly to alter the nuclear size. Spillage of solution into the cytoplasm during a nuclear injection is generally apparent and often unavoidable.

If the micrometer assembly is used to create pressure during injection a steady flow of sample is produced and the micropipette is moved from cell to cell without pressure adjustment. This results in loss of sample into the medium. Rapid and nearly uniform injections are readily performed in this way, and one hand is free for stage adjustments. A micropipette filled with the sample by suction through the orifice contains an insufficient amount of sample for this type of injection.

The glass syringe allows pressure to be created only when the orifice is within the cell. Even with this approach it is likely that some of the sample escapes into the medium. If the micrometer assembly is used to create pressure only when the orifice is within the cell, caution must be used to ensure that when the pressure is released between injections negative pressure is not created to draw growth medium back into the micropipette. If a micropipette has a tip perpendicular to the coverslip, injections can be made with an orifice slightly larger than the cell thickness at the site of injection. With a properly designed micropipette 1000 injections can be made in less than 2 hr.

whether the cell divides. A more rapid analysis is possible under phase-contrast optics. A cell only mildly affected by an injection is unaltered in appearance. Even if the cell only exhibits a granular cytoplasmic appearance or withdraws its cellular processes, it is likely to survive. If the phase-contrast cytoplasmic appearance of an injected cell lightens permanently, the cytoplasmic granules exhibit Brownian motion, or the nuclear boundary becomes unusually sharp, it is almost certain the cell will not survive.

The average injection volume can be determined by performing several hundred injections of concentrated [^{125}I]BSA (New England Nuclear, Boston, Massachusetts) solution, washing the cells several times over the next hour, and then measuring the radioactivity within the cell in relation to the specific activity of the injected solution. It is not as yet possible to inject repeatedly a strictly uniform or a predetermined volume.

If injected cells must be identified after 1–2 days, radiolabeled, fluorescent labeled, or antigenic markers may be added to the sample. Alternatively, the position of injected cells may be marked with an object marker or all the cells within a given area can be injected.

Clogging of the Micropipette. This is a frustrating problem. The first step in clearing a clogged orifice is to increase the pressure as much as possible with the micrometer device and wait 1–2 min. If this fails to clear the orifice, the tip may be broken off against the coverslip in the hope that the resulting orifice will not be too large for injection. A micropipette tip that is perpendicular to the coverslip is best suited for this procedure. Wall thickness at the orifice may be a factor in recurrent problems along with debris in the sample or in the micropipette prior to loading of the sample. In the latter case the sample filling techniques described above may eliminate the problem.

A variety of cells in monolayer culture have been microinjected. In general, established lines withstand injection better than primary cultures. Nonadherent cells may be attached to a coverslip with antibodies or lectin.[7] Alternatively they may be embedded in a thin layer of agar. Graessmann has increased the size of target cells by fusion to form a syncytium.[8] Even with a large multinuclear mass, indiscriminately large injections will have cytopathic effects at the injection site. Phase-contrast optics may not be preferable in injections of rounded cells.

[7] A. Graessmann, H. Wolf, and G. W. Bornkamm, *Proc. Natl. Acad. Sci. U.S.A.* **77**, 433 (1980).

[8] A. Graessmann, M. Graessmann, and C. Mueller, *Biochem. Biophys. Res. Commun.* **88**, 428 (1979).

[13] Assay of Human Interferon Messenger RNA by Microinjection into Mammalian Cells

By CHIH-PING LIU, DORIS L. SLATE, and FRANK H. RUDDLE

Microinjection of DNA and RNA into intact eukaryotic cells provides an efficient method for studying transcription and translation of specific genes. The most widely exploited microinjection system employs the *Xenopus laevis* oocyte, in which both transcription and translation of injected nucleic acid have been described.[1,2] The pioneering work of Diacumakos[3] extended the microinjection procedure to individual mammalian cells that have approximately 1/100,000 the volume of *Xenopus* oocytes. Stacey and Allfrey[2] used microinjection to transfer duck globin mRNA into HeLa cells and observed efficient translation of this mRNA.

Several laboratories have reported the translation of interferon mRNA in intact heterologous cells,[4-6] cell-free extracts,[7-9] and *Xenopus laevis* oocytes.[7,10] We have studied interferon production in mouse LTK⁻ cells after microinjection of mRNA extracted from human fibroblasts that had been superinduced[11,12] to synthesize interferon.

Superinduction of Human Interferon mRNA

FS11 human diploid fibroblast cells were treated with poly(I) · poly(C) (100 μg/ml, P-L Biochemicals), DEAE-dextran (20 μg/ml, Pharmacia), cycloheximide (50 μg/ml, Sigma), and actinomycin D (2 μg/ml, Calbiochem) as described by Slate and Ruddle.[13] Briefly, cells were ex-

[1] J. E. Mertz and J. B. Gurdon, *Proc. Natl. Acad. Sci. U.S.A.* **74**, 1502 (1977).
[2] D. W. Stacey and V. G. Allfrey, *Cell* **9**, 725 (1976).
[3] E. G. Diacumakos, *Methods Cell Biol.* **7**, 287 (1973).
[4] F. H. Reynolds and P. M. Pitha, *Biochem. Biophys. Res. Commun.* **59**, 1023 (1974).
[5] J. De Maeyer-Guignard, E. De Maeyer, and L. Montagnier, *Proc. Natl. Acad. Sci. U.S.A.* **69**, 1203 (1972).
[6] L. H. Kronenberg and T. Friedmann, *J. Gen. Virol.* **27**, 225 (1975).
[7] F. H. Reynolds, Jr., E. Premkumar, and P. M. Pitha, *Proc. Natl. Acad. Sci. U.S.A* **72**, 4881 (1975).
[8] S. Pestka, J. McInnes, E. A. Havell, and J. Vilček, *Proc. Natl. Acad. Sci. U.S.A.* **72**, 3898 (1975).
[9] M.-N. Thang, D. C. Thang, E. De Maeyer, and L. Montagnier, *Proc. Natl. Acad. Sci. U.S.A.* **72**, 3975 (1975).
[10] R. L. Cavalieri, E. A. Havell, J. Vilček, and S. Pestka, *Proc. Natl. Acad. Sci. U.S.A.* **74**, 3287 (1977).
[11] Y. H. Tan, J. A. Armstrong, Y. H. Ke, and M. Ho, *Proc. Natl. Acad. Sci. U.S.A.* **67**, 464 (1970).
[12] P. B. Sehgal and I. Tamm, *Virology* **92**, 240 (1979).
[13] D. L. Slate and F. H. Ruddle, *Cell* **16**, 171 (1979).

METHODS IN ENZYMOLOGY, VOL. 79

posed to phosphate-buffered saline containing poly(I) · poly(C), DEAE dextran, and cycloheximide for 2 hr, washed, and refed with serum-free medium containing cycloheximide. After incubation for 3 hr, the translation inhibitor was washed out, and serum-free medium containing actinomycin D was added for 1 hr. After washing, cells were fed medium supplemented with 2% fetal bovine serum and incubated for 2 hr. This incubation medium was assayed for interferon activity; typical yields were 2 to 5 × 10^4 units/ml. Cells were then scraped from roller bottles, washed three times in physiological saline, and pelleted for mRNA extraction.

Preparation of mRNA

Total cellular RNA from superinduced FS11 or control cells (FS7 fibroblasts or HeLa cells) was prepared by the guanidine · HCl technique of Deeley et al.[14] Cell pellets were resuspended in 20 ml of 6 M guanidine · HCl, 10 mM dithiothreitol, and 25 mM NaOAc, pH 5.0, per gram wet weight of cells at room temperature. Cells were disrupted by gentle pipetting, and the extract was overlaid on 0.2 volume of 5.7 M CsCl, 0.1 M EDTA-Na_2, pH 7.5, and centrifuged for 18 hr at 25,000 rpm in a Beckman SW27 rotor. All the liquid was removed, and the tip of the tube was cut off. The gelatinous pellet was collected and dissolved in a small amount of water. RNA was precipitated by the addition of 0.1 volume of 2 M NaOAc, pH 5.0, and 2 volumes of ice-cold absolute ethanol. The RNA was pelleted and washed twice in 20 parts of ethanol 9 parts of water, and 1 part of 2 M NaOAc, pH 5.0, and once in absolute ethanol. The final pellet was dissolved in a small amount of 0.5 M NaCl, 0.5% sodium dodecyl sulfate (SDS).

The RNA preparation was chromatographed on oligo(dT)-cellulose (type 2, Collaborative Research, Inc.) as described by Deeley et al.[14] Unbound material [poly (A)$^-$-RNA] was washed from the column with 0.5 M NaCl, 0.5% SDS; the bound fraction [poly(A)$^+$-RNA] was eluted with water. Each fraction was then precipitated and washed as above. The pellets were dissolved in water to a final concentration of 1.2 $A_{260 nm}$ units/ml for poly(A)$^-$-RNA and 1 $A_{260 nm}$ units/ml for poly(A)$^+$-RNA.

Sucrose Gradient Fractionation of mRNA

Poly(A)$^+$-RNA (0.5 $A_{260 nm}$ unit) from FS11 superinduced fibroblasts was loaded on a 15 to 40% linear sucrose gradient in 0.01 M Tris · HCl, pH 7.5, and centrifuged in a Beckman SW40 rotor at 35,000 rpm at 4° for 13

[14] R. G. Deeley, J. I. Gordon, A. T. H. Burns, K. B. Mullinix, M. Bina-Stein, and R. F. Goldberger, *J. Biol. Chem.* **252**, 8310 (1979).

hr. One-milliliter fractions were collected, and the RNA in each fraction was precipitated with ethanol and then dissolved in 0.1 ml of water. Double-stranded, labeled reovirus RNAs (provided by E. Slattery, Yale University) were used as size markers in a parallel run.

Microinjection Procedure

Microinjection of RNA solutions into individual mouse LTK⁻ cells was performed following the method of Diacumakos[3] as modified by Stacey and Allfrey[1,2] (see this volume [12]). The average volume injected in each cell was about 3×10^{-11} ml; this represents about 6% of the volume of the injected cell.

Assay for Human Interferon Translation

Small sections of coverslip containing 10–100 microinjected cells were placed cell side up in a well of a 96-well microtiter tray (Linbro) containing a preseeded monolayer of human FS7 fibroblasts in 50 μl of medium. After incubation for 18 hr, the medium was removed by vigorous inversion of the microtiter tray, and 0.1 PFU/cell of vesicular stomatitis virus was added in 100 μl of medium. After 24–30 hr, cytopathic effect was scored in sample wells, as in cell, virus, and interferon control wells.

For quantitation of interferon yields, 100 injected cells were incubated in 100 μl of medium for 18 hr, and then this medium was diluted serially on monolayers of FS7 cells. Antibody neutralizations of these interferon preparations were done by incubating anti-leukocyte or anti-fibroblast human interferon sera (provided by Dr. J. K. Dunnick, National Institute of Allergy and Infectious Disease) with 50 μl of interferon for 1 hr at 37° before assay on FS7 cells.

Summary of Translation Experiments

Since the mouse and human interferons produced by the LTK⁻ and FS11 cells used in this study do not exhibit cross-species activity, it was possible to assay for the production of human interferon in mouse cells microinjected with FS11 mRNA. Poly(A)⁺-RNA from superinduced, but not uninduced, human fibroblasts, can direct the synthesis of human interferon in injected mouse cells. Poly(A)⁻-RNA is not translated into detectable human interferon. When cycloheximide is added to the medium just after microinjection of the LTK⁻ cells, the translation of poly(A)⁺-RNA into interferon is blocked. The size of the interferon mRNA molecule is approximately 12 S, as determined by sucrose density gradients. The human interferon made in response to poly(A)⁺-RNA from superinduced FS11 fibroblasts is of the fibroblast antigenic type.

The microinjection assay method provides greatly enhanced sensitivity in detecting translation products from extremely small amounts of mRNA. We could detect interferon translation using as few as 13 injected mouse cells. In our quantitation assays, 100 injected cells were found to produce 4–8 units of human interferon. We have also used modifications of the microinjection mRNA assay to detect human thymidine kinase, adenine, and hypoxanthine phosphoribosyltransferases, and propionyl-CoA carboxylase[15] in microinjected cells.

Acknowledgments

We wish to thank Dr. Elaine Diacumakos for instruction in the microinjection procedure and Frances Lawyer for excellent technical assistance. This work was supported by NIH Grant GM 09966.

[15] C.-P. Liu, D. L. Slate, R. Gravel, and F. H. Ruddle, *Proc. Natl. Acad. Sci. U.S.A.* **76**, 4503 (1979).

[14] Isolation and Translation of Mouse Interferon mRNAs

By M. N. THANG

The wheat germ cell-free protein-synthesizing system was observed to translate mouse interferon mRNA to protein that was active in inhibiting viral multiplication in mouse cells.[1] The easy availability of the wheat embryo and the simplicity of the procedure for the preparation of the extract make this material a good choice for studies of protein synthesis. Mouse interferon mRNAs isolated by the two methods described in this section are translated in the wheat germ extract with high efficiency. The fidelity of the translation is reflected in the production of the biologically active product.

Induction of Cells for mRNA Extraction

Mouse C-243 cells, a line transformed by murine sarcoma virus, are induced with poly(I) · poly(C) in large petri dishes (150 mm in diameter); they received 10 ml of minimum Eagle's medium (MEM), without serum, containing 30 μg of poly(I) · poly(C) and 100 μg of DEAE-dextran per milliliter.[1] The inducer is left for 2 hr on the cells and then removed; the

[1] M.-N. Thang, D. C. Thang, E. De Maeyer, and L. Montagnier, *Proc. Natl. Acad. Sci. U.S.A.* **72**, 3975 (1975).

cells are washed once with warm MEM, and 30 ml of MEM containing 3% calf serum are then added. Nine hours after the onset of induction, the cells are harvested. If extraction of RNA is not to be performed immediately, the cells can be frozen at $-70°$ until the time of RNA extraction.

C-243 cells are also obtained by culture in suspension according to Tovey et al.[2] MEM medium (500 ml) containing 10% fetal calf serum is inoculated with 2×10^5 cells/ml and grown to about 1×10^6 cells/ml. Mouse interferon with specific activity of 10^6 units/mg is added to make 20 units/ml for priming during 16 hr. The cells are then centrifuged and resuspended in 5 ml of medium without serum, and Newcastle disease virus (10 PFU/ml) is added. After 1 hr of incubation, the cells are centrifuged to separate the virus and resuspended in 50 ml of medium for the superinduction incubation in the presence of 10 μg/ml of cycloheximide for 3 hr and an additional 2 hr in the presence of 1 μg/ml of actinomycin D per milliliter.

The cells are centrifuged again, and new medium (500 ml) is added to make up the initial cell concentration ($\approx10^6$ cells/ml). After 9–10 hr of incubation, the cells are harvested by centrifugation.

Interferon Assay

Routinely, mouse interferon titer is determined by inhibition of vesicular stomatitis virus (VSV)-induced cytopathogenicity in L-929 cells by reference to a National Institutes of Health (Bethesda) mouse interferon reference standard included in each titration.

A more accurate determination, when it is needed, is performed by a plaque reduction method in mouse embryo fibroblasts with VSV as challenge (for details, see Section IV of this volume).

Preparation of Mouse Interferon mRNA

Two Step Extraction according to Montagnier et al.[3] (Fig. 1). To induced cells harvested at time indicated, a solution of sodium ethylenediaminetetraacetate (EDTA), pH 7, 0.04 M, containing 1% sodium dodecyl sulfate (SDS) and 0.1% sodium deoxycholate is added to produce a homogeneous cell suspension. Usually, 10 ml are needed for about 5×10^7 cells. To this suspension, an equal volume of saturated phenol solution is added. The mixtures, pooled in an Erlenmeyer flask are stirred gently on a magnetic stirrer for 15 min at room temperature.

[2] M. J. Tovey, J. Degon-Louis, and I. Gresser, *Proc. Soc. Exp. Biol. Med.* **146**, 809 (1974).
[3] J. De Maeyer-Guignard, E. De Maeyer, and L. Montagnier, *Proc. Natl. Acad. Sci. U.S.A.* **69**, 1203 (1972).

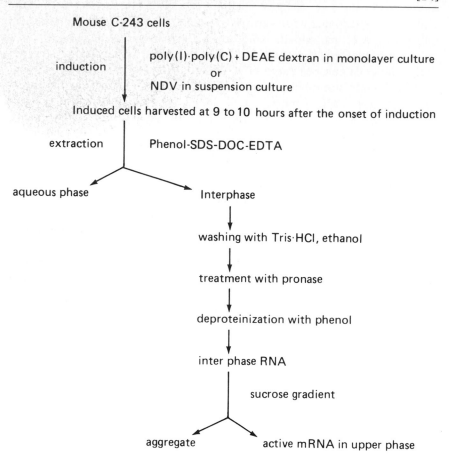

FIG. 1. Induction of cells and two-step extraction of mRNA.

The suspension is then centrifuged to separate the phenol. The upper aqueous phase as well as the phenol in the lower phase are removed by careful pipetting. The interphase containing insolubilized proteins and RNA is homogenized in 10 ml of 10 mM Tris · HCl, pH 7, plus 25 mM EDTA. After mixing with 20 ml of ethanol, the precipitate is recovered by centrifugation at 3500 g for 10 min. The precipitate is rewashed once as above. The protein–RNA mixture is then suspended in 2 ml of the following solution: 50 mM EDTA, pH 7; 0.25% SDS; Pronase, 1 mg/ml (sometimes, 0.5 mg of proteinase K is added per milliliter). The mixture is incubated at 37° overnight. The proteins are then extracted twice with an equal volume of phenol. The RNA is recovered by ethanol precipitation and dissolved in 100 mM Tris · HCl, pH 7.5; 100 mM NaCl, and 1 mM EDTA. The RNA fraction is further purified by centrifugation in a 10 to

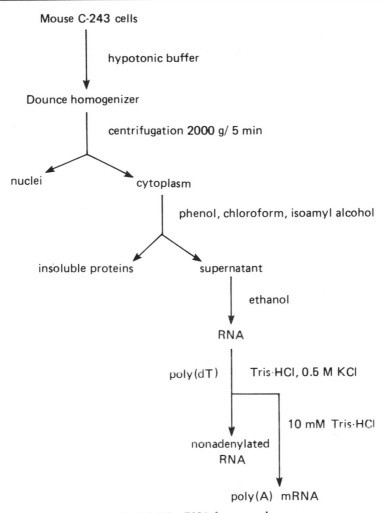

FIG. 2. Poly(A) mRNA from cytoplasm.

20% sucrose gradient for 30 min at 110,000 g in a Spinco SW41 rotor. The upper half of the gradient contains the active mouse interferon mRNA.

This procedure yields an mRNA highly active in the wheat germ cell-free system. However, the nature of the RNA components is not well defined, and the yield of RNA is not always reproducible.

Extraction of Cytoplasmic mRNA (Fig. 2)

A more classical technique can be used to obtain active mouse interferon mRNA. The method consists of separating nuclei from the cytoplasm before extraction.[4] Cells are allowed to swell in 5 volumes of hypotonic buffer (10 mM HEPES, pH 7.5, 20 mM KCl, 2.5 mM MgCl$_2$, and 2 mM dithiothreitol) at 4° for 10 min. The swollen cells are broken in a Dounce homogenizer. Cytoplasm is recovered by centrifugation at 2000 g for 5 min and deproteinized by a mixture of phenol, chloroform, and isoamyl alcohol (v/v/v, 25 : 25 : 1, pH 9). After centrifugation, RNA in the supernatant is precipitated by 2 volumes of ethanol. The precipitate is dissolved in 10 mM Tris · HCl, pH 7.5, in the presence of 0.5 M KCl. This solution is applied to an oligo(dT)-cellulose column. The nonadenylated RNA passes through the column. After washing with 5 volumes of Tris · HCl, 0.5 M KCl, the poly(A)-mRNA is eluted with 10 mM Tris · HCl, pH 7.5.[5] The RNA is concentrated by precipitation with 2 volumes of ethanol and redissolved in a minimum volume of 10 mM Tris · HCl.

Preparation of Wheat Germ Extracts[1,6]

Commercial wheat germ, mechanically extracted and not toasted, is prewashed with the grinding buffer: 20 mM N-2-hydroxyethylpiperazine-N'-2-ethanesulfonic acid (HEPES), pH 7.5; 100 mM potassium chloride; 1 mM magnesium acetate; 2 mM calcium chloride; and 6 mM 2-mercaptoethanol. The wheat germ is then ground in the presence of the buffer in a chilled mortar with an equal weight of acid-washed Fontainebleau sand. The proportion germ : sand : buffer is 1 g/1 g/5 ml. The grinding is performed gently in a discontinuous fashion for 20 min. The sand and intact cells are eliminated by two centrifugations at 500 g for 10 min. The active extract is obtained by centrifugation at 30,000 g for 10 min. The supernatant is removed carefully, with a pipette, without the upper lipid phase. The Mg^{2+} concentration of this extract is raised to 3.5 mM, and the following components are added: to make 1 mM ATP, 20 μM GTP, 8 mM creatine phosphate, 40 μg of creatine kinase per milliliter, and 2 mM dithiothreitol. The extract is incubated for 15 min at 30° before filtration through Sephadex G-25. For an extract prepared with 4–5 g of wheat germ, for example, a column of 160–200 ml of Sephadex G-25 coarse (2 cm × 30 cm) can be used. The Sephadex is equilibrated with the buffer: 20 mM HEPES, pH 7.6; 100 mM KCl; 5 mM magnesium acetate; and 2 mM dithiothreitol. The extract, in a volume about one-tenth that of the column, is filtered through the Sephadex at a rate around 1 ml/min.

[4] B. Lebleu, E. Hubert, J. Content, L. De Wit, I. A. Braude, and E. De Clercq, *Biochem. Biophys. Res. Commun.* **82**, 665 (1978).
[5] H. Aviv and P. Leder, *Proc. Natl. Acad. Sci. U.S.A.* **69**, 1408 (1972).
[6] B. E. Roberts and B. M. Paterson, *Proc. Natl. Acad. Sci. U.S.A.* **70**, 2330 (1973).

TABLE I

PREPARATION OF INCUBATION MIXTURE FOR PROTEIN SYNTHESIS[a]

Stock reagents	Volume (μl) for preparing 1 ml (final) of incubation mixture	
	Mix A	Mix B
HEPES, 1 M, pH 7.6	20	20
Dithiothreitol, 0.1 M	20	20
ATP, 25 mM, pH 7	40	40
GTP, 1.2 mM	20	20
Phosphocreatine, 0.5 M	15	15
Phosphocreatine kinase, 2 mg/ml	10	10
Mixture of all amino acids, 0.1 mM each	30	30
Spermine, 10 mM	10	10
KCl, 1 M	40	60
MgCl$_2$, 0.1 M	—	10
H$_2$O	95	65
	300	300

[a] Add 30 μl of mix A to 100 μl of incubation mixture containing 40 μl of S$_{00}$ extract, or 30 μl of mix B to mixture containing 20 μl of S$_{30}$. The final volume is completed with mRNA and H$_2$O.

The proteins are followed by the absorbance at 280 nm. From a column of 160 ml, the proteins are excluded in 60 ml. The fractions (30 ml) containing usually 15–20 mg of proteins per milliliter, are pooled. This extract, called S$_{30}$, is stored at −80° in small aliquots. It retains full activity for several months and remains active for over a year.

Assay of the Translation System

The incubation mixture contains 20 mM HEPES, pH 7.5; 2 mM dithiothreitol; 1 mM ATP, 25 μM GTP, 7.5 mM creatine phosphate, 20 μg of creatine kinase per milliliter; all essential amino acids (30 μM); 2 mM magnesium acetate, 80 mM potassium chloride, S$_{30}$ extract, 4 mg of protein/ml, 0.1 mM spermine; and mRNA, as indicated. Since S$_{30}$ extract is made in buffer containing KCl, MgCl$_2$, and DTT, the mixes should be made in such a way that the added volume of S$_{30}$ gives the correct final concentrations of K$^+$ and Mg^{2+}. A typical protocol for the preparation of assay medium is shown in Table I.

The efficiency of the translation system can be easily checked by the incorporation of radioactive phenylalanine directed by poly(U). The fidel-

TABLE II
EFFECT OF SPERMINE ON THE TRANSLATION OF MOUSE INTERFERON mRNA[a]

	Spermine (mM)					
	0	0.05	0.10	0.15	0.20	0.25
Interferon (units/ml)	<100	800	3800	4500	1800	800

[a] Five micrograms of interphase mRNA are added in 100 μl of the translation mixture described in the text. The preparation is incubated at 25° for 120 min. The interferon titer is expressed in international units.

ity of the translation by the wheat germ extract can be measured by the use of commercially available plant viral RNA, such as tobacco mosaic virus (TMV), turnip yellow mosaic virus (TYMV), or even alfalfa mosaic virus (AMV) RNAs. It is noteworthy that the endogenous incorporation of amino acids without added messagers is very low.

As discussed previously,[7] the translation of many plant viral RNAs in the wheat germ extract, prepared as described above, is stimulated by addition of polyamines such as spermine or spermidine. It is thus important to maximize the translation system for both the ionic requirements and the polyamine concentration. The use of a plant viral RNA for this purpose is convenient.

Translation of Mouse Interferon mRNA

As shown previously,[1] a high efficiency of translation of mouse interferon mRNA to biologically active protein in the wheat germ cell-free system needs the presence of polyamines. The dependency is illustrated in Table II.

At optimum ionic and spermine concentrations, the synthesis of interferon by this system is linear for about 90 min. From many experiments performed in this laboratory, the optimum incubation time to obtain a high yield of active interferon is around 90–120 min.

It also seems fruitful to use high concentrations of S_{30} proteins. The maximum yield of interferon was obtained with 4 mg of protein per milliliter (Table III).

Characterization of the Translation Product

The product of translation of mouse mRNA in the wheat germ extract is characterized as mouse interferon by (a) its inhibition of virus multipli-

[7] M.-N. Thang, L. Dondon, and E. Mohier, FEBS Lett. 61, 85 (1976).

TABLE III

EFFECT OF CONCENTRATION OF S_{30} EXTRACT ON THE
BIOSYNTHESIS OF MOUSE INTERFERON[a]

	S_{30} extract (mg protein/ml)		
	1.25 mg	2.5 mg	4 mg
Interferon (units/ml)	450	2200	4500

[a] Same conditions as described for Table II except for the concentrations of S_{30} extracts per milliliter of reaction mixture as indicated.

cation in homologous cells; (*b*) its absence of activity in heterologous cells; (*c*) its neutralization by anti-mouse interferon antiserum; (*d*) the absence of inhibitory effects on viral multiplication by the translation products obtained from mRNA isolated from noninduced cells.

[15] Translation of Mouse Interferon mRNA in Heterologous Cells

By EDWARD DE MAEYER

Heterologous cells are by definition derived from an animal species different from the species from which the interferon (IF) mRNA is extracted. The method is based on the relative species-specificity of the antiviral action of IF and consists of the translation of exogenous IF mRNA by whole cells, in tissue culture. It has been successfully used for the translation of human, monkey, mouse, and chick IF mRNA in a variety of cells as summarized in the table.[1-5] The procedure described here has been used at Orsay for mouse IF mRNA in chick embryo fibroblasts and Vero cells.

[1] J. De Maeyer-Guignard, E. De Maeyer, and L. Montagnier, *Proc. Natl. Acad. Sci. U.S.A.* **69**, 1203 (1972).

[2] F. H. Reynolds and P. M. Pitha, *Biochem. Biophys. Res. Commun.* **59**, 1023 (1974).

[3] T. G. Orlova, I. I. Georgadze, A. I. Kognovitskaya, and V. D. Solovyov, *Acta Virol.* (*Engl. Ed.*) **18**, 210 (1974).

[4] L. H. Kronenberg and T. Friedmann, *J. Gen. Virol.* **27**, 225 (1975).

[5] J. J. Greene, C. W. Dieffenbach, and P. O. P. Ts'o, *Nature* (*London*) **271**, 81 (1978).

HETEROLOGOUS CELL SYSTEMS THAT HAVE BEEN USED TO
TRANSLATE INTERFERON (IF) mRNA[a]

IF mRNA	Donor cells	Recipient cells
Human	Fibroblasts	Chick embryo fibroblast (CEF)
	Fibroblasts (HF-926)	Hamster: Syrian hamster embryo cells
Monkey	ESI cells (*Maccacus cynomolgus*)	CEF, monkey Vero (primate)
Mouse	L cells	CEF, monkey Vero
	Mouse embryo fibroblast	CEF
	C-243 cells	CEF
	Bone marrow	Chick bone marrow
Chicken	CEF	Mouse bone marrow
	Bone marrow	Mouse bone marrow

[a] See references cited in text footnotes 1–5.

Preparation of Cell Cultures

Chick Embryo Cells. We have used Brown Leghorn chick embryos. Individual cells are obtained by cutting 8- or 9-day-old embryos into 2–3 mm³ fragments, which are exposed to a solution of 0.25% trypsin and hyaluronidase (Wydase, Wyeth), 150 N.F. units per 30 ml, in PBS, at 37°. Primary cell cultures can be made, for example, in 35 mm diameter plastic petri dishes (Falcon) by seeding 250,000 cells/ml in a total volume of 2 ml of minimal essential medium (MEM) containing 15% fetal calf serum. Under these conditions, the cultures reach confluency 48–72 hr later and should be used for translation at this time.

Vero Cells. A continuous line of African green monkey cells is maintained in our laboratory in MEM containing 10% newborn calf serum. They can be seeded at a density of 125,000 cells/ml, for example, in plastic petri dishes 35 mm in diameter, and are confluent and ready to be used 48 hr later. The rationale for using Vero cells as a translational system is their incapacity to synthesize interferon upon induction,[6] which provides an additional safeguard against the possibility of interferon induction versus translation. However, since the incapacity to synthesize interferon is not a property of all Vero cell lines, each individual line should be checked for interferon production before being selected for translation of primate or human IF mRNA. This criterion is not important when Vero cells are used to translate RNA of other species.

Treatment of Cultures with Actinomycin D

We have consistently found that treatment of recipient cells with actinomycin D enhances their translational efficiency (Fig. 1), an observa-

[6] J. Desmyter, J. L. Melnick, and W. E. Rawls, *J. Virol.* 2, 955 (1968).

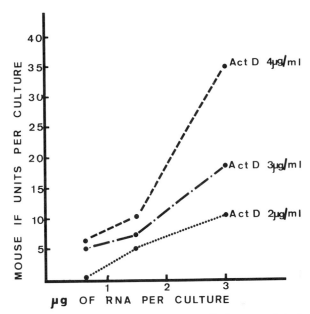

FIG. 1. Increase of translational efficiency of Vero cells for mouse interferon (IF) mRNA as a function of actinomycin D concentration. The mRNA was extracted from Newcastle disease virus-induced C-243 cells.

tion confirmed by other workers except for Kronenberg and Friedmann,[4] who reported that treatment with actinomycin D did not enhance translational capacity of Vero cells for mouse L cell IF mRNA.

Six hours before the RNA is added to the cells, the growth medium is aspirated from the culture and replaced by MEM with 3% calf serum containing actinomycin D (0.2 μg/ml for CEF and 2–4 μg/ml for Vero cells). At the end of the 6-hr treatment, the culture medium is aspirated and the cell monolayers are washed once with 2 ml of PBS at 37°.

Addition of Facilitators of Uptake of RNA

We have employed DEAE-dextran, but, according to Greene et al.,[5] one can also use $CaCl_2$. Mixtures of RNA and DEAE-dextran can be added to the cells, or the DEAE-dextran solution can be added first, followed by the RNA. We have routinely used the latter procedure, which is described below.

After washing the cultures once with warm PBS, 0.5 ml of a DEAE–dextran solution (75 μg/ml) in PBS is added to all the culture plates; then 0.25 ml of the RNA solution at three times the final concentration desired to test for messenger activity is placed in each dish. This brings the final

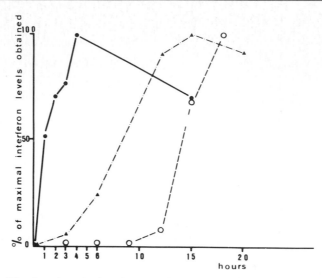

FIG. 2. Kinetics of mouse interferon production by mRNA-treated chick cells as compared to induced mouse cells. The figure clearly demonstrates the much earlier appearance of mouse interferon in mRNA-treated chick cells, versus the slower kinetics of an interferon induction in mouse cells. ●——●, Mouse interferon made by mRNA-treated chick cells; ▲–·–·▲, interferon induction with Newcastle disease virus (NDV) in mouse embryo fibroblast cultures; ○----○, inteferon induction with poly(I) · poly(C) also in mouse embryo fibroblast cultures.[7]

concentration of DEAE-dextran during incubation with the RNA to 50 μg/ml, in a volume of 0.75 ml.

The RNA–DEAE dextran mixture is left on the cells at 37° for 45 min and then aspirated. The cells are washed once with warm PBS, 2 ml of culture medium containing 3% of calf serum are added, and dishes are incubated at 37°.

Since the heterologous interferon made by the cells is secreted and accumulates in the culture medium, no extraction step is needed. Translation is essentially terminated 5 hr after removal of nonadsorbed RNA (Fig. 2), and the culture medium can be harvested and stored at 4° or −20° until interferon titrations.

The whole procedure, from pretreatment of the cells until harvesting of the media requires at least 12 hr. If necessary, the culture media can be harvested after overnight incubation rather than after 5 hr. However, this causes a small loss of interferon activity.

The composition of the PBS used throughout this method is as follows: NaCl, 8 g/liter; KCl, 0.2 g/liter; $Na_2HPO_4 \cdot 12 H_2O$, 2.88 g/liter; KH_2PO_4, 0.2 g/liter.

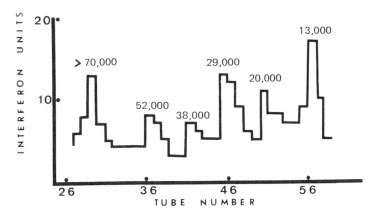

FIG. 3. Molecular sizes of mouse interferon translated in chick cells. The mRNA was obtained from poly(I) · poly(C) induced L cells and partially purified by sucrose gradient centrifugation. The chick cells used for translation received RNA collected from the 7 S to 10 S region. The interferon made by these cells was concentrated 10-fold before being filtered on Sephadex G-75.[7]

Fidelity of Translation. The translational product is measured by its biological antiviral activity, which implies that it has been transcribed correctly. Furthermore, it is neutralized by anti-type I mouse-interferon serum, indicating that at least some antigenic sites are shared with normally induced mouse interferon. In addition, the four molecular species described for mouse interferon are also obtained after translation of mouse interferon mRNA in CEF, plus two larger species, probably resulting from complex formation with some other proteins present in the culture medium[7] (Fig. 3).

Linearity of the Assay. Probably the main disadvantage of the method is that uptake of RNA by the cells is not well controlled and may well vary from one experiment to another, no matter how much one tries to standardize the conditions. Linearity of the assay (i.e., correlation between amount of mRNA put on the cells and interferon activity measured at the end) is a function of the degree of purification of the mRNA, and improves with increasing specific activity of the mRNA. Several authors have found the response to be linear within a certain range of RNA concentrations.[4,5,8]

[7] E. De Maeyer, L. Montagnier, J. De Maeyer-Guignard, and H. Collandre, *in* "Effects of Interferon on Cells, Viruses and the Immune System" (A. Geraldes, ed.), p. 191 Academic Press, New York, 1975.

[8] L. Montagnier, H. Collandre, J. De Maeyer-Guignard, and E. De Maeyer, *Biochem. Biophys. Res. Commun.* **59**, 1031 (1974).

[16] Translation of Exogenous Interferon mRNA in Intact Mammalian Cells

By JAMES J. GREENE, CARL W. DIEFFENBACH, and
PAUL O. P. Ts'o

Exogenous interferon (IF) mRNA has been translated in oocytes, cell-free lysates, and intact mammalian cells.[1] While each of these translational systems can provide quantitative measurement of IF mRNA, the whole-cell translational system offers one distinct characteristic. When the IF mRNA is being introduced into and translated within a mammalian cell, the recipient cell can be subjected to a variety of manipulations, and its response on translation of the exogenous IF mRNA can be observed. This approach can provide unique information on the cellular elements affecting the stability and translation of IF mRNA.

Principle

Whole-cell translation of IF mRNA was first reported by De Maeyer-Guignard *et al.*[2] for the translation of mouse and simian IF mRNA in avian and simian cells. Since then a variety of other whole-cell translation systems have been described (summarized in Table I).[2-6] In all these systems, the uptake and subsequent translation of exogenous IF mRNA requires the application of the RNA together with another substance—a facilitator. The facilitator acts to increase the quantity of exogenous RNA taken up by the cells as well as to protect the RNA from nuclease damage. Quantitative measurement of IF mRNA translation is by a bioassay based on the species-specific antiviral properties of IF. When exogenous IF mRNA usually of different species is translated by the recipient cells, the IF produced is active only on cells of the same species as the IF mRNA. The titer of the antiviral activity produced in the culture medium is a measure of the translation of IF mRNA.

[1] W. E. Stewart, II, "The Interferon System," pp. 90–97, Springer-Verlag, Vienna and New York, 1979.
[2] J. De Maeyer-Guignard, E. De Maeyer, and L. Montagnier, *Proc. Natl. Acad. Sci. U.S.A.* **69**, 1203 (1972).
[3] F. H. Reynolds and P. M. Pitha, *Biochem. Biophys. Res. Commun.* **59**, 1023 (1974).
[4] L. H. Kronenberg and T. Friedmann, *J. Gen. Virol.* **27**, 225 (1975).
[5] J. J. Greene, C. W. Dieffenbach, and P. O. P. Ts'o, *Nature (London)* **271**, 81 (1978).
[6] J. J. Greene, C. W. Dieffenbach, L. C. Yang, and P. O. P. Ts'o, *Biochemistry* **19**, 2485 (1980).

TABLE I

WHOLE-CELL TRANSLATIONAL SYSTEMS

Reference	Inducer	mRNA source	Cells used for translation	Facilitator used
De Maeyer-Guignard et al.[2]	Poly(I)·poly(C) NDV	Mouse embryo fibroblasts Simian (*Maccacus cynomolgus*) cell line	Chicken embryo fibroblasts Vero	DEAE-dextran
Reynolds and Pitha[3]	Poly(I)·poly(C)	Human fibroblasts	Chicken embryo fibroblasts	DEAE-dextran
Kronenberg and Friedmann[4]	Poly(I)·poly(C) and NDV	Mouse L cells	Vero, BSC-1, HeLa	DEAE-dextran
Greene et al.[5,6]	Poly(I)·poly(C) and NDV	Human fibroblasts and Syrian hamster fibroblasts	Syrian hamster and human fibroblasts	$CaCl_2$

Procedures

Induction of IF mRNA Synthesis

Synthesis of IF mRNA can be induced by viruses or by synthetic compounds such as polyinosinic acid · polycytidylic acid [poly(I) · poly(C)] and other double-stranded RNAs. For poly(I) · poly(C) induction of human foreskin fibroblasts, confluent cultures in 850 cm² Corning roller bottles (approximately 4×10^7 cells) are treated with 10 ml of 0.3 mM (100 μg/ml) poly(I) · poly(C) in PBS or in Dulbecco's modified minimum essential medium (DMEM). After incubation for 1 hr at 37°, the cultures are washed four times with PBS or DMEM to remove any residual poly(I) · poly(C). Then, DMEM supplemented with 2% fetal bovine serum is added. Viral induction of IF mRNA synthesis is conducted similarly to poly(I) · poly(C) induction except that virus replaces poly(I) · poly(C) during the 1-hr incubation. Hamster cells are refractory to poly(I) · poly(C) for IF synthesis, so they must be induced to synthesize IF mRNA by virus. Inactivated Newcastle disease virus (NDV) is an excellent inducer of IF synthesis in hamster cells. The procedure for ultraviolet (UV) inactivation of NDV and induction with NDV has been described in this series.[7]

Extraction of RNA

Total cytoplasmic RNA is extracted by a modified Penman procedure.[8] The modifications are designed (*a*) to keep DNA contamination of the preparation to a minimum by adjusting the detergent concentrations so that the nuclei and mitochondria remain intact; (*b*) to minimize the length of time between cell lysis and the addition of the phenol/CHCl₃ solution to the cell extract; and (*c*) to adjust conditions for maximal RNase inhibition. For each cell type and for each induction protocol, the kinetics of IF production should be determined to extrapolate the optimum time after induction for IF mRNA extraction. The maximal rate of IF synthesis is at 3–4 hr after induction for poly(I) · poly(C)-induced human fibroblasts whereas for NDV-induced Syrian hamster embryo (SHE) cells, it is 16–19 hrs after induction. At appropriate times after induction, the cells are washed three times with cold PBS to remove serum nucleases. The extraction and purification of the RNA are done at 5°. The cells are lysed in a two-step procedure by treatment with lysis buffer (0.5 M KCl, 0.01 M Tris · HCl, 0.005 M MgCl₂, 0.5% Nonidet P-40, saturated with dextran

[7] R. Friedman, this series, Vol. 58, p. 292.
[8] S. Penman, *in* "Fundamental Techniques in Virology" (K. Habel and N. P. Salzman, eds.), pp. 35–48. Academic Press, New York, 1969.

sulfate, pH 8.0; 8.0 ml/roller bottle) for 10 min followed by the addition of Tween 40/deoxycholate (stock solution: 10% (v/v) Tween 40, 10% (w/v) sodium deoxycholate) to the final concentration of 0.1% Tween 40 and 0.1% deoxycholate. Lysis is allowed to continue for an additional 5 min, after which the lysates are transferred from roller bottles to 30-ml Corex tubes and centrifuged at 10,000 g for 10 min to pellet the nuclei and mitochondria. The supernatant is carefully transferred to 250-ml Corning centrifuge tubes, where it is adjusted to 5 mM EDTA. The cytosol is extracted three times with equal volumes of a solution containing a mixture of phenol, chloroform, and isoamyl alcohol in the ratio 25 : 24 : 1 (v/v/v) followed by one final extraction with chloroform/isoamyl alcohol (24 : 1). The final extraction with chloroform/isoamyl alcohol is primarily to remove small quantities of phenol dissolved in the aqueous phase. Elimination of phenol is critical, since contamination of the RNA with even minute quantities of phenol will prevent translation of IF mRNA due to the cytotoxic effects of phenol on the recipient cells. The final aqueous phase is mixed with two volumes of ethanol and stored at $-20°$ overnight to precipitate the RNA.

The proportion of IF mRNA in the population of cytoplasmic RNA in poly(I) · poly(C)-induced human fibroblasts is sufficiently high to detect IF upon addition of unfractionated cytoplasmic RNA to the recipient cells. However, the concentration of IF mRNA in NDV-induced SHE cytoplasmic RNA is not sufficiently high for its direct use in translation. Consequently, the IF mRNA in the SHE cytoplasmic RNA must first be enriched by oligo(dT)-cellulose chromatography before use.

Purification on Oligo(dT)-Cellulose Columns

The precipitated RNA is dissolved in 0.4 M NaCl, 10 mM Tris · HCl (pH 7.6) and applied to an oligo(dT)-cellulose (Collaborative Research T-4 or Boehringer Mannheim T-1) column. After extensive washing of the column with 0.4 M NaCl, 10 mM Tris · HCl (pH 7.6), the bound RNA is eluted with 10 mM Tris · HCl (pH 7.6) and precipitated with ethanol at $-20°$.

Translation in Intact Whole Cells

The RNA precipitate is washed extensively with ethanol to remove all traces of phenol (phenol contamination is much less for oligo(dT)-cellulose-purified RNA). The washed RNA is dissolved in HEPES-PBS buffer (0.5% w/v HEPES in calcium-free PBS, pH 6.6–6.8) to a concentration of 50 μg/ml as determined by optical absorbance using a molar extinction coefficient of 7400 per nucleotide residue at 260 nm. Each translation requires 5 ml (250 μg of RNA) of the RNA solution. CaCl$_2$ (1 M solution) is added to each 5-ml RNA aliquot to a final concentration of 30 mM, forming a calcium–phosphate–RNA coprecipitate. Recipient cells are grown to confluency in 100-mm dishes or 75-cm^2 flasks. Prior to the addition of RNA, the recipient cells are depleted of calcium by incubation in calcium-free HEPES-PBS buffer for 10 min. The calcium-free buffer is aspirated from the recipient cells, and the precipitated RNA is vortexed and applied immediately to the calcium-depleted cells. The cultures are then incubated undisturbed for 45 min to allow the precipitate to adhere to the cells. After incubation, the HEPES–PBS buffer is removed and replaced with 5–10 ml of DMEM supplemented with 2% fetal bovine serum and incubated at 37°. Maximal translation is achieved within 5 hr.

At the conclusion of translation, the media are removed from the cultures for titering of IF activity. Carryover of the calcium–phosphate–RNA coprecipitate into the collected medium is unavoidable, but it can be removed by gentle centrifugation (500 g for 5 min) or by allowing the precipitate to settle overnight. This procedure can be used for homologous translations (the species of origin of the exogenous IF mRNA is the same as the species of the recipient cell) as well as heterologous translations. When assaying IF yield from homologous translation of IF mRNA, particular care must be taken to ensure that no endogenous IF production occurs simultaneously; if this is not possible, the endogenous translation as background should be assessed.

The whole-cell translation procedure just described has been optimized for translation of exogenous human and hamster IF mRNAs in human fibroblasts and in SHE cells. Translation of exogenous IF mRNA in mammalian cells is affected by a variety of factors which appears to be more dependent upon the recipient cell type than on the type of IF mRNA (i.e., species of origin). Consequently, some of these factors, discussed below, must be considered when optimizing conditions for cells other than hamster or human.

Factors Affecting Whole-Cell Translation of IF mRNA

The specific factors that influence the yield of IF are those that affect the uptake of RNA by cells and those that affect the translational capacity of the recipient cells.

Factors That Affect the Uptake of RNA

1. Facilitator. Both CaCl$_2$ and DEAE-dextran can be used to enhance RNA uptake by cells. As shown in Table II, CaCl$_2$ is more effective than DEAE-dextran in mediating the translation of human IF mRNA in SHE cells. An important requisite for either facilitator to be effective is that the RNA precipitate must adhere tightly to the cell monolayer.

TABLE II
EFFECTS OF CaCl₂, DEAE-DEXTRAN, CALCIUM DEPLETION, AND ACTINOMYCIN D
TREATMENT ON HUMAN INTERFERON (IF) PRODUCTION BY SHE CELLS[a]

Type of facilitator	Prior calcium depletion	Actinomycin D (0.01 mg/ml) treatment	IF titer (units/ml)
DEAE-dextran	None	Treated, 5 hr	4
CaCl₂	None	None	13
CaCl₂	None	Treated, 5 hr	22
CaCl₂	None	Treated, 15 hr	41
CaCl₂	Calcium depletion	None	76

[a] Production of human IF is dependent on addition of mRNA from human cells that were in the process of producing human IF.

2. Calcium depletion. Depleting the recipient cells of calcium prior to the application of the RNA–calcium–phosphate coprecipitate increases the yield of IF produced during translation. As shown in Table II, calcium depletion results in a fourfold increase in translation.

3. pH optimum. The CaCl₂-mediated translation of exogenous IF mRNA is highly dependent upon the pH of the buffer containing the RNA. The pH dependence of human IF production by SHE cells following the addition of calcium-precipitated IF mRNA is shown in Fig. 1. The pH range 6.6 to 7.0 is optimal. Outside this range, the translational efficiency is greatly reduced. When exogenous IF mRNA is translated in human fibroblasts, the optimum pH for translation is 0.2 unit lower than for translation in SHE cells. This strong dependency of translation on pH may be related to the adherence of the RNA–calcium–phosphate coprecipitate

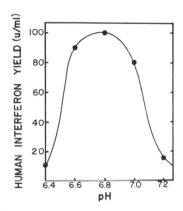

FIG. 1. Dependence of the translation of human IF mRNA in SHE cells on the pH of the application buffer. Reprinted, with permission, from Greene et al.[5]

to the cells since the coprecipitate adheres most strongly to the cells at the optimum pH. We have noted that every cell line has a specific pH optimum for adherence. Therefore, the pH optimum must be determined for each different cell type used for translation.

Factors Affecting Translational Capacity. Pretreatment of the cell monolayer with actinomycin D increases the IF yield during translation of exogenous IF mRNA. The degree of the actinomycin D effect, however, varies from system to system. While some systems require prior treatment with actinomycin D to achieve reproducible translation, others do not. Prior treatment with actinomycin D is not required for CaCl$_2$-mediated translation of IF mRNA, although it does significantly improve translation (Table II). This enhancement is presumed to be the consequence of endogenous mRNA degradation resulting in a more favorable competition of exogenous IF mRNA for the translational machinery.

Quantitation of the Translational System

Accurate quantitation of IF mRNA is dependent upon the proportional response of the translational system to the amount of added IF mRNA. Two types of response can be expected depending on whether or not the amount of RNA applied is a saturating dose. As seen in Fig. 2A, the yield of human IF by calcium-depleted SHE cells is dose dependent until 50 μg of RNA are applied. At this dose, the translational system saturates and

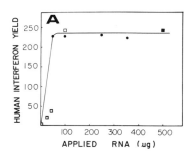 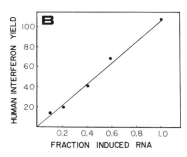

FIG. 2. Dose response for the heterologous translation of human IF mRNA in SHE cells. (A) RNA extracted from induced human cells was diluted to various concentrations before application to SHE cells to determine the absolute dosage dependency of translation. Five milliliters of each solution were applied for translation. The IF yields are expressed as units per milliliter. ●, RNA concentrations of 10 μg/ml, 20 μg/ml, 50 μg/ml, 100 μg/ml; □, RNA concentrations of 1 μg/ml, 5 μg/ml, 10 μg/ml, 20 μg/ml, 100 μg/ml. (B) Cytoplasmic RNA extracted from induced cells was mixed with cytoplasmic RNA from mock-induced cells, at the ratios shown, to determine the relative dosage dependency of translation. The total concentration of RNA in each sample was 50 μg/ml. Five milliliters of each sample were applied to SHE cells and measured for translation of human IF mRNA. All points represent the average of duplicate assays.

becomes independent of RNA dose. Under saturating conditions, the SHE translational system responds proportionally not to the absolute amount of IF mRNA, but rather to its relative concentration within the RNA preparation (Fig. 2B). The dose of RNA recommended above (50 μg/ml, 250 μg/translation) is saturating, since translations under saturating conditions yield more reproducible results than under subsaturating conditions.

Characterization of the Antiviral Activity Produced during Translation

The antiviral activity produced by the recipient cells following the application of IF mRNA can only be *presumed* to result from the translation of the exogenous IF mRNA until certain criteria have been established. These criteria include the following ones.

1. Demonstration that the production of the antiviral activity is the result of a metabolic process (i.e., prior treatment of the recipient cells with actinomycin D will increase the amount of antiviral activity produced whereas prior treatment with inhibitors of protein synthesis will abolish production of the antiviral activity).
2. The antiviral activity is produced only from poly(A)-containing RNA extracted from appropriately induced cells.
3. Verification that the physical and biological properties of the antiviral product corresponds to that of the IF that is produced by the cell from which the IF mRNA was extracted.

[17] Procedures to Estimate the Size of Interferon mRNA and Use of UV Irradiation to Estimate the Size of Its Primary Transcript[1]

By PRAVINKUMAR B. SEHGAL[2]

The polyadenylated RNA fraction extracted from induced human or murine cells contains interferon mRNA as shown by translation into biologically active interferon.[3] Microinjection of interferon mRNA into oocytes of *Xenopus laevis* is a sensitive and reliable assay for this molecule in biologically active form.[4] Interferon mRNA can be sized by sedimentation

[1] Supported by Grant AI-16262 from the NIAID, DHEW.
[2] Recipient of a Junior Faculty Research Award from the American Cancer Society.
[3] R. McCandliss, A. Sloma, and S. Pestka, this volume [8].
[4] A. Sloma, R. McCandliss, and S. Pestka, this volume [10].

of denatured mRNA through sucrose gradients[5] as well as by electrophoresis in denaturing gels.

Estimation of the DNA target size for inactivation of the synthesis of an mRNA species is a simple and convenient means of determining the size of the transcription unit from which a particular mRNA is derived. The size of the transcription unit gives an estimate of the length of the primary transcript from which the mRNA is processed. The lengths of all of the primary transcripts of eukaryotic mRNA species that have been investigated thus far are larger than those of the mature mRNA species.[6] As an example, the primary transcript for human fibroblast interferon mRNA (size of mature mRNA: 10.8 S, 700 residues long) is transcribed from a DNA segment of 4000–11,000 base pairs, as estimated by UV-target size analysis.[7]

Sizing of Interferon mRNA by Sucrose Gradient Sedimentation

Sucrose gradient analysis provides a convenient and simple way to estimate the size of interferon mRNA. Denatured polyadenylated interferon mRNA preparations are sedimented through an appropriate sucrose gradient followed by an estimation of the interferon mRNA content of each fraction of the gradient by means of a translation assay. Marker RNA species of known size are analyzed in parallel in order to assign sedimentation velocity and RNA length values to interferon mRNA.

Materials and Reagents

All glassware used is acid-cleaned and sterilized by dry heating. The following stock solutions are used: Tris · HCl, 1 M, pH 7.4; NaCl, 5 M; sodium dodecyl sulfate (SDS), 20%; ethylenediaminetetraacetic acid (EDTA), 0.5 M, pH 7.4; and LiCl, 2 M. These solutions are made up in distilled water in acid-cleaned glassware. To dissolve EDTA, it is necessary to stir the solution continuously and to adjust the pH to 7.4 using NaOH pellets. All solutions are autoclaved before use. The stock solutions can be used to make up the desired buffers by dilution with autoclaved, distilled water, preferably in disposable plastic ware. Sucrose solutions are made up as 15% and 30% (w/v) solutions in 0.05 M NETS (0.05 M NaCl; 0.01 M EDTA; 0.01 M Tris · HCl, 7.4; 0.2% SDS). Analytical grade or, preferably, ribonuclease-free sucrose is used for this purpose. The sucrose solutions are treated with diethylpyrocarbonate (Sigma; or Baycovin, Naftone, Inc.) to inactive completely any ribonucleases that may be present. The sucrose solution is made 0.1% (v/v) with diethyl-

[5] P. B. Sehgal, D. Lyles, and I. Tamm. *Virology* **89**, 189 (1978).

[6] W. Sauerbier and K. Hercules, *Annu. Rev. Genet.* **12**, 329 (1978).

[7] P. B. Sehgal and I. Tamm, *Virology* **102**, 245 (1980).

pyrocarbonate and is heated just short of boiling with continuous stirring in a fume hood for 2–3 hr until there is no detectable odor of diethyl-pyrocarbonate. Autoclaved distilled water is used to readjust for any volume loss due to evaporation, and the solution is stored in acid-cleaned, sterile glassware at room temperature.

Dimethyl sulfoxide (DMSO, 99%) contains 0.01 M in Tris · HCl, pH 7.4 and 0.001 M in EDTA. This solution can be stored at room temperature in the dark.

Ribosomal RNA and transfer RNA (4, 18, and 28 S) to be used as sedimentation markers are conveniently prepared by labeling growing cells (HeLa, L, etc.) with [^{3}H]uridine or ^{32}PO$_4$ for 24 hr and extracting the cytoplasmic RNA. Most of the incorporated radioactivity after a long labeling period is in the rRNA and tRNA species. The total cytoplasmic RNA can be used directly. Alternatively, the RNA can be sedimented through a 15 to 30% sucrose gradient, and the 4 S, 18 S, and 28 S RNA species can be isolated, ethanol-precipitated, and then used as sedimentation markers.

Centrifugation is carried out in polyallomer tubes using SW27 or SW41 rotors. New polyallomer tubes are nuclease-free. Corex (15 ml or 30 ml) glass tubes or Eppendorf (1.5 ml) disposable tubes are siliconized prior to use. This is best done by pouring a solution of dichlorodimethylsilane (Eastman, 5% v/v) in chloroform into a tube, pouring it out, inverting the tube, and letting it dry for a few minutes. The siliconized tube can then be rinsed with sterile distilled water and used directly or first autoclaved and dried prior to use.

Disposable glass or plastic pipettes are recommended for use throughout.

A peristaltic pump system (e.g., Pharmacia P-3), preferably but not necessarily equipped with a UV-absorbance monitor, is used to fractionate gradients. A 50-μl disposable Yankee Micropet (Clay-Adams; this is nuclease-free if used directly from carton), connected to the peristaltic pump via appropriate tubing, can be inserted into a gradient from the top, its tip gently resting at the bottom of the gradient. The sucrose solution can then be pumped out and sampled from the bottom of the gradient.

Carrier RNA, which does not interfere with the translation of interferon mRNA, is prepared from commercial yeast tRNA (e.g., soluble yeast ribonucleic acid, A grade, Calbiochem) by a further round of phenol extraction and ethanol precipitation. It is dissolved in sterile distilled water at 10 mg/ml and is stored in small aliquots at $-20°$.

Procedure

Polyadenylated interferon mRNA preparation obtained by poly(U)-Sepharose or oligo(dT)-cellulose chromatography is dissolved in 50 μl of

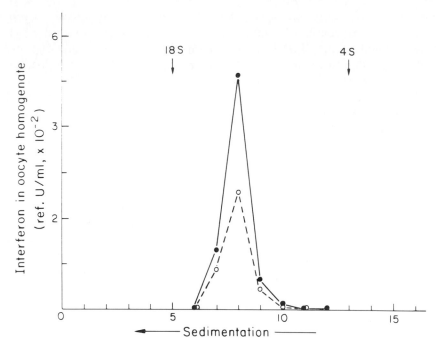

FIG. 1. Sucrose gradient sedimentation of human fibroblast interferon mRNA. Polyadenylated RNA extracted from FS-4 cell cultures grown in 150-mm dishes and induced with poly(I) · poly(C) in the presence of cycloheximide for 4 hr (46 μg of RNA from 14 cultures) was sedimented through a 15 to 30% sucrose gradient in an SW41 rotor at 32,000 rpm for 18 hr. RNA from each fraction was precipitated with ethanol, dissolved in 5 μl of sterile distilled water, and a 2-μl aliquot of each was assayed for interferon mRNA content using the *Xenopus* oocyte assay (●). The remaining 3-μl samples of fractions 8, 9, and 10 were pooled and sedimented through a second sucrose gradient identical to the first. RNA from each fraction of the second sucrose gradient was also dissolved in 5 μl of sterile distilled water, and a 2-μl aliquot of each was assayed for interferon mRNA content (○).

ETS (0.01 M EDTA; 0.01 M Tris, pH 7.4; 0.2% SDS) in a siliconized tube, mixed with 200 μl of 99% DMSO (0.001 M EDTA; 0.01 M Tris · HCl, pH 7.4), heated at 65° for 2 min, and rapidly cooled on ice. Then 500 μl of 0.05 M NETS (ETS buffer with 0.05 M NaCl) is added to dilute the DMSO (otherwise the sample will sink into 15% sucrose), and the solution (at room temperature) is then carefully layered onto a 15 to 30% preformed sucrose gradient in an SW41 (11 ml gradient) or SW27 (36 ml gradient) polyallomer tube using a disposable Pasteur pipette. Vortexing of the RNA solution after addition of DMSO is to be avoided. A 50-μl aliquot of marker RNA (4 and 18 S) is also denatured and layered onto a parallel gradient. A suitable aliquot of marker RNA can also be mixed in with the interferon mRNA sample and both sedimented through the same gradient.

In our experience sucrose gradients can be reproducibly fractionated into an identical number of equal fractions.

Gradients loaded with RNA are centrifuged at 20° such that 18 S marker RNA is sedimented close to the bottom of the tube (e.g., SW41: 32,000 rpm for 18.5 hr; SW27: 25,500 rpm for 28 hr). Each gradient is then fractionated into approximately 20 fractions of equal volume. Fractions of the marker RNA gradient or small aliquots thereof are precipitated with trichloroacetic acid (TCA) by the addition of 50 μl of 0.1% bovine serum albumin solution followed by 1–2 ml of 10% trichloroacetic acid, and the precipitated RNA samples are collected on GF/C glass fiber filters and counted in a liquid scintillation spectrometer.

Gradient fractions likely to contain interferon mRNA (10–12 S) are transferred to siliconized Corex tubes, and the following are added: LiCl (2 M) to a final concentration of 0.1 M, carrier RNA (5–10 μg) per tube, and finally 2.5 volumes of ethanol. The RNA is precipitated overnight at $-20°$, and the precipitate is pelleted by centrifugation at 8000 rpm for 30–45 min in a Sorvall SS-34 rotor. The RNA pellet is washed with 0.5–1 ml of ethanol and centrifuged again at 8000 rpm for 30 min; the ethanol is drained, and the pellet is dried at room temperature in an air stream. The mRNA precipitated from each gradient fraction can then be dissolved in 5 μl of sterile distilled water and assayed for interferon mRNA activity by translation in an appropriate assay system.

Human fibroblast interferon mRNA sediments as a sharp peak in the 10–12 S region (Fig. 1). The sedimentation (s) value is arrived at by interpolation between the 4 S and 18 S markers. The length of interferon mRNA can be computed from the s value using the empirical formula: number of residues $= 4.7 \times s^{2.1}$.

Sizing of Interferon mRNA by Electrophoresis in 2.5% Agarose–6 M Urea Gels

Electrophoresis of RNA in an agarose–6 M urea gel provides high resolution of RNA species under partially denaturing conditions. It provides RNA size information that, although not completely accurate, is superior to that obtained from sucrose gradients. Even greater accuracy can be achieved by electrophoresis of RNA under fully denaturing conditions. Agarose–methyl mercury hydroxide gels may prove to be superior for the purpose of sizing interferon mRNA. However, an agarose–urea gel is superior in its resolving power to a gel run under fully denaturing conditions. The following description is based on the method of Locker,[8] as used in our laboratory.

[8] J. Locker, *Anal. Biochem.* **98**, 358 (1979).

Materials and Reagents

Electrophoresis of human fibroblast interferon mRNA is carried out in cylindrical gels cast in glass tubes (0.6 cm in internal diameter and 13 cm long) in a GT5/6 electrophoresis unit (Hoefer Scientific Instruments, USA). Agarose (electrophoresis grade, standard low molecular weight, Bio-Rad), urea (Eastman), glycerol saturated with bromophenol blue, iodoacetic acid, as well as many of the reagents described in the section on sizing by gradient sedimentation, are necessary. The running buffer (half-strength Loening's buffer) is 0.02 M Tris · HCl, 0.01 M sodium acetate, 0.001 M EDTA, pH 7.8. A stock solution of this buffer (5 × strength) is made up in distilled water and autoclaved before use. ^{32}P-labeled marker rRNA is prepared as indicated in the section on gradient sedimentation sizing.

We have used an aluminum gel holder to slice the gel manually into 1 mm-thick slices. The gel holder is washed thoroughly and autoclaved prior to use. New single-edge razor blades are used each time for slicing gels. Forceps, Eppendorf pipette tips, Eppendorf tubes (1.5 ml capacity), glass scintillation vials and their caps are all autoclaved and dried prior to use. Disposable plastic gloves are worn throughout the procedure.

Preparation of the Tube Gel

Appropriate amounts of agarose (to make 2.5%), urea (to make 6 M), 5× concentrated running buffer, and distilled water are mixed in an acid-cleaned, autoclaved beaker. Iodoacetic acid is added to make 15 mM. *This appears to be essential for the success of this procedure.* The solution is boiled briefly to dissolve all components and is poured, while still hot, into acid-cleaned and autoclaved glass tubes whose one end is closed with parafilm. Owing to the viscosity of the solution, a sterile syringe equipped with a long sterile cannula is helpful in pouring gels free of air bubbles. The gels are allowed to set overnight at 4°, both ends sealed with parafilm. We have stored gels in this manner for up to 3 weeks in the cold room. Just prior to use, the parafilm covering the lower end of the tube is perforated (but not removed), that covering the upper end is removed, the tube is tilted to let the gel protrude out of the tube a short distance, and the top surface of the gel is sliced flat using a razor blade.

Electrophoresis of the mRNA Sample

Each tube gel is loaded with a volume of 50 μl containing urea (18 mg), appropriate aliquots of interferon mRNA in distilled water and ^{32}P-labeled marker rRNA in ETS buffer (see section on sizing by sucrose gradient sedimentation), 5 μl of glycerol saturated with bromophenol blue (autoclaved prior to use), and 0.5 μl of 5 M NaCl. The components are mixed in an Eppendorf tube. The RNA is denatured at 95° for 2 min and then cooled on ice. Electrophoresis is carried out at 4°. The running buffer is made up fresh by dilution of a 5× stock solution with sterile distilled water. The tube gel is prerun for 15–20 min at 5 mA per tube. We perforate and retain the parafilm at the lower end of the tube in order to prevent slippage of the gel and yet establish electrical continuity. Electrophoresis is performed at 5 mA per tube until the dye front is close to the bottom of the tube (3–4 hr).

Location and Extraction of Interferon mRNA from Gel Slices

The gel is sliced manually and the ^{32}P content of each slice is determined by Cerenkov counting. Figure 2 illustrates the usual migration of 28 S, 18 S, 5 S, and 4 S RNA in a 2.5% agarose–6 M urea gel. Gel slices corresponding to the 10–12 S region are each placed in Eppendorf tubes, melted in 0.2 ml of 0.1 M NETS buffer (95° for 2 min), mixed, and allowed to resolidify. The samples are then frozen at $-70°$ for at least 2 hr, thawed, and then centrifuged in an Eppendorf centrifuge at top speed for 15 min at 4°. The agarose collapses into a pellet. The supernatant, which contains 60–80% of the RNA that was present in the slice, is removed into a new tube, phenol extracted, and the aqueous phase ethanol precipitated after addition of 5 μg of carrier RNA. The RNA pellet can be washed with ethanol, dried, and dissolved in 5 μl of sterile distilled water; an aliquot of this sample is assayed for interferon mRNA activity (Fig. 2). The apparent length of interferon mRNA (700 residues; 10.8 S) is estimated graphically on the basis of the migration of the markers.

Sizing of Primary Transcripts of Interferon mRNA by UV Irradiation

Rationale

Irradiation of cells with UV light induces lesions such as thymidine–thymidine dimers in DNA in a random manner.[6] RNA transcription is terminated, and incomplete RNA chains are released at sites of UV-induced lesions in the DNA molecule. There is little effect on the initiation of RNA synthesis. The longer an RNA transcription unit, the greater is the likelihood of a UV-induced lesion in that unit. Thus, the synthesis of promoter-distal RNA is more sensitive to inhibition by UV-irradiation than promoter-proximal. The UV sensitivity of the synthesis of RNA reflects the distance of that RNA sequence from its promoter.

The extent of inactivation of RNA synthesis observed after UV irradiation is usually calibrated in terms of the DNA target using the ribosomal

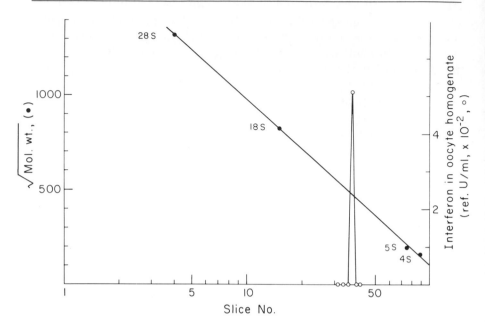

F<small>IG</small>. 2. Electrophoresis of human fibroblast interferon mRNA (○) in a 2.5% agarose–6 M urea tube gel. Polyadenylated RNA (45 μg) from induced FS-4 cultures was mixed with ^{32}P-labeled HeLa cell rRNA (●) and electrophoresed as described in the text. RNA from adjacent slices (31 to 42) was pooled prior to microinjection into oocytes. (A. Sagar and P. B. Sehgal, unpublished data.)

RNA transcription unit.[6] The 18 S and 28 S rRNA species can be readily separated and quantitated by sucrose gradient centrifugation or by gel electrophoresis. The 3' ends of 18 S and 28 S RNA are approximately 4 and 11 kibobases from the promoter.[9,10] Thus, the extent of inhibition of the labeling of 18 S and 28 S RNA by UV irradiation corresponds to DNA targets of 4 and 11 kbp (kilobase pairs), respectively. Since the major effect of UV irradiation is to generate lesions in DNA with little effect on RNA polymerases themselves, a calibration of UV-target size with reference to a transcription unit transcribed by RNA polymerase I can be used for estimating the size of transcription units for mammalian mRNA species transcribed by RNA polymerase II. The validity of this calibration procedure has been adequately documented in several instances by comparison of the UV target and the DNA sequence information for specific

[9] P. K. Wellauer and I. B. Dawid, *Proc. Natl. Acad. Sci. U.S.A.* **70**, 2827 (1973).
[10] I. B. Dawid and P. K. Wellauer, *Cell* **8**, 443 (1976).

mRNA species.[6] Repair of UV-induced lesions in DNA is minimized by handling irradiated cultures in the dark.

The inhibition in the rate of mRNA synthesis after UV irradiation is ideally estimated by pulse-labeling the cells after a 15–20 min "runoff" period with labeled precursors such as [³H]uridine and monitoring the appearance of label in a specific mRNA species. The inhibition of ribosomal RNA and mRNA synthesis has been monitored in this manner for up to 90 min after irradiation of cells with satisfactory results.[6] Secondary radiation-induced effects become apparent thereafter. The pulse-labeling procedure cannot be used for interferon mRNA, an inducible mRNA, at the present time. For inducible mRNA species, two alternative procedures can be used to monitor the inactivation of mRNA synthesis. The mRNA synthesized within a period of ≤90 min after irradiation and induction can be extracted and assayed by translation in an appropriate system. Alternatively, irradiated cells can be pulse-labeled with labeled amino acids and label incorporation into specific proteins quantitated. This alternative at present is not practical for interferon. Furthermore, since UV irradiation can inhibit protein synthesis directly,[11] a measurement of interferon production by irradiated cultures could provide inaccurate data.

In poly(I) · poly(C)-induced FS-4 cells it is possible to irradiate cultures before induction and then analyze interferon mRNA levels within 90 min of irradiation. In systems where transcription of interferon mRNA takes place several hours after exposure to an inducer (e.g., most viral inductions) it may be necessary to irradiate induced cells at a time such that sufficient levels of interferon mRNA are present 90 min later.

An important consideration in the use of the UV procedure to size the primary transcripts for inducible mRNA species is the possible involvement of several genes in series in the induction process. In such cases the UV target estimated by irradiating cultures prior to induction may reflect the aggregate target of all the genes involved, not the size of the primary transcript for the mRNA under consideration. This possibility can be tested by irradiating cells before induction as well as just after transcription of the induced mRNA has begun. However, irradiation should not be delayed to such an extent that mRNA synthesized prior to irradiation would interfere with subsequent quantitation of the effects of irradiation. If the target sizes estimated by irradiating cultures before and after induction are similar, then it is reasonable to conclude that the observed target size reflects the size of the primary transcript per se.

[11] J. R. Greenberg, *Nucleic Acids Res.* 6, 715 (1979).

Materials and Reagents

A high intensity shortwave (254 nm) UV lamp with the emission tube large enough to uniformly irradiate petri dish cultures is essential. We have used a model R-52 Mineralight (Ultraviolet Products, Inc., U.S.A.) placed approximately 49 cm above the surface of monolayer cultures in 150-mm petri dishes. At this distance the UV dose corresponds to 6 erg mm^{-2} sec^{-1}. Protective eyeglasses are a necessary precaution. A tight-fitting Dounce homogenizer is used for cell fractionation. Most of the reagents required for UV inactivation experiments are similar to those required for sucrose gradient analyses of RNA. In addition, sodium acetate, 1 M, pH 5.1, sterilized by autoclaving and stored at 4°, is a convenient stock solution that can be used to make up acetate–SDS buffer (0.1 M sodium acetate, pH 5.1; 0.01 M EDTA; 0.2% SDS), which is used for hot acid–phenol extraction of cellular RNA. Phenol (commercial phenol can be used directly if it is colorless) saturated with acetate buffer (without SDS) is stored at 4° in a dark bottle. Other solutions required are: phosphate-buffered saline (containing per liter of solution: NaCl, 8 g; KCl, 0.2 g; Na_2HPO_4, 1.15 g; $MgCl_2 \cdot 6\ H_2O$, 0.1 g; KH_2PO_4, 0.2 g; $CaCl_2$, 0.1 g, pH 7.4; abbreviated PBS), PBS without $CaCl_2$ and $MgCl_2 \cdot 6\ H_2O$ (PBS-def); trypsin–EDTA (trypsin, 5 g; and EDTA 1 g dissolved at 56° in 2 liters of PBS-def); reticulocyte standard buffer, RSB (0.01 M NaCl; 0.01 M Tris, pH 7.4; 1.5 mM $MgCl_2$) and reagents for poly(U)-Sepharose or oligo(dT)-cellulose chromatography.

Procedure

The experiment is carried out in two parts. First, the extent of inactivation of rRNA synthesis is monitored at various UV doses. Second, the extent of inactivation of interferon mRNA synthesis is monitored at various UV-doses and is related to the size of the transcription unit using the calibration curves derived from the inactivation of rRNA synthesis.

The following description applies to cells in monolayer cultures. Cells in suspension culture can be handled in a similar fashion in a small volume of medium placed in a petri dish.

Inactivation of Ribosomal RNA Synthesis. Most cell culture media contain phenol red, which strongly absorbs UV irradiation. Hence, cells need to be washed well with warm PBS to eliminate phenol red. Photoinduced DNA repair is minimized by irradiating cultures in a dark room. Cultures can be seen and handled in the faint bluish light emitted by the UV lamp. After irradiation, phenol red-containing media can be added back and the cultures wrapped in aluminum foil to prevent exposure to visible light. These can then be removed from the dark room and incubated at 37°. Ideally, the preparation of cultures, their irradiation and subsequent incubation should be carried out at 37° (e.g., in a warm room). However, the irradiation step can also be carried out at room temperature without the introduction of significant error. Ultraviolet-inactivation data are most accurate in the range 0 to 67% inhibition.[6] Hence, the UV lamp should be placed at such a distance from the cultures that the inhibition of RNA synthesis falls within this range when cultures are irradiated for short periods of varying length.

Irradiation of monolayer FS-4 cultures in 150-mm petri dishes for 0, 5, 10, 20, and 40 sec generates data illustrated in Fig. 3. In this experiment,

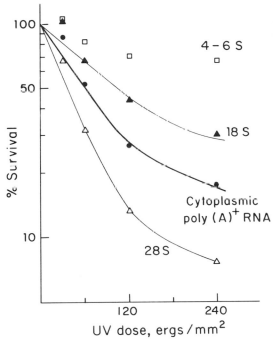

FIG. 3. Ultraviolet (UV) inactivation of ribosomal and polyadenylated messenger RNA synthesis in confluent, poly(I) · poly(C)-treated monolayer cultures of human diploid fibroblasts (FS-4) grown in 150-mm petri dishes. The 100% levels of ^{3}H incorporation correspond, per culture, to 6704 cpm for cytoplasmic polyadenylated RNA (●), 1953 cpm for 28 S RNA (△), 5457 cpm for 18 S RNA (▲), and 1741 cpm for 4–6 S RNA (□). See text for details. From Sehgal and Tamm.[7]

confluent FS-4 cultures were rinsed twice with warm PBS and the medium was decanted. At this time the room was darkened and the UV lamp (6 erg mm^{-2} sec^{-1} at 49 cm distance) was turned on for a few minutes. The duration of UV exposure was controlled by rapidly placing a culture on the bench top (with a nonreflecting or dark surface) below the UV lamp and removing it after a suitable interval or by covering a culture with aluminum foil, placing it on the bench and removing the foil from over the culture for an appropriate period of time. One culture was used per time interval, whereas two cultures were separately monitored as unirradiated controls. Ten milliliters of warm Eagle's MEM containing 200 μCi [^{3}H]uridine (New England Nuclear, 27.9 Ci/mmol) were then added to each culture. The cultures were wrapped in aluminum foil and incubated at 37° for 90 min. The cells were harvested by trypsinization (rinse once with cold PBS-def, add 3 ml of the trypsin-EDTA solution per culture, incubate at room temperature for 3 min, remove cells by vigorous pipetting, add serum to stop trypsin action, pellet cells by centrifugation). They were then rinsed twice with ice-cold PBS-def and resuspended in 1.5 ml of one-third strength reticulocyte standard buffer and kept on ice for 15–20 min. An equal aliquot of FS-4 cells labeled overnight with [^{14}C]uridine (New England Nuclear, 52.4 mCi/mmol; 10 μCi per 40 ml of medium per culture) was added to each of the ^{3}H-labeled cell suspensions. The cells were broken by 40 strokes in a tight-fitting Dounce, and cytoplasmic RNA was extracted. Poly(A)-free and poly(A)-containing cytoplasmic RNA were prepared by poly(U)-Sepharose chromatography. The poly(A)-free fraction was ethanol precipitated, dissolved in 0.5 ml of buffer containing 0.05 M NaCl; 0.01 M EDTA; 0.01 M Tris, pH 7.4; and 0.2% SDS (0.05 M NETS); 0.3 ml of each sample was sedimented through a 15 to 30% sucrose gradient in 0.05 M NETS in a Beckman SW41 rotor at 25,000 rpm for 19.5 hr. Each gradient fraction was precipitated with 10% trichloroacetic acid (TCA), and incorporation into 4–6 S, 18 S, and 28 S RNA was monitored. Aliquots of total cytoplasmic poly(A)-containing RNA were also precipitated with TCA (10%) and counted. The recovery of ^{14}C radioactivity in each RNA sample was used to correct for variations in the recovery of ^{3}H-labeled 28 S, 18 S, 4–6 S RNA and polyadenylated RNA in different samples. An alternative procedure to correct for variations in the recovery of RNA is to label cells with [^{14}C]uridine overnight and to UV-irradiate ^{14}C-labeled cultures followed by [^{3}H]uridine labeling after irradiation. It should be noted that in the experiments in Fig. 3 we did not allow RNA polymerase distal to sites of UV-induced lesions to run off (15–20 min at 37°) prior to labeling of cells with [^{3}H]uridine, since we labeled cells for a long time (90 min) because interferon mRNA synthesis was also monitored 90 min after irradiation.

Figure 3 summarizes our data for rRNA and mRNA inactivation in FS-4 cells. The initial sections (up to 120 ergs/mm^2) of the lines describing the inactivation of 18 S and 28 S rRNA are straight and are taken to represent first-order (single-hit) curves describing the loss of rRNA synthesis as given by the equation $R/R_0 = 1 - e^{-kt}$, where R is the level of residual transcription product after UV irradiation for time t (sec), R_0 is the level without irradiation, and k is a constant consisting of the target size and the frequency of transcription-terminating hits per unit length of DNA per unit dose of UV.[6] The slope (k) of the line log (R/R_0) versus UV dose is directly proportional to the size of the DNA target. A simple procedure to relate DNA target sizes is to calculate the UV dose at which RNA synthesis is reduced to 37% of that in unirradiated controls (D_{37}). The D_{37} values for 18 S and 28 S rRNA from Fig. 3 are 162 and 57 erg

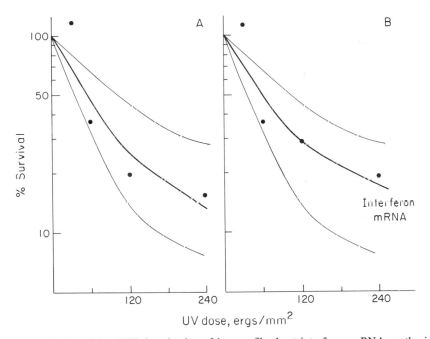

FIG. 4. Ultraviolet (UV) inactivation of human fibrobast interferon mRNA synthesis. Confluent monolayers of FS-4 cultures were irradiated prior to (panel A) or 30 min after (panel B) induction with poly(I) · poly(C) in the presence of cycloheximide. The cells were harvested 90 min after irradiation, and interferon mRNA levels were estimated using the oocyte microinjection assay. Data on interferon mRNA levels (●) from four experiments are pooled in panel A and from two experiments in panel B. The geometric mean interferon titer of the unirradiated controls (100% level) was 60 reference units/ml in both panels A and B. The fine lines correspond to the 18 S and 28 S ribosomal RNA lines in Fig. 3. See text for details. From Sehgal and Tamm.[7]

mm^{-2}, respectively. These values correspond to 4 and 11 kbp DNA lengths. Thus, the size of the DNA target is inversely proportional to the D_{37} value.

UV Inactivation of Human Fibroblast Interferon mRNA Synthesis. Confluent monolayers of FS-4 cells in 150-mm petri dishes (4 or 5 per group) were either (*a*) irradiated as described in the section on rRNA labeling and immediately induced for 90 min at 37° with poly(I) · poly(C) (20 μg/ml, P-L Biochemicals) and cycloheximide (50 μg/ml, Polysciences, Inc.) in Eagle's MEM, 10 ml per culture (Fig. 4A) or (*b*) first induced with poly(I) · poly(C) and cycloheximide for 30 min, rinsed with cold PBS, UV-irradiated, replenished with warm Eagle's MEM containing 2% fetal bovine serum (10 ml medium/culture), and incubated for a further 90 min at 37° (Fig. 4B). Cycloheximide is included in the medium in order to superinduce interferon mRNA synthesis maximally so that the superinducing effect of UV irradiation does not mask its inhibitory effect.[7] Cultures were irradiated 30 min after induction because we estimated that commitment of FS-4 cells to make interferon is established within 1–5 min of exposure to poly(I) · poly(C) and that *transcription* of interferon mRNA begins 10–20 min later.[7] Considerable levels of interferon mRNA are observed by 60 min after induction. Two groups of 4–5 cultures each were usually used as unirradiated controls in each experiment. Polyadenylated RNA from harvested cells was prepared and dissolved in 5 μl of sterile, distilled water. Aliquots of 2 μl were assayed for interferon mRNA by microinjection into *Xenopus* oocytes.

The scatter in the UV-inactivation data is largely due to the twofold error in interferon assays. This error can be minimized by assaying each RNA sample twice and by pooling data from several experiments. Furthermore, the slight leakiness of cycloheximide in preventing the superinducing effect of UV-irradiation[7] probably explains the high values for interferon mRNA synthesis at a UV dose of 30 erg mm^{-2}.

The data in Fig. 4 suggest that the size of the transcription unit for human fibroblast interferon mRNA is in the range 4 to 11 kbp. The scatter in the data precludes a better estimate.

Acknowledgment

I thank Dr. Igor Tamm for helpful discussions and a critical review of this manuscript.

[18] Procedures for the Measurement of Interferon mRNA Distribution in Induced Mouse Cells

By JEAN CONTENT, LUK DE WIT, MARGARET I. JOHNSTON, and ERIK DE CLERCQ

Mouse interferon mRNA can be translated successfully in several cell-free systems, including the wheat germ system[1] (see this volume [14]) and the micrococcal nuclease-treated rabbit reticulocyte lysate system.[2]

We have shown that total cytoplasmic RNA from Newcastle disease virus (NDV)-induced L-929 cells can program the synthesis of interferon in *Xenopus laevis* oocytes. Moreover, if this RNA is further purified by sucrose-formamide gradient centrifugation, a linear dose-response relationship is obtained between the amount of RNA injected into the oocyte and the amount of interferon translated.[2] A similar quantitative relationship has also been established for the oocyte and wheat germ systems programmed with human interferon mRNA.[3-5] These observations have facilitated a study of the mechanism of superinduction of human fibroblast interferon production.[6-8]

When studying the mechanism of interferon priming on interferon production (see Volume 78 [1]) in NDV-induced L-929 cells, we took advantage of this simple quantitative assay to study the kinetics and intracellular distribution of interferon mRNA (IF mRNA). The conclusion of this study was that priming by interferon did not affect IF mRNA transcription or stability, but rather its rate of translation in NDV-induced mouse L cells.[9,10]

[1] M.-N. Thang, D. C. Thang, E. De Maeyer, and L. Montagnier, *Proc. Natl. Acad. Sci. U.S.A.* **72**, 3975 (1975).

[2] B. Lebleu, E. Hubert, J. Content, L. De Wit, I. A. Braude, and E. De Clercq, *Biochem. Biophys. Res. Commun.* **82**, 665 (1978).

[3] P. B. Sehgal, B. Dobberstein, and I. Tamm, *Proc. Natl. Acad. Sci. U.S.A.*, **74**, 3409 (1977).

[4] R. L. Cavalieri, E. A. Havell, J. Vilček, and S. Pestka, *Proc. Natl. Acad. Sci. U.S.A.* **74**, 3287 (1977).

[5] N. B. K. Raj and P. M. Pitha, *Proc. Natl. Acad. Sci. U.S.A.* **74**, 1483 (1977).

[6] P. B. Sehgal, D. S. Lyles, and I. Tamm, *Virology* **89**, 186 (1978).

[7] P. B. Sehgal and I. Tamm, *Virology* **92**, 240 (1979).

[8] R. L. Cavalieri, E. A. Havell, J. Vilček, and S. Pestka, *Proc. Natl. Acad. Sci. U.S.A.* **74**, 4415 (1977).

[9] J. Content, M. I. Johnston, L. De Wit, J. De Maeyer Guignard and E. De Clercq, *Biochem. Biophys. Res. Commun.* **96**, 415 (1980).

[10] S. L. Abreu, J. C. Bancroft, and W. E. Stewart, II, *J. Biol. Chem.* **254**, 4114 (1979).

METHODS IN ENZYMOLOGY, VOL. 79

We present here the methodology used to extract IF mRNA, partially purify IF mRNA, and assay IF mRNA by translation in *Xenopus laevis* oocytes (see also this volume [10]).

Materials

Frogs and Oocytes. Xenopus laevis were imported from the South African Snake Farm, P.O. Box 6, Fish Hoek, Cape Province, South Africa. Frogs are anesthesized by immersion for 10 min in MS 222 (Sandoz) 1% (w/v) in water. Pieces of one ovary are removed, teared, and free individual oocytes are collected and kept in Barth solution until injection. Frogs are usually sewn and can be used a second time within 1 or 2 weeks.

Micropipettes. Drummond microcaps (2 μl) are stretched to 10–30 μm diameter in a Narishige automatic glass microelectrode puller PN-3 (Narishige Sci. Inst. Lab., Tokyo, Japan).

Buffers and Reagents

Barth medium (modified by Gurdon[11]): Tris · HCl, 2.0 mM; NaCl, 88.0 mM; KCl, 1.0 mM; Ca(NO$_3$)$_2$ · 4 H$_2$O, 0.33 mM; CaCl$_2$ · 2 H$_2$O, 0.41 mM; MgSO$_4$ · 7 H$_2$O, 0.82 mM; NaHCO$_3$, 2.4 mM; penicillin, 0.1 g/liter; streptomycin, 0.01 g/liter

Phosphate-buffered saline (PBS): NaCl, 136 mM; Na$_2$HPO$_4$ · 12 H$_2$O, 8 mM; KCl, 2.7 mM; KH$_2$PO$_4$, 1.47 mM, pH 7.2

Triton lysing medium (TLM): Tris · HCl, 10 mM, pH 7.5; KCl, 10 mM; MgCl$_2$, 1.5 mM; 2-mercaptoethanol, 6.5 mM; Triton X-100, 0.13% (w/v); sucrose, 13% (w/v)

Sucrose lysing medium (SLM)[11]: Tris · HCl, 50 mM, pH 7.6; KCl, 100 mM; MgCl$_2$, 5 mM; sucrose, 27.4% (w/v). LiCl stock solution (10 M) is used after filtration on a 0.45 μm Millipore filter.

Formamide stock: formamide (J. T. Baker Chemicals, Reagent grade) is deionized by stirring 2 hr with 4% (w/v) Bio-Rad mixed-bed resin and stored frozen.

Sucrose formamide gradient solution: Sucrose (ultracentrifugation grade, Merck) is dissolved at 5% or 20% (w/v) in a solution containing 50% deionized formamide, 5 mM EDTA, 0.01 M Tris · HCl, pH 7.5, 0.1 M LiCl, and 0.2% SDS (sodium dodecyl sulfate "specially pure," BDH).

Methods

Cells and Viruses

L-929 mouse cells are grown in Eagle's minimal essential medium (MEM) supplemented with 10% fetal calf serum. The Kumarov strain of

---

[11] J. B. Gurdon, *J. Embryol. Exp. Morphol.* **20**, 401 (1968).

NDV is used for interferon induction. Stock virus is propagated in 10-day-old chick embryos (see also Volume 78 [42]). Vesicular stomatitis virus (VSV) (Indiana strain) serves as the challenge virus for interferon assays. Stock virus is propagated in BS-C-1 cells.

Induction of Interferon

L-929 cells are grown in half-gallon roller bottles; 5–12 bottles are used for one typical RNA preparation. When confluent, the cells are inoculated with NDV at approximately 1 PFU per cell. After 1 hr of incubation at 37°, the virus inoculum is removed; the cells are further incubated with MEM for 12 hr, at which time the cells are harvested, fractionated, and extracted. The *in vivo* interferon production is usually controlled by assaying the cell culture fluid for interferon content at regular times (i.e., 4, 8, 12, 16, 20, and 24 hr) after NDV infection (see "Interferon Assay" below).

Cell Harvesting and Fractionation

For optimal IF mRNA activity, cells are extracted 12 hr after NDV injection. The cells are first washed with ice-cold PBS and then scraped off with a rubber policeman (rubber bottle scraper Bellco, 7731-22222). All subsequent operations are done at 4°. The harvested cells are centrifuged for 2 min at 3000 rpm (1200 g) in a clinical centrifuge and washed three times with ice-cold PBS. The cell suspension is then divided into two parts (A and B). Part A is disrupted and extracted in the presence of detergent and is used to prepare fractions 1, 2, and 3. Part B is disrupted without detergent and is used to prepare fractions 4 and 5 (the whole procedure is summarized in Fig. 1).

Part A. The cell suspension is centrifuged in ice-cold PBS for 5 min at 3000 rpm in a graduated tube, and the packed cell volume is measured. The cells are suspended at 4° in 2.5 volumes of a buffer containing 0.13% Triton X-100 (TLM). After 10 min the suspension is disrupted by 5 strokes of a tight-fitting Dounce glass homogenizer ("B" pestle). Nuclei are removed by centrifugation for 2 min at 2000 rpm (500 g) in an SS34 Sorvall rotor. A portion of the $S_{0.5}$ supernatant is immediately extracted with phenol and further purified by LiCl precipitation and formamide-sucrose gradient centrifugation (see details below); this constitutes *fraction 1* (*total cytoplasmic RNA*). Fractions 2 and 3 are obtained by processing the rest of the $S_{0.5}$ supernatant for the precipitation of polysomes. It is first centrifuged for 10 min at 12,000 rpm in a SS34 Sorvall rotor (17,300 g). The supernatant (S_{17}) is made up to 30 mM MgCl$_2$ (by addition of 1 M MgCl$_2$), incubated for 2 hr at 4°, and centrifuged for 20 min at 2500 rpm (755 g) in a SS34 Sorvall rotor. The pellet from this centrifugation is washed two times with 30 mM MgCl$_2$, redissolved in H$_2$O, and phenol-extracted (see

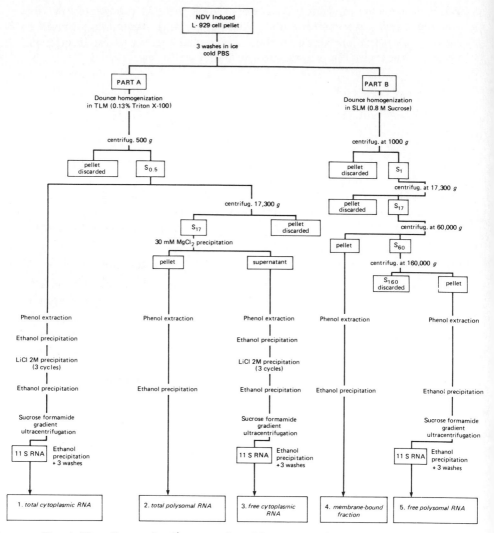

FIG. 1. Flow diagram for the preparation of five classes of RNA by cell fractionation. NDV, Newcastle disease virus; TLM, Triton lysing medium; SLM, sucrose lysing medium; centrifug., centrifugation.

below). This RNA constitutes *fraction 2* (*total polysomal RNA*). The supernatant from the last centrifugation is phenol-extracted and further purified by LiCl precipitation and formamide-sucrose gradient centrifugation (see below) and yields *fraction 3* (*free cytoplasmic RNA*).

Part B. The washed cells are lysed by 15 strokes of a tight-fitting

Dounce homogenizer in a buffer containing $0.8\,M$ sucrose (SLM).[12] The lysate is centrifuged for 5 min at $1000\,g$, and the resulting supernatant is centrifuged for 10 min at 12,000 rpm $(17,300\,g)$. The resulting S_{17} is layered onto 2 ml of SLM and centrifuged at $60,000\,g$ for 20 min in the SW41 rotor of a Beckman ultracentrifuge. The pellet is phenol-extracted and constitutes *the membrane-bound fraction* (*fraction 4*). The supernatant (S_{60}) is further centrifuged for 2 hr at $160,000\,g$ in a Beckman type 65 or 75 Ti fixed-angle rotor. The resulting pellet is phenol-extracted and further purified by formamide-sucrose gradient centrifugation (see below). This RNA constitutes *fraction 5* (*free polysomal RNA*).

Interferon mRNA Extraction

Fractions 1–5 described above are deproteinized immediately by three extractions with an equal volume of phenol, chloroform, and isoamyl alcohol [25 : 25 : 1 (v/v/v)] in 100 mM Tris HCl, pH 8.9, and 1% SDS. The resulting aqueous phase is further extracted with chloroform, isoamyl alcohol [25 : 1 (v/v)]; after addition of $0.2\,M$ NaCl, the RNA is precipitated with two volumes of ethanol at $-20°$. When required, the RNA is redissolved in water and precipitated in $2\,M$ LiCl at $4°$ for 6–16 hr. The LiCl precipitate is redissolved in water, and the precipitation is repeated 2 or 3 times. After the last precipitation the RNA is precipitated with 2 volumes of ethanol, washed twice in 75% ethanol, and dried under vacuum before it is finally dissolved in water.

Sucrose Gradient Purification of IF mRNA

IF mRNA samples from fractions 1, 3, and 5, containing 150–800 μg of RNA, are first dissolved in water, then brought to 50% formamide, incubated for 2 min at $37°$, and layered onto a linear 5 to 20% w/v sucrose gradient in 5 mM EDTA, 0.01 M Tris HCl, pH 7.5, 0.1 M LiCl, 0.2% SDS, and 50% formamide. The gradients are centrifuged at $20°$ for 16 hr at 40,000 rpm in a Beckman SW60 Ti rotor. [3H]Uridine-labeled cytoplasmic mouse cell RNA is analyzed in a parallel gradient in order to localize the position of the major (18 S and 4 S) RNA components and to deduce the position of the IF mRNA peak. The latter is reproducibly found at approximately 11 S.[2] Usually, each gradient is divided into 18 fractions. Fractions corresponding to the 9–12 S region, containing most of the IF mRNA activity, are pooled, made up to $0.2\,M$ NaCl, precipitated with two volumes of ethanol, incubated for at least 16 hr at $-20°$, centrifuged, washed 3 times with 75% ethanol, dried, and finally redissolved in water at 2 mg of RNA per milliliter.

[12] G. E. Sonenshein and G. Brawerman, *Biochemistry* 15, 5501 (1976).

TYPICAL DISTRIBUTION OF RNA AND IF mRNA ACTIVITY IN NEWCASTLE DISEASE
VIRUS-INDUCED L-929 CELLS

	Fraction 1: total cytoplasmic RNA	Fraction 2: total polysomal RNA	Fraction 3: free cytoplasmic RNA	Fraction 4: membrane-bound RNA	Fraction 5: free polysomal RNA
Amount of RNA (μg) obtained from 1 ml of packed cells	148	354	141	11	30
Amount of interferon (units/ml) translated in oocytes when injected with 100 ng of RNA per oocyte	1584	500	316	22[a]	158

[a] Upon injection of 50 ng of RNA per oocyte.

Translation of Interferon mRNA in Xenopus laevis Oocytes

Xenopus laevis oocytes are microinjected with the mRNA preparations as described previously.[13] RNA samples are dissolved in water at 0.5–2 mg/ml. This corresponds to 25–100 ng of RNA per oocyte, for each oocyte is injected with 50 nl. Within this range of RNA concentrations the amount of biologically active interferon translated is directly proportional to the amount of RNA injected per oocyte, thus allowing a direct estimation of IF mRNA content in a given RNA preparation.[2] This quantitative relationship is also observed for both polysomal and membrane-bound RNA (fractions 2 and 4, respectively). In contrast, fractions 1, 3, and 5 require the sucrose formamide gradient purification step to show the same linear response.[9] Ten oocytes are injected for every determination of IF mRNA activity. After injection they are incubated in 0.2 ml of Barth's medium in 1.5-ml Eppendorf plastic tubes for 16 hr at 19°. After 120 μl of the incubation medium have been discarded, the oocytes are homogenized in the remaining medium with a small Teflon pestle. The homogenates are centrifuged at 10,000 g for 2 min in an Eppendorf microcentrifuge. The supernatants are carefully removed (to avoid contamination by the fat layer) and assayed for interferon either directly or after storage at −70°.

[13] J. B. Gurdon, C. D. Lane, H. R. Woodland, and G. Marbaix, *Nature (London)* 233, 177 (1971).

Interferon Assay

Mouse interferon is assayed by inhibition of VSV-induced cytopathogenicity in L-929 cells. The assay is done in duplicate in 96-well microtiter plates. A standard calibrated against the National Institutes of Health mouse interferon reference standard G-002-904-511 is included in each assay.

Results

The results of a representative experiment are presented in the table. They indicate the amounts of RNA obtained for each of the five fractions as well as the amounts of interferon translated in oocytes upon injection of a standardized amount of RNA.

Acknowledgments

These investigations were supported by grants from the Belgian FRFC (Fonds de la Recherche Fondamentale Collective), FGWO (Fonds voor Geneeskundig Wetenschappelijk Onderzoek), and Geconcerteerde Onderzoeksacties, and by a U.S. National Research Service Award (1 F32 A105662 to M. I. J.).

Section III

Methods to Study Mode of Interferon Action

A. General Methodology and Considerations
Articles 19 and 20

B. Phosphorylation and cAMP in Interferon Action
Articles 21 through 23

C. Synthesis and Assay of (2'-5')-Oligoadenylic Acid and Related Events
Articles 24 through 35

D. Effect of Interferon on Protein Synthesis and Other Cellular Processes
Articles 36 through 42

[19] Enzymology of Interferon Action—A Short Survey

By P. Lengyel

Multiplicity of Interferons and Complexity of Interferon Action

One organism can produce several different interferons.[1-3] Thus, three classes of human interferons have been described: leukocyte, fibroblast, and immune types. Moreover, the leukocyte class was shown to consist of several species.[4,5]

When discovered in 1957, the interferons were considered to be antiviral agents.[6] Studies in the last two decades, however, revealed that they also affect a large variety of seemingly diverse biological phenomena. These include cell motion, cell proliferation, and several immunological processes (e.g., the antibody response, delayed hypersensitivity, histocompatibility antigen expression, graft rejection, natural killer-cell recruitment, and macrophage activation.[2,3,7-9] Considering the multiplicity of interferons together with the diversity of their actions, it should not be surprising that the biochemical basis of interferon action appears to be complex.

The early attempts to understand this biochemistry aimed to discover the mechanism(s) by which the replication of viruses is impaired by interferons. The question usually asked was: Which step in the replication of a particular virus is blocked, or which is (or are) the viral component(s)

[1] Nomenclature of interferons and enzymes: $(2'-5')(A)_n$ is also designated as oligoisoadenylate and is sometimes abbreviated as 2-5A. The enzyme synthesizing $(2'-5')(A)_n$ is designated by different authors as $(2'-5')(A)_n$ synthetase or $(2'-5')$-oligo(A)polymerase or oligoisoadenylate synthetase E. The endoribonuclease that can be activated by $(2'-5')(A)_n$ is designated as RNase L or RNase F. The protein kinase that can be activated by double-stranded RNA is also designated as protein kinase PK-i, or eIF-2 kinase. The phosphodiesterase degrading $(2'-5')(A)_n$ is also designated as phosphodiesterase 2'-PDi.

[2] "The Interferon System" (S. Baron and F. Dianzani, eds.), *Tex. Rep. Biol. Med.* **35**, 1–573 (1977).

[3] W. E. Stewart, II, "The Interferon System." Springer-Verlag, Vienna and New York, 1979.

[4] M. Rubinstein, S. Rubinstein, P. C. Familletti, R. S. Miller, A. A. Waldman, and S. Pestka, *Proc. Natl. Acad. Sci. U.S.A.* **76**, 640 (1979).

[5] N. Mantei, M. Schwarzstein, M. Streuli, S. Panem, S. Nagata, and C. Weissmann, *Gene* **10**, (1980).

[6] A. Isaacs and J. Lindenmann, *Proc. R. Soc. London, Ser. B* **147**, 258 (1957).

[7] E. De Maeyer and J. De Maeyer-Guignard, *Compr. Virol.* **15**, 205 (1979).

[8] I. Gresser, *Cell. Immunol.* **34**, 406 (1977).

[9] Conference on regulatory functions of interferons (J. Vilček, ed.), *Ann. N. Y. Acad. Sci.* **350** (1980).

METHODS IN ENZYMOLOGY, VOL. 79

or virus-specific molecule(s) whose accumulation is diminished in cells treated with interferon? Many of the results obtained pointed toward viral RNA and protein accumulation as likely targets of interferon action.[10–12] This caused investigators to compare cell-free extracts from interferon-treated cells with those from control cells for their capacity to process, translate, and cleave viral and host mRNAs. These comparisons resulted in the findings that (a) protein synthesis is more sensitive to inhibition by double-stranded RNA in extracts from interferon-treated cells than in those from control cells[13]; (b) the treatment of cells with interferons increases the level of several enzymes.[14–18] Some (but not all) of these enzymes remain latent unless activated by double-stranded RNA and ATP. The most likely rationale for the involvement of double-stranded RNA in the interferon response is that double-stranded RNA might be formed as a side product or intermediate in the replication of various viruses[19,20] and thus might be a signal for the presence of replicating viruses in cells.

The establishment of an increase in the level of several enzymes in cells treated with interferon prompted investigations on the activities of interferons as inducers of mRNAs and proteins.[21–27]

Studies on the biochemical basis of the cell growth-inhibitory and immunoregulatory actions of interferons have also been initiated.[7,28–33]

[10] R. M. Friedman. *J. Gen. Physiol.* **56**, 149s (1970).

[11] J. A. Sonnabend and R. M. Friedman, *in* "Interferons and Interferon Inducers" (N. B. Finter, ed.), pp. 201–239. North-Holland Publ., Amsterdam, 1973.

[12] D. H. Metz, *Cell* **6**, 429 (1975).

[13] I. M. Kerr, R. Brown, and L. A. Ball. *Nature (London)* **250**, 57 (1974).

[14] A. G. Hovannessian, R. E. Brown, and I. M. Kerr, *Nature (London)* **268**, 537 (1977).

[15] B. Lebleu, G. C. Sen, S. Shaila, B. Cabrer, and P. Lengyel, *Proc. Natl. Acad. Sci. U.S.A.* **73**, 3107 (1976).

[16] A. Zilberstein, A. Kimchi, A. Schmidt, and M. Revel, *Proc. Natl. Acad. Sci. U.S.A.* **75**, 4734 (1978).

[17] W. K. Roberts, A. Hovanessian, R. E. Brown, M. J. Clemens, and I. M. Kerr, *Nature (London)* **264**, 477 (1976).

[18] A. Schmidt, Y. Chernajovsky, L. Shulman, P. Federman, H. Berissi, and M. Revel, *Proc. Natl. Acad. Sci. U.S.A.* **76**, 4788 (1979).

[19] C. Colby and P. H. Duesberg, *Nature (London)* **222**, 940 (1969).

[20] R. F. Boone, R. P. Parr, and B. Moss, *J. Virol.* **30**, 365 (1979).

[21] L. A. Ball, *Proc. Natl. Acad. Sci. U.S.A.* **75**, 1167 (1978).

[22] E. Knight, Jr. and B. D. Korant, *Proc. Natl. Acad. Sci. U.S.A.* **76**, 1824 (1979).

[23] P. J. Farrell, R. J. Broeze, and P. Lengyel, *Nature (London)* **279**, 523 (1979).

[24] M. De Ley, A. Billiau, and P. De Somer, *Biochem. Biophys. Res. Commun.* **89**, 701 (1979).

[25] S. L. Gupta, B. Y. Rubin, and S. L. Holmes, *Proc. Natl. Acad. Sci. U.S.A.* **76**, 4817 (1979).

[26] P. J. Farrell, R. J. Broeze, and P. Lengyel, *Ann. N. Y. Acad. Sci.* **350**, 615 (1980).

[27] B. Y. Rubin and S. L. Gupta, *J. Virol.* **34**, 446 (1980).

[28] P. Lindahl, P. Leary, and I. Gresser, *Proc. Natl. Acad. Sci. U.S.A.* **70**, 2785 (1973).

Enzymes of Posttranscriptional Control in Extracts from
Interferon-Treated Cells

The $(2'-5')$ $(A)_n$ Synthetase-RNase L System

This system was discovered serendipitously in a comparison of the
rates of cleavage of reovirus mRNAs in extracts from interferon-treated
cells with those in extracts from control cells. The faster cleavage in the
extract from interferon-treated cells turned out to depend on the presence
of genomic double-stranded RNA from reovirions contaminating the
reovirus mRNA preparation. After the removal of the double-stranded
RNA, the rates of cleavage in the two types of extracts became identical;
the readdition of double-stranded RNA [from reovirions or of
poly(I) · (poly (C))] accelerated the cleavage only in the extract from
interferon-treated cells, and thereby restored the original difference.[34]
Subsequent studies revealed that in addition to double-stranded RNA,
ATP is also required for accelerating RNA cleavage, and both of these
compounds are needed only for the activation of an endoribonuclease
system (designated as endoribonuclease$_{INT}$), not for its action.[33] The
endoribonuclease$_{INT}$ system consists of at least two enzymes. In the presence of double-stranded RNA, the first of these catalyzes the production
of a small thermostable product from ATP. This in turn activates the
second enzyme, a latent endoribonuclease that was designated as RNase
L.[35] The small thermostable product was identified as $(2'-5')$-
oligo$(A)_n$,[1-17,35-38] a set of compounds discovered earlier as inhibitors of
protein synthesis that are formed from ATP in extracts from interferon-
treated cells if the extracts are supplemented with double-stranded
RNA.[14,39,40]

---

[29] M. Degré and T. Hovig, *Acta Pathol. Microbiol. Scand. Sect. B* **84B**, 347 (1976).

[30] D. Brouty-Boyé and M. G. Tovey, *Intervirology* **9**, 243 (1978).

[31] M. G. Tovey, C. Rochette-Egly, and M. Castagna, *Proc. Natl. Acad. Sci. U.S.A.* **76**, 3890 (1979).

[32] A. Kimchi, H. Shure, and M. Revel, *Nature (London)* **282**, 849 (1979).

[33] G. C. Sen, B. Lebleu, G. E. Brown, M. Kawakita, E. Slattery, and P. Lengyel, *Nature (London)* **264**, 370 (1976).

[34] G. E. Brown, B. Lebleu, M. Kawakita, S. Shaila, G. C. Sen, and P. Lengyel, *Biochem. Biophys. Res. Commun.* **69**, 114 (1976).

[35] L. Ratner, R. Wiegand, P. Farrell, G. C. Sen, B. Cabrer, and P. Lengyel, *Biochem. Biophys. Res. Commun.* **81**, 947 (1978).

[36] C. Baglioni, M. A. Minks, and P. A. Maroney, *Nature (London)* **273**, 684 (1978).

[37] M. J. Clemens and B. R. G. Williams, *Cell* **13**, 565 (1978).

[38] D. A. Epstein and C. E. Samuel, *Virology* **89**, 240 (1978).

[39] I. M. Kerr, R. E. Brown, and A. G. Hovanessian, *Nature (London)* **268**, 540 (1977).

[40] I. M. Kerr and R. E. Brown, *Proc. Natl. Acad. Sci. U.S.A.* **75**, 256 (1978).

(2'-5')(A)$_n$ Synthetase. The enzyme synthesizing this set of compounds [(2'-5')(A)$_n$ synthetase] has been found so far in human,[36] mouse,[14] rabbit,[41] and chicken cells.[21] Treatment with interferon boosts the level of the latent enzyme 10- to 100-fold.[42] Leukocyte and fibroblast interferons are equally effective as inducers.[43] The (2'-5')(A)$_n$ synthetase was purified to homogeneity from interferon-treated mouse Ehrlich ascites tumor cells (EAT).[44]

The molecular weight of the enzyme is 105,000 as determined by gel electrophoresis in the presence of sodium dodecyl sulfate and 85,000 as determined by sedimentation through a glycerol gradient.

In the presence of double-stranded RNA, the purified enzyme can convert the large majority (over 97%) of the ATP added to (2'-5')(A)$_n$ (n extends from 2 to about 15) and pyrophosphate, although it does not cleave pyrophosphate.[45] The stoichiometry of the reaction can be formulated as

$$n + 1 \text{ ATP} \rightarrow (2'\text{-}5')\text{pppA(pA)}_n + n \text{ pyrophosphate}$$

The designation (2'-5')pppA(pA)$_n$ stands for the same series of compounds that are abbreviated otherwise as (2'-5')(A)$_n$. The different abbreviation is used here to indicate that the 5' terminus of (2'-5')(A)$_n$ is a triphosphate. Among the products of the enzyme usually di-, tri-, and tetraadenylates are the most abundant. The percentage of longer oligomers diminishes with increasing chain length.[44,46] The proportion of different oligomers appears to reflect an increasing probability of release from the enzyme as the oligomers are elongated. Dimers added to the enzyme are substrates for adenylate addition, but two dimers do not become linked to each other.[47]

(2'-5')(A)$_n$ synthetase was reported to be capable of adding adenylate residues to a variety of compounds with 3'-terminal adenylate moieties such as NAD, ADPribose, and AppppA.[48] The affinity of the enzyme for ATP is low: the rate of the reaction increases when the ATP concentration

[41] A. G. Hovanessian and I. M. Kerr, *Eur. J. Biochem.* **81**, 149 (1978).
[42] G. R. Stark, W. J. Dower, R. T. Schimke, R. E. Brown, and I. M. Kerr. *Nature (London)* **278**, 471 (1979).
[43] R. J. Broeze, J. P. Dougherty, and P. Lengyel, *Fed. Proc., Fed. Am. Soc. Exp. Biol.* **39**, 2205 (1980).
[44] J. P. Dougherty, H. Samanta, P. J. Farrell, and P. Lengyel, *J. Biol. Chem.* **255**, 3813 (1980).
[45] H. Samanta, J. P. Dougherty, and P. Lengyel, *J. Biol. Chem.* **255**, 9807 (1980).
[46] E. M. Martin, N. J. M. Birdsall, R. E. Brown, and I. M. Kerr, *Eur. J. Biochem.* **95**, 295 (1979).
[47] M. A. Minks, S. Benvin, and C. Baglioni, *J. Biol. Chem.* **255**, 5031 (1980).
[48] L. A. Ball and C. N. White, *in* "Regulation of Macromolecular Synthesis by Low Molecular Weight Mediators" (G. Koch and D. Richter, eds.), pp. 303–317. Academic Press, New York, 1979.

is increased from 5 mM to 10 mM. The enzyme is maximally active when the concentration of double-stranded RNA (w/w) is about half of that of the enzyme.[45] Synthetic double-stranded RNA shorter than 30 base pairs does not activate the enzyme, whereas double-stranded RNA longer than 65–80 base pairs causes maximal activation.[49]

Phosphodiesterase Degrading $(2'-5')(A)_n$. An enzyme catalyzing the cleavage of $(2'-5')(A)_n$ into ATP and AMP was partially purified from L cells.[18] The level of this enzyme increases four- to fivefold after the treatment of the cells with interferons. The enzyme can also cleave dinucleoside-monophosphates. Its activity is generally higher on $(2'-5')$ than on $(3'-5')$ phosphodiester bonds. The presence of a phosphate at the 5' terminus of an oligonucleotide does not affect the rate of enzymic cleavage, whereas the presence of a phosphate at the 3' end impairs the cleavage. These results suggest that the enzyme may attack oligonucleotides starting at their 2'-hydroxyl end. The enzyme can also remove the CCA termini from tRNA (see below the section on impairment of exogenous mRNA translation).

RNase L. This latent endoribonuclease, which can be activated by $(2'-5')(A)_n$, was partially purified from EAT cells that had been treated with interferon[35,50] and also from L cells.[51] The interferon treatment raises the level of the enzyme only slightly (1.5- to 2-fold).[52] The enzyme purified from EAT cells is low in $(2'-5')(A)_n$-independent nuclease activity.[50] The $(2'-5')(A)_n$ activating the enzyme must be the trimer or longer and must have either a tri- or a diphosphate moiety at its 5' terminus.[46]

The activation of the enzyme by $(2'-5')(A)_n$ is reversible. Upon removal of $(2'-5')(A)_n$ from the activated enzyme (by gel filtration), the enzyme reverts to the latent state. Readdition of $(2'-5')(A)_n$ reactivates the enzyme.[50]

Consistent with the possibility that the activation of RNase L may involve the binding of $(2'-5')(A)_n$ to the enzyme are the findings that (a) a partially purified RNase L preparation retains $(2'-5')(A)_n$ on nitrocellulose filters[50]; (b) the agent retaining $(2'-5')(A)_n$ copurifies with RNase L both during ion exchange chromatography and gel filtration[50]; (c) the treatment of the enzyme with N-ethylmaleimide abolishes both its activatability by $(2'-5')(A)_n$ and its ability to retain $(2'-5')(A)_n$ on nitrocellulose filters.[53]

[49] M. A. Minks, D. K. West, S. Benvin, and C. Baglioni, *J. Biol. Chem.* **254**, 10180 (1979).
[50] E. Slattery, N. Ghosh, H. Samanta, and P. Lengyel, *Proc. Natl. Acad. Sci. U.S.A.* **76**, 4778 (1979).
[51] M. Revel, *in* "Interferon" (I. Gresser, ed.), Vol. 1, pp. 101–163. Academic Press, New York, 1979.
[52] L. Ratner, Ph.D. dissertation, Yale University, New Haven, Connecticut, 1979.
[53] E. Slattery, N. Ghosh, H. Samanta, and P. Lengyel, *in* "Interferon: Properties and Clinical Uses (A. Khan and G. L. Dorn, eds.), pp. 521–528. Wadley Institutes of Molecular Medicine, Dallas, Texas, 1979.

The molecular weight of the enzyme is 185,000 as estimated by gel filtration. The activation does not seem to result in a large size change of the enzyme. This and other data make it unlikely that the activation should involve the binding or the release of a protein.[50]

When activated, RNase L (partially purified from mouse Ehrlich ascites tumor cells) cleaves only poly(U) of the four ribohomopolymers poly(A), (C), (G) or (U).[53a] In natural RNAs RNase L cleaves at the 3' side of UA, UG, and UU sequences, yielding 3' phosphate-terminated product.[53a,b]

The following observations are in line with a possible involvement of the endonuclease$_{INT}$ (i.e., the $(2'-5')(A)_n$ synthetase-RNase L) system in mediating at least some of the effects of interferon action.

1. The replication of reovirus in mouse L-929 cells is inhibited by interferon, and reovirus mRNAs are degraded faster in interferon-treated cells than in control cells.[54]

2. $(2'-5')(A)_n$ occurs in intact cells; its level is higher in virus-infected, interferon-treated cells than in cells treated only with interferon, or only infected with virus, or in control cells.[55]

3. The introduction into intact cells of $(2'-5')(A)_n$ or of $(2'-5')(A)_n$ core [obtained by treating $(2'-5')(A)_n$ with alkaline phosphatase to remove its 5' terminal triphosphate] results in a transient impairment of protein synthesis and virus replication and also in an accelerated cleavage of RNA.[56]

In vitro, the double-stranded RNA-activatable endonuclease$_{INT}$ system appears to cleave both viral and host RNAs without discrimination.[36,57] *In vivo,* however, at least in the case of some virus-cell systems (e.g., reovirus and L cells), viral protein synthesis is inhibited preferentially above host protein synthesis.[58] An intriguing series of experiments[59] reveals that, at least in principle, the endonuclease$_{INT}$ system is capable of such discrimination. For these experiments the poly(A) segment of a viral

[53a] G. Floyd Smith, E. Slattery, and P. Lengyel, *Science* **212**, 1030 (1981).

[53b] D. H. Wreschner, J. W. McCauley, J. J. Skehel, and I. M. Kerr, *Nature (London)* **289**, 414 (1981).

[54] P. Lengyel, R. Desrosiers, R. Broeze, E. Slattery, H. Taira, J. Dougherty, H. Samanta, J. Pichon, P. Farrell, L. Ratner, and G. C. Sen, in "Microbiology 1980" (D. Schlessinger, ed.), pp. 219–226. Am. Soc. Microbiol., Washington, D.C., 1980.

[55] B. R. G. Williams, R. R. Golgher, R. E. Brown, C. S. Gilbert, and I. M. Kerr, *Nature (London)* **282**, 582 (1979).

[56] A. G. Hovanessian and J. N. Wood, *Virology* **101**, 81 (1980).

[57] L. Ratner, G. C. Sen, G. E. Brown, B. Lebleu, M. Kawakita, B. Cabrer, E. Slattery, and P. Lengyel, *Eur. J. Biochem.* **79**, 565 (1977).

[58] S. L. Gupta, W. D. Graziadei, III, H. Weideli, M. L. Sopori, and P. Lengyel, *Virology* **54**, 49 (1974).

[59] T. W. Nilsen and C. Baglioni, *Proc. Natl. Acad. Sci. U.S.A.* **76**, 2600 (1979).

RNA (from vesicular stomatitis virus) was annealed to poly(U), resulting in an mRNA in which a single-stranded segment is covalently linked to a double-stranded segment. In an extract from interferon-treated cells supplemented with ATP, this mRNA covalently bound to a double-stranded RNA segment is degraded faster than an identical mRNA not bound to double-stranded RNA. This result is consistent with the possibility that in interferon-treated cells infected by an RNA virus the RNase L activity may be localized in a region near the partially double-stranded replicative intermediate of the virus. This localization might be a consequence of (a) the localized synthesis of $(2'-5')(A)_n$ by the $(2'-5')(A)_n$ synthetase bound to the partially double-stranded viral replicative intermediate; and/or (b) the decrease in the concentration of $(2'-5')(A)_n$ away from its site of formation that is caused by cleavage by phosphodiesterase.

Protein Kinase

The need for double-stranded RNA and ATP for the activation of the endonuclease$_{INT}$ system in extracts from interferon-treated cells, the knowledge that several proteins are activated or inactivated by phosphorylation,[60] and other considerations, resulted in tests on the effect of double-stranded RNA on protein phosphorylation in extracts from interferon-treated and control cells. The addition of double-stranded RNA [from reovirus or of poly(I) · poly(C)] to an extract from interferon-treated EAT cells or L cells (but not, or only to a much lesser extent, to an extract from control cells) was found to cause the phosphorylation of at least two proteins: P_1 (67,000 daltons) and P_2 (37,000 daltons).[15-17] Subsequently, a double-stranded RNA-activatable protein kinase system was purified several thousandfold from L cells and from EAT cells treated with interferon.[61,62] The extent of induction by interferon is 3- to 10-fold. Leukocyte, fibroblast, and immune interferons can all induce the enzyme.[43,63]

The purified kinase preparation is essentially free of double-stranded RNA-independent kinase activity and is similar to a double-stranded RNA-dependent protein kinase in reticulocyte lysates.[64] $(2'-5')(A)_n$ does not substitute for double-stranded RNA in activating the enzyme, and $(2'-5')(A)_n$ is not synthesized by the activated enzyme.[65]

[60] C. S. Rubin and O. M. Rosen, *Annu. Rev. Biochem.* **44**, 831 (1975).
[61] G. C. Sen, H. Taira, and P. Lengyel, *J. Biol. Chem.* **253**, 5915 (1978).
[62] A. Kimchi, A. Zilberstein, A. Schmidt, L. Shulman, and M. Revel, *J. Biol. Chem.* **254**, 9846 (1979).
[63] A. G. Hovanessian, E. Meurs, O. Aujean, C. Vaquero, S. Stefanos, and E. Falcoff, *Virology* **104**, 195 (1980).
[64] P. J. Farrell, K. Balkow, T. Hunt, and R. J. Jackson, *Cell* **11**, 187 (1977).
[65] P. J. Farrell, G. C. Sen, M.-F. Dubois, L. Ratner, E. Slattery, and P. Lengyel, *Proc. Natl. Acad. Sci. U.S.A.* **75**, 5893 (1978).

The activation of a partially purified protein kinase by double-stranded RNA and ATP was reported to be enhanced by an acidic protein designated as factor A. The level of this factor in the cell is not affected by treatment with interferon.[62]

The P_2 protein that can be phosphorylated by the kinase is the small subunit of the peptide chain initiation factor eIF-2.[65,66] The identity of P_1 has not been definitely established. However, P_1 copurifies with the double-stranded RNA-activatable protein kinase system throughout a several thousandfold purification, and the most highly purified kinase preparation available at present consists of P_1 as the major protein and two minor components.[67] It remains to be established whether or not P_1 is identical with the protein kinase. The purified enzyme, if activated, can also phosphorylate some histones. Phosphoprotein phosphatase(s), which can remove phosphate moieties from both phosphorylated eIF-2 and phosphorylated P_1, is (are) present in cell extracts.[62] This dephosphorylating activity is impaired by double-stranded RNA.[68]

The addition of the activated protein kinase preparation to a cell-free protein synthesizing system from EAT cells or reticulocyte lysates results in the inhibition of peptide chain initiation. The inhibition can be overcome by the addition of further eIF-2[65] and is apparently the consequence of the phosphorylation of eIF-2.[64]

The finding that double-stranded RNA added to intact, interferon-treated cells (but not to control cells) results in the phosphorylation of P_1 protein indicates that the activation and action of the protein kinase is not an artifact of the *in vitro* system.[69]

Double-Stranded RNA Does Not Have to Be "Free" to Activate the Latent Enzymes

The presence of double-stranded RNA (or at least of complementary RNA strands) has been established in vaccinia virus-infected cells[19,20] and in cells infected with viruses with a double-stranded RNA genome, such as reovirus. In the latter case, however, the double-stranded RNA is apparently always packaged in a protein coat.[70]

This fact prompted tests aiming to establish whether the double-stranded RNA has to be "free" to activate the endonuclease$_{INT}$ system

---

[66] C. E. Samuel, *Proc. Natl. Acad. Sci. U.S.A.* **76**, 600 (1979).

[67] P. Lengyel, H. Samanta, J. Pichon, J. Dougherty, E. Slattery, and P. Farrell, *Ann. N. Y. Acad. Sci.* **350**, 441 (1980).

[68] D. A. Epstein, P. F. Torrence, and R. M. Friedman, *Proc. Natl. Acad. Sci. U.S.A.* **77**, 107 (1980).

[69] S. L. Gupta, *J. Virol.* **29**, 301 (1979).

[70] W. K. Joklik, *Compr. Virol.* **1**, 231 (1974).

and the protein kinase. This is not the case. Reovirions (which contain genomic double-stranded RNA), reovirus cores (i.e., reovirions partially uncoated by treatment with chymotrypsin), and reovirus subviral particles (formed from reovirions in infected cells by cleavage and removal of some outer coat proteins) purified by centrifugation through CsCl gradients can replace double-stranded RNA in promoting RNA degradation in an extract from interferon-treated cells. However, in enhancing RNA cleavage the reovirions are only 4% as efficient (per milligram of double-stranded RNA) as free double-stranded RNA. Furthermore, reovirions treated with RNase III (an enzyme specific for cleaving double-stranded RNA), under conditions in which most free double-stranded RNA but little or no double-stranded RNA in virions is degraded, are no longer able to promote RNA cleavage. This suggests that double-stranded RNA on the surface of the reovirions may be responsible for the activation of the endonuclease$_{INT}$ system.[52,54] It remains to be established whether the double-stranded RNA on the surface of the virions is of cellular origin or is an exposed segment of the genomic RNA.

It is in accord with the above findings that reovirus subviral particles (isolated by CsCl density gradient centrifugation) from an extract of cells that had been treated with a partially purified interferon preparation manifest *in vitro* an endonuclease activity and produce mainly reovirus mRNAs shorter than full size.[71] It is probable that it is $(2'-5')(A)_n$ synthetase that is binding to and becoming activated by the double-stranded RNA of such subviral particles, and it is RNase L that degrades the mRNAs. Reovirus subviral particles isolated from cells not treated with interferon synthesize full-size reovirus mRNAs and have little or no endonuclease activity.

Possible Rationale for the Multiple Roles of Double-Stranded RNA in Interferon Induction and Action

It remains to be understood why double-stranded RNA has become an important modulator (i.e., both inducer and enzyme activator) of the interferon system. It is conceivable that double-stranded RNA serves as a signal revealing the presence of replicating viruses in the cell. This might be the case for viruses in the replication of which double-stranded RNA (even if partially coated) is a side product or intermediate.

As described in previous sections, the treatment of cells with interferon induces, among others, two latent enzyme systems. The activation of these requires double-stranded RNA in some form. This puzzling complexity might be rationalized in the following hypothesis.[55,57] Localized

[71] R. J. Galster and P. Lengyel, *Nucleic Acids Res.* 3, 581 (1976).

virus infection results in the synthesis of interferon in the infected cells and its secretion and spreading in the body. In the cells exposed to interferon the two enzyme systems are induced. Since they are latent, they do not impair cell metabolism. When the cells previously exposed to interferon become infected with a virus, however, this might result (in the formation of some double-stranded RNA and perhaps thereby) in the activation of the two enzyme systems. These in turn impair protein synthesis in the virus-infected cells. Whether or not the activator of these enzyme systems in intact cells is double-stranded RNA remains to be established.

Two biochemical phenomena elicited by interferon treatment of cells manifested in the cell extracts, even without double-stranded RNA, are discussed in the following sections.

Impairment of Reovirus mRNA Methylation

The 5' termini of most eukaryotic mRNAs, including those of reovirus mRNAs, are capped and methylated. The discovery that methylated, capped reovirus mRNAs are usually more efficient as messengers *in vitro* than unmethylated capped reovirus mRNAs[72] prompted a comparison of the process of mRNA methylation in an extract from interferon-treated cells with that in an extract from control cells. It was established that the methylation of unmethylated, capped reovirus mRNAs by the cellular methylating enzymes is impaired in extracts from interferon-treated cells.[73] The inhibitor is a macromolecule that is inactivated during the incubation. The impairment is not due to cleavage or irreversible inactivation of reovirus mRNA, to depletion of the methyl donor (*S*-adenosylmethionine), or to the irreversible inactivation of the methylating enzymes. It is possible that the inhibitor acts by binding to the unmethylated capped reovirus mRNAs and making the cap region unavailable for the methylating enzymes. It is conceivable that the impairment of methylation is only one of the manifestations of the "inhibitor(s) of methylation" and may not even be the most important one. Thus it is imaginable that the same agent(s) might also interfere with the process of attachment of mRNA to ribosomes. The impairment of methylation is apparently not affected by double-stranded RNA.[74]

The finding of an impairment of reovirus mRNA methylation *in vitro* prompted a comparison of the extent of cap methylation of reovirus mRNAs in intact interferon-treated cells and control cells. It was estab-

---

[72] A. J. Shatkin and G. W. Both, *Cell* 7, 305 (1976).
[73] G. C. Sen, B. Lebleu, G. E. Brown, M. A. Rebello, Y. Furuichi, M. Morgan, A. J. Shatkin, and P. Lengyel, *Biochem. Biophys. Res. Commun.* 65, 427 (1975).
[74] G. C. Sen, S. Shaila, B. Lebleu, G. E. Brown, R. C. Desrosiers, and P. Lengyel, *J. Virol.* 21, 69 (1977).

lished that the percentage of reovirus mRNAs with cap 2 termini (i.e., those carrying three methyl groups attached to the cap) is 35–50% lower in interferon-treated cells than in control cells.[75] The conversion of cap 1 into cap 2 termini (i.e., the attachment to the cap of the third methyl group) is catalyzed *in vivo* by cellular enzymes. It remains to be seen whether the decrease in the proportion of reovirus mRNAs with cap 2 termini in interferon-treated cells *in vivo* is mediated by the same inhibitor(s) that impair(s) methylation of reovirus mRNA caps *in vitro*. The methylation of the cap of vaccinia virus mRNAs is also impaired in cells treated with interferon.[76]

Impairment of Exogenous mRNA Translation: The tRNA Effect

This impairment is most pronounced in an extract from interferon-treated cells that had been "preincubated" to decrease protein synthesis directed by endogenous mRNA and passed through Sephadex G-25 to remove small molecules. It is mediated by an inhibitor loosely attached to ribosomes. The impairment is manifested in the cessation of translation of exogenous mRNA in an extract from interferon-treated cells at a time when the translation in an extract from control cells still continues. It is apparently a consequence of a block in peptide chain elongation, but is not due to the cleavage of the mRNA.

After the translation in the extract from interferon-treated cells has ceased, it can be restored by adding tRNA.[77] [79] In the course of studying the basis for the need for added tRNA, it was established that, in the extract from interferon-treated EAT cells which had been preincubated and passed through Sephadex G-25, certain species of aminoacyl tRNAs (e.g., leucyl-tRNAs) are inactivated faster than in a corresponding extract from control cells. This result makes it conceivable that the requirement for added tRNA to restore translation in an extract from interferon-treated cells may be due to a faster tRNA inactivation.[80,81] As noted in an earlier section, the treatment of cells with interferon increases the level of an enzyme that cleaves $(2'-5')(A)_n$; it was reported that the same enzyme also inactivates tRNAs by removal of their 3' terminal CCA residues.

[75] R. C. Desrosiers and P. Lengyel, *Biochim. Biophys. Acta* 562, 471 (1979).
[76] H. Kroath, H. J. Gross, G. Jungwirth, and G. Bodo, *Nucleic Acids Res.* 5, 2441 (1978).
[77] S. L. Gupta, M. L. Sopori, and P. Lengyel, *Biochem. Biophys. Res. Commun.* 57, 763 (1974).
[78] J. Content, B. Lebleu, A. Zilberstein, H. Berissi, and M. Revel, *FEBS Lett.* 41, 125 (1974).
[79] A. Zilberstein, B. Dudock, H. Berissi, and M. Revel, *J. Mol. Biol.* 108, 43 (1976).
[80] G. C. Sen, S. L. Gupta, G. E. Brown, B. Lebleu, M. A. Rebello, and P. Lengyel, *J. Virol.* 17, 191 (1976).
[81] R. Falcoff, E. Falcoff, J. Sanceau, and J. A. Lewis, *Virology* 86, 507 (1978).

Thus it is conceivable that it is the increase in the level of this enzyme that may account for the faster tRNA inactivation in extracts from interferon-treated cells.[18] It should be noted, however, that so far no significant difference was detected in the amount of tRNA needed to overcome the impairment of mRNA translation in the extracts of interferon-treated cells between tRNAs from interferon-treated and control cells.[80,82] Thus the *in vivo* relevance of these findings remains to be seen.

mRNAs and Proteins Induced by Interferon

The increase in the level of various enzymes after exposing cells to interferon, together with the fact that the establishment in cells of the antiviral state by interferons requires a functioning nucleus,[83] RNA synthesis[84] and protein synthesis, prompted investigators to test whether interferons induce the synthesis of new mRNAs and new proteins. The treatment of various human, mouse, and chicken cells with homologous interferons was found to induce several mRNAs and the corresponding proteins.[21-27] The correspondence was established by translating the mRNAs into the proteins *in vitro*.[23,26] The induction of both the mRNAs and proteins could be inhibited by actinomycin D indicating *de novo* synthesis. Short-term labeling at different times after treatment with interferons indicated that the mRNAs and proteins were induced within a few hours after the beginning of the treatment with interferon. The synthesis of the different proteins peaks at different times and thereafter declines.[27] The results make it likely that the interferon-induced mRNAs have short half-lives. The functions and identities of the interferon-induced proteins and their relationship to the enzymes whose level is increased by interferon treatment remain to be established. It is of interest that some of the proteins induced can be retained on poly(I) · poly(C) agarose columns indicating their high affinity to double-stranded RNA.[25]

Discussion and Conclusions

As described in preceding sections, most of our knowledge about the enzymology of interferon action is based on the comparison of enzyme activities in extracts from interferon-treated cells and extracts from control cells. This approach led to the following major conclusions. Interferon treatment of cells enhances the level of two distinct enzymic pathways

---

[82] C. Colby, E. E. Penhoet, and C. E. Samuel, *Virology* **74**, 262 (1976).

[83] K. L. Radke, C. Colby, J. R. Kates, H. M. Krider, and D. M. Prescott, *J. Virol.* **13**, 623 (1974).

[84] J. Taylor, *Biochem. Biophys. Res. Commun.* **14**, 447 (1964).

that remain latent unless activated by double-stranded RNA. One of the pathways, if activated, results in the accelerated cleavage of mRNA. This pathway involves two latent enzymes: $(2'-5')(A)_n$ synthetase and RNase L. If activated by double-stranded RNA, $(2'-5')(A)_n$ synthetase produces $(2'-5')(A)_n$ that in turn activates RNase L. The activation of RNase L ceases upon removal of $(2'-5')(A)_n$. The level of phosphodiesterase that can degrade $(2'-5')(A)_n$ is also raised by interferon treatment.

The second pathway of activation leads to the impairment of peptide chain initiation. This is apparently a consequence of the phosphorylation of chain initiation factor eIF-2. It remains to be established if the protein kinase system catalyzing this pathway consists of one or several enzymes. The significance of the fact that the same enzyme system can phosphorylate some histones is unclear and so is the relationship of the kinase to P_1 protein. This protein is present even in the most purified kinase preparations and becomes phosphorylated in the presence of double-stranded RNA. The impairment of peptide chain initiation resulting from the action of the kinase system can be reversed by a protein phosphatase that can remove phosphate from phosphorylated eIF-2.

The treatment of cells with interferons also boosts pathways whose functioning does not appear to depend on double-stranded RNA. Thus the methylation of the cap structure of mRNA is impaired in an extract from interferon-treated cells. The lability of the agents responsible for this impairment makes their characterization difficult.

Finally, in extracts from interferon-treated cells, especially if these have undergone gel filtration, the inactivation of some tRNA species is accelerated. This happens presumably as a consequence of the increase in the level of the phosphodiesterase, which in addition to cleaving $(2'-5')(A)_n$ can also remove the CCA termini from tRNA.

Some of the data that are in accord with a role of the two double-stranded RNA-dependent pathways and of the inhibition of mRNA cap-methylation in mediating some of the antiviral effects of interferons were outlined in the earlier sections. A definitive test for the function of each of the above pathways would require mutants that are (at least conditionally) defective in the pathway.

The following features of interferon-inducible enzymes and of their actions need further comments: Agents other than interferon can also affect the level of the enzymes that can be induced by interferons. Thus withdrawal of estrogen from chick oviducts results in a severalfold increase in $(2'-5')(A)_n$ synthetase.[42] This observation, together with the fact that the level of RNase L is high even in cells not treated with interferon, makes it probable that the $(2'-5')(A)_n$ synthetase–RNase L system has functions besides that of mediating interferon action.

The levels of both protein kinase and $(2'-5')(A)_n$ synthetase are remarkably high in reticulocytes from rabbits not treated with interferon.[41,64] It remains to be seen whether or not this is a consequence of an exposure of these cells to interferon in the bone marrow.

At present the only characterized activity of $(2'-5')(A)_n$ in cell-free systems is the activation of RNase L. However, it is not known whether or not RNase L activation can account for the impairment of DNA synthesis in mitogen-stimulated lymphocytes by $(2'-5')(A)_n$ and by $(2'-5')(A)_n$ from which the 5'-terminal triphosphate moiety had been removed[32] (see also footnote 56).

As noted earlier the increase in the activity of the various enzymes upon interferon treatment is most likely a consequence of the induction of the synthesis of mRNAs and proteins. However, neither the mechanism of this induction nor the details of the early steps of interferon action have been elucidated. The earliest effect of interferon treatment of cells reported so far is a short-lived increase in the level of cyclic GMP. This takes place within 5–10 min after exposing the cells to interferon.[31] Neither the significance nor the mechanism of this increase has been established.

It is likely that interferons must interact with the outside of the cell surface to initiate the conversion of cells into the antiviral state.[85,86] So far, however, no interferon receptor has yet been isolated although genetic and biochemical data point to its existence.[87-89] Furthermore, it is not clear whether interferons act only at the cell surface or whether they must enter the cells to exert some or all of their functions.[90]

Acknowledgment

Studies in the author's laboratory were supported by research grants from the U.S. National Institutes of Health.

[85] V. E. Vengris, B. D. Stollar, and P. M. Pitha, *Virology* **65**, 410 (1975).

[86] P. Lebon, M.-C. Moreau, L. Cohen, and C. Chany, *Proc. Soc. Exp. Biol. Med.* **149**, 108 (1978).

[87] M. Revel, D. Bash, and F. Ruddle, *Nature (London)* **260**, 139 (1976).

[88] F. Besancon and H. Ankel, *Nature (London)* **250**, 784 (1974).

[89] C. Chany, *Biomedicine* **24**, 148 (1976).

[90] R. M. Friedman, *in* "Interferon" (I. Gresser, ed.), Vol. 1, pp. 53–74. Academic Press, New York, 1979.

[20] Interferon-Induced Enzymes: Microassays and Their Applications; Purification and Assay of (2'-5')-Oligoadenylate Synthetase and Assay of 2'-Phosphodiesterase

By M. REVEL, D. WALLACH, G. MERLIN, A. SCHATTNER, A. SCHMIDT, D. WOLF, L. SHULMAN, and A. KIMCHI

At least three enzymic activities are elevated in cells that have been exposed to interferon: a double-stranded (ds) RNA-activated protein kinase (PK-i) phosphorylating initiation factor eIF-2's small subunit, the dsRNA-dependent (2'-5')-oligoadenylate synthetase (E), and a 2'-phosphodiesterase (2'-PDi), which splits the (2'-5')ppp(A2'p)$_n$A oligonucleotide formed by the synthetase.[1,2] Microassays have been developed for these enzymes and have become important tools for interferon research.[1,3-5] The specificity of the induction process allows the elevation in these enzymes to be used as an assay for interferon. The capacity of various cells to respond to a given interferon can be studied as well. Some cells have a higher basal level of these enzymes than others, and the enzyme level can also vary with the physiological state of the culture or of the organism from which the cells are taken. In our work, we have found several applications for these enzyme measurements; these are summarized in this section with a detailed description of the assay procedures used. The experiments described here are concerned with the (2'-5')-oligoadenylate synthetase, which shows the largest variations, and with the 2'-phosphodiesterase. Uses of the protein kinase PK-i can be found in previous publications.[1,6]

Assays of the (2'-5')-Oligoadenylate Synthetase

The (2'-5')-oligoadenylate produced by this enzyme inhibits protein synthesis in cell-free systems by activating a specific ribonuclease, F.[7-9]

[1] A. Kimchi, L. Shulman, A. Schmidt, Y. Chernajovsky, A. Fradin, and M. Revel, *Proc. Natl. Acad. Sci. U.S.A.* **76**, 3208 (1979).

[2] M. Revel, in "Interferon" (I. Gresser, ed.), Vol. 1, pp. 101–163. Academic Press, New York, 1979.

[3] G. R. Stark, W. J. Dower, R. T. Schimke, R. E. Brown, and I. M. Kerr, *Nature (London)* **278**, 471 (1979).

[4] C. Baglioni, P. A. Maroney, and D. K. West, *Biochemistry* **18**, 1765 (1979).

[5] M. A. Minks, S. Benvin, P. A. Maroney, and C. Baglioni, *J. Biol. Chem.* **254**, 5058 (1979).

[6] A. Kimchi, A. Zilberstein, A. Schmidt, L. Shulman, and M. Revel, *J. Biol. Chem.* **254**, 9846 (1979).

[7] I. M. Kerr and R. E. Brown, *Proc. Natl. Acad. Sci. U.S.A.* **75**, 256 (1978).

This property can be used to assay the synthetase. Only trimers, pppA2'-p5'A2'p5'A, or longer chains, however, have biological activity. When small amounts of synthetase are used, the products of the reaction are often mainly dimers pppA2'p5'A, which would not be detectable by the biological assay.[8] We have, therefore, used direct biochemical assays of the oligoadenylate produced to measure the level of the synthetase with a minimum number of cells. When microcultures of 20,000–30,000 cells are used, the best results are obtained if the cells are lysed with a nonanionic detergent such as Nonidet P-40 (NP40). Less enzyme activity is recovered if NP40 is omitted, possibly because some of the enzyme is membrane-bound.

Principle

The incorporation of $[\alpha\text{-}^{32}P]ATP$ into pppA2'p5'A or longer chains, is measured. The enzyme in crude cell extracts is subjected to a one-step purification by adsorption to poly(rI) · poly(rC)-agarose beads.

Materials

Microtest plate for cell culture (96 wells)
Wash buffer: 35 mM Tris · HCl, pH 7.6; 140 mM KCl; 3 mM magnesium acetate
Buffer B: 20 mM HEPES buffer, pH 7.5; 120 mM KCl; 5 mM magnesium acetate; 7 mM β-mercaptoethanol; 10% glycerol
Lysis buffer: 0.5% Nonidet P-40 in buffer B
Agarose-polyribonosinic-polyribocytidylic acid (AG poly(I) · poly(C) type 6; P-L Biochemicals)
Polyethyleneimine-cellulose thin-layer plates (Schleicher and Schuell)
$[\alpha\text{-}^{32}P]ATP$ (The Radiochemical Centre, Amersham, UK)
Alumina powder: acid WAI (Sigma)
Alumina column buffer: 1 M glycine, 0.9 N HCl, pH 2
Electrophoresis paper: Whatman 3 MM
A2'p5'A (Sigma)

Procedure

Cell Cultures. Mouse L cells or human fibroblasts (or any other suitable cell line) are grown as confluent monolayers in microtest plates (2 to 3

[8] A. Schmidt, A. Zilberstein, L. Shulman, P. Federman, H. Berissi, and M. Revel, *FEBS Lett.* **25**, 257 (1978).
[9] E. Slattery, N. Ghosh, H. Samanta, and P. Lengyel, *Proc. Natl. Acad. Sci. U.S.A.* **76**, 4778 (1979).

$\times 10^4$ cells in 0.1 ml of minimal essential medium with 2 mM L-glutamine, 5–10% fetal calf serum, and 100 units of penicillin, 0.1 mg of streptomycin and 0.4 mg of gentamycin per milliliter) at 37° in a 95% air–5% CO_2 incubator. Medium is removed from each well by gentle suction and replaced by 0.075–0.1 ml of fresh medium containing mouse or human interferon (e.g., 10 units/ml) or any solution in which the presence of interferon is to be detected. The plates are returned to the 37°/CO_2 incubator for at least 6 hr (usually 8 hr for mouse cells and 14 hr for human cells).

Extraction. Plates are cooled to 4° and inverted on blotting paper to remove medium. Each well is washed 3 times with wash buffer. The plate is drained at a 45° angle for 5 min to remove the last drops of wash buffer. Lysis buffer (0.03–0.05 ml) is added to each well. The plates are covered with an adhesive plastic sheet to seal each well, and a few hand-strikes are applied on the side of the plate to detach the cells. After 5–10 min, the plates are centrifuged in an International centrifuge with plate adaptors, at 2000 rpm (r = 18 cm). The supernatant from each well is transferred to Eppendorf microtest 1.5-ml tubes, which already contain poly(rI) · poly(rC)-agarose beads (see below). Each well of the plate is washed with lysis buffer, according to the above procedure, and the second supernatant is combined with the first.

Adsorption to Poly(rI) · poly(rC)-Agarose Beads. Before use, the beads (suspended in 0.02% azide) are washed three times with buffer B. Introduce 25 μl of the 1 : 1 (v/v) bead suspension in buffer B into Eppendorf microtest tubes. Centrifuge and remove the remaining buffer solution with a capillary. To the packed beads add the NP40 extracts prepared as above. Mix well and leave 10 min at 30°. Add 1 ml of buffer B to each tube, mix well, and centrifuge in an Eppendorf microfuge (8000 g) for 2 min. Remove completely the supernatant from the packed beads. The washing with buffer B may be repeated.

Assay of Oligoadenylate Synthesis. To the packed beads in each tube, a 10-μl reaction mixture is added that contains 5 mM [α-^{32}P]ATP (100 μCi/μmol), 2.5 mM dithiothreitol, 10–20 mM magnesium acetate, and 25 mM KCl. The tubes are sealed and incubated at 30° for 2–20 hr (depending on the activity expected; in practice we use 18 hr for cells treated with 1–10 units of interferon per milliliter). To each tube, 8 μl of 20 mM Tris · HCl, pH 8.0, are added, and the beads are sedimented by centrifugation. A 10-μl aliquot of the supernatant is treated with 4 μl of bacterial alkaline phosphatase (100 units/ml in 0.17 M Tris base) for 30 min at 37°. A second 30-min treatment with an additional 2 μl of phosphatase is recommended. Synthetic (2'-5')ApA and ApApA(30 μg) are added as carrier, and the reaction mixture is applied to Whatman 3 MM paper sheets (80 × 46 cm). High-voltage electrophoresis is carried out at 3000 V for 4 hr in a Savant varsol refrigerated tank. The spots of (2'-5')-oligo(A) markers are visualized under ultraviolet light, cut out, and counted. Alternatively, the electrophoresis paper can be subjected to autoradiography before counting. A typical separation is shown in Fig. 1.

When high enzyme activity is expected, the phosphatase-treated products may be separated by thin-layer chromatography on polyethyleneimine-cellulose plates with 1 M acetic acid for 1 hr at room temperature. The dimers run faster than trimers and longer oligonucleotides.

Assay of Oligoadenylate with Alumina Columns. For assays involving a large number of samples to be analyzed, we have developed a rapid assay on alumina columns. The assay is based on the fact that with poly(rI) · poly(rC)-bound enzyme preparations, obtained as described above, the only phosphatase-resistant material is the mixture of

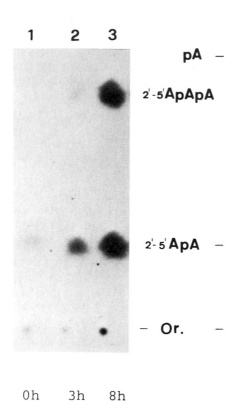

FIG. 1. Assay of (2'-5')-oligoadenylate synthetase. Mouse L cells were exposed to 200 units of mouse interferon per milliliter (10^7 units per milligram of protein) for 0, 3, and 8 hr (lanes 1–3). The assay was carried out as described and the ^{32}P-labeled, phosphatase-treated products were analyzed by high-voltage electrophoresis at pH 3.5. Left: autoradiography; right: ultraviolet absorbance. Or. = origin.

oligoadenylates. Alumina powder binds strongly free phosphate[10] and has been used for the fractionation of various cyclic and oligonucleotides.[11]

In the present assay, aliquots of 2–10 μl of the alkaline phosphatase-treated reaction products[12] (described in the preceding section) are applied on 0.3-ml alumina columns, previously washed with 1 ml of 1 M glycine · HCl buffer, pH 2. To the columns, 3 ml of the same buffer are applied and the liquid effluent is collected in scintillation vials and counted in the ^{3}H channel of a scintillation counter (Cerenkov radiation). With properly packed columns, elution can be completed within 5 min and a large number of samples can be processed together.

At pH 2, between 95 and 100% of the oligoadenylate dimer and trimer cores are recovered, whereas only 0.1–0.3% of the free radioactive phosphate is eluted. Up to 20% of the tetramer and pentamer cores may be retained on the column under these conditions, but since in most reactions with small amounts of enzyme the long oligomers represent only a small fraction of the products, these losses do not impair the assay. Analysis of the oligoadenylate products, before and after the column, is illustrated in Fig. 2. Examples of assays with the alumina column procedure are given in Table I. The assay is very sensitive, and treatment of cells with 0.1 unit of interferon per milliliter increases the synthetase level enough to give a significant signal in this system.

Special Applications

Comparison of Diploid and Trisomic 21 Human Fibroblasts

Human fibroblasts trisomic for chromosome 21 are more sensitive to the antiviral action of interferon, presumably because chromosome 21 carries a gene for the human interferon receptor.[13] With a trisomic 21 cell line such as GM-2504 (from the human genetic mutant cell repository, Institute for Medical Research, Camden, New Jersey), the microassay for (2′-5′)-oligoadenylate synthetase can be carried out exactly as described above for diploid fibroblasts. Enzyme induction is much higher in trisomic cells than in diploid cells, in particular at low concentrations of interferon (Table II).

Assay of in Vitro Translation Products of Interferon mRNA

The activity of *in vitro* synthesized interferon can be assayed by measuring induction of the (2′-5′)-oligoadenylate synthetase. Translation

---

[10] H. Kubli, *Helv. Chim. Acta* **30**, 453 (1947).
[11] J. Ramachandran, *Anal Biochem.* **43**, 227 (1971).
[12] G. Merlin, M. Revel, and D. Wallach, *Anal. Biochem.* **110**, 190 (1981).
[13] M. Revel, D. Bash, and F. H. Ruddle, *Nature (London)* **260**, 139 (1976).

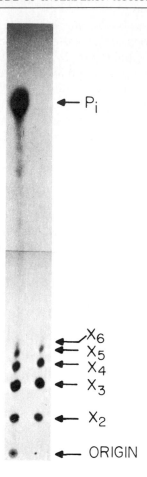

FIG. 2. Purification of (2'-5')-oligoadenylates on alumina columns. The ^{32}P-labeled, phosphatase-treated products of a (2'-5')-oligoadenylate synthetase assay with human Namalva cell extracts were subjected directly to paper electrophoresis and autoradiography (lane 1) or passed through an alumina column prior to electrophoresis (lane 2). X_2–X_6 designates $(A2'p)_n A$ oligomers where n is 1–5. P_i is inorganic phosphate.

reactions in reticulocyte lysates are carried out in 25 or 100 μl as described by Weissenbach *et al.*[14] After 60–90 min of incubation, the translation reaction mixture is centrifuged for 10 min at 4° in an Eppendorf microfuge. An aliquot of 20 μl is diluted 10- to 15-fold in minimum essential medium with 10% fetal calf serum, and 100 μl are applied in duplicate to the cell cultures in the wells of a microtest plate as described above. It

[14] J. Weissenbach, M. Zeevi, T. Landau, and M. Revel, *Eur. J. Biochem.* **98**, 1 (1979).

TABLE I
ASSAY OF (2'-5')-OLIGOADENYLATE SYNTHETASE BY ALUMINA PROCEDURE AND
PAPER ELECTROPHORESIS

| | | (2'-5')-Oligoadenylate synthetase activity $[^{32}P](A2'p)_n5'A$ | |
Cells	Human leukocyte interferon treatment (units/ml)	Alumina procedure (cpm)	Paper electrophoresis (cpm)
Bovine MDBK cells[a]	None	130	180
	50	3850	3710
	250	5210	4430
Human Namalva cells[b]	None	1180 ± 40	
	0.1	2750 ± 50	
	1	4120 ± 440	
	10	7300 ± 490	
	100	14,020 ± 240	
	1000	23,910 ± 910	

[a] Confluent monolayers exposed for 18 hr to human leukocyte interferon as described in the text.
[b] Lymphoblastoid Namalva cells (10^6 cells in 1 ml) exposed 24 hr to interferon and extracted in 0.1 ml of lysis buffer.

is recommended to use the high-voltage electrophoresis procedure for this assay. A standard solution of interferon (10 units/ml) is used as reference.

The same assay can be performed with *Xenopus laevis* oocyte extracts that have been injected with interferon mRNA. After incubation of 10 oocytes for 40 hr in 0.1 ml of Barth's medium, as described by Colman and

TABLE II
(2'-5')-OLIGOADENYLATE SYNTHETASE ASSAY IN HUMAN DIPLOID AND
TRISOMIC 21 FIBROBLASTS[a]

| Human fibroblast interferon (units/ml) | (2'-5')Oligoadenylate synthetase $[^{32}P](A2'p)_n5'A$ | |
	Diploid cells (FS11) (cpm)	Trisomic 21 cells (GM2504) (cpm)
0	200	260
1	790	2350
5	1760	3310
10	2460	5300
40	3300	5400

[a] Monolayers of 30,000 cells in microtest plates were used for both cell types. The high-voltage electrophoresis procedure was used in this experiment.

TABLE III

INDUCTION OF (2'-5')-OLIGOADENYLATE SYNTHETASE BY *in Vitro* TRANSLATION
PRODUCTS OF INTERFERON MESSENGER RNA[a]

Expt. No.	Addition to cell cultures (human fibroblast FS-11)	(2'-5')-Oligoadenylate synthetase [^{32}P](A2'p)$_n$5'A (cpm)
1	Reticulocyte lysate translation products (diluted 1 : 15).	
	Without mRNA	75
	With poly A$^+$-mRNA from induced cells	2,630
	With poly A$^+$-mRNA from uninduced cells	150
	Human fibroblast interferon, 10 units/ml	13,850
2	*Xenopus* oocyte incubation medium (diluted 1 : 60).	
	Injected with Barth medium	0
	Injected with poly A$^+$ mRNA from induced cells	2,070
	Human fibroblast interferon 10 units/ml	4,570

[a] A background value for untreated cells of 885 cpm in Expt. 1 and of 235 cpm in Expt. 2 was subtracted.

Morser,[15] the Barth medium is recovered and the oocytes are homogenized in 0.1 ml of minimum essential medium with 10% fetal calf serum. An aliquot of oocyte incubation medium and of oocyte homogenate is diluted 15-fold or more and added to the microtest plate cultures as above. Examples of such assays are given in Table III.

Assay of (2'-5')-Oligoadenylate Synthetase in Human Blood Samples

Mononuclear Cell Extracts (Lymphocytes). Venous blood samples of 2 ml are collected with heparin, diluted with an equal volume of PBS (0.15 M NaCl, 0.015 M phosphate buffer, pH 7.0) and applied on a 3-ml cushion of Ficoll–Paque (Pharmacia Fine Chemicals). After centrifugation for 30 min at 1600 rpm, the cells at the interphase are collected, washed twice with 5 ml of PBS, and counted. Approximately 10^7 cells are transferred to an Eppendorf microtest tube, and the pelleted cells are lysed in 1 ml of buffer B with 0.5% NP40, at 4°. The cell extract is recovered by centrifugation for 6 min at 8000 g.

Polymorphonuclear Cell Extracts (Granulocytes). The cells at the bottom of the Ficoll–Paque centrifugation (see above) are resuspended in 1 volume of 0.15 M NaCl; a 0.25-volume of 6% (w/v) dextran (T250, Pharmacia Fine Chemicals) in 0.15 M NaCl is added, and erythrocytes are sedimented at 1 g for 30 min at room temperature. The supernatant is

[15] A. Colman and J. Morser, *Cell* **17**, 517 (1979).

centrifuged at 250 g for 10 min, and the cell pellet is suspended in 34 mM NaCl for 45 sec to lyse residual erythrocytes. Isotonicity is restored with an equal volume of 0.27 M NaCl, the cells are centrifuged at 250 g for 10 min, then washed twice in PBS and lysed as described for lymphocytes above.

Assay of Oligoadenylate Synthetase in Blood Cells. The procedure is similar to that for tissue culture cells. Poly(rI) · poly(rC)-agarose beads (10 μl for each reaction tube) are washed with 1 ml of 10 mM HEPES buffer, pH 7.5, 50 mM KCl, 5 mM MgCl$_2$, 7 mM dithiothreitol, and 20% glycerol (buffer C). Aliquots (10 μl) of cell extracts, or appropriate dilutions in buffer B, are mixed with the pelleted beads and incubated for 15 min at 30°. The beads are washed with 1 ml of buffer C, and all liquid is carefully removed. A reaction mixture of 10 μl containing 10 mM HEPES buffer pH 7.5, 5 mM MgCl$_2$, 7 mM dithiothreitol, 10% glycerol, 2.5 mM [α-^{32}P]ATP (100–300 μCi/μmol), 3 mg of creatine kinase per milliliter, 10 mM creatine phosphate, and 40 μg of poly(rI) · poly(rC) per milliliter are added. After incubation for 8 hr or 20 hr at 30°, 10 μl of bacterial alkaline phosphatase (100 units/ml in 140 mM Tris base) are added directly, and incubation is continued for 1 hr at 37°. Water (20 μl) is added to each tube, and, after centrifugation in an Eppendorf microfuge, 10–20 μl of the supernatant are applied to the alumina columns as described above.

Human peripheral lymphocytes have a high (2'-5')-oligoadenylate synthetase activity. With 50 μg of protein from crude extract, 2.5 nmol of ATP are converted into oligoadenylate in 8 hr. This value, can, however, increase 5- to 10-fold in patients receiving interferon injections, or in various viral diseases, whereas the values are decreased in some cases of acute lymphatic leukemias (unpublished data). Granulocytes have a low enzyme activity. Determination of the (2'-5')-oligoadenylate synthetase in blood cells may become an important diagnostic tool with widespread medical applications.

Purification of (2'-5')-Oligoadenylate Synthetase

An outline for the purification of the enzyme from human lymphoblastoid Namalva cells is given. Cells are taken after Sendai virus infection, which induces interferon and the interferon-induced enzymes. Homogenization of cells is performed in 0.05% NP40, 20 mM HEPES buffer, pH 7.5, 1 mM dithiothreitol (DTT), 10% glycerol. Centrifugation at 10,000 g for 60 min yields a supernatant (S$_{10}$) that is centrifuged at 100,000 g for 4–5 hr to obtain an S$_{100}$ supernatant. The ribosomal pellet is resuspended in 0.5 M KCl, 20 mM HEPES buffer, pH 7.5, 1 mM DTT, and 10% glycerol with vigorous mixing for 4 hr. After sedimentation of ribosomes as above, the ribosomal wash is added to the S_{100}, and the solution is diluted to 50 mM KCl in the same buffer.

DEAE-Cellulose Step. A W-52 DEAE-cellulose column (Whatman) is used. The nonadsorbed fraction is recovered.

Phosphocellulose Step. After adjusting the pH to 6.7, the fraction is absorbed onto a PC-11 (Whatman) column, which is washed with 20 mM HEPES buffer, pH 6.7, 150 mM KCl, 1 mM DTT, and 5% glycerol. Elution is carried out with a 0.3 to 0.75 M KCl gradient, the enzyme eluting around 350 mM KCl.

Blue Sepharose Step. The fraction is diluted to 150 mM KCl with 20 mM HEPES buffer, pH 7.5, 1 mM DTT, and 5% glycerol and applied to a Cibacron Blue Sepharose column (Pharmacia Fine Chemicals). Elution is carried out with a 0.15 to 0.8 M KCl gradient, the enzyme eluting around 350 mM KCl.

Gel Filtration. For this step, the protein may be concentrated by adsorption to a small phosphocellulose column and eluted with a minimal volume of 20 mM HEPES buffer, pH 7.5, 500 mM KCl, 1 mM KCl, 1 mM DTT, and 5% glycerol. This is applied to a P-100 column (Bio-Rad) or a Sephacryl S-200 column (Pharmacia Fine Chemicals) equilibrated in the same buffer with 150 mM KCl. The activity from Namalva cells elutes from the gel as a 30,000 molecular weight (MW) protein. With low protein concentration, we have observed some retardation of the enzyme on Sephacryl S-200. With mouse L cells, the activity is excluded from P-100 and behaves as a 90,000 MW protein. With HeLa cells, the enzyme gives two peaks of 70,000 and 30,000 MW.

The purification data are summarized in Table IV.

Assay of the Enzyme during Purification

The assay is done in solution, in a final volume of 0.02 ml with 15 mM Hepes buffer, pH 7.5, 90 mM KCl, 20 mM MgCl$_2$, 1 mM dithiothreitol, 8% glycerol, 1 mM [α-^{32}P]ATP (150 μCi/μmol) and 250 μg of poly(rI) · poly(rC) per milliliter. After incubation for 1–3 hr at 30°, the reaction is treated with alkaline phosphatase and analyzed by thin-layer chromatography on polyethyleneimine cellulose as described above.

Assay of 2′-Phosphodiesterase

This enzyme degrades the 2′-5′-oligoadenylate, yielding ATP and 5′-AMP:

$$pppA2'p5'A2'p5'A \rightarrow pppA + (pA)_2$$

It requires a free 3′ end; it shows a marked preference for 2′-5′-oligonucleotides of various base composition, but cleaves also 3′-5′-

TABLE IV

PURIFICATION OF (2'-5')-OLIGOADENYLATE SYNTHETASE E FROM INTERFERON-TREATED MOUSE L CELLS AND HUMAN NAMALVA CELLS INDUCED FOR INTERFERON PRODUCTION[a]

Step	Interferon-treated L cells	Interferon-induced Namalva cells		
	Specific activity (nmol AMP polymerized/ mg protein/hr)	Specific activity (nmol AMP polymerized/ mg protein/hr)	Total protein (mg)	Yield of activity (%)
1. S$_{10}$	0	11	44,000	100
2. S$_{100}$	0.7	18	36,000	134
3. DEAE-cellulose	30	120	15,080	377
4. Phosphocellulose	340	875	1,020	184
5. Gel filtration	14,100	50,900	17.1	180
6. Blue Sepharose	N.D.	91,665	4.8	91

[a] The mouse enzyme was assayed with 1 μg of poly(rI) · poly(rC) per milliliter; the human enzyme, with 250 μg/ml. The increase in the total activity at step 3 is due to the removal of inhibitors of the synthetase.

TABLE V
2'-PHOSPHODIESTERASE ACTIVITY[a]

Expt. No.	Cell source	Conditions	2'-Phosphodiesterase (units/mg protein)
1	Monkey kidney cells BSC-1	Growing cells	43
		Confluent, serum starved	10
		Confluent, serum stimulated	25
2	Mouse L cells	Untreated monolayers	36
		At 8 hr after mouse interferon, 5 units/ml	182
		At 8 hr after mouse interferon, 125 units/ml	230

[a] In experiment 1, [^{32}P]pppA2'p5'A2'p5'A was used as substrate; in experiment 2, unlabeled A2'p5'A was used.

oligonucleotides. The enzyme removes the CCA terminus of tRNAs, but does not degrade the tRNA itself. Purification and properties of this enzyme were described by Schmidt *et al.*[16] We describe here the methods we use to assay this enzyme in crude cell extracts. These methods established that the 2'-phosphodiesterase increases in cells treated with interferon[1,16] and is 5 times higher in fast growing cells or mitogen-stimulated lymphocytes than in resting cells[17,18] (see Table V).

Principle

The degradation of enzymically synthesized [^{32}P]pppA2'p5'A2'p5'A or of chemically synthesized A2'p5'A is measured. Alternatively, the release of 5'-AMP from these substrates is assayed.

Procedure

Preparation of Radioactive Substrate. (2'-5')-Oligoadenylate synthetase E purified as outlined in the preceding section is used. The reaction mixture contains in 0.15 ml, 12 μg of enzyme E, 17 mM HEPES buffer, pH 7.5, 100 mM KCl, 4 mM magnesium acetate, 1 mM dithiothreitol, 8% glycerol, 1 mM [α-^{32}P]ATP (150 μCi/μmol), and 20 μg of poly(rI) · poly(rC) per milliliter. After 20 hr of incubation at 30°, the reaction is stopped by heating for 10 min to 60°. Aliquots (10 μl) of the incubation mixture [containing approximately 400 pmol or 20,000 cpm of an equimo-

[16] A. Schmidt, Y. Chernajovsky, L. Shulman, P. Federman, H. Berissi, and M. Revel, *Proc. Natl. Acad. Sci. U.S.A.* **76**, 4788 (1979).
[17] A. Kimchi, H. Shure, and M. Revel, *Nature (London)* **282**, 849 (1979).
[18] A. Kimchi, H. Shure, and M. Revel, *Eur. J. Biochem.* **114**, 5 (1981).

lar mixture of (2'-5')pppApA and pppApApA] can be used as substrates for the 2'-phosphodiesterase. It is, however, as easy to separate the different oligomers by paper electrophoresis at pH 3.5 (as described above), and to elute the oligomers by a simple procedure. The spots are cut out from the paper and introduced in an Eppendorf 1.5-ml microtest tube with a hole pierced at the bottom. The tube is fitted on top of a second tube, the paper is wetted with water, and the two tubes are centrifuged. The procedure is repeated three times, and the eluate in the bottom tube is lyophilized.

Preparation of Crude Cell Extracts. For fibroblasts (e.g., monkey BSC-1 cells), monolayer cultures containing 6×10^6 cells are washed twice with 35 mM HEPES buffer, pH 7.5, 140 mM KCl, 3 mM MgCl$_2$, and then lysed with 1 ml of 20 mM HEPES buffer, pH 7.5, 120 mM KCl, 5 mM MgCl$_2$, 1 mM dithiothreitol, 10% glycerol containing 0.5% NP40. The lysates are centrifuged for 6 min in an Eppendorf microfuge (8000 g), and the supernatants are stored in liquid air. For lymphocytes, similar extracts were made with 5 to 10×10^8 cells for 1 ml of lysate.

Assay of 2'-Phosphodiesterase. Reaction mixtures (20 μl) containing 2–20 μg of protein from crude cell extracts and the radioactive substrates (prepared as above) with 15 mM HEPES buffer, pH 7.5, 90 mM KCl, 3.5 mM MgCl$_2$, 0.7 mM dithiothreitol, 7% glycerol, are incubated at 30° for 1 hr. To 10 μl, 0.2 unit of bacterial alkaline phosphatase is added, with 30 mM Tris base, and incubation is continued at 37° for 30 min. Samples are then subjected to thin-layer chromatography on polyethyleneimine cellulose sheets with 1 M acetic acid. After autoradiography, the spots of (2'-5')ApA and ApApA are cut out and counted. The residual amount of oligonucleotide is subtracted from that of a control reaction without 2'-phosphodiesterase.

One unit of 2'-phosphodiesterase is defined as the degradation of 1 nmol of (2'-5')pppApA or conversion of 1 nmol of (2'-5')-pppApApA into (2'-5')pppApA in 60 min at 37°.

If chemically synthesized unlabeled A2'p5'A (1–2 mM) is used as substrate, the activity is measured by the production of 5'-AMP. After incubation, the reaction is analyzed by thin-layer chromatography on polyethyleneimine cellulose with 0.32 M LiCl. The 5'-AMP spots (R_f 0.25 relative to A2'p5'A) are cut and eluted with 0.7 M MgCl$_2$ in 20 mM Tris-HCl, pH 7.5, and the absorbance is determined at 260 nm. Examples of assays are given in Table V.

[21] Assays and Procedures to Study Changes in cAMP Concentrations after Interferon Treatment of Cells

By R. K. Maheshwari, P. M. Lad, and Robert M. Friedman

Several laboratories have examined the possible connection between cAMP and interferon (IF) action. Although the original report[1] that cAMP potentiates interferon's antiviral activity has been confirmed by other workers,[2-4] the detailed mechanistic connections between cAMP and IF action are unclear. For example, cAMP could act by increasing IF production[5] or by affecting pathways of viral uncoating or viral assembly.[6] The cell surface is important in several phases of interferon action.[7] When a cell is treated with an IF to which it is sensitive, an intracellular antiviral state develops. During this process IF appears to bind to specific receptors on the cell surface. This binding is necessary, but it is not sufficient for the development of antiviral activity. Interferon binding in turn appears to bring about chemical, immunological, physical, and morphological alterations in the plasma membrane, the meaning of which is presently unclear; however, these may be related to various aspects of the antiviral activity of IF. The expression of some phases of antiviral activity itself seems to be regulated at the cell surface. While the inhibition of replication of most viruses takes place at the level of virus-directed translation, the inhibition of murine leukemia viruses and also vesicular stomatitis virus, both of which bud from the plasma membrane, seems to be distinctly different: in IF-treated cells, protein synthesis directed by these viruses does not appear to be inhibited, and either virus release by cells was inhibited or virus of markedly decreased infectivity was produced.[6]

Possibly connected with the effects mentioned above is the stimulation of cAMP by IF.[8] Yet no definitive effects of IF on adenylate cyclase, the enzyme that produces cAMP from MgATP, have been reported. The need

[1] I. Pastan and R. M. Friedman, *Biochem. Biophys. Res. Commun.* **36**, 735 (1969).

[2] J. M. Weber and R. B. Stewart, *J. Gen. Virol.* **28**, 363 (1975).

[3] L. B. Allen, N. C. Eagle, J. H. Huffman, D. A. Shuman, R. B. Meyer, Jr., and R. W. Sidwell, *Proc. Soc. Exp. Biol. Med.* **146**, 580 (1974).

[4] I. Mecs, M. Kasler, A. Lozsa, and G. Kovacs, *Acta Microbiol. Acad. Sci. Hung.* **21**, 265 (1974).

[5] R. M. Friedman, *Bacteriol. Rev.* **41**, 543 (1977).

[6] R. M. Friedman, R. K. Maheshwari, F. Jay, and C. Czarniecki, *Ann. N. Y. Acad. Sci.* **350**, 533 (1980).

[7] R. M. Friedman, *Pharmacol. Ther. Part A* **2**, 725 (1978).

[8] R. M. Friedman and L. Kohn, *Biochem. Biophys. Res. Commun.* **70**, 1078 (1970).

for such studies is made even more noticeable by findings that IF causes changes in the physical properties of the plasma membrane. Thus, the effects, stimulatory as well as inhibitory, on cAMP production could arise from a nonspecific perturbation of the plasma membrane. Alternatively, the possibility exists that IF may be regulating cyclase and cAMP production in a receptor-dependent fashion, as has been reported in great detail for both the beta[9] and glucagon receptors.[10] Functional analogies between IF and other known receptor-related regulatory patterns would in turn imply a cell surface receptor type of mechanism for IF action with all its attendant implications.[11] As direct, functionally related binding studies of IF with the cell surface have not yet been possible, such studies on the cyclase would seem to be all the more essential.

In this chapter we describe procedures for the determination of cAMP and adenylate cyclase that would be required in such studies. We have also outlined those types of experimental approaches that would be useful in establishing or dismissing connections between cAMP and IF action.

Procedure for cAMP Radioimmunoassay[12]

The principle of this procedure is the competitive interaction of labeled and unlabeled cAMP with antibody to cAMP. The reagents for this assay, the buffers, iodinated modified cAMP, and antibody to cAMP are available from NEN (New England Nuclear). A protocol for the use of these reagents in cAMP radioimmunoassay is provided below.

A curve for the standard is constructed by the following steps.

1. Use 50 μl of cAMP standard (0–1000 fmol of cAMP of per 50 μl).
2. Mix 5 μl of triethylamine–acetic anhydride (TME) (2 : 1) at room temperature.
3. Add 50 μl of [^{125}I]cAMP (10,000 cpm per assay tube) in 50 mM sodium acetate buffer, pH 6.3.
4. Add 200 μl of diluted antiserum in 50 mM sodium acetate buffer, pH 6.3.
5. Incubate overnight at 4°.
6. Add 1 ml of sodium acetate buffer (50 mM, pH 6.3) and mix.
7. Centrifuge in a table-top centrifuge for 20 min.
8. Decant tube, and count pellet in tube in a gamma counter. Repeat procedure for cAMP unknown.

[9] P. M. Lad, T. B. Nielson, M. S. Preston, and M. Rodbell, *J. Biol. Chem.* **255**, 988 (1980).
[10] P. M. Lad, A. F. Welton, and M. Rodbell, *J. Biol. Chem.* **252**, 5942 (1977).
[11] M. Rodbell, *Nature (London)* **284**, 17 (1980).
[12] G. Brooker, J. F. Harper, W. L. Terasaki, and R. D. Moylan, *Adv. Cyclic Nucleotide Res.* **10**, 1 (1979).

The cAMP may be extracted from the tissue or cell suspension by boiling the sample in water, centrifuging to remove debris, and using the supernatant in the assay as described above. Extraction with trichloroacetic acid or diethyl ether saturated with water is equally efficient.

Analysis of Data

A standard curve is constructed of $\log B/B_0$ vs log cAMP. The cAMP in the unknown is read from these standard curves. B_0 represents the amount bound in the absence of any standard; B represents the amount bound in the presence of standard. Statistical analysis of the data for linearized logit/log plots is also available.[13]

Membrane Preparations. L cells were grown in 150 cm² petri plates to confluency (2×10^7) and treated with IF in Eagle's MEM for different periods of time. The media were then aspirated, cells were washed with PBS twice and scraped, and membrane was prepared. Numerous methods for membrane preparation now in use include (*a*) homogenization of cell or tissue; (*b*) nitrogen cavitation; (*c*) osmotic lysis.

The lytic procedure has to be separately designed for each type of cell. Generally, cells that prefer solid substrates for attachment and growth, and are small and lenticular in shape, present problems. Homogenization is probably best for such cells. For larger cells that can be grown in suspension, osmotic lysis or nitrogen cavitation would be the preferred methods. These are not general prescriptions, but pointers from our experience.

Sucrose gradients (discontinuous, continuous, or flotation type) have also been employed. In general, a pilot study of sedimentation at different times to equilibrium will show which type of separation is required.

Interferon treatment is generally carried out in the presence of a complex growth medium that contains hormones and could activate or inhibit adenylate cyclase. Thus, the effect of IF on cAMP could be due to impairment or stimulation by these hormones and may not reflect the primary action of IF itself. For these reasons, the diagnostic tests of adenylate cyclase activities mentioned below might be relevant.

Adenylate Cyclase Assay[14]

Adenylate cyclase activity is measured by monitoring the conversion of $[\alpha\text{-}^{32}\text{P}]\text{ATP}$ to cAMP. The cAMP is separated from ATP by chromatography on Dowex and alumina.

[13] D. Rodbard and D. M. Hutt, *Radio-Immunoassay and Related Procedures in Medicine* **1**, 165 (1974).

[14] Y. Salomon, D. Londos, and M. Rodbell, *Anal. Biochem.* **58**, 541 (1974).

Solutions

Assay cocktail: 0.2 mM [³H]ATP; 40 mM Tris · HCl, pH 7.5; 20 mM MgCl₂; 2 mM DTT. This solution can be made up in advance in volumes up to 125 ml, dispensed in 2.5-ml aliquots into glass tubes, and frozen. Each tube is sufficient for 50 assays. Prior to assay a regenerating system, 9 mg of creatine phosphate (Sigma) and 1 mg of creatine kinase (Sigma) are added to each tube.

cAMP standard: A mixture of 2.5 ml of 1 M Tris (pH 7.5), 1 mM cAMP, and [³H]cAMP is diluted to 500 ml. The solution should contain 20,000 cpm/50 μl and is stored frozen.

Stopping solution: 8 g of sodium dodecyl sulfate (SDS), 10 g of ATP (Grade II Sigma), and 170 mg of cAMP in 400 ml. Adjust to pH 7.6 with 2 M Tris. This solution should be kept frozen.

Chromatography

Dowex AG-50 W-X4, is washed with water until the effluent is color free. A volume of water equal to the resin volume is added: 2 ml of the slurry are poured into glass wool-stoppered columns. The columns are washed with 2 ml of 1 N HCl. Alumina columns are prepared by pouring 0.5 g of alumina with a measuring spoon into glass wool-stoppered columns. Columns are washed with 25 ml of 0.1 M imidazole HCl buffer, pH 7.3.

Incubations

The assay is carried out by the addition of 50 μl of assay cocktail solution I, 10 μl of required ligand, and 40 μl of membrane suspension to a glass tube. The protein concentration in the assay can vary from as little as 1 μg (e.g., the adipocyte membrane) to 30 μg (rat liver plasma membrane). All additions may be carried out at 0°. The reaction is performed at either 24° or 30° for 10–15 min. The reaction is stopped by adding 100 μl of stopping cocktail to each tube. The tubes are then boiled for 1 min and cooled; 1 ml of water is added to each tube. The tubes are then decanted onto Dowex columns. The tubes are removed. Each column is washed twice with 1 ml of water. The Dowex column rack is mounted on the alumina rack and 3 ml of water are added to the Dowex columns. After the 3 ml of water have passed through the alumina columns, the Dowex column rack is removed. The alumina rack is placed on top of a tray of scintillation vials containing 12 ml of scintillant (Aquasol or Hydrofluor). The columns are eluted with 4 ml of 0.1 M imidazole. The vials are capped, thoroughly shaken until a solid gel is formed, and then counted in a liquid scintillation spectrometer. The Dowex columns are regenerated with 2 ml of 0.1 N HCl. The alumina columns are washed with 8 ml of 0.1 M imidazole, pH 9.3.

Calculations

The amount of cAMP produced is calculated by subtracting the machine background and blank values (zero time values) from each count. The counts are then normalized for recovery by multiplying each count by the ratio of the ^{3}H counts in the standard to the ^{3}H counts in the sample. The normalized counts are then converted into picomoles by dividing with the specific activity of the ATP used in the assay (generally 10,000 cpm/pmol). Protein determinations are carried out by the Lowry method. Activities are expressed as picomoles of cAMP in 15 min per milligram of membrane protein.

General Precautions

Temperature and Divalent Cation Concentration.[15] The preferred temperature range when hormones that inhibit cyclase are being studied is 16° to 24°. Higher temperatures eliminate these inhibitory effects; 30° is preferred for ligands such as hormones, guanine nucleotides, and fluoride ion. A divalent cation concentration of 10 mM Mg^{2+} is adequate. For assays of intrinsic catalytic unit activity 1 mM Mn^{2+} is used. Concentrations of metal ions higher than these will lead to complex modulations of nearly all the kinetic responses that have been studied.

Chelators.[15] These should be excluded from the assay wherever possible. EDTA as well as EGTA produce dramatic and ill-understood effects on adenylate cyclase. If a chelator is present in the stock protein solution, it should be removed either by washing the membranes or, if a solubilized preparation is used, by dialysis.

Substrate.[16] ATP from equine muscle contains appreciable amounts of GTP. Purification of this ATP on Dowex 50, is strongly advised. Alternatively, ATP synthesized from adenosine can be used. Low concentrations of substrate (10–100 μM) are recommended.

Impurities in Membrane Preparations. A number of hormones can be present bound as isolated to their receptors in membranes. A quick diagnostic test is a high "basal" activity relative to the activity in the presence of fluoride ion. Pretreatment of membrane with GTP to dissociate bound hormone may be necessary. A purified membrane preparation is advised.

Adenosine.[17] ATP can be converted to adenosine, which is a powerful modulator of several cyclase systems. The deoxy ATP system should be used if such a problem is suspected. (Deoxyadenosine does not produce the same modulatory effects as does adenosine.) A simple test for whether such an assay is required is to examine the effect of any added theophylline or adenosine deaminase.[17]

[15] H. Yamamura, P. M. Lad, and M. Rodbell, *J. Biol. Chem.* **252**, 7964 (1977).
[16] M. C. Lin, A. F. Welton, and M. F. Berman, *J. Cyclic Nucleotide Res.* **4**, 159 (1978).
[17] D. M. F. Cooper and C. Londos, *J. Cyclic Nucleotide Res.* **5**, 289 (1979).

Analysis of Adenylate Cyclase Activity

It has been shown that the adenylate cyclase system consists of at least three components: the hormone receptor R, the nucleotide component N, which binds guanine nucleotides with the high affinity, and the catalytic unit C. The guanine nucleotide unit N regulates both R and C and is in turn regulated by these components. Although it is not necessary to describe the entire functioning of this system here, a simple version of this system and its use in analyzing plasma membrane events are described below.

Hormones activate the enzyme adenylate cyclase through their receptors by modifying the interactions of guanine nucleotides at the N unit. Hormones alone do not produce any activation, but hormone and GTP produce maximal activation. The detailed mechanistic events have been analyzed by several workers.[18] From these studies, it is now clear that the various parameters that can be measured can be assigned to appropriate structures; selective perturbations of certain types of activities can therefore be of diagnostic value. In addition, it should be mentioned that if interferon is acting through a receptor mechanism even vaguely analogous to other receptors, two simple predictions should hold: (a) the time course of cAMP production in the intact cell should be compatible with the stimulation of the enzyme adenylate cyclase; and (b) interferon stimulation of the enzyme in the isolated membrane should be GTP dependent.

Listed in the tabulation are the various cyclase activities that are frequently measured and the structures that they probably represent in the membrane.[11]

Activity measured	Probable structure
a. GTP regulation of hormone binding	RN
b. Hormone displacement of guanine nucleotide	RN
c. Gpp(NH)p stimulation of enzyme	NC
d. Fluoride stimulation of enzyme	NC
e. Mn ATP basal activity	C
f. Hormone + GTP activation	RN/NC
g. Mg basal activity	RN/NC
h. Hormone + Gpp(NH)p activation	RN/NC
i. Hormone binding	R
j. Capacity to confer activation on acceptor system, by fluoride and Gpp(NH)p	N
k. Labeling by cholera toxin	N

[18] P. M. Lad, M. S. Preston, A. F. Welton, T. B. Nielson, and M. Rodbell, *Biochim. Biophys. Acta* 551, 368 (1979).

Two types of tests are required to establish the correlation between interferon and its stimulation of cAMP in the cell: (*a*) direct effects by known receptor mechanisms in activating cyclase; (*b*) indirect effect in perturbing other hormone-mediated responses on cyclase.

In addition, the effects of viral assembly or disassembly in the plasma membrane on the organization of the cyclase also need to be investigated. Unless correlations such as these are made, and a close connection between cAMP enhancement and influences on cyclase is established, the exact place of cAMP in the overall regulatory pattern cannot be clearly understood. The methodology and the comments in this chapter may be useful to workers interested in the relationship (if any) between interferon, cAMP, and adenylate cyclase.

[22] Procedures for Measurement of Phosphorylation of Ribosome-Associated Proteins in Interferon-Treated Cells[1]

By CHARLES E. SAMUEL

Posttranslational modifications such as protein phosphorylation provide an important mechanism by which the functional activity of proteins can be controlled and, hence, biological processes regulated.[2-4] Interferon inhibits the replication of a wide range of animal viruses[5]; interferon treatment also mediates an enhanced selective protein phosphorylation of at least two ribosome-associated proteins.[6-9] In many virus–host systems, the level of genome expression inhibited as a result of interferon treatment

[1] This work was supported in part by research grants from the National Institute of Allergy and Infectious Diseases (AI-12520) and the American Cancer Society (MV-16), and by a Research Career Development Award from the U.S. Public Health Service.
[2] E. G. Krebs and J. A. Beavo, *Annu. Rev. Biochem.* **48**, 923 (1979).
[3] P. Greengard, *Science* **199**, 146 (1978).
[4] G. Taborsky, *Adv. Protein Chem.* **28**, 1 (1974).
[5] W. E. Stewart, II, "The Interferon System." Springer-Verlag, Vienna and New York, 1979.
[6] A. Zilberstein, P. Federman, L. Shulman, and M. Revel, *FEBS Lett.* **68**, 119 (1976).
[7] C. E. Samuel, D. A. Farris, and D. A. Epstein, *Virology* **83**, 56 (1977).
[8] B. Lebleu, G. C. Sen, S. Sharla, B. Cabrer, and P. Lengyel, *Proc. Natl. Acad. Sci. U.S.A.* **73**, 3107 (1976).
[9] W. K. Roberts, A. Hovanessian, R. E. Brown, M.-J. Clemens, and I. M. Kerr, *Nature (London)* **264**, 477 (1976).

is the accumulation or translation, or both, of viral messenger RNAs.[10,11] The inhibition of viral protein synthesis observed in cell-free extracts prepared from interferon-treated murine and human cells is due, at least in part, to the presence of a ribosome-associated inhibitor of translation.[12–17] The ribosome-associated proteins selectively phosphorylated in response to interferon treatment may play a role in regulating the initiation of translation of viral mRNAs in interferon-treated cells.[10,11]

A cyclic AMP (cAMP)-independent protein kinase present in the translational inhibitory ribosomal salt-wash fractions prepared from interferon-treated cells catalyzes the $[\gamma\text{-}^{32}P]$ATP-mediated phosphorylation of the 38,000-dalton subunit of protein synthesis initiation factor 2 (eIF-2α) as well as the phosphorylation of a ribosome-associated protein designated P_1.[18,19] Phosphorylation of protein P_1 is one of the most sensitive biochemical markers of interferon action and is significantly enhanced in interferon-treated as compared to untreated cells.[6–9,18–23] The apparent molecular weight of P_1 from interferon-treated human cells (amnion U[19] and foreskin fibroblast,[20] ~69,000 daltons) is generally slightly greater than that from interferon-treated murine cells (L929,[19] CCL1,[6] and L$_e$[21] fibroblast, ~67,000 daltons; Ehrlich ascites tumor,[8] ~64,000 daltons). The interferon-mediated phosphorylation of both eIF-2α and P_1 is dependent upon the presence of RNA with double-stranded character.[6–9,18–23] However, double-stranded RNA (dsRNA) also affects the phosphorylation of at least two additional proteins, designated P_{ds}, that are not dependent upon interferon treatment; the P_{ds} phosphorylations are observed in untreated human and murine cells as well as interferon-treated cells.[18,19] In addition to P_1 and eIF-2α, interferon-treated human cells also possess a ribosome-associated phosphorylated protein of relatively low molecular

[10] C. E. Samuel, S. M. Kingsman, R. W. Melamed, D. A. Farris, M. D. Smith, N. G. Miyamoto, S. R. Lasky, and G. S. Knutson, *Ann. N. Y. Acad. Sci.*, **350**, 473 (1980).

[11] C. Baglioni, *Cell* **17**, 255 (1979).

[12] C. E. Samuel, this volume [37].

[13] C. E. Samuel and W. L. Joklik, *Virology* **58**, 476 (1974).

[14] E. Falcoff, R. Falcott, B. Lebleu, and M. Revel, *J. Virol.* **12**, 421 (1973).

[15] S. L. Gupta, M. L. Sopori, and P. Lengyel, *Biochem. Biophys. Res. Commun.* **54**, 777 (1973).

[16] C. E. Samuel and D. A. Farris, *Virology* **77**, 556 (1977).

[17] U. Mayr, C. Parajsz, C. Jungwirth, and G. Bodo, *Virology* **88**, 55 (1978).

[18] C. E. Samuel, *Proc. Natl. Acad. Sci. U.S.A.* **76**, 600 (1979).

[19] C. E. Samuel, *Virology* **93**, 281 (1979).

[20] J. Wérenne and M. Revel, *Arch. Int. Physiol. Biochem.* **86**, 471 (1978).

[21] D. A. Epstein, P. F. Torrence, and R. M. Friedman, *Proc. Natl. Acad. Sci. U.S.A.* **77**, 107 (1980).

[22] G. C. Sen, H. Taira, and P. Lengyel, *J. Biol. Chem.* **253**, 5915 (1978).

[23] A. Kimchi, L. Shulman, A. Schmidt, Y. Chernavorsky, A. Fradin, and M. Revel, *Proc. Natl. Acad. Sci. U.S.A.* **76**, 3208 (1979).

weight, about 16,000, designated P_f. This protein is not observed in untreated human cells or in interferon-treated murine cells; P_f phosphorylation is not affected by dsRNA.[18,19]

Thus, four classes (I–IV) of cAMP-independent protein phosphorylations may be observed when the $[\gamma\text{-}^{32}P]ATP$-mediated reactions are catalyzed by preparations from untreated as compared to interferon-treated cells: (I) The net level of phosphorylation is significantly enhanced in extracts from interferon-treated as compared to untreated cells, and the enhancement is dependent upon the presence of dsRNA; examples are eIF-2α and P_1. (II) The net level of phosphorylation is significantly enhanced in extracts from interferon-treated as compared to untreated cells, and the enhancement is dsRNA-independent; example is P_f. (III) The net level of phosphorylation is comparable in extracts from interferon-treated as compared to untreated cells, but is significantly enhanced in the presence of dsRNA; examples are at least two P_{ds} species. (IV) The net level of phosphorylation is comparable in extracts from interferon-treated as compared to untreated cells, both in the presence and in the absence of dsRNA. Procedures are described below for the measurement of these phosphorylations of ribosome-associated proteins in interferon-treated as compared to untreated murine and human cells.

Materials

Reagents. All chemicals used in the following procedures are of reagent grade. Buffers are prepared, and the pH is measured at 25° unless otherwise indicated.

$[\gamma\text{-}^{32}P]ATP$, 15–1000 Ci/mmol (New England Nuclear, or prepared by the method of Glynn and Chappell[24] as modified by Maxam and Gilbert[25])

Reovirus genome double-stranded RNA, 25 μg/ml (prepared from purified reovirions by the method of Ito and Joklik[26])

Reovirus single-stranded RNA, 1 mg/ml (prepared as the *in vitro* transcription product by the method of Skehel and Joklik[27] and purified as described by Samuel and Joklik[13])

Mouse fibroblast interferon (prepared from mouse L-929 fibroblast cells induced with Newcastle disease virus as described by Samuel and Joklik[13])

Human leukocyte interferon (supplied by Dr. K. Cantell and prepared from leukocytes induced with Sendai virus[28])

[24] I. M. Glynn and J. B. Chappell, *Biochem. J.* 90, 147 (1964).
[25] A. M. Maxam and W. E. Gilbert, this series, Vol. 65 [57].
[26] Y. Ito and W. K. Joklik, *Virology* 50, 189 (1972).
[27] J. J. Skehel and W. K. Joklik, *Virology* 32, 822 (1969).
[28] K. Cantell, this series, Vol. 78 [4] and [41].

Protein synthesis initiation factor eIF-2 (preparations supplied by Drs. W. C. Merrick and I. M. London, and purified as previously described.[29])

N-2-Hydroxyethylpiperazine-N'-2-ethanesulfonic acid (HEPES), 1.0 M, pH 7.5

Tris · HCl, 1.5 M, pH 8.8, and 1.0 M, pH 6.8

KCl, 3 M

Mg(OAc)$_2$, 0.5 M

Dithiothreitol (DTT), 0.1 M (Sigma)

ATP, 0.1 M (Sigma or P-L Biochemicals)

Glycerol

Nonidet P-40 (NP40) (Shell)

2-Mercaptoethanol (Mallinckrodt)

Sodium dodecyl sulfate (SDS), 10% w/v (Matheson, Coleman and Bell; recrystallized from ethanol and washed with ether prior to use)

Ammonium persulfate, 30 mg/ml (solution made just prior to use)

Acrylamide (Eastman; recrystallized from chloroform prior to use)

N,N'-Methylene bisacrylamide (Eastman, recrystallized from acetone prior to use)

N,N,N',N'-Tetramethylethylenediamine (Eastman or Sigma)

Trichloroacetic acid, 50% w/v

Isopropanol

Acetic acid

Cell Culture, Interferon Treatment, and Cell Harvest. Clone L-929 mouse fibroblast cells and human amnion U cells are routinely grown in roller culture (1330-cm^2 growth area/glass bottle) in Eagle's MEM containing 7% fetal calf serum or, more recently, 5.6% newborn calf serum and 1.4% fetal calf serum. Confluent cells are treated at 37° with 300 units of interferon per milliliter for 18 hr before harvesting and harvested without trypsinization by scraping into cold (0–4°) isotonic buffer (35 mM Tris · HCl, pH 6.8 at 25°; 146 mM NaCl; 11 mM glucose). Cells are then washed three times in isotonic buffer and, after the last wash, are suspended in a small volume of isotonic buffer, transferred to a graduated conical centrifuge tube, and pelleted at 800 g for 10 min in an IEC CRU-5000 refrigerated centrifuge. The volume of the cell pellet is recorded, and then as much buffer as possible is carefully removed with a pipette.

Preparation of Cell-Free S$_{10}$ Extracts by Dounce Homogenization. All operations, unless otherwise indicated, are carried out at 0–4°; preparations should be kept on ice whenever possible, and ice-cold, degassed buffers prepared with deionized, glass-distilled water should be utilized in all steps. Suspend the cell pellet in three packed-cell volumes of hypotonic

[29] W. C. Merrick, this series, Vol. 60 [8].

buffer (10 mM HEPES, pH 7.5; 15 mM KCl; 1.5 mM Mg(OAc)$_2$; 1 mM DTT), allow to swell for 10 min, and then disrupt with 30–50 strokes of a tight-fitting Dounce homogenizer. Immediately add 0.1 volume of 10× buffer (100 mM HEPES, pH 7.5; 1050 mM KCl; 35 mM Mg(OAc)$_2$; 10 mM DTT), and centrifuge the homogenate at 5000 g for 5 min in a Sorvall RC-5 centrifuge and an SS-34 rotor; discard the pellet, and centrifuge the supernatant solution at 10,000 g for 10 min. The resulting supernatant solution, the S$_{10}$ extract, can be stored at $-90°$ in small propylene vials for subsequent phosphorylation analysis or, alternatively, can be used directly for the preparation of ribosome and cell-sap fractions.

Preparation of Ribosome, Ribosomal Salt-Wash, and Cell-Sap Fractions. Ribosome and cell-sap fractions are separated by centrifugation of non-preincubated S$_{10}$ extracts at 150,000 g for 2.5 hr in a DuPont Sorvall OTD-50 ultracentrifuge and a T865.1 rotor. The cell-sap fraction (the upper two-thirds of the supernatant solution remaining after the 150,000 g centrifugation) is passed through a Sephadex G-25 (medium mesh, Pharmacia Fine Chemicals, Inc.) column equilibrated with 20 mM HEPES buffer, pH 7.5, containing 120 mM KCl, 1 mM Mg(OAc)$_2$, and 1 mM DTT, and stored in small portions at $-90°$.

The crude, translucent ribosome pellet is gently suspended in 20 mM HEPES buffer, pH 7.5, containing 120 mM KCl, 5 mM Mg(OAc)$_2$, and 1 mM DTT, stirred slowly for 1 hr, and centrifuged for 2.5 hr at 150,000 g; the resulting supernatant solution is defined as the 0.12 M ribosomal salt-wash. The 0.12 M salt-washed ribosomal pellet is suspended, stirred, and centrifuged as before except that the buffer is modified to contain 0.3 M KCl; the resulting supernatant solution is defined as the 0.3 M ribosomal salt-wash. The 0.3 M salt-washed ribosomal pellet is suspended and stirred for 1 hr as before except that the buffer is modified to contain 0.8 M KCl; the suspension is centrifuged overnight at 120,000 g. The resulting supernatant solution, defined as the 0.8 M ribosomal wash, as well as the 0.12 M and 0.3 M ribosomal salt-washes, are all centrifuged for 4 hr at 120,000 g and then dialyzed against 250 volumes of 20 mM HEPES buffer, pH 7.5, containing 50 mM KCl, 1 mM Mg(OAc)$_2$, 1 mM DTT, and 10% (v/v) glycerol. The dialysis is routinely carried out for 18 hr with two changes of buffer. After centrifugation of the dialyzed ribosomal washes at 5000 g for 5 min, the resulting supernatants are removed, divided into small aliquots, and stored at $-90°$.

Preparation of Cell-Free S$_{10}$ Extracts by Nonidet P-40 Lysis. The preparation of cell-free S$_{10}$ extracts by NP40 lysis rather than by Dounce homogenization is relatively more convenient when large numbers of samples and small numbers of cells in monolayer culture are involved. As before, all extraction and fractionation procedures are carried out at $0-4°$

unless otherwise indicated. Cells in monolayer culture are washed three times in isotonic buffer; particular care should be taken consistently to remove all isotonic buffer from the tissue culture (T.C.) dish following the third wash. The T.C. dish is then floated on an ice-water bath, and the cells are disrupted by addition of NP40 lysis buffer [20 mM HEPES, pH 7.5; 120 mM KCl; 1 mM DTT; 1 mM Mg(OAc)$_2$; 10% (v/v) glycerol; and 0.5% (v/v) NP40], 1.6 ml per 150-mm T.C. dish and 0.3 ml per 60-mm T.C. dish. After agitation for 10 min, the T.C. dish is scraped and the disrupted cells are transferred to a conical centrifuge tube and centrifuged at 800 g for 10 min in an IEC CRU-5000 refrigerated centrifuge. The supernatant fraction is then centrifuged for 10 min at 10,000 g in a Sorvall RC-5 centrifuge and a SS-34 rotor. The resulting supernatant solution, the S$_{10}$ extract, is stored in small aliquots at $-90°$.

Assay for Phosphorylation of Ribosome-Associated Proteins

In vitro [γ-^{32}P]ATP-mediated phosphorylation of endogenous ribosome-associated proteins and of purified protein synthesis factor eIF-2 catalyzed by protein kinases present in untreated and interferon-treated murine and human cells is determined by analyzing ^{32}P-labeled products by SDS–polyacrylamide slab-gel electrophoresis and autoradiography.

Reaction Mixture. The standard reaction mixture (26 μl) contains 20 mM HEPES (pH 7.5), 100 mM KCl, 4 mM Mg(OAc)$_2$, 1.5 mM DTT, 100 μM ATP containing 5–10 μCi of [γ-^{32}P]ATP, and, as indicated, reovirus genome dsRNA (1 μg/ml), 0.8 M ribosomal salt-wash fraction or S$_{10}$ cell-free extract possessing the interferon-induced kinase(s) (routinely 2.5–10 μg of protein), and purified protein synthesis initiation factor eIF-2 (0.5 μg of protein). After incubation at 30° for the time indicated, the reaction mixture is stopped by addition of 30 μl of 2× extraction buffer (0.125 M Tris · HCl, pH 6.8; 2.5% SDS; 0.42 M 2-mercaptoethanol; 25% glycerol; and 0.001% bromophenol blue) and boiled for 2 min.

Analysis of ^{32}P-Labeled Protein Products. The ^{32}P-labeled products of the *in vitro* protein phosphorylation reactions are analyzed by electrophoresis followed by autoradiography. Standard conditions for electrophoresis using a discontinuous Tris · glycine-buffered SDS–polyacrylamide gel system are essentially as described by Laemmli[30] and modified by Studier.[31] Slab gels, 0.75 mm thick, with a 5% acrylamide stacking layer containing 10–25 sample wells and a 10% acrylamide running layer are routinely used. Electrophoresis is performed at 20 mA,

[30] U. K. Laemmli, *Nature (London)* 227, 680 (1970).
[31] F. W. Studier, *J. Mol. Biol.* 79, 237 (1973).

normally for approximately 3 hr. After electrophoresis, gels are fixed in a mixture of 10% trichloroacetic acid–25% isopropanol, washed with 7% acetic acid, and dried onto Whatman filter paper under vacuum and over steam heat. Autoradiography is then performed with DuPont Cronex or Kodak X-Omat RP or SB5 film, routinely without the use of an intensifying screen.

The relative amount of [^{32}P]phosphate transfered to the various proteins present in the reaction mixtures can be quantitated either by densitometer scanning of a series of autoradiographs of varying exposure times or by measuring the radioactivity content of the individual bands excised from the gel with the film as a template.

Results and Discussion

Results of typical reactions illustrating the effect of interferon treatment and RNA with double-stranded character on the phosphorylation of endogenous ribosome-associated proteins as well as purified protein synthesis initiation factor eIF-2 are shown in Figs. 1–3.

As shown in Fig. 1, phosphorylation of a component designated P_1 present in the 0.8 M ribosomal wash fraction prepared both from human amnion U and from mouse L-929 cells treated with interferon for 18 hr is greatly enhanced as compared to the phosphorylation obtained with the comparable wash fractions prepared from untreated cells. The relative mobility on SDS–polyacrylamide gels of P_1 from U cells is slightly slower than that of P_1 from L-929 cells. Interferon-treated human cells also possess a smaller ribosome-associated phosphorylated protein, designated P_f, that is not observed either in untreated human cells or in interferon-treated murine cells. The phosphorylation of both human and murine P_1 is dependent upon the presence of dsRNA; phosphorylation of human P_f is not affected by dsRNA. Reovirus genome dsRNA at a concentration of 1 μg/ml also, however, enhances the phosphorylation of two additional ribosome-associated proteins designated P_{ds} that are not dependent upon interferon treatment.

When cells are rapidly chilled before harvesting, and when extracts are kept ice cold and not preincubated prior to ribosome isolation, essentially all (>90%) of the detectable P_1 protein is ribosome-associated; the apparent amounts of P_1 in the cell-sap fraction are minor when compared to the ribosome salt-wash fractions. Most of the protein P_1 is dissociated from ribosomes with salt concentrations between 0.3 and 0.8 M KCl, although the distribution of P_1 in the various ribosomal salt-washes between 0.3 and 0.8 M (for example, 0.3 to 0.5 M as compared to 0.5 to 0.8 M) varies between preparations and depends upon the concentration of ribosomes during washing, efficiency of suspension prior to stirring, and length of time stirred.

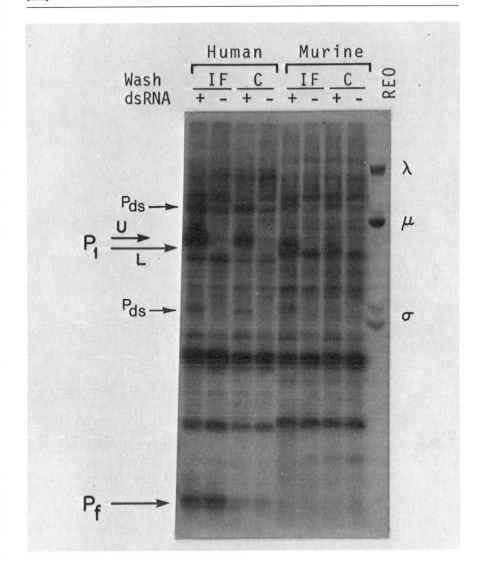

FIG. 1. Autoradiogram showing the effect of interferon treatment on $[\gamma\text{-}^{32}P]$ATP-mediated phosphorylation of 0.8 M KCl ribosomal salt-wash fractions (wash) prepared from untreated (C) and interferon-treated (IF) human amnion U and mouse fibroblast L-929 cells. The *in vitro* phosphorylation assay was carried out in the presence (+) or the absence (−) of 1 μg of reovirus genome dsRNA per milliliter. The incubation mixture (26 μl) contained 7.5 μg of protein of the indicated ribosomal salt wash. Incubation was for 30 min at 30°. U and L indicate phosphoprotein P_1 in human U and murine L-929 cells, respectively. Arrows indicate the positions of $[^{32}P]$phosphorylated proteins P_1 and P_f (interferon-dependent) and P_{ds} (interferon-independent), REO, $[^{14}C]$leucine-labeled reovirion (γ, μ, σ) marker proteins. From Samuel.[19]

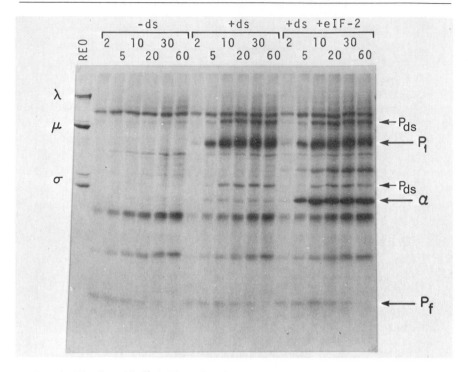

FIG. 2. Kinetics of [γ-^{32}P]ATP-mediated phosphorylation of proteins P_1 and P_f and of eIF-2α catalyzed by ribosomal salt-wash fraction prepared from interferon-treated human cells. The reaction mixture contained 0.8 M ribosomal salt-wash (12.5 μg of protein), reovirus genome dsRNA (1 μg/ml), and eIF-2 (2.5 μg of protein) as indicated. Reaction volume, 125 μl; 15-μl aliquots were analyzed after incubation at 30° for 2, 5, 10, 20, 30, or 60 min. Symbols are as described for Fig. 1. From Samuel.[18]

The effect of incubation time of the reaction mixture on the ability of the interferon-mediated human kinase activity to catalyze the phosphorylation of endogenous proteins P_1 and P_f in the presence and the absence of dsRNA is shown in Fig. 2. Phosphorylation of human P_1 is clearly dsRNA dependent; in the absence of dsRNA, no P_1 phosphorylation is detectable after 60 min of incubation. In the presence of dsRNA, the phosphorylation of P_1 and of purified eIF-2α show similar kinetics and are maximal by 10 min of incubation under standard reaction conditions; the amount of radioactivity associated with P_1 and eIF-2α does not change significantly between 20 and 60 min. By contrast, the accumulation of phosphorylated P_f is generally more rapid than P_1 and eIF-2α, and the amount of radioactivity associated with P_f decreases appreciably between 20 and 60 min both in the presence and in the absence of dsRNA.

The major phosphoester linkage for the human, as well as murine, phosphoprotein P_1 is O-phosphoserine.[18] In addition, the tryptic

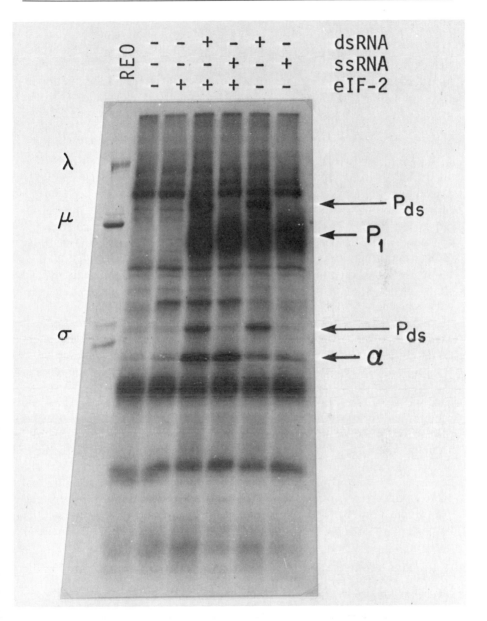

FIG. 3. Effect of reovirus dsRNA and ssRNA on phosphorylation of eIF-2α and of proteins P_1 and P_{ds} present in the 0.8 M ribosomal salt-wash fraction. The incubation mixture contained 2.7 μg of wash fraction protein from interferon-treated human cells and, where indicated, eIF-2 (0.5 μg of protein), reovirus genome dsRNA (1 μg/ml), and reovirus ssRNA (60 μg/ml). Incubation was for 15 min at 30°. Symbols are as described for Fig. 1. From Samuel.[18]

[^{32}P]phosphopeptide pattern observed for eIF-2α phosphorylated by the interferon-mediated human kinase is indistinguishable from the pattern obtained for eIF-2α phosphorylated by the hemin-regulated rabbit reticulocyte kinase.[18] The phosphorylation of eIF-2α, as well as the phosphorylation of the endogenous ribosome-associated protein P$_1$, is dependent upon the duration of interferon treatment,[7,19,23] correlates with the appearance of the ribosome-associated inhibitor of translation,[7,19] and shows a dose dependence similar to that for the development of the antiviral state.[23]

The interferon-mediated phosphorylation of protein P$_1$ and of eIF-2α shows an apparent dependence upon dsRNA (Fig. 2); poly(I) · poly(C), available commercially from P-L Biochemicals, can also be used in place of reovirus genome RNA as a source of dsRNA for the interferon-dependent dsRNA-dependent phosphorylation of P$_1$.[5,8] Reovirus dsRNA at a concentration near 0.2 μg/ml activates the phosphorylation of P$_1$ by 50%; we routinely use a concentration of 1 μg/ml of dsRNA, although concentrations as high as 10 μg/ml do not inhibit P$_1$ phosphorylation.[32] As shown in Fig. 3, the *in vitro* activation of the interferon-mediated kinase by reovirus genome dsRNA is also obtained with reovirus ssRNA purified by repeated LiCl precipitation. At a concentration of ssRNA typical of that used in cell-free protein synthesizing systems, 60 μg/ml,[7,13] the phosphorylation of protein P$_1$ and of eIF-2α is comparable to that observed in the presence of reovirus dsRNA. However, the interferon-independent P$_{ds}$ protein phosphorylations are observed only in the presence of reovirus genome dsRNA, not in the presence of reovirus ssRNA (Fig. 3). Thus, the requirement for dsRNA per se may not be absolute in that different RNAs possessing presumably different effective concentrations of double-stranded character—for example, synthetic poly(I) · poly(C), reovirus genome dsRNA, and transcribed reovirus ssRNA—appear to affect the net measurable level of [^{32}P]ATP-mediated phosphorylation of individual proteins differently in preparations from untreated and interferon-treated cells (Figs. 1 and 3). The different effects of reovirus dsRNA and ssRNA on protein phosphorylation (Fig. 3) suggest that either different kinase or phosphatase activities, or both, may be affected by the viral RNAs. *In vivo,* the apparent requirement for RNA with double-stranded character could conceivably be fulfilled by double-stranded regions of viral mRNAs resulting from the secondary structure of the ssRNAs or, alternatively, by dsRNA per se synthesized during replication of RNA virus genomes or by symmetrical transcription of dsDNA.

---

[32] S. R. Lasky and C. E. Samuel, unpublished observations, 1979.

[23] Measurement of Inhibition of Phosphoprotein Phosphatase Activity by Poly(I) · Poly(C) in Extracts of Interferon-Treated Cells

By David A. Epstein, Paul F. Torrence, and Robert M. Friedman

Double-stranded RNA (dsRNA) is a potent inhibitor of protein synthesis in reticulocyte lysates[1] as well as in extracts of interferon (IF)-treated cells.[2] A dsRNA-activated protein kinase that catalyzes the phosphorylation of the eIF-2 α-subunit, as well as the concomitant phosphorylation of a ribosome-associated protein, P_1, of molecular weight 67,000 has been partially purified from reticulocyte lysates[3] and IF-treated cells.[4] The function of P_1 remains unclear.

Previously reported work from this laboratory has shown that inhibition by a polynucleotide depends on strandedness, the nature of the polynucleotide sugar moiety, and the nature of the substitutions made on the polynucleotide bases.[5] Although poly(I) · poly(C) can both stimulate interferon production and inhibit protein synthesis, some dsRNAs are ineffective or affect only one activity. For example, poly(I) · poly(br5C) is a good inducer of interferon, but it does not inhibit protein synthesis in extracts of interferon-treated cells[5] or in reticulocyte lysates.[6]

In reticulocyte lysates the extent of phosphorylation of the eIF-2 α-subunit is regulated in part by a phosphoprotein phosphatase.[7] Kimchi et al.[8] have reported the existence of an activity in extracts of control and IF-treated cells that dephosphorylates P_1 and the eIF-2 α-subunit. They reported that the addition of this phosphoprotein phosphatase activity restored protein synthesis to that of the control incubated in the absence of dsRNA. In this chapter, a qualitative method is described for detecting the presence in preparations of interferon-treated mouse L cells of a phosphoprotein phosphatase activity, which is inhibited by poly(I) · poly(C) but not by poly(I) · poly(br5C).

[1] E. Ehrenfeld and T. Hunt, *Proc. Natl. Acad. Sci. U.S.A.* **68**, 1075 (1971).

[2] I. M. Kerr, R. E. Brown, and L. A. Ball, *Nature (London)* **250**, 57 (1974).

[3] R. S. Ranu and I. M. London, *Proc. Natl. Acad. Sci. U.S.A.* **73**, 4349 (1976).

[4] A. Kimchi, A. Zilberstein, A. Schmidt, L. Shulman, and M. Revel, *J. Biol. Chem.* **254**, 9846 (1979).

[5] P. F. Torrence and R. M. Friedman, *J. Biol. Chem.* **254**, 1259 (1979).

[6] J. Content, B. Lebleu, and E. De Clercq, *Biochemistry* **17**, 88 (1978).

[7] B. Safer and R. Jagus, *Proc. Natl. Acad. Sci. U.S.A.* **76**, 1094 (1979).

[8] A. Kimchi, L. Shulman, A. Schmidt, Y. Chernajovsky, A. Fradin, and M. Revel, *Proc. Natl. Acad. Sci. U.S.A.* **76**, 3208 (1979).

Principle

The method for demonstrating the inhibition of the P_1 phosphoprotein phosphatase is based on the observation that poly(I) · poly(br⁵C) stimulates P_1 phosphorylation under certain conditions, but is unable to do so under conditions approximating those of protein synthesis. Thus, by labeling P_1 in the presence of poly(I) · poly(br⁵C) and then transferring the labeled extract to conditions approximating those of protein synthesis, it is possible to follow (*a*) the decay of label from P_1 or (*b*) in the presence of an inhibitor of the P_1 phosphoprotein phosphatase, the stabilization of label in P_1.

Materials

Reagents

[γ-³²P]Adenosine triphosphate
Poly(I) · poly(C): s_{20} for poly(I) = 9.4; poly(C) = 10.0
Poly(I) · poly(br⁵C): s_{20} for poly(I) = 9.4; poly(br⁵C) = 6.2
Mouse interferon ($\approx 2 \times 10^7$ N.I.H. units/mg protein)
L cells. These cells (originally obtained from Dr. I. Kerr) are maintained in spinner culture. They are grown in minimal Eagle's medium supplemented with 10% fetal calf serum, 1× Eagle's vitamin mixture, 1× Eagle's nonessential amino acid mixture, 2 mM glutamine, and 50 µg/ml of gentamicin. Cells are treated with 30 units of interferon per milliliter for 14–18 hr.
Rabbit muscle creatine phosphokinase, 163 units per milligram of creatine phosphate substrate (Calbiochem)
Yeast hexokinase, 200 units/mg (P-L Biochemicals)
Escherichia coli alkaline phosphatase 38 units/mg (Worthington Biochemical Corp.)
Molecular weight protein markers: phosphorylase *b* (93,000 daltons), bovine serum albumin (68,000 daltons); ovalbumin (45,000 daltons), carbonic anhydrase (30,000 daltons), soybean trypsin inhibitor (21,000 daltons), and lysozyme (14,400) daltons. (Supplied as a mixture by Bio-Rad.)

Buffers

Washing buffer (buffer A): 0.034 M Tris · HCl, pH 7.5, 0.14 M NaCl
Lysis buffer (buffer B): 20 mM HEPES (adjusted to pH 7.5 with KOH), 10 mM KCl, 1.5 mM Mg(OAc)₂, 0.5 mM dithiothreitol.[9]
Labeling buffer (buffer C): 20 mM HEPES (adjusted to pH 7.5 with

---

[9] L. A. Weber, E. R. Feman, and C. Baglioni, *Biochemistry* **14**, 5315 (1975).

KOH), 10 mM KCl, 60 mM KOAc, 0.5 mM dithiothreitol, 25 mM Mg(OAc)$_2$

Mock protein synthesis buffer (buffer D): 30 mM HEPES (adjusted to pH 7.5 with KOH), 110 mM KCl, 7 mM 2-mercaptoethanol, 1 mM adenosine triphosphate, 0.1 mM guanosine triphosphate, 0.6 mM cytosine triphosphate, 10 mM creatine phosphate, 40 μM amino acids (except leucine), 0.16 mg of creatine kinase (163 units/mg)

Stop buffer (buffer E): 60 mM Tris $\cdot$ HCl, pH 7.5, 1% (v/v) 2-mercaptoethanol, 1% (w/v) sodium dodecyl sulfate, 30% (w/v) glycerol, 0.05% (w/v) bromophenol blue[10]

Preparation of P_{100} Fraction. Mouse L suspension culture cells are grown and maintained as described above. After treatment of the cells (final concentration of 5 to 6 $\times$ 10^5 cells per milliliter) with 30 units of mouse interferon per milliliter, the cells are collected by centrifugation at 500 g for about 5 min and washed three times with buffer A. The volume of the packed cells is measured, then the cells are resuspended and allowed to swell in approximately 1.25 packed cell volumes of buffer B. The cells are then homogenized by 10–12 strokes of a Dounce homogenizer with a type A pestle. The homogenate is centrifuged at 10,000 g for 20 min. The 10,000 g supernatant (S_{10}) is then centrifuged at 100,000 g for 2 hr. The supernatant (S_{100}) is frozen in small portions in Dry Ice. The pellet (P_{100}-IF) is resuspended in 0.75 packed cell volumes. Control P_{100} fractions are prepared from cells not treated with interferon. The P_{100} fractions are routinely stored in liquid nitrogen vapor.

Labeling of the P_{100}-IF fraction with $[\gamma\text{-}^{32}P]ATP$. The labeling of the P_{100} fraction with $[\gamma\text{-}^{32}P]ATP$ is carried out in a 30-μl reaction volume as follows. Six microliters of 10 mM ATP and 5 μl of $[\gamma\text{-}^{32}P]ATP$ (4 mCi/ml, specific activity 2600 Ci/mmol) are mixed and lyophilized in a 1.5-ml polypropylene conical centrifuge tube with attached cap (Eppendorf). To the dried $[\gamma\text{-}^{32}P]ATP$ is added 12 μl of 0.25 M Mg(OAc)$_2$, 3.6 μl of 2 M KOAc, and 32 μl of buffer B. Twelve microliters of this mixture are distributed into three separate 1.5-ml centrifuge tubes. To each of the tubes are added respectively 3 μl of (a) 2 $\times$ 10^{-4} M_p poly(I) $\cdot$ poly(C) (70 μg/ml); (b) 2 $\times$ 10^{-4} M_p poly(I) $\cdot$ poly(br^{5}C) (ca. 70 μg/ml); or (c) buffer B. Fifteen microliters of P_{100}-IF extracts (A_{260}/ml, 110; A_{280}/ml, 66) are added to each of the tubes, placed in a 30° water bath, and incubated for 10 min.

Demonstration of the Inhibition by Poly(I) $\cdot$ Poly(C) of P_1 Phosphoprotein Phosphatase Activity. After the 10-min incubation period the tubes are removed from the 30° bath and placed in ice for subsequent dilution into mock protein synthesis buffer. This buffer is prepared as follows. A master mix is prepared consisting of 110 μl of a mixture of 10 mM ATP, 1 mM

[10] B. Lebleu, G. C. Sen, S. Shaila, B. Cabrer, and P. Lengyel, *Proc. Natl. Acad. Sci. U.S.A.* **73**, 3107 (1976).

GTP, 6 mM CTP, and 100 mM creatine phosphate; 110 μl of a buffer containing 300 mM HEPES (adjusted to pH 7.5 with KOH), 1.1 M KCl, 70 mM 2-mercaptoethanol; 55 μl of an amino acid mixture consisting of all the standard amino acids except leucine at a concentration of 1 mM; 55 μl of creatine kinase (3.2 mg/ml); and 110 μl of sterile double-distilled water. Two hundred microliters of this master mix are added to a mixture of 55 μl of buffer B and 160 μl of water. One hundred fifteen microliters of the diluted master mix are distributed into three tubes to which are added, respectively, 14 μl of 2 $\times$ 10^{-4} M_p poly(I) $\cdot$ poly(C), 2 $\times$ 10^{-4} M_p poly(I) $\cdot$ poly(br^{5}C), or buffer B. Eleven microliters of P_{100}-IF previously labeled in the presence of poly(I) $\cdot$ poly(C), poly(I) $\cdot$ poly(br^{5}C), or no dsRNA are added to the mock protein synthesis buffer, which contains the same dsRNA present in the labeled extracts. The diluted labeled P_{100}-IF fractions are returned to the 30° water bath, and 20-μl samples are removed at intervals over an 80-min period. The samples are added to 20 μl of buffer E that has been equilibrated at 100°. The extract is denatured by boiling for 5 min, cooled to room temperature, and centrifuged to collect condensed water from the cap and sides of the centrifuge tube. The entire sample can be loaded onto a 10% SDS polyacrylamide gel. *It is extremely important throughout these procedures to use tubes that can be securely capped in order to prevent radioactive contamination of equipment and personnel.*

Polyacrylamide gel electrophoresis of the samples is carried out according to the procedure of Laemmli[11] as modified by Studier.[12] The formulation per 10 $\times$ 14 cm of gel is as follows: 10.4 ml of H_2O; 2.5 ml of 1.5 M Tris $\cdot$ HCl, pH 8.8; 0.2 ml of 0.2 M EDTA, pH 7.5; 0.2 ml of 10% (w/v) sodium dodecyl sulfate (SDS); 6.7 ml of acrylamide : N,N'-bismethylene acrylamide (30 : 0.8%, w/v); 0.01 ml of N,N,N',N'-tetramethylethylenediamine (TEMED); and 0.065 ml of 10% (w/v) of ammonium persulfate. All reagents except TEMED and SDS are mixed prior to degassing. After adding TEMED and SDS, the gel solution is poured and deionized water is layered over it. Polymerization occurs in about 30 min. The water covering the polymerized resolving gel is removed, then a stacking gel solution is poured over the polymerized resolving gel. The composition of the stacking gel solution per gel is: 7.13 ml of H_2O; 1.25 ml of 1.5 M Tris $\cdot$ HCl, pH 6.8; 0.1 ml of 0.2 M EDTA; 0.1 ml of 10% SDS; 1.33 ml of acrylamide : bisacrylamide (30 : 0.8); 0.005 ml of TEMED; and 0.1 ml of 10% ammonium persulfate. A well-mold is inserted in the stacking gel solution, which polymerizes in about 20 min. The composition per liter of running buffer is as follows: 6 g of Tris base, 28.8 g of glycine, 1 g of SDS,

[11] U. K. Laemmli, *Nature (London)* 227, 680 (1970).
[12] F. W. Studier, *J. Mol. Biol.* 79, 237 (1973).

10 ml of 0.2 M EDTA. Appropriate markers should be included in every polyacrylamide gel analysis. The gels are generally run for 16–18 hr at 30 V constant voltage. The voltage can be raised to about 100 V to accelerate the movement of the bromophenol blue dye to the bottom of the gel at the end of the run. After completion of the run, the gels are stained for 20–30 min with 0.25% Coomassie Blue dissolved in 50% (v/v) methanol and 9.3% (v/v) acetic acid. The gels may be rapidly destained in 55% (v/v) methanol and 7.5% (v/v) acetic acid at 55° for 30 min. The destaining medium is changed at least once. Since gels tend to shrink under these conditions, they are incubated for an additional hour in 5% methanol and 7.5% acetic acid. The gels are removed from the methanol–acetic acid bath and placed on heavy filter paper to which they are bonded by heating under vacuum with a gel drier (Bio-Rad). The dried gels are placed against Kodak X-Omat X-ray film. The exposure time needed to obtain a satisfactory autoradiogram is significantly reduced by sandwiching the X-ray film between the gel and a DuPont Cronex intensifying screen at $-70°$. The developed autoradiogram will show the progressive decrease in ^{32}P label in P_1 as a function of time in the sample incubated with poly(I) · poly (br^{5}C). Little or no change in the sample labeled in the presence of poly(I) · poly(C) can be observed. Moreover, if poly(I) · poly(C) is added to the mock protein synthesis buffer containing the P_{100}-IF fraction, which was labeled initially in the presence of poly(I) · poly(br^{5}C) alone, the decrease in ^{32}P-labeled P_1 is inhibited.

The inhibition by poly(I) · poly(C) of the P_1 phosphoprotein phosphatase can also be demonstrated under conditions of ATP depletion. This depletion is carried out under the same buffer conditions used during the initial labeling procedure, in order to avoid using the ATP-generating system of the mock protein synthesis buffer. The hexokinase–glucose ATP depletion system is formulated as follows. To 140 μl of buffer B are added 7.5 μl of 2 M KOAc, 24.5 μl of 0.25 M Mg(OAc)$_2$, 2.5 μl of 100 mM glucose, and 2.5 μl of 500 units of hexokinase per milliliter. Of this buffer, 51 μl are distributed into three tubes containing 7 μl of $2 \times 10^{-4} M_p$ poly(I) · poly(C), $2 \times 10^{-4} M_p$ poly(I) · poly(br^{5}C), or buffer B. Approximately 5.5 μl of the P_{100}-IF fractions labeled in the presence of poly(I) · poly(C), poly(I) · poly(br^{5}C), or no dsRNA are added to the buffer containing glucose and hexokinase. The reaction is carried out at 30°. The reactions are assayed for P_1 dephosphorylation by polyacrylamide gel electrophoresis and autoradiography and also for ATP depletion by thin-layer chromatography on PEI-cellulose coated plates run in 1 M LiCl.

An alternative ATP-depletion system utilizes *Escherichia coli* alkaline phosphatase. After centrifugation to remove most of the 2.6 M ammonium sulfate in which it is suspended, the alkaline phosphatase is diluted to a concentration of 200 units/ml with buffer B. The alkaline phosphatase ATP-depletion buffer is prepared by mixing 135 μl of buffer B, 7.4 μl of 2 M KOAc, 24.5 μl of 0.25 M Mg(OAc)$_2$, and 7.0 μl of alkaline phosphatase (200 units/ml). Precisely the same proportions of this buffer, dsRNA, and labeled P$_{100}$-IF are mixed together as in the hexokinase-glucose depletion. It is perhaps noteworthy that alkaline phosphatase as well as hexokinase–glucose have no apparent effect on the ^{32}P label incorporated into P$_i$ in the presence of poly(I) · poly(C). However, ^{32}P label incorporated into P$_i$ rapidly decreases. Because little if any ^{32}P-labeled ATP remains after less than 20 min, depletion of the labeled samples of ATP should give a reasonably good measure of P$_i$ phosphoprotein phosphatase activity, in the absence of protein kinase activity.

This assay should be generally applicable to the detection of any inhibitor of the P$_i$ phosphoprotein phosphatase.

[24] Enzymic Synthesis, Purification, and Fractionation of (2'-5')-Oligoadenylic Acid

By A. G. HOVANESSIAN, R. E. BROWN, E. M. MARTIN, W. K. ROBERTS, M. KNIGHT, and IAN M. KERR

A series of 2'-5'-linked oligoadenylic acid 5'-triphosphate inhibitors of protein synthesis (collectively referred to as 2-5A) are formed from ATP by an enzyme (2-5A synthetase) induced by interferon and activated by double-stranded RNA (dsRNA).[1,2] The major biologically active species is the trimer (pppA2'p5'A2'p5'A) but most enzymically synthesized preparations also contain the corresponding dimer (pppA2'p5'A), tetramer [ppp(A2'p)$_3$A], pentamer [ppp(A2'p)$_4$A] and higher oligomers in decreasing amounts.[2,3] The trimer, tetramer, and pentamer are biologically active at similar subnanomolar levels in the inhibition of cell-free protein synthesis. The dimer (and trimer in the rabbit reticulocyte system[4]) is less potent if active at all.[2] The inhibitory activity of 2-5A in such systems is medi-

[1] A. G. Hovanessian, R. E. Brown, and I. M. Kerr, *Nature (London)* **265**, 537 (1977).

[2] I. M. Kerr and R. E. Brown, *Proc. Natl. Acad. Sci. U.S.A.* **75**, 256 (1978).

[3] The terms dimer, trimer, tetramer, and pentamer will be used to denote preparations consisting largely of the 5'-triphosphates, but also containing smaller amounts of the 5'-di- and monophosphates. The more complete nomenclature, e.g., trimer triphosphate, will be used to denote the individual species.

[4] B. R. G. Williams, R. R. Golgher, R. E. Brown, C. S. Gilbert, and I. M. Kerr, *Nature (London)* **282**, 582 (1979).

ated, in large part at least, through the activation of a nuclease that degrades mRNA.[5-7]

The 2-5A synthetase is widely distributed in very different amounts in a variety of cells and tissues.[8] The level increases between 10- and 10,000-fold in response to interferon treatment.[1,8,9] Assay of these different synthetase levels and of the product 2-5A are dealt with elsewhere (Stark *et al.* and Williams *et al.*, this volume [25] and [26], respectively). Here we discuss the optimal synthesis of 2-5A. Methods to optimize the yield of 2-5A or to obtain radioactive material of known specific activity are described with the 2-5A synthetases from three sources containing relatively large amounts of the enzyme: interferon-treated mouse L or human HeLa cells and rabbit reticulocytes.[1,8,10,11] In addition, methods for the purification and fractionation of 2-5A on a preparative scale (10–50 mg) are described.

Materials

Dialysis buffer with glycerol (DBG): 20 mM HEPES buffer, pH 7.5; 1.5 mM MgAc$_2$; 7 mM 2-mercaptoethanol; 20% glycerol (v/v); and KCl as specified

Poly(I) · poly(C) (P-L Biochemicals or Sigma): Stock solutions (10^{-4} g/ml) in saline or in 20 mM HEPES buffer, pH 7.5, and 90 mM KCl were stored at $-20°$ (prolonged storage in water or low-salt buffer can result in denaturation of the dsRNA and loss of activity)

Mouse L and human HeLa cells were treated for 17 hr at 37° with 100–400 reference units of mouse or human interferon, respectively. Treatment with higher concentrations of interferon were without additional effect on 2-5A synthetase levels.

Cell extracts were prepared from the interferon-treated L or HeLa cells or rabbit reticulocyte lysates by the same methods used for the preparation of such material for use in cell-free protein-synthesizing systems.[12,13] In our experience the use of postmicrosomal supernatant

[5] M. J. Clemens and B. R. G. Williams, *Cell* 13, 565 (1978).

[6] C. Baglioni, M. A. Minks, and P. A. Maroney, *Nature (London)* 273, 684 (1978).

[7] B. R. G. Williams, I. M. Kerr, C. S. Gilbert, C. N. White, and L. A. Ball, *Eur. J. Biochem.* 92, 455 (1978).

[8] G. R. Stark, W. J. Dower, R. T. Schimke, R. E. Brown, and I. M. Kerr, *Nature (London)* 278, 471 (1979).

[9] L. A. Ball, *Virology* 94, 282 (1979).

[10] M. A. Minks, S. Benvin, P. A. Maroney, and C. Baglioni, *J. Biol. Chem.* 254, 5058 (1979).

[11] A. G. Hovanessian and I. M. Kerr, *Eur. J. Biochem.* 84, 149 (1978).

[12] I. M. Kerr, R. E. Brown, M. J. Clemens, and C. S. Gilbert, *Eur. J. Biochem.* 69, 551 (1976).

[13] T. Hunt, G. Vanderhoff, and I. M. London, *J. Mol. Biol.* 66, 471 (1972).

fractions (S100s) can give better results in the "in solution" assays, but postmitochondrial supernatant fractions (S10s) are equally effective when preparing poly(I) · poly(C)-Sepharose or paper-bound enzymes. The 2-5A synthetases in the extracts from all these types of cells are stable both to storage at −70° for many months and to multiple cycles of freezing and thawing.

Several groups have found commercially available [^{3}H]- or [^{32}P]ATP from a variety of suppliers to be inhibitory when used at higher concentrations (e.g., ≥0.1 mCi/ml) in the synthetase reaction. Most of this inhibitory activity can be removed by stepwise chromatography on DEAE-Sephacel (Pharmacia). For example, 5 mCi of [^{3}H]ATP (1 mCi/ml, 22 Ci/mmol in 50% ethanol from the Radiochemical Centre, Amersham, England), from which the ethanol was removed under vacuum, was applied to a 1-ml column of DEAE-Sephacel in a 1-ml pipette equilibrated with 10 mM NH$_4$HCO$_3$. The column was washed with 10 ml each of 10 mM and 50 mM NH$_4$HCO$_3$. The [^{3}H]ATP was eluted with 250 mM NH$_4$HCO$_3$ (all of the radioactivity was recovered in 400 μl) and the NH$_4$HCO$_3$ removed by lyophilization (overall recovery ≥ 90%).

Synthesis of 2-5A

2-5A can be synthesized "in solution" by incubation of appropriate crude cell extracts with ATP and dsRNA. Alternatively, the 2-5A synthetase can be substantially purified by binding it to dsRNA, itself bound to a solid support [such as poly(I) · poly(C)-Sepharose or paper], and the "insoluble" enzyme employed. With crude extracts "in solution," the synthesis of 2-5A is balanced by degradation and net accumulation does not normally occur for more than 2 hr at 30° (Fig. 1). In addition, when synthesizing radioactive 2-5A the specific activity of the radioactive ATP, and hence of the product 2-5A, will not be known unless the size of the cold ATP pool has been determined. Nevertheless, with extracts containing high levels of synthetase 2-5A can be rapidly and conveniently synthesized in this way. Accordingly, examples of the use of this method are given below. The use of the partially purified enzyme bound to poly(I) · poly(C)-Sepharose[1] or poly(I) · poly(C) paper[8] largely circumvents the above problems and is, therefore, generally to be preferred, although incomplete removal of the enzymes (nuclease and phosphatase) responsible for the degradation of 2-5A can result in some residual breakdown.

For convenience, the concentration of 2-5A (trimer and higher oligomers) will be given in AMP equivalents throughout.

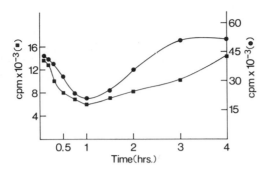

FIG. 1. Kinetics of 2-5A formation in reticulocyte lysates. A rabbit reticulocyte lysate (80 μl), which had been supplemented with 20 μM hemin, was incubated with poly(I) · poly(C) (1 μg/ml) and 1 mM ATP at 30° without further addition in a final volume of 100 μl. Aliquots (5 μl) were removed at the times indicated (abscissa), diluted 10-fold with water, and heated at 95° for 5 min. The denatured proteins were removed by centrifugation, and the supernatants were assayed for 2-5 A at final dilutions of 2500-fold for their ability to inhibit amino acid incorporation in the reticulocyte lysate (●) or EMC RNA programmed L cell-free system (■). Results are given as the incorporation of [¹⁴C]leucine (●) or of a mixture of ¹⁴C-labeled amino acids (■) per 10 μl of the appropriate cell-free system. From Hovanessian and Kerr.[11]

Synthesis of 2-5A "in Solution"

For the interferon-treated HeLa or L cell extracts (in 20 mM HEPES buffer, pH 7.5; 90 mM KCl; 1.5 mM MgAc₂; 7 mM 2-mercaptoethanol) typical incubations contain 20 μl of cell extract; 1.5–30 mM MgAc₂; 10^{-5} to 10^{-6} g/ml of poly(I) · poly(C), and 1–10 mM ATP (plus [³H]ATP—0.1 mCi/ml, 22 Ci/mmol—if a radioactivity assay is to be employed) in a final volume of 26 μl. The optimum concentrations of cell extract and of the other components should be determined for individual extracts. In general, the higher the ATP concentration, the higher the Mg^{2+} required. In our experience optimal conversion of input ATP to 2-5A ($\geq 30\%$) was obtained with a 10-fold dilution (in DBG 50 mM KCl) of extracts from interferon-treated HeLa cells incubated with 15 mM MgAc₂, 10^{-5} g/ml of poly(I) · poly(C), and 4 mM ATP. For interferon-treated L cell extracts, the corresponding figures were 8.5 mM MgAc₂, 10^{-5} g/ml of poly(I) · poly(C), and 3 mM ATP. For reticulocyte lysates ($\pm 20 \mu M$ hemin), 100 mM ATP and 10^{-4} g of poly(I) · poly(C) per milliliter were added to final concentrations of 1 mM and 10^{-6} g/ml, and the lysate was incubated without further addition.

In all the above, incubations were routinely for 60–120 min at 30°. Longer incubations resulted in decreased yields (e.g., Fig. 1). Dilution to obtain optimum results was necessary only with the HeLa cell extracts.

For example, 10 ml of a 10-fold, 3 ml of a 3-fold, and 1 ml of undiluted extract incubated for 120 min at 30° with 3, 10, and 10 mM ATP, respectively, yielded the equivalent of 13, 10, and 3 ml of 1 mM 2-5A in AMP equivalents. For large-scale preparations, however, the disadvantage of increased volume on dilution could outweigh the advantage of a significantly increased yield. For convenience we use commercially available poly(I) · poly(C). Both reovirion and *Penicillium chrysogenum* phage dsRNAs are also highly effective at similar concentrations. The effects of variation in the concentration of the different components in the HeLa and reticulocyte systems has been described in detail elsewhere.[10,11]

Synthesis of 2-5A with the Synthetase Bound to
Poly(I) · Poly(C)-Sepharose or Poly(I) · Poly(C) Paper

Poly(I) · poly(C)-Sepharose can be prepared as described previously[11] or obtained from P-L Biochemicals. The former has been more satisfactory in our experience. The bound enzyme is obtained by the very slow (30 min) passage of the cell extract (2 volumes) supplemented with 20% glycerol (v/v) through a column of poly(I) · poly(C)-Sepharose (one volume) equilibrated with DBG (50 mM KCl) in a plastic syringe. The column-bound enzyme is extensively washed with DBG (90 mM KCl) to remove unbound protein and nucleotides. It is then equilibrated with the incubation medium to be used in the synthesis of 2-5A. The Sepharose is then removed from the column and incubated in two volumes of this medium.

In our experience, the use of poly(I) · poly(C) paper has proved to be more convenient than poly(I) · poly(C)-Sepharose, and this is now our method of choice. The preparation of poly(I) · poly(C) paper and the bound synthetase is described elsewhere (Stark *et al.*, this volume [25]). Optimum conditions for the synthesis of 2-5A with the synthetase from interferon-treated L cells or rabbit reticulocyte lysates bound to either support have proved to be remarkably similar. [The 2-5A synthetase from both types of extract have been extensively compared on both types of support with very similar results, but that from HeLa cells has been tested in this laboratory only with poly(I) · poly(C) paper.] Accordingly, the conditions described below are suitable for either enzyme with either support. Typically, incubation of the synthetase from 1 ml of reticulocyte lysate bound to 3 cm^2 of poly(I) · poly(C) paper (10–25 μg of poly(I) · poly(C)/cm^2) in 0.5 ml of DBG (50 mM KCl) supplemented with 3 mM ATP and MgAc$_2$ to 8.5 mM yielded the equivalent of 1.0 ml of 0.3–0.6 mM 2-5A for each 24 hr of incubation at 30°. The medium can be harvested each day and replaced with fresh medium, and the incubation can be continued in this way for 6–12 days before significant loss of enzyme activity occurs.

With interferon-treated HeLa extracts, higher yields have been obtained. For example, incubation of the synthetase from 1.0 ml of extract bound to 10 cm^2 of poly(I) · poly(C) paper for 20 hr at 30° in 1.25 ml of DBG (50 mM KCl) supplemented with 30 or 50 mM ATP and 22.5 or 50 mM MgAc$_2$, respectively, yielded the equivalent of 1.0 ml of 13 and 15 mM 2-5A.

Remarkably similar yields have been obtained in all of the above procedures at 25, 30, or 37°, but the column- or paper-bound enzymes appear to be more stable to prolonged incubation at the lower temperatures. The MgAc$_2$ and ATP concentrations given throughout are intended as guidelines. It is important to optimize these for individual extracts and batches of poly(I) · poly(C) paper or Sepharose. In all the above, for convenience, [^{3}H]ATP (1 mCi/ml, 22 Ci/mmol) can be added to a final concentration of 0.1 mCi/ml to monitor its conversion to 2-5A (Williams et al., this volume [26]).

Purification and Fractionation of 2-5A

2-5A synthesized enzymically, whether "in solution" or with the 2-5A synthetase bound to a solid support, is a mixture of oligomers. The relative amounts of the various oligomers observed with different enzyme preparations and in different laboratories vary. Products up to 15 residues in length have been informally reported by a number of groups. Nevertheless, the factors controlling their distribution are not yet understood. The products can be analyzed by HPLC (Brown et al., this volume [27]), by chromatography on DEAE-Sephadex[2] or cellulose[14] in the presence of 7 M urea, or by TLC or PEI-cellulose with or without prior treatment with bacterial alkaline phosphatase.[2,15] Here methods are described that we have found useful for the purification and fractionation of 2-5A when the objective is to obtain large amounts (≥10 mg) of material. Depending on the objectives, two methods have been used. In the first, residual ATP and related by-products are removed from the 2-5A by simple two-step elution from DEAE–Sephacel. This yields material consisting of unfractionated 2-5A, which is suitable for many experiments concerning its biological activity. Much of the work concerning its mechanism of action has been carried out with material of this type. Alternatively, for experiments requiring the individual oligomers, these can be obtained by fractionation on DEAE–Sephadex.[15]

[14] L. A. Ball and C. N. White, Proc. Natl. Acad. Sci. U.S.A. 75, 1167 (1978).
[15] E. M. Martin, N. J. M. Birdsall, R. E. Brown, and I. M. Kerr, Eur. J. Biochem. 95, 295 (1979).

Milligram quantities of 2-5A can be readily synthesized with the reticulocyte or HeLa 2-5A synthetase bound to poly(I) · poly(C) paper. In a typical experiment, the enzyme from 30 ml of reticulocyte lysate bound to poly(I) · poly(C) paper was incubated at 30° for 10 days in successive daily 15-ml batches of medium as described above. To obtain unfractionated 2-5A, the pooled supernatants (150 ml) were applied to a column of DEAE–Sephacel (20 cm × 4 cm diameter) washed and equilibrated with water. (The DEAE–Sephacel from Pharmacia is available as a slurry in 50% ethanol and should be washed with water and allowed to settle; the "fines" are decanted before use). Residual ATP and related by-products were eluted with a solution consisting of 125 mM KCl, 1.5 mM MgAc$_2$, and 10 mM HEPES buffer, pH 7.5 and the column washed with water to remove the KCl. The product 2-5A was eluted with 450 mM NH$_4$HCO$_3$ (200 ml), lyophilized, and taken up in 3.0 ml of 10 mM HEPES buffer, pH 7.5. It had an absorbance at 260 nm of 565 units/ml corresponding to a total yield of approximately 35 mg of 2-5A. Subsequent HPLC analysis showed it to contain approximately 50% trimer, 40% dimer, 10% tetramer, and 2% pentamer (by weight). It was contaminated with approximately 10% (by weight) of ATP. There was no significant loss of biological activity throughout the procedure.

To obtain the individual oligomers, in an experiment similar to that described above, the products from 6 days of incubation were pooled (90 ml), diluted with an equal volume of water, loaded (1 ml/min) onto a column (50 cm × 2.3 cm in diameter) of DEAE-Sephadex A25 (Pharmacia), washed and equilibrated with 0.2 M NH$_4$HCO$_3$. The column was washed with 500 ml of 0.2 M NH$_4$HCO$_3$ prior to elution (1 ml/min) with a gradient (500 ml/500 ml) of 0.2 M to 0.6 M NH$_4$HCO$_3$ followed by further elution with 0.6 M NH$_4$HCO$_3$. Elution of the individual components was monitored by their absorbance at 260 nm (A_{260}) (Fig. 2). Appropriate fractions were pooled and lyophilized to remove the NH$_4$HCO$_3$. The individual peaks were identified by the inclusion of small amounts of appropriate radioactive markers ([32P]ATP and 3H-labeled trimer and tetramer, data not shown). Identification was confirmed by subsequent analysis by HPLC or TLC. The yields obtained were 510, 320, and 145 A_{260} units corresponding to approximately 12, 7, and 3 mg of trimer, dimer, and tetramer, respectively. For all of these, the 5′-triphosphate was the major component (80%), but each contained smaller amounts of the 5′-di- and traces of the 5′-monophosphates. As first noted by L. A. Ball (personal communication), the 5′-diphosphates of the trimer, tetramer, and pentamer have essentially the same specific biological activity as the triphosphates.[2,15] The monophosphates are essentially inactive.[2,15] The degree of phosphorylation of the different components can be most conveniently

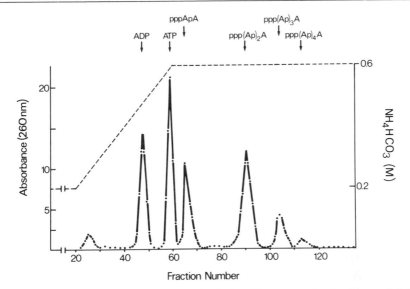

FIG. 2. Fractionation of 2-5A on DEAE-Sephadex. The preparation, loading, and elution of the column were carried out as described in the text; 20-ml (1 to 60) and 10-ml (61 to 120) fractions were collected. The absorbance (units/ml) at 260 nm of the individual fractions is presented (●, left ordinate). The right-hand ordinate gives the NH_4HCO_3 concentration (---). All of the oligonucleotides identified (arrows) were 2'-5'-linked, although this is not indicated on the figure.

determined analytically by TLC or HPLC (Brown *et al.*, this volume [27]). Fractionation of the 5' mono-, di-, and triphosphorylated trimer can be carried out on a preparative scale by chromatography on DEAE–Sephacel as described previously.[15]

Synthesis of Radioactive 2-5A of Known Specific Activity

This can be most conveniently done with the 2-5A synthetase bound to poly(I) · poly(C) paper (from which all the unlabeled ATP in the cell extract has been removed by washing). Incubations are carried out in DBG (50 mM KCl) supplemented with $MgAc_2$ to 8.5 mM and appropriate concentrations (0.1–1 mM) of radioactive ATP. The specific activity that can be achieved is limited by the progressively decreasing percentage of conversion of ATP to 2-5A at lower ATP concentrations (Fig. 3). For example, undiluted [^{3}H]ATP at a specific activity of 20 Ci/mmol is approximately 50 μM at 1 mCi/ml, yielding only 2–3% conversion to 2-5A (Fig. 3). The use of higher concentrations ($\geq$2 mCi/ml) of radioactive ATP has proved to be inhibitory. Accordingly, the maximum specific activity that can readily be achieved while retaining a reasonable conversion (5%) of

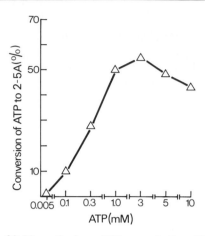

Fig. 3. Dependence of 2-5A synthesis on ATP concentration. The 2-5A synthetase from interferon-treated HeLa cell extracts (20 μl of a threefold dilution) bound to 0.1 cm² of poly(I) · poly(C) paper was incubated for 18 hr at 30° in 25 μl of reaction mixtures containing 20 mM HEPES buffer, pH 7.5; 50 mM KCl; ATP as indicated on the abscissa: 8.5 (for incubations containing 3 mM ATP), 10 (for 5 mM ATP) and 15 (for 10 mM ATP) mM MgAc₂; and 20% glycerol. [³H]ATP (22 Ci/mmol) was added to 0.1 mCi/ml. The results are presented (ordinate) in terms of the percentage conversion of the [³H]ATP to 2-5A (see Williams et al., this volume [26]).

the input $[^3\text{H}]$- or $[\alpha\text{-}^{32}\text{P}]$ATP is about 10 Ci/mmol (AMP equivalents, 30 Ci/mmol trimer) involving the utilization of 100 μM ATP at 1 mCi/ml. Simple variation in the conditions of incubation have not allowed any improvement of this. A similar ATP concentration dependence is seen in the "in solution" assays. The synthesis of very high specific activity 2-5A therefore remains a problem. On the other hand, the very high levels of 2-5A synthetase in the interferon-treated HeLa cell (and chick embryo fibroblast: L. A. Ball, personal communication) extract allow the very efficient (≥50%) utilization of input radioactive ATP at specific activities of ≤1000 (Fig. 3). Similar results to those in Fig. 3 have been obtained with the reticulocyte or L cell synthetases under the same conditions, but the maximum conversion of ATP to 2-5A obtained was lower (10–20%, and frequently less than this), presumably reflecting the lower levels of synthetase in these extracts.

Comments

In our experience, the synthesis of 2-5A with the synthetase bound to a solid support has been more satisfactory than incubations "in solution" giving markedly higher yields of 2-5A per unit of extract, reflecting the removal (in large part) of degradative enzymes and the consequent ability to maintain the accumulation of the 2-5A products over a period of 6–12 days. There is, in addition, the obvious advantage of the use of the partially purified column- or paper-bound enzyme free of any unlabeled nucleotide pool, both with respect to the purity of product and the synthesis of radioactive 2-5A. Once the purified synthetase is more widely available, however, its use in solution may well be preferred. The "in solution" synthesis with crude extracts is more convenient if relatively lower yields of crude material will suffice.

Throughout the above, the use of an ATP regenerating system based on creatine phosphate and creatine kinase identical to that used in many cell-free protein synthesizing systems[12] has frequently resulted in slightly increased yields of 2-5A. This may be particularly true in the use of 2-5A synthetase from HeLa cell extracts where there appears to be significant contamination of the poly(I) · poly(C) paper-bound synthetase with phosphatase, with the consequent breakdown of ATP on prolonged incubation. Its inclusion may be particularly helpful when low concentrations of high specific activity [^{3}H]ATP are to be employed. Similarly, there is in cell extracts a nuclease capable of degrading 2-5A to AMP (or ATP in the presence of an ATP regenerating system).[7,16] AMP (5 mM) has been found to stabilize 2-5A in extracts from chick embryo fibroblasts (L. A. Ball, personal communication).

Of the extracts for which data is presented above, that from the interferon-treated HeLa cells in use in this laboratory[17] contained by far (10- to 20-fold) the greatest amount of 2-5A synthetase activity. Similar levels appear to be present in extracts of specially aged interferon-treated chick embryo fibroblasts.[9] Although richer sources may be found in due course, these extracts appear to contain the highest levels reported to date. The data for the more widely available reticulocyte lysate may, however, be of value to those who have neither interferon nor tissue culture facilities readily available.[18]

Acknowledgments

A. G. H. was the recipient of a Wellcome Trust Fellowship, and W. K. R. of Grant No. NP200 from the American Cancer Society.

[16] A. Schmidt, Y. Chernajovsky, L. Shulman, P. Federman, H. Berissi, and M. Revel, *Proc. Natl. Acad. Sci. U.S.A.* **76**, 4788 (1979).
[17] The HeLa cells used here were HeLa S3 from Flow Laboratories, Irvine, Scotland. Minks *et al* [10] also used HeLa cells and appeared to obtain high levels of synthetase. Not all strains of HeLa cells, however, contain these high levels after interferon treatment (A. G. Hovanessian, unpublished work, 1980).
[18] For improved methodology see this volume [25, footnote 10].

[25] Assay of (2′-5′)-Oligoadenylic Acid Synthetase Levels in Cells and Tissues: A Convenient Poly(I) · Poly(C) Paper-Bound Enzyme Assay

By G. R. STARK, R. E. BROWN, and IAN M. KERR

The 2-5A synthetase responsible for the synthesis of the unusual oligonucleotide pppA2′p5′A2′p5′A (and related oligomers collectively referred to as 2-5A[1]) was discovered[2] on the basis of its role in the interferon-mediated inhibition of cell-free protein synthesis by double-stranded RNA (dsRNA). This latter occurs by two mechanisms: (*a*) inactivation of initiation factor eIF-2 by dsRNA-mediated phosphorylation; (*b*) degradation of mRNA by a nuclease(s) activated by subnanomolar concentrations of 2-5A (reviewed by Baglioni[3]). The subsequent demonstration that the 2-5A synthetase is widely distributed in very different amounts in a variety of cells and tissues and that its level varies with both growth and hormone status, as well as in response to interferon, has raised the possibility that the 2-5A system may be of wider significance than simply in the antiviral action of interferon.[4] Its assay in crude extracts is, therefore, of interest both in studies on the mechanism of action of interferon (or concerning the clinical effectiveness of interferon in reaching a particular tissue) and on the possible wider significance of the 2-5A system.

Assay of the synthetase, which both binds to and requires dsRNA for activity, depends on measurement of 2-5A and is complicated in crude extracts by the presence of enzymes degrading 2-5A. Here we describe a convenient assay based on the activity of the partially purified enzyme adsorbed from crude extracts onto poly(I) · poly(C) paper, which normally circumvents this difficulty. The crude extract is added to the poly(I) · poly(C) paper, which is then washed to remove low molecular weight material and most of the proteins (>90%) that do not bind. The partially purified poly(I) · poly(C) paper-bound enzyme is then incubated with ATP, and the product 2-5A is assayed. The method is equally amenable for use on the 10-μl or ≥10-ml scale. Accordingly, it can be applied to the assay of synthetase levels in cells and tissues even when only rela-

[1] I. M. Kerr and R. E. Brown, *Proc. Natl. Acad. Sci. U.S.A.* **75**, 256 (1978).

[2] A. G. Hovanessian, R. E. Brown, and I. M. Kerr, *Nature (London)* **265**, 537 (1977).

[3] C. Baglioni, *Cell* **17**, 255 (1979).

[4] G. R. Stark, W. J. Dower, R. T. Schimke, R. E. Brown, and I. M. Kerr, *Nature (London)* **278**, 471 (1979).

tively small amounts of material are available. The lower limits are set by the minimum volume of tissue that can be conveniently handled (≤ 50 μl), the level of 2-5A synthetase present,[4] and the sensitivity of the assay for the 2-5A product employed (Williams *et al.* and Knight *et al.*, this volume [26] and [28]).

Preparation of Poly(I) · Poly(C) Paper

Materials

Poly(I) · poly(C) from P-L Biochemicals
Poly(C) from Sigma
Poly(I) · poly(C) paper (10–25 μg of poly(I) · poly(C) per cm^2) is prepared by a modification of the method of Alwine *et al.*[5] To a 20 × 10 cm sheet of diazobenzyloxymethyl paper are added 5 ml of a mixture of 0.5 ml of poly(I) · poly(C) (10 mg/ml), 0.9 ml of 25 mM sodium phosphate buffer, pH 6.0, and 3.6 ml of dimethyl sulfoxide. After incubation overnight at room temperature, the following are added to allow the denatured poly(I) · poly(C) to rehybridize: 3 ml of bovine serum albumin (2 mg/ml, pretreated with 0.1% diethylpyrocarbonate for 1 hr at room temperature to inactivate nucleases); 1.5 ml of 10% sodium dodecyl sulfate; 25 ml of 2 × SSC (300 mM NaCl, 30 mM trisodium citrate); 30 ml of formamide; and 8.5 μl of poly(C) (5 mg/ml). After 24 hr at room temperature with gentle rocking, the paper is washed thoroughly three times with 2 × SSC, blotted dry, and stored at $-70°$. (A detailed account of the methodology and use of nucleic acids bound to paper is given in this series.[6])

Preparation of Cell Extracts

Reagents

HEPES-buffered saline (HBS): 20 mM HEPES buffer, pH 7.5; 140 mM NaCl
Lysis buffer: 10 mM HEPES buffer, pH 7.5; 10 mM KCl; 1.5 mM MgAc$_2$; 7 mM 2-mercaptoethanol
10× buffer: 100 mM HEPES buffer, pH 7.5; 500 mM KCl; 15 mM MgAc$_2$; 70 mM 2-mercaptoethanol
Nonidet P-40 (NP40) lysis buffer: 20 mM HEPES buffer, pH 7.5; 120 mM KCl; 5 mM MgAc$_2$; 1 mM dithiothreitol; 10% glycerol (v/v); 0.5% NP40 (w/v)

[5] J. C. Alwine, D. J. Kemp, and G. R. Stark, *Proc. Natl. Acad. Sci. U.S.A.* **74**, 5350 (1977).
[6] M. L. Goldberg *et al.*, this series, Vol. 68 [14] and J. C. Alwine *et al.*, this series, Vol. 68 [15].

Procedure. Samples of cells grown in suspension culture are centrifuged at 2000 g for 3 min at 4°, resuspended in HBS, and recentrifuged to obtain packed cells (50–100 μl, approximately 1 to 2 × 10⁷ cells depending on cell type). These can be stored at −70° or processed immediately. In either case the cell pellet is resuspended in 1.5 volumes of lysis buffer for 2 min at 0° and homogenized vigorously in a glass Dounce homogenizer; 0.1 volume of 10× buffer is added prior to centrifugation at 4° for 10 min at 10,000 g. The supernatant (S_{10}) can be assayed immediately or after storage at −70°.

Solid tissues (e.g., chick oviduct, guinea pig mammary gland[4]) are first frozen at −70°. Samples are "minced" with sharp-pointed scissors while still frozen, immediately prior to thawing, suspended in 1.5 volumes (or a volume equivalent to 1.5 times the weight of the sample) of lysis buffer, and homogenized; S_{10} is isolated as described above.

With monolayer cultures of cells the method is applicable to small numbers of cells, provided these contain relatively high levels of synthetase. For example, Kimchi *et al.*[7] have described the assay of 2-5A synthetase from cell monolayers of 96-well microtiter plates. After washing in buffered saline, the cells in each well were lysed with NP40 lysis buffer. The microtiter plate was centrifuged in an international swinging rotor at 3600 rpm for 15 min, and the supernatants (10–15 μl) were collected for assay with the original poly(I) · poly(C)-Sepharose method.[2] With larger cultures, the cells can be scraped off the bottle or plate, washed in HBS and the extract prepared as described above for suspension cultures.

The absorbance of the extract at 260 nm (A_{260}) after dilution in 0.4 M NaOH provides a convenient basis for comparison of the very large differences in synthetase levels (units/A_{260}) in extracts from different sources.[4]

Synthesis of 2-5A

Reagents

> Dialysis buffer glycerol (DBG): 10 mM HEPES buffer, pH 7.5; 90 mM KCl; 1.5 mM MgAc₂; 7 mM 2-mercaptoethanol; 20% glycerol (v/v)
>
> DBG (50 KCl, 8.5 MgAc₂, 3 ATP): 10 mM HEPES buffer, pH 7.5; 50 mM KCl; 8.5 mM MgAc₂; 7 mM 2-mercaptoethanol; 3 mM ATP (Sigma); 20% glycerol (v/v)
>
> [³H]ATP, 1 mCi/ml, 22 Ci/mmol (The Radiochemical Centre, Amersham, U.K.)
>
> *Procedure.* The 2-5A synthetase present in different extracts is bound

<hr>

[7] A. Kimchi, L. Shulman, A. Schmidt, Y. Chernajovsky, A. Fradin, and M. Revel, *Proc. Natl. Acad. Sci. U.S.A.* **76**, 3208 (1978).

to poly(I) · poly(C) paper by incubating 20 μl of extract with 0.1 cm^2 of paper for 1 hr at 25°. The paper is washed thoroughly with DBG prior to its incubation for 30 min at 25° with 1 ml of the same buffer and then for 18 hr at 30° in 25 μl of DBG (50 KCl, 8.5 MgAc$_2$, 3 ATP) to synthesize 2-5A. The yield of 2-5A can be assayed either by the sensitive biological, radioimmune, or radiobinding[8] assays or, for samples containing higher levels of synthetase, by [^{3}H]ATP incorporation into the product 2-5A (Williams *et al.* and Knight *et al.*, this volume [26] and [28]).

In this assay 0.1 cm^2 of poly(I) · poly(C) paper is at least five times more than enough to bind all of the 2-5A synthetase available in 20 μl of extract from rabbit reticulocytes, interferon-treated L cells, or chick oviducts.[4] There is a linear relationship between the amount of extract used in the assay and the amount of 2-5A synthesized. For example, if different amounts of extract from interferon-treated cells are mixed with extracts from control cells before adsorption of the synthetase to poly(I) · poly(C) paper, the yield of 2-5A is proportional to the amount of interferon extract employed (Fig. 1A and B). The level of ATP is not limiting, and synthesis of 2-5A continues for at least 40 hr (Fig. 1B). On occasion, for example, with extracts from HeLa cells, we have obtained evidence for residual phosphatase adsorbed to the poly(I) · poly(C) paper, emphasizing the necessity to check the linearity of the assay. The effect of this, if it occurs, can be countered by the inclusion of an ATP regenerating system of the type routinely employed in assays of cell-free protein synthesis.[9]

The incubation conditions given above are optimal for the synthetase from interferon-treated mouse L cells, rabbit reticulocytes, and chick oviduct and were used successfully in a survey of 2-5A levels in a variety of cells and tissues.[4] With extracts containing exceptionally high levels of synthetase, however, it is possible that higher ratios of poly(I) · poly(C) paper per unit of extract and higher concentrations of ATP may be required to maintain the linearity of the assay.

Comments

Assays of this general type employing the 2-5A synthetase bound to a solid support were first reported[2] with poly(I) · poly(C) bound to Sepharose rather than paper, and such assays continue to be used successfully.[7] The 2-5A synthetase from interferon-treated mouse L cells and rabbit reticulocytes (and presumably, therefore, from other tissues) is

[8] M. Knight, D. H. Wreschner, R. H. Silverman, and I. M. Kerr, this volume [28] and M. Knight, P. J. Cayley, R. H. Silverman, D. H. Wreschner, C. S. Gilbert, R. E. Brown, and I. M. Kerr, *Nature (London)* **288**, 189 (1980).
[9] L. Villa-Komaroff, M. McDowell, D. Baltimore, and H. F. Lodish, this series, Vol. 30, p. 709.

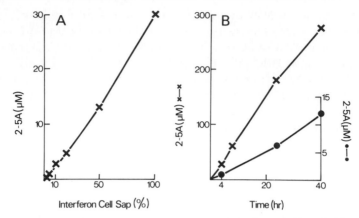

FIG. 1. Synthesis of 2-5A with the 2-5A synthetase bound to poly(I) · poly(C) paper. (A) Dependence upon 2-5A synthetase concentration. The ordinate gives the yield of 2-5A at 20 hr (×) obtained on incubation of the poly(I) · poly(C) paper-bound enzyme prepared from extracts from interferon-treated (high 2-5A synthetase) and control (negligible 2-5A synthetase) L cells (Hovanessian *et al.*, this volume [24]) mixed in the proportions shown on the abscissa from 0% interferon (100% control) on the left to 100% interferon (0% control) on the right. (B) Kinetics of 2-5A synthesis. The ordinates give the yields of 2-5A with time (abscissa) obtained with the poly(I) · poly(C) paper-bound enzyme from chick oviducts either 1 hr (●, low 2-5A synthetase) or 10 days (×, high 2-5A synthetase) after withdrawal from hormonal stimulation (Stark *et al.*[4]). Preparation of the extracts and of the poly(I) · poly(C) paper-bound enzymes and the incubation conditions for the synthesis of 2-5A are given in the text. The 2-5A product was assayed by its inhibitory effect on cell-free protein synthesis in comparison with suitable standards (Williams *et al.*, this volume [26]). The concentrations of 2-5A are given in AMP equivalents.

equally stable and active when bound to either poly(I) · poly(C)-Sepharose or poly(I) · poly(C) paper.[4] The poly(I) · poly(C) paper, however, has a number of advantages, especially reproducibility of preparation, high capacity (10–25 μg of poly(I) · poly(C)/cm²), stability on storage (−70°), and convenience in use, particularly in the routine assay of multiple small-volume samples. There is very little loss of poly(I) · poly(C) from the paper during prolonged incubation with bound enzyme, allowing the same paper to be reused several times. This is of particular value for the large-scale preparation of 2-5A when 100 cm² or more of the paper is used. The paper-bound synthetase retains full activity on storage for at least 6 weeks at −70°.

Accepting the linearity of the assay in proportion to the 2-5A synthetase content of known extracts (Fig. 1), it is reasonable to assume that the assay is suitable for the determination of available synthetase in extracts. We cannot yet be certain, however, that preactivated synthetase already bound to dsRNA (or an alternative activator) might not be sequestered and be unavailable for binding to poly(I) · poly(C). On the other hand, in control experiments the deliberate addition of dsRNA (10^{-6} g/ml) to interferon-treated L cell extracts prior to this type of assay failed to affect the results obtained, presumably reflecting the very high concentration and competitive advantage of the paper-bound poly(I) · poly(C).[10]

Acknowledgments

We are indebted to Drs. W. J. Dower and R. T. Schimke for the provision of the chick oviducts from which the extracts described in Fig. 1 were prepared. G. R. S. was recipient of a Josiah Macy Jr. Foundation grant.

[10] J. Wells and G. R. Stark have found that some batches of glycerol can cause inactivation of the synthetase, probably due to the presence of aldehydes, since small amounts of periodate-oxidized glycerol inactivate very effectively. They have also found that poly(I) · poly(C) attached after periodate oxidation to a hydrazide derivative of finely divided cellulose is superior to poly(I) · poly(C) paper as a support for synthetase. It gives tenfold higher rates of synthesis and much less nonspecific binding of other proteins.

[26] Assay of (2'-5')-Oligo(A) Synthesized *in Vitro* and the Analysis of Naturally Occurring (2'-5')-Oligo(A) from Intact Cells

By B. R. G. WILLIAMS, R. E. BROWN, C. S. GILBERT, R. R. GOLGHER, D. H. WRESCHNER, W. K. ROBERTS, R. H. SILVERMAN, and IAN M. KERR

Assay of 2-5A[1] synthesized in crude extracts or with the partially purified 2-5A synthetase can be carried out by (a) the incorporation of radioactive ATP into 2-5A; (b) "biological" assay of the product 2-5A by its ability to inhibit cell-free protein synthesis or to activate a nuclease(s); or (c) radioimmune or radiobinding assay. In general, the biological assays are less convenient but more than 1000-fold more sensitive than those based on the incorporation of radioactive ATP. The radioimmune and radiobinding assays remain to be firmly established but promise to combine the sensitivity of the biological with convenience approaching that of the radioactive ATP incorporation assays.[2] Here we present the methods used in our laboratory for radioactive ATP incorporation and "biological"

[1] The series of 2'-5'-linked 5'-triphosphorylated oligoadenylic acids collectively referred to as 2-5A is described in detail by Hovanessian *et al.* (this volume [24]). The terms trimer, tetramer, and pentamer will be used to denote preparations consisting largely of the 5'-triphosphates but also containing smaller amounts of the corresponding di- and monophosphates. The more complete nomenclature, e.g., trimer triphosphate, will be used to identify the individual species.

[2] M. Knight, D. H. Wreschner, R. H. Silverman, and I. M. Kerr, this volume [28], and M. Knight, P. J. Cayley, R. H. Silverman, D. H. Wreschner, C. S. Gilbert, R. E. Brown, and I. M. Kerr, *Nature (London)* **288**, 189 (1980).

assays. Similar but slightly different assays that have been developed by others may prove to be more attractive to individual users depending on their circumstances.[3,4]

The wide distribution of the 2-5A synthetase in a variety of cells and tissues and its variation with growth and hormone status have emphasized the wider possible significance of the 2-5A system and the desirability of assay methods for the routine estimation of naturally occurring preformed 2-5A.[5] Accordingly, the methods that have been used successfully in the initial experiments on the recovery of 2-5A from intact cells[6] are also presented.

Materials

DEAE paper disks (DE-81, 2.3 cm in diameter) were from Whatman, and thin-layer chromatography (TLC) plates of polyethyleneimine cellulose (PEI cellulose, Machery Nagel Polygram Cel 300 PEI) were from Camlab. Bacterial alkaline phosphatase (BAP; EC 3.1.3.1; 250–450 units/ml; 20–30 units/mg protein) was from Sigma (product 4252).

Vesicular stomatitis virus (VSV) mRNA labeled with [^{3}H]UTP was prepared *in vitro* and purified on oligo(dT)-cellulose.[7] In the gel nuclease assays (see below) similar results are obtained using ^{32}P-labeled polyadenylated mRNA or ribosomal RNA from Ehrlich ascites tumor (EAT) cells[8] or ribosomal RNA labeled at the 3' terminus with [5'-^{32}P]pCp.[9] The last is by far the most convenient to prepare. Ribosomal RNA (10 μg, 18 or 28 S) is incubated with 3 μl of T4 RNA ligase (EC 6.5.1.3), P-L Biochemicals product (0880; 3700 units/ml; 2055 units/mg protein) and 20 μCi of [5'-^{32}P]pCp (2 to 3 × 10^6 Ci/mol, Radiochemical Centre, Amersham, U.K.; lyophilized and redissolved in the reaction mix) in 40 μl (final volume) of 50 mM HEPES buffer, pH 7.6, 7.5 mM MgCl$_2$; 3.3 mM dithiothreitol and 10% dimethyl sulfoxide (v/v) for 16 hr at 2–4°. The ^{32}P-labeled RNA is separated from residual [^{32}P]pCp and ^{32}P$_i$ by passage through a column (7 cm × 0.7 cm in diameter) of Sephadex G-100 in H$_2$O. One-drop fractions are collected. The [^{32}P]RNA that elutes in the void volume is lyophilized and resuspended in H$_2$O prior to use (yield: 150 μl of 60,000 cpm/μl and 10^6 cpm/μg RNA representing 25% utilization of the [^{32}P]pCp).

Interferon treatment of cells, preparation of cell extracts, and purifica-

[3] C. Baglioni, M. A. Minks, and P. A. Maroney, *Nature (London)* 273, 686 (1978).
[4] M. A. Minks, S. Benvin, P. A. Maroney, and C. Baglioni, *J. Biol. Chem.* 254, 5058 (1979).
[5] G. R. Stark, W. J. Dower, R. T. Schimke, R. E. Brown, and I. M. Kerr, *Nature (London)* 278, 471 (1979).
[6] B. R. G. Williams, R. R. Golgher, R. E. Brown, C. S. Gilbert, and I. M. Kerr, *Nature (London)* 282, 582 (1979).
[7] L. A. Ball and C. N. White, *Virology* 84, 479 (1978).
[8] B. R. G. Williams, R. R. Golgher, and I. M. Kerr, *FEBS Lett.* 105, 47 (1979).
[9] T. E. England and O. C. Uhlenbeck, *Nature (London)* 275, 560 (1978).

tion of radioactive ATP are as described by Hovanessian *et al.* (this volume [24]).

Assay of 2-5A

The synthesis of 2-5A in crude extracts or with the partially purified synthetase is described by Hovanessian *et al.* (this volume [24]). Where necessary, for example with material synthesized with crude extracts, the incubation mixtures (diluted 10-fold in the case of the reticulocyte lysate) are heated to 90° for 5 min and centrifuged to remove denatured protein prior to the assay of the product 2-5A.

Assay by Incorporation of Radioactive ATP. If [^{3}H]ATP is used in the synthetase reaction its conversion to 2-5A can be readily monitored, as first described by Ball,[10] by digestion of the products with BAP to convert [^{3}H]2-5A to "core" (A2'p)$_n$A (where $n \geq 1$), which is negatively charged and will bind to DEAE paper or a PEI-cellulose TLC plate and residual [^{3}H]ATP to A, which is uncharged and is readily removed by washing from either of these supports. In typical assays containing 1 mM ATP, 10-μl samples of the reaction product are incubated with 1 μl of BAP (diluted to 0.025 unit/μl) for 1 hr at 37°. For higher concentrations of ATP longer incubations or higher concentrations of BAP should be used [1 μl of undiluted BAP (0.25 units/μl) is sufficient to digest 10 μl of 10 mM ATP in 1 hr at 37°]. In the latter case, the BAP supplied as a suspension in saturated ammonium sulfate should be recovered by centrifugation and dissolved in 20 mM Tris · HCl, pH 8.0, prior to use, as otherwise the ammonium sulfate will interfere with the subsequent TLC analysis of the products. Samples of the BAP digests are then assayed either directly on DEAE paper or after chromatography on PEI-cellulose.

In the DEAE paper procedure, 2-μl samples are applied to numbered DEAE paper disks that can be washed together in a beaker five times with water with gentle agitation because wet DEAE paper is very fragile. The disks are dried and counted in a toluene-based scintillant.

For chromatographic analysis 1-μl samples of each BAP digest plus 1 μl of marker mix [10 mM of each of (A2'p)$_n$A, where n = 0 to 3] are applied at 1-cm intervals 1 cm from the foot of a strip of PEI-cellulose 7 to 10 cm high, and the plate is developed (30 min) in 0.1 M NH$_4$HCO$_3$ or 1 M acetic acid.[11] The dimer, trimer, and tetramer "core" areas are located under ultraviolet (UV) light, marked, cut out, and placed in scintillation vials. The "cores" are eluted in 200 μl of 0.5 M NH$_4$HCO$_3$ for 5 min at room temperature prior to mixing with 2 ml of methoxyethanol and 3 ml of toluene-based scintillant for counting.

Appropriate controls for the complete digestion of the [^{3}H]ATP to [^{3}H]A and for the efficiency of counting must be included with both of the

[10] L. A. Ball, *Virology* **94**, 282 (1979).
[11] I. M. Kerr and R. E. Brown, *Proc. Natl. Acad. Sci. U.S.A.* **75**, 256 (1978).

above methods as well as for determination of background levels of residual [³H]A after washing the DEAE paper and of the recovery (usually 70%) of "cores" on elution after the chromatographic analysis.

Without chromatography, the DEAE paper BAP "core" assay detects all species of 2-5A including the biologically inactive dimer. Moreover, it provides no identification of the product which with crude extracts could contain [³H]A-labeled products other than 2-5A. Nevertheless, it is extremely convenient and is recommended when the latter can be excluded by experience. The chromatographic analysis takes surprisingly little longer, however, and in this way quantitation of the yields, identification of the 2-5A, and an analysis of its heterogeneity are all obtained.

In both of these assays, $\geq 2\%$ conversion of the input [³H]ATP is required for a clear signal above background. The absolute amount of product 2-5A detectable depends, therefore, on the concentration of ATP employed, for example, 20 or 100 μM 2-5A (in AMP equivalents) with 1 and 5 mM ATP, respectively. At the limit of detection with high specific activity [³H]ATP (20 Ci/mmol) at 1 mCi/ml, 2% conversion would correspond to 1 μM 2-5A. The reluctance of the synthetase to utilize ATP at concentrations of ≤ 100 μM (Hovanessian et al., this volume [24], Fig. 3) makes radioactivity assays based on [³²P]ATP subject to a similar limitation in sensitivity. The main value of these radioactivity-based assays, therefore, lies in monitoring the synthesis of 2-5A with systems containing relatively high concentrations of enzyme and consequently not requiring the much greater sensitivity of the biological assays.

Assay by Inhibition of Cell-Free Protein Synthesis. The preparation of suitable cell-free systems from rabbit reticulocytes or EAT, L, and HeLa cells and the assay of protein synthesis programmed with exogenous mRNA have been described in detail elsewhere.[12-14] In the case of the reticulocyte lysate, the use of the micrococcal nuclease-treated system[12] with exogenous mRNA is recommended because synthesis on endogenous mRNA occurs rapidly in the untreated native system, whereas the inhibition by 2-5A operating through the activation of a nuclease takes a significant time to develop.[15] It is, therefore, best seen in systems in which amino acid incorporation proceeds linearly for at least an hour, preferably longer. With any of the above systems suitable dilutions of the product 2-5A are added at the beginning of the incubation and the effect on amino acid incorporation is compared at or over a period of 120 min with that of a similar series of dilutions of a standard preparation of 2-5A added in

[12] H. R. B. Pelham and R. J. Jackson, *Eur. J. Biochem.* **67**, 247 (1976).

[13] L. Villa-Komaroff, M. McDowell, D. Baltimore, and H. F. Lodish, this series, Vol. 30, p. 709.

[14] I. M. Kerr, R. E. Brown, M. J. Clemens, and C. S. Gilbert, *Eur. J. Biochem.* **69**, 551 (1976).

[15] B. R. G. Williams, C. S. Gilbert, and I. M. Kerr, *Nucleic Acids Res.* **6**, 1335 (1979).

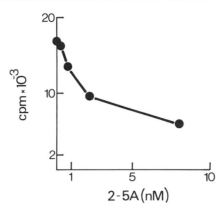

Fig. 1. Inhibition of cell-free protein synthesis by (2'-5')-oligo(A) (2-5A). A standard preparation of reticulocyte 2-5A was added to the final concentrations indicated (AMP equivalents, abscissa) to an encephalomyocarditis (EMC) RNA-programmed cell-free protein synthesizing system from Ehrlich ascites tumor cells. The ordinate gives the ^{14}C-labeled amino acid incorporation per 10-μl sample.

parallel. Inhibition of amino acid incorporation in such systems normally occurs at final 2-5A concentrations of 1–10 nM in AMP equivalents (Fig. 1). In the reticulocyte system, in contrast to those from L, EAT, or HeLa cells, the trimer is essentially inactive, the higher oligomers being required to activate the nuclease[6] (see below). These assays thus are sensitive and reproducible, but are not capable of accurately measuring small differences ($\leq$2-fold) in 2-5A concentration. A significant disadvantage is that such systems are rather sensitive to nonspecific inhibition by, for example, changes in salt concentration. It is normally essential, therefore, to dilute crude 2-5A preparations $\geq$50-fold overall in order to remove any possibility of such nonspecific effects. The minimum concentration of 2-5A required in a crude extract for it to be detectable after a 50-fold dilution with this type of assay is, therefore, of the order of 50 nM. From this point of view the alternative biological assay by nuclease activation is preferable, being significantly (at least 10-fold) less sensitive to such nonspecific inhibition.

Assay by Activation of a Nuclease. In these assays activation of a nuclease(s) by 2-5A in cell-free protein synthesizing systems is measured by the degradation of added ^{3}H- or ^{32}P-labeled messenger or ribosomal RNA. Depending on the system, breakdown of the RNA is best monitored either by its conversion to trichloroacetic acid-soluble material or by electrophoretic analysis on SDS–polyacrylamide gels (gel nuclease assay).[6,15,16] The overall rate and extent of breakdown of the RNA in these

[16] B. R. G. Williams, I. M. Kerr, C. S. Gilbert, C. N. White, and L. A. Ball, *Eur. J. Biochem.* **92**, 455 (1978).

systems reflects the combined effects of the 2-5A-activated nuclease and other endogenous nucleases. The effect of the latter are counteracted to different extents in different systems by endogenous nuclease inhibitors. Accordingly, with reticulocyte lysates, which display a low level of net endogenous nuclease activity, we have found electrophoretic analysis on gel to be necessary.[6,15] Conversion to trichloroacetic acid-soluble material has been used with EAT, L, and HeLa cell extracts.[6,16,17] The reticulocyte "gel nuclease assay" may appear at first sight the more cumbersome, but it has the advantage of a clearer signal-to-background ratio and has been the preferred assay in this laboratory. It has consistently proved to be as sensitive as any available. With most lysates the limits of detection for tetramer [ppp(A2′p)$_3$A] and pentamer [ppp(A2′p)$_4$A] are in the range 0.1–2.4 nM, averaging about 0.8 nM (in AMP equivalents). This compares with values from about 0.5–8 nM for L and EAT cell extracts assayed either by activation of a nuclease or inhibition of protein synthesis. In the light of this, it is unfortunate that the 2-5A activatable nuclease in the reticulocyte lysate, in contrast to that in EAT or L cells, requires the tetramer or pentamer (higher oligomers not yet tested) for activation, the trimer being essentially inactive in this particular system.[6]

It is clear from the above that the choice of assay is determined by the objectives and circumstances. Accordingly, procedures for both the gel and trichloroacetic acid assay are presented.

In the assay of nuclease activity in the rabbit reticulocyte lysate system 3–300 ng (equivalent to 3000–5000 cpm at a specific activity of 10^4 to 10^6 cpm/μg of ^{32}P-labeled mRNA or ribosomal RNA are normally added per 12.5-μl assay. The assays are set up and incubated for 60 min at 30° exactly as for the measurement of cell-free protein synthesis, except that no radioactive amino acid is added.[13,15] An equal volume of a solution containing 4% SDS (w/v), 140 mM 2-mercaptoethanol, 125 mM Tris · HCl, pH 6.8, and 40% glycerol (v/v) is added; the mixture is heated at 90° for 2 min and loaded onto 7.5% polyacrylamide gels containing 0.1% SDS prepared as described by Laemmli[18] for the analysis of proteins. Electrophoresis is routinely for 2 hr at 150 V (approximately 40 mA per 14 cm × 10 cm slab gel). The gels can then be wrapped in cling film and autoradiographed using an intensifier screen (e.g., Fast Tungstate from Ilford) and pre-exposed film to enhance the sensitivity,[19] or dried and subjected to autoradiography in the routine way. The concentration of 2-5A in an unknown sample is determined by comparison of the effect of

[17] M. A. Minks, P. A. Benvin, P. A. Maroney, and C. Baglioni, *Nucleic Acids Res.* **6**, 767 (1979).
[18] U. K. Laemmli, *Nature (London)* **227**, 680 (1970).
[19] R. A. Lasky and A. D. Mills, *FEBS Lett.* **82**, 314 (1977).

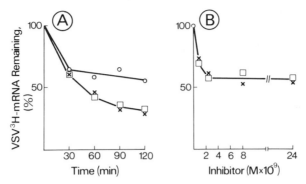

Fig. 2. Activation of a nuclease by the trimer and tetramer components of reticulocyte (2'-5')-oligo(A) (2-5A). Breakdown of ^{3}H-labeled vesicular stomatitis virus (VSV) mRNA on incubation in an EAT cell-free system: (A) alone (○) or with 8 nM trimer (×) or tetramer (□) or (B) with different concentrations (AMP equivalents, abscissa) of trimer (×) or tetramer (□) for the times indicated in (A) and for 120 min in (B). The ordinates give the residual acid-insoluble radioactivity as a percentage of the zero time value (2600 cpm per 4-μl sample) in (A) and of the 120-min control (no inhibitor) value (3500 cpm per 10-μl sample) in (B). From Williams et al.[6]

suitable dilutions with that of appropriate standards. The results of a typical assay are presented below (see Fig. 3) in relation to the detection of naturally occurring 2-5A in extracts from interferon-treated, virus-infected L cells.

With EAT, L, or HeLa cell extracts, a TCA assay can be used.[6,16,17] In a typical assay comparing the activity of reticulocyte trimer and tetramer (Fig. 2), ^{3}H labeled VSV mRNA (4 × 10⁵ cpm/μg) was added to an EAT cell-free system to a final concentration of 2 μg/ml and incubated at 30° under normal protein synthesis conditions.[6,14,16] Of the 2-5A solution under test, 2 μl were added per 25-μl assay. At appropriate times (Fig. 2) samples (4 μl) were transferred to glass fiber filters (Whatman GF/C) that were soaked in 5% trichloroacetic acid and air-dried prior to use. The filters were placed in 5% trichloroacetic acid (w/v) for 10 min at 0° and washed three times with ice cold 5% trichloroacetic acid, then with ethanol and acetone. They were dried and counted in a liquid scintillation counter with a toluene-based scintillant. The ordinates (Fig. 2) give the residual acid-insoluble [^{3}H]RNA remaining as a percentage of the zero time value for systems (Fig. 2A) incubated in the absence of 2-5A or with 8 nM trimer or tetramer. The concentration of 2-5A in an unknown sample is determined by comparison of the effect of a series of dilutions with those of a range of suitable standards. Typical concentration curves obtained with trimer and tetramer are presented in Fig. 2B.

Analysis of Naturally Occurring 2-5A from Intact Cells

The 2-5A is effective at nanomolar concentrations in cell-free systems. Assuming that naturally occurring 2-5A is effective at similar concentrations in the intact cell, methods capable of detecting such levels are ideally required. Even with ^{32}P, specific activities in excess of 100–1000 Ci per mole of nucleotide cannot normally be achieved in the intact cell, and even the higher specific activity would be barely sufficient for the characterization of 2-5A present at nanomolar concentrations in 1 ml of packed cells. Moreover, labeling in this way would not be possible for the detection of 2-5A in tissue samples in animal experiments. Although convenient routine methods for the detection of 2-5A in such samples have yet to be developed, naturally occurring, preformed, 2-5A has been detected in extracts from interferon-treated, encephalomyocarditis (EMC) virus-infected mouse L cells with the "biological" assays described above.[6] In six different experiments it was recovered in amounts corresponding to concentrations of 20–200 nM in the intact cell. The detection of 2-5A at concentrations lower than this has been complicated by the nonspecific inhibitory effects of the crude extracts in the "biological" assay employed, the sensitivity of the assays, and the amounts of starting material readily available. However, radioimmune and radiobinding assays are greatly facilitating analyses of this type. It is to be hoped that with these alone, or in combination with high-performance liquid chromatography (HPLC) or more trivial partial purification procedures, the routine analysis of cells and tissues for nanomolar levels of 2-5A will be possible. Meanwhile, naturally occurring 2-5A can be detected in crude trichloroacetic acid-soluble extracts that are simple to prepare from intact cells. Recoveries with these simple procedures appear to be good (>50% on the basis of control experiments in which 200 nM 2-5A was added to extracts at the time of extraction).

Extraction Procedure. Mouse L cells (300 ml, 0.75×10^6 cells/ml) were treated for 17 hr at 37° with 400 reference (20 effective) units per milliliter of partially purified (5×10^7 reference units/mg protein) mouse interferon, infected with 20 PFU per cell of EMC virus and, after a 30-min adsorption period, incubated for 4 hr at 37°. Control cells with or without interferon treatment and EMC infection were incubated in parallel. Cells were pelleted by centrifugation at 300 g for 10 min, resuspended in 10 ml of medium, and recentrifuged at 1000 g to obtain the packed cells (about 1 ml). These were lysed (with or without prior storage at −70°) by addition of an equal volume of a solution containing 0.5% NP40 (v/v), 10 mM HEPES buffer, pH 7.5, 90 mM KCl, and 1.5 mM MgAc$_2$. Then 50% trichloroacetic acid was immediately added to a final concentration of 5%, and the precipitated protein was removed by centrifugation for 10 min at 10,000 g at 0–5°. The supernatant solutions were extracted six times at room tem-

FIG. 3. Recovery of (2'-5')-oligo(A) (2-5A) from intact cells. The 2-5A was assayed by its ability to activate a nuclease and cause the degradation of added [^{32}P]RNA in rabbit reticulocyte lysates under the conditions described in the text. An autoradiograph of the electrophoretic analysis on a 14 × 10 cm SDS–polyacrylamide slab gel of the [^{32}P]RNA is shown. Comparison of 2-5A activity isolated from control (A), encephalomyocarditis (EMC)-infected (B), interferon-treated (C), and interferon-treated, EMC virus-infected (D–H) cells at an overall dilution of 6-fold (A–D) and 18-, 54-, 162- and 486-fold (E–H), respectively; I, control (no inhibitor); J–M, plus 8, 2.4, 0.8, and 0.24 nM (AMP equivalents) reticulocyte 2-5A. The results shown are for an early experiment in which a similar but lengthier extraction procedure was used in the recovery of 2-5A from intact cells (Williams *et al.*[6]). Essentially identical results have since been obtained with the simpler extraction procedure described in the text. Undegraded [^{32}P]RNA remains at the top of the gel.

perature with water-saturated ether to remove the trichloroacetic acid (raising the pH to ≥5), lyophilized, and taken up in 20–200 µl (per milliliter or original packed cells) of 20 mM HEPES buffer, pH 7.5, and adjusted to pH 7 with NH$_4$OH prior to analysis by the reticulocyte "gel nuclease assay" described above. Typical results are shown in Fig. 3. That the biological activity detected in such extracts is indeed 2-5A can be established by its further analysis by HPLC in comparison with suitable 2-5A standards.[6] In one such analysis the triphosphorylated trimer, tetramer, and pentamer were all present (the dimer and higher oligomers would not have been detected). Smaller amounts of the corresponding diphosphates were also detected, and it is possible that they are present also in the intact cell, but it is equally likely that they reflect the partial breakdown of the triphosphates during extraction. The amine columns used in the initial HPLC studies proved to be unsatisfactory, being unstable under the experimental conditions required to obtain good separations.[6] Alternative methods with octadecylsilyl-silica columns (e.g., μBondapak C_{18}, Waters Associates) are now being developed and are described by Brown et al. (this volume [27]).

Acknowledgments

R. R. G. was the recipient of a Fellowship from the Conselho Nacional de Desenvolvimento Cientifico e Tecnologico, Brazil; and W. K. R. of a grant from the American Cancer Society.

[27] Analysis of (2'-5')-Oligo(A) and Related Oligonucleotides by High-Performance Liquid Chromatography

By R. E. BROWN, P. J. CAYLEY, and IAN M. KERR

(2'-5')-Oligo(A) (2-5A) is a complex mixture of the 5'-triphosphorylated oligoadenylic acids $ppp(A2'p)_nA$ (where n = 1 to $\geq$4). Lesser amounts of the corresponding 5'-di- and monophosphates are also normally present. It can be synthesized in vitro with the 2-5A synthetase (Hovanessian et al., this volume [24]) and occurs naturally in intact cells (Williams et al. this volume [26]).[1–4] Resolution of the major components of 2-5A can be achieved by chromatography on DEAE-cellulose or DEAE-Sephadex in the presence[1,2] or the absence[4] (this volume [24]) of urea. These methods, however, are time consuming and inconvenient for the routine analysis of multiple samples. On the other hand, thin-layer chromatography (TLC) is ideal for the latter, but no single TLC system

[1] I. M. Kerr and R. E. Brown, Proc. Natl. Acad. Sci. U.S.A. **75,** 256 (1978).

[2] L. A. Ball and C. N. White, Proc. Natl. Acad. Sci. U.S.A. **75,** 1167 (1978).

[3] B. R. G. Williams, R. R. Golgher, R. E. Brown, C. S. Gilbert, and I. M. Kerr, Nature (London) **282,** 582 (1979).

[4] E. M. Martin, N. J. M. Birdsall, R. E. Brown, and I. M. Kerr, Eur. J. Biochem. **95,** 295 (1979).

has yet been reported that is capable of resolving all the different components of interest.

Here we report high-performance liquid chromatography (HPLC) systems that we have found to be invaluable in the analysis of 2-5A and related products, such as those synthesized by the 2-5A synthetase with nicotinamide-adenine dinucleotide (NAD) as primer (NAD-2-5A) first reported by Ball and White.[5] Among the advantages of these HPLC systems are their high resolution and high sensitivity: 100 pmol in AMP equivalents can be readily detected by absorbance at 260 nm. In addition, the fractionated products are obtained in a highly purified state in solvents that do not appear to interfere with their subsequent assay (except in cell-free protein synthesizing systems in which some dilution of the salt is essential) and from which they can be readily recovered. Each run takes less than 30 min, allowing the convenient analysis of large numbers of samples.

In our initial analysis of 2-5A occurring naturally in interferon-treated, encephalomyocarditis virus-infected L cells, amine columns (isopropylamine-bonded phase) from a number of suppliers were used.[3] These columns gave very good separation of the individual components present in 2-5A (Fig. 1), but have proved to be unstable in the phosphate buffers essential to achieve these separations.[6] The use of C_{18} columns (octadecylsilyl-bonded phase) was therefore investigated. These columns give highly reproducible separations of the various components of interest (Figs. 2–5) and have proved to be completely stable under the different conditions used in several hundred runs over more than 6 months.

Apparatus and Materials

The HPLC system comprised a μBondapak C_{18} column, two 6000A pumps, a 660 solvent programmer, a U6K injector valve (all from Waters Associates) and a fixed-wavelength (254 nm) spectrophotometric detector (Altex Model 153) linked to a Phillips (PM 8252) dual-pen recorder.

Distilled water was further purified using a Vaponics unit (VLT1) with

---

[5] L. A. Ball and C. N. White, in "Regulation of Macromolecular Synthesis by Low Molecular Weight Mediators (G. Koch and D. Richter, eds.), pp. 303–317. Academic Press, New York, 1980.

[6] In our experience, amine columns from more than one supplier showed variable progressive "breakdown" (apparent loss of the bonded phase) after use as in Fig. 1 with phosphate buffers. Breakdown continued on subsequent storage in methanol. Accordingly, even during the relatively short useful life of the column, constant readjustments of pH and ionic strength were frequently required to maintain good separations. On widespread inquiry, it appears that such "breakdown" of amine columns in phosphate buffers is a recognized but little publicized phenomenon, the basis for which is not understood.

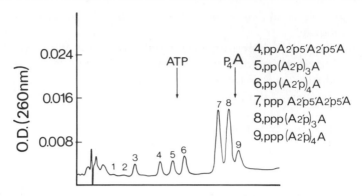

FIG. 1. High-performance liquid chromatography (HPLC) analysis of (2'-5')-oligo(A) (2-5A) on an amine column. Reticulocyte 2-5A containing the trimer, tetramer, and pentamer 5'-triphosphates (peaks 7–9, respectively) and lesser amounts of the corresponding 5'-di- (peaks 4–6) and mono- (peaks 1–3) phosphates was analyzed on a LiChrosorb 10 NH_2 column with a linear gradient of 40 to 200 mM ammonium phosphate, pH 7.2. In a parallel run, ATP and p_4A were included with the above components, and the positions at which they eluted are indicated by arrows. From Williams et al.[3]

an MR01 cartridge and VMF7.2.22 membrane filter. Organic solvents were HPLC grade from Rathburn Chemicals (Walkerburn) Ltd., Walkerburn, Peeblesshire, Scotland. Ammonium phosphate (AR grade, Hopkin and Williams) solutions were made up each day, and the pH was adjusted with ammonium hydroxide. Potassium phosphate (AR) was from BDH. All solutions were deaerated and passed through a Millipore filter prior to use.

The 2-5A was prepared, purified, and fractionated as described by Hovanessian et al. (this volume [24]). The NAD 2-5A was prepared in an essentially identical way with the poly(I) · poly(C) paper-bound 2-5A synthetase from interferon-treated HeLa cells, but with 10 mM NAD in addition to 5 mM ATP in the reaction. The ATP, NAD, and 3'-5'-linked ApA, $(Ap)_2A$ and $(Ap)_3A$ were obtained from Sigma and Boehringer. Residual particulate material ("lint") in the samples to be analyzed by HPLC was removed by centrifugation at 10,000 g.

HPLC Analysis of 2-5A and Related Nucleotides

In all cases the 2-5A or related products are eluted with a gradient of methanol: H_2O (1 : 1,v/v) using different buffer systems. The first system[7] employing 50 mM ammonium phosphate, pH 7, separates the individual

[7] Based on the systems described by F. S. Anderson and R. C. Murphy [*J. Chromatogr.* **121**, 251 (1976)] for the separation of purine nucleosides and nucleotides.

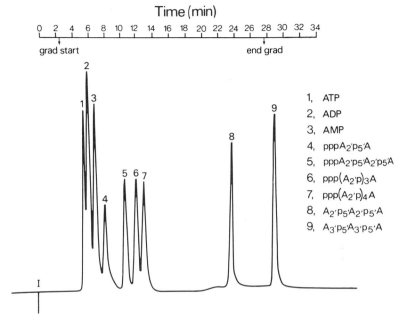

FIG. 2. High-performance liquid chromatography (HPLC) analysis of (2'-5')-oligo(A) (2-5A) in ammonium phosphate at pH 7. Reticulocyte 2-5A containing the trimer, tetramer, and pentamer 5'-triphosphates and lesser amounts of the corresponding di- and monophosphates was analyzed together with AMP, ADP, ATP, A2'p5'A2'p5'A, and A3'p5'A3'p5'A on a μBondapak C_{18} column. Pump A delivered 50 mM ammonium phosphate, pH 7. Elution (1 ml/min) was with a linear gradient (program 6) of from zero to 50% of 1:1 methanol:H_2O (pump B) in 25 min. The injection point (I), start (allowing for the void volume), end of the gradient, and position of elution of the individual components are shown on the figure. The 5'-mono-, di-, and triphosphates of the individual oligomers of 2-5A are not resolved in this system.

2-5A oligomers [ppp(A2'p)$_n$A where n = 1–4; higher oligomers not yet tested] from each other and from the AMP, ADP, and ATP that are normally present as "contaminants" in enzymically synthesized 2-5A (Fig. 2). The nonphosphorylated "cores" of 2-5A [(A2'p)$_n$A, where n = 1–3, higher oligomers not tested] are not resolved in this system but elute as a single peak (peak 8, Fig. 2) with a greater retention time than the phosphorylated components and are clearly separated from the corresponding 3'-5'-linked oligomers [(A3'p)$_n$A where n = 1–3] (peak 9, Fig. 2). These conditions are, therefore, ideally suited for an initial analysis of enzymically synthesized or naturally occurring 2-5A, in which the resolution of the 5'-mono-, di-, and triphosphorylated components of the individual oligomers and of the nonphosphorylated "cores" is not required. An analysis of the mono-, di-, and triphosphate components of the trimer,

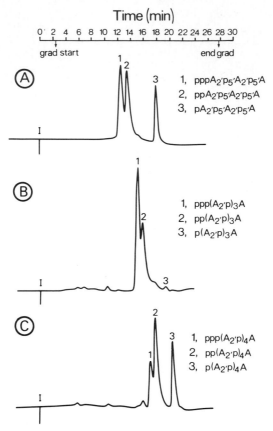

FIG. 3. High-performance liquid chromatography (HPLC) analysis of (2'-5')-oligo(A) (2-5A) in ammonium phosphate at pH 6. (A) Chemically synthesized 5'-tri-, di-, and monophosphorylated A2'p5'A2'p5'A (peaks 1–3, respectively; Jones and Reese.[8]) (B) and (C) Enzymically synthesized ppp(A2'p)₃A and ppp(A2'p)₄A, respectively, containing the corresponding di- and monophosphates. (D) Recombined mixture of enzymically synthesized 2-5A. Separation was on a µBondapak C_{18} column. Peaks corresponding to the individual components, the injection point (I), the start (allowing for the void volume), and end of the gradients are shown. In (D) the positions at which ATP, ADP, and AMP eluted in a parallel run are shown on the figure. Pump A delivered 50 mM ammonium phosphate, pH 6. Elution (1 ml/min) was with a linear gradient (program 6) from zero to 50% of 1:1 methanol:H_2O (pump B) in 25 min.

tetramer, and pentamer (Figs. 3A, B, and C, respectively) can be achieved in the same solvents but at pH 6. The trimer standard (Fig. 3A) was a recombined mixture of chemically synthesized[8] pppA2'p5'A2'p5'A (peak 1), ppA2'p5'A2'p5'A (peak 2), and pA2'p5'A2'p5'A (peak 3). In Figs. 3B and C, enzymically synthesized tetramer (B) and pentamer (C)

[8] S. S. Jones and C. B. Reese, *J. Am. Chem. Soc.* **101**, 7399 (1979).

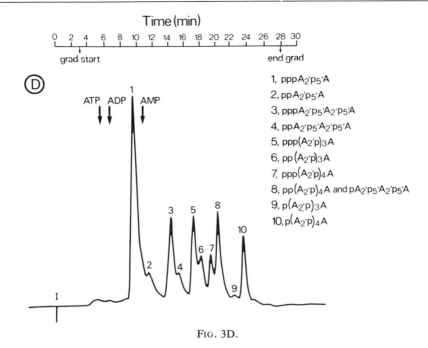

Time (min)

0 2 4 6 8 10 12 14 16 18 20 22 24 26 28 30

grad start   end grad

D

ATP ADP | AMP

1, pppA$_2$'p$_5$'A
2, ppA$_2$'p$_5$'A
3, pppA$_2$'p$_5$'A$_2$'p$_5$'A
4, ppA$_2$'p$_5$'A$_2$'p$_5$'A
5, ppp(A$_2$'p)$_3$A
6, pp(A$_2$'p)$_3$A
7, ppp(A$_2$'p)$_4$A
8, pp(A$_2$'p)$_4$A and pA$_2$'p$_5$'A$_2$'p$_5$'A
9, p(A$_2$'p)$_3$A
10, p(A$_2$'p)$_4$A

FIG. 3D.

were employed and the individual peaks (1 to 3) corresponding to the 5'-tri', di- and monophosphates, respectively, were identified by TLC. [Most preparations of ppp(A2'p)$_n$A contain lesser amounts of the corresponding 5'-di- and monophosphates, which are thought to be breakdown products of the major triphosphorylated components.[2-4]]. An analysis of a recombined mixture of enzymically synthesized dimer, trimer, tetramer, and pentamer under the same pH 6.0 conditions is shown in Fig. 3D. As would be expected from Fig. 3A–C, analysis in this system provides information concerning the degree of phosphorylation of the individual oligomers that is not provided by analysis at pH 7 (Fig. 2), but the pattern obtained is correspondingly more complex, with the disadvantage that there is some overlap between the monophosphorylated components of the lower oligomers with the di- and triphosphorylated higher oligomers; for example, the trimer monophosphate (pA2'p5'A2'p5'A) is not resolved from the pentamer diphosphate [pp(A2'p)$_4$A] (peak 8, Fig. 3D).

For a more detailed analysis of the nonphosphorylated "cores" of 2-5A, 4 mM potassium phosphate, pH 6.5, is employed. Once again elution is with a gradient of methanol. Under these conditions the 2'-5'-linked dimer, trimer, and tetramer "cores" are resolved and well separated from the corresponding 3'-5'-linked molecules (Fig. 4).

Ball first showed that an enzyme that is almost certainly the 2-5A

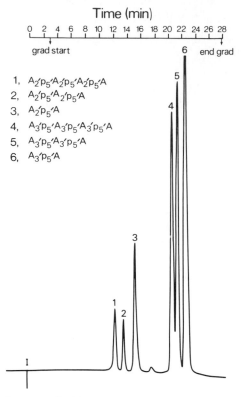

FIG. 4. High-performance liquid chromatography (HPLC) analysis of the nonphosphorylated "cores" of (2'-5')-oligo(A) (2-5A). Enzymically synthesized 2-5A oligomers from which the terminal phosphates had been removed by digestion with bacterial alkaline phosphatase [$(A2'p)_nA$, $n = 1-3$] were analyzed together with commercial A3'p5'A, (A3'p)$_2$A, and (A3'p)$_3$A on a μBondapak C$_{18}$ column. Peaks corresponding to the individual components, the injection point (I), the start (allowing for the void volume), and end of the gradient are shown on the figure. Pump A contained 4 mM potassium phosphate, pH 6.5. Elution (1 ml/min) was with a linear gradient (program 6) from zero to 30% of 1 : 1 methanol : H$_2$O (pump B) in 25 min.

synthetase is capable of adding one or more AMP residues from ATP to an NAD primer.[5] Nuclear magnetic resonance analysis of the product[9] has confirmed the 2'-5' nature of the additional linkages and indicated that the first additional AMP is linked to the AMP, rather than the nicotinamide ribose moiety of the NAD molecule. The NAD, NAD2'p5'A, and NAD 2'p5'A2'p5'A can also be resolved in the ammonium phosphate, pH 7, system described above (Fig. 5). The ammonium phosphate system at pH 7 also, incidentally, gives excellent separation of 2', 3', and 5' AMP.

[9] P. J. Cayley, unpublished work, 1979.

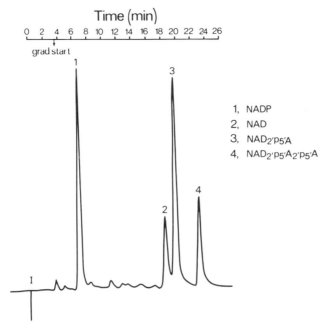

FIG. 5. High-performance liquid chromatography (HPLC) analysis of NAD-2-5A. Enzymically synthesized NAD-2-5A was analyzed together with NAD and NADP on a μBondapak C$_{18}$ column. Peaks corresponding to the individual components, the injection point (I), and the start (allowing for the void volume) of the gradient are shown on the figure. Pump A delivered 50 mM ammonium phosphate, pH 7. Elution (1 ml/min) was with a linear gradient (program 6) from zero to 20% of 1 : 1 methanol water (pump B) in 30 min.

With all of the above systems that employ a gradient for elution, it is essential to reequilibrate the column with the starting buffer for 10–20 min between runs.

Recoveries with the above procedures are routinely in excess of 80% even with picomole amounts of material, and there is no detectable loss of biological activity on the columns. We have, however, noticed, particularly with relatively small (≤10 μCi) amounts of ^{32}P-labeled nucleotides at specific activities in excess of 10^6 Ci/mol, that considerable losses can occur if polyvinylchloride (to which the nucleotides stick) rather than Teflon tubing is used between the detector and fraction collector.

Concluding Remarks

Although the data presented (Figs. 1–5) were obtained with recombined mixtures of purified components, we have found HPLC with these systems to be successful and invaluable in the analysis of enzymically synthesized 2-5A.[10] Of the above systems we have found the pH 7 buffer (Fig. 2) ideal for the initial analysis of material of this type. More detailed information can be obtained with the aid of the systems described in Figs. 3 and 4. In addition, very high resolution of the 5'-mono-, di-, and triphosphates of any or all of the oligomers from dimer to pentamer can be achieved on the amine columns as originally described.[3] In general, however, until a method can be found to stabilize these latter columns to phosphate buffers (or alternative buffer systems devised) their use cannot be recommended. Meanwhile, we have obtained separations of the type shown in Fig. 2 with C_{18} columns from two different manufacturers. The characteristics of the columns do appear to vary slightly, however, and it is clear that the separations achieved are very sensitive to small variations in pH and ionic strength (compare Figs. 2 and 3), the effects of which are not always easy to predict. For example, AMP elutes more rapidly than pppA2'p5'A at pH 7 but more slowly at pH 6 (Figs. 2 and 3). The conditions described should, therefore, be taken as a basis for the determination of optimum conditions for individual columns rather than as absolute standards. The use of cyanopropyl-bonded phase columns could give better resolution than that reported here. This has not yet been investigated.

Acknowledgment

We are indebted to Simon Jones and Colin Reese for the gift of the chemically synthesized 2-5A trimer 5'-mono-, di-, and triphosphates, and to R. A. Harkness for suggesting the used of 4 mM potassium phosphate, pH 6.5, in the HPLC analysis of cores.

[10] These methods have been used successfully also for the analysis of naturally occurring 2-5A in cell extracts made in the absence of detergents. M. Knight, P. J. Cayley, R. H. Silverman, D. H. Wreschner, C. S. Gilbert, R. E. Brown, and I. M. Kerr, *Nature* (*London*) **288**, 189 (1980).

[28] Radioimmune and Radiobinding Assays for A2'p5'A2'p5'A, pppA2'p5'A2'p5'A, and Related Oligonucleotides

By M. KNIGHT, D. H. WRESCHNER, R. H. SILVERMAN, and IAN M. KERR

(2'-5')-Oligoadenylic acid synthetase, the enzyme that synthesizes pppA2'p5'A2'p5'A and related oligonucleotides collectively referred to as 2-5A,[1] has been found in a wide variety of cells in amounts varying with the state of growth and hormone status.[2] Moreover, the possible role for

[1] The series of 2'-5'-linked 5'-triphosphorylated oligoadenylic acids collectively referred to as 2-5A is described in detail by Hovanessian *et al.* (this volume [24]). The terms trimer,

2-5A in mediating the antiviral[3] and anticellular[4-6] effects of interferon has made the development of a convenient assay highly desirable. To date assays for 2-5A per se have involved either (*a*) inhibition of cell-free protein synthesis or (*b*) direct assay of the activation of the 2-5A-dependent RNase (Williams *et al.*, this volume [26]). Although these methods have been successfully used to estimate 2-5A levels in cell extracts,[3] they are inconvenient and not, therefore, suitable for the routine assay of large numbers of samples. In the case of the nonphosphorylated "core" [$(A2'p)_nA$, where $n = 2$–6] of 2-5A, the only assays available have been based on its anticellular activity.[4-6] These are relatively insensitive, e.g., $\geq 1 \mu M$ "core" is needed to inhibit DNA synthesis and cell growth in Daudi cells, and significant volumes ($\geq 200 \mu$l) of material are required[4] (see also Martin *et al.*, this volume [35]). Here we describe convenient, sensitive, and relatively simple radioimmune and radiobinding assays that are capable of detecting physiological concentrations (i.e., nanomolar) of "core" and 2-5A, respectively. These assays are based on the high affinities of "core" for antibody prepared against $A2'p5'A2'p5'A$ bound to bovine serum albumin (BSA) in the radioimmune (RI) assay, and of 2-5A for the 2-5A-dependent RNase in the radiobinding (RB) assay.

The major difficulty in the development of these assays lay in the synthesis of radioactive 2-5A of sufficiently high specific activity to give the required sensitivity. Relatively high concentrations of ATP ($\geq 100 \mu M$) are required for its efficient conversion to 2-5A by the 2-5A synthetase, thus limiting the specific activity that can conveniently be achieved in this way (Hovanessian *et al.*, this volume [24]). Modified 2-5A or core of sufficiently high specific activity can, however, be obtained by the 3'-terminal addition of cytidine 3'-5'-[5'-^{32}P]bisphosphate (pCp) using T4 RNA ligase (EC 6.5.1.3).[7] The ^{32}P-labeled pCp derivatives of 2-5A and

tetramer, and pentamer will be used to denote preparations consisting largely of the 5'-triphosphates but also containing smaller amounts of the corresponding di- and monophosphates. The more complete nomenclature, e.g., trimer triphosphate, will be used to identify the individual species. The tri- and diphosphates, but not the monophosphates, are biologically active in the activation of a nuclease in cell extracts.

[2] G. R. Stark, W. J. Dower, R. T. Schimke, R. E. Brown, and I. M. Kerr, *Nature* (*London*) **278**, 471 (1978).

[3] B. R. G. Williams, R. R. Golgher, R. E. Brown, C. S. Gilbert, and I. M. Kerr, *Nature* (*London*) **282**, 582 (1979).

[4] D. M. Reisinger and E. M. Martin, Reported at the Proceedings of the Second International Workshop on Interferons.

[5] A. Kimchi, H. Shure, and M. Revel, *Nature* (*London*) **282**, 849 (1979).

[6] R. R. Golgher, B. R. G. Williams, C. S. Gilbert, R. E. Brown, and I. M. Kerr, *Proc. N. Y. Acad. Sci.* **350**, 448 (1980).

[7] R. Silber, V. G. Malathi, and J. Hurwitz, *Proc. Natl. Acad. Sci. U.S.A.* **69**, 3009 (1972).

"core" bind to the 2-5A-dependent nuclease and antibody, respectively, and are suitable for use as "probes" in the RB and RI assays. These are competition or displacement assays in which suitable dilutions of test samples compete with a fixed amount of labeled probe for binding to either the nuclease or the antibody. The concentrations of the unknown samples are obtained by comparison with the displacement observed with standard preparations of 2-5A or "core." The binding of the probe to antibody or RNase is most conveniently determined by a nitrocellulose filter assay in which bound probe is retained with the protein on the nitrocellulose filter, whereas free probe does not bind and is removed by washing.

The RB assay is specific for the 5'-phosphorylated components of 2-5A. The RI assay provides a simple and sensitive assay for preexisting "core" or for 2-5A converted to "core" with bacterial alkaline phosphatase (BAP, EC 3.1.3.1). It is also capable of detecting higher concentrations, in decreasing order of sensitivity, of the 5'-mono-, di-, and triphosphates of 2-5A. 2-5A concentrations are expressed in AMP equivalents throughout.

Preparation of High Specific Activity ^{32}P-Labeled (A2'p)$_2$A3'p5'C3'p and Related Oligonucleotides with [5'-^{32}P]pCp and T4 RNA Ligase

T4 RNA ligase[7,8] is used to link [5-^{32}P]pCp covalently to the 3'-OH of (A2'p)$_2$A for use in the RI assay and to ppp(A2'p)$_2$A, or ppp(A2'p)$_3$A for use in the RB assay. The reaction for (A2'p)$_2$A is

$$(A2'p)_2A + ppp5'A + p5'C3'p \xrightarrow{Mg^{2+}} (A2'p)_2A3'p5'C3'p + p5'A + PP_i$$

[^{32}P]pCp is available at specific activities of 2 to 3 × 10^6 Ci/mol. Any pure product of pCp addition will have the same high specific activity. After synthesis, however, it is necessary to purify the 2-5A-pCp to remove the unreacted 2-5A and pCp. The 2-5A would bind to the antibody or nuclease, making a sensitive assay for 2-5A impossible; the pCp would raise the background radioactivity to unacceptable levels. Purification of 2-5A-pCp is achieved by high-performance liquid chromatography (HPLC).

Materials

Ligase buffer: 100 mM HEPES buffer, pH 7.6; 15 mM MgCl$_2$; 6.6 mM dithiothreitol; and 20% dimethyl sulfoxide (v/v)

pCp: [5'-^{32}P]pCp (2 to 3 × 10^6 Ci/mol) (Radiochemical Centre, Amersham, U.K.)

Unlabeled pCp (P-L Biochemicals, Inc.)

[8] T. E. England and O. C. Uhlenbeck, *Nature (London)* 275, 560 (1978).

T4 RNA ligase: EC 6.5.1.3 (P-L Biochemicals, product 0880), 3700 units/ml containing 50% glycerol; 20 mM Tris · HCl, pH 7.5; 5 mM dithiothreitol; and 50 μM ATP

2-5A and "core": A2'p5'A2'p5'A was synthesized chemically.[9] ppp(A2'p)$_n$A (n = 2 and 3) were synthesized enzymically as described previously[9] and by Hovanessian et al., this volume [24].

Thin-layer chromatography (TLC) plates of polyethyleneimine (PEI) cellulose (Macherey-Nagel Polygram Cel 300 PEI) were from Camlab, Cambridge, U.K.

Method

$(A2'p)_2A3'p5'C3'p$. High-specific activity $(A2'p)_2$ApCp is synthesized in a final volume of 50 μl containing 100 μCi of [5'-^{32}P]pCp (lyophilized); 5 μl of T4 RNA ligase; 25 μl of ligase buffer; and $(A2'p)_2$A at a final concentration of 100 μM; 5 μM ATP is contributed to the reaction mixture by the enzyme solution. Increasing the ATP concentration does not improve the yield (typically 50 to 70%, but see "Phosphatase in Ligase" below). Unlabeled $(A2'p)_2$ApCp can be synthesized for use as an optical density marker in 50-μl reactions containing 3 μl of T4 RNA ligase, 5 μM ATP, 25 μl of ligase buffer, 400 μM unlabeled pCp, and 100 μM $(A2'p)_2$A. About 50% conversion of $(A2'p)_2$A to $(A2'p)_2$A pCp is obtained. In both of the above, the reactions are incubated at 2–4° for 16 hr, heated to 90° for 5 min, and centrifuged for 5 min at 12,000 g to remove denatured protein.

The HPLC system used in the purification is that using a gradient of methanol in ammonium phosphate buffer, pH 7, described in detail by Brown et al. (this volume [27]). Typical results for optical density amounts of unlabeled products are shown in Fig. 1A in which $(A2'p)_2$ApCp (peak 3) is well separated from residual $(A2'p)_2$A, pCp, and AMP (peaks 4, 1, and 2, respectively). Additional unidentified peaks in Fig. 1A are contaminants from the enzyme preparation.

When ^{32}P-labeled material of high specific activity is being prepared, the identity of the appropriate peak [corresponding in retention time to the unlabeled material, (peak 3, Fig 1A)] is confirmed by spotting fractions from the HPLC separation on PEI-cellulose TLC plates and developing in 0.5 M potassium phosphate, pH 3.4, in comparison with suitable markers.[10]

Verification of the transfer of pCp to the 3' terminus of $(A2'p)_2$A can be readily obtained by digestion with alkali and with T2 RNase. The $(A2'p)_2$A-[^{32}P]pCp is digested with 0.3 N KOH at 37° for 17 hr to yield a

[9] E. M. Martin, N. J. M. Birdsall, R. E. Brown, and I. M. Kerr, Eur. J. Biochem. 95, 295 (1979).
[10] R. H. Silverman, D. H. Wreschner, C. S. Gilbert, and I. M. Kerr, Eur. J. Biochem. 115, 49 (1981).

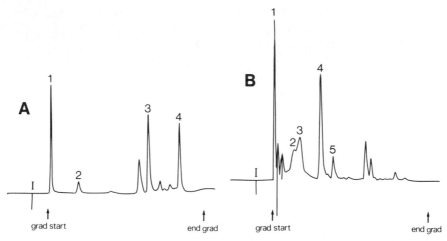

FIG. 1. High-performance liquid chromatography (HPLC) purification of $(A2'p)_2ApCp$ and $ppp(A2'p)_3ApCp$. The T4 RNA ligase products (see the text) were separated on a μBondapak C_{18} column with the methanol gradient: 50 mM ammonium phosphate, pH 7, system described by Brown *et al.* (this volume [27]). (A) Purification of $(A2'p)_2ApCp$. Peaks: 1, pCp; 2, AMP; 3, $(A2'p)_2ApCp$; and 4, $(A2'p)_2A$. (B) Purification of $ppp(A2'p)_3ApCp$. Peaks: 1, pCp; 2, ATP; 3, AMP; 4, $ppp(A2'p)_3ApCp$; and 5, $ppp(A2'p)_3A$. Pump A delivered 50 mM ammonium phosphate, pH 7. Elution (1 ml/min) was with a linear gradient from zero to 50% of 1:1 methanol: H_2O (pump B) in 25 min. The injection points (I), start, and finish of the gradients are shown.

mixture of $[^{32}P]A2'p$ and $A3'p$ as the only radioactive products. These can be identified in comparison with suitable standards by TLC on PEI plates with 1 M acetic acid as solvent.[10] On digestion with T2 RNase (to which $3'$-$5'$ linkages are sensitive, but $2'$-$5'$ linkages are resistant[11]), the only radioactive product obtained is $(A2'p)_2A$-$[^{32}P]p$, which can be similarly identified by TLC on PEI cellulose in 0.175 M ammonium bicarbonate.[10] As expected, digestion with T2 RNase also renders the ^{32}P in $(A2'p)_2A$-$[^{32}P]pCp$ sensitive to digestion with BAP.[10]

 $ppp(A2'p)_2A3'p5'C3'p$ *and* $ppp(A2'p)_3A3'p5'C3'p$. These derivatives are synthesized and purified as for the $(A2'p)_2ApCp$ with the appropriate substitution of 100 μM $ppp(A2'p)_2A$ or $ppp(A2'p)_3A$ for $(A2'p)_2A$ in the RNA ligase reaction. In the preparation of material of high specific activity, 60–75% conversion of the input $[5'$-$^{32}P]pCp$ to 2-5A-pCp has been observed. Typical results obtained on HPLC fractionation of an unlabeled preparation of $ppp(A2'p)_3ApCp$ (peak 4) are presented in Fig. 1B.

Phosphatase in the Ligase. Some preparations of T4 RNA ligase were found to contain a phosphatase activity that degraded $[^{32}P]pCp$ by 50–90%

---

[11] I. M. Kerr and R. E. Brown, *Proc. Natl. Acad. Sci. U.S.A.* **75**, 256 (1978).

on incubation overnight at 2–4° under the above conditions, reducing the yield of product by as much as sixfold. Under these circumstances, much improved yields can usually be obtained by increasing the ATP concentration to 100 μM or by the addition of 100 μM TTP.

Radioimmune Assay of 2-5A "Core"

Ribonucleotides and ribonucleosides conjugated to protein carriers have been shown to elicit antibodies specific for the nucleotide determinant group.[12,13] (2'-5')-Oligoadenylic acid-specific antibodies are prepared by injecting rabbits with $(A2'p)_2A$ conjugated to BSA. Dilutions of the antisera so obtained are used without further treatment in the RI assay with ^{32}P-labeled $(A2'p)_2ApCp$, prepared as described above, as a probe. Essentially identical results are obtained with or without the removal of the 3'-phosphate from the probe by digestion with BAP.

Preparation of $(A2'p)_2A$ Conjugated to Bovine Serum Albumin

Materials

Bovine serum albumin (BSA), fraction V (Sigma)
Sodium periodate, 0.1 M
Potassium carbonate, 5%
Sodium cyanoborohydride (Sigma)
$(A2'p)_2A$: chemically synthesized and purified as described previously[9]
$[^3H](A2'p)_2A$: synthesized as described by Williams *et al.*, this volume [26]

Method

One milligram of $(A2'p)_2A$ containing 20,000 cpm of $[^3H]$ $(A2'p)_2A$ as marker in 60 μl is added to 50 μl of 0.1 M sodium periodate and incubated at 0° for 30 min in the dark. The mixture is added with stirring to 1.0 ml of 0.9–2.8 mg of BSA per milliliter in saline adjusted to pH 9–9.5 with 5% potassium carbonate. After stirring for 45 min on ice while maintaining the pH at 9–9.5 with the potassium carbonate, 1.0 ml of 1.5 mg/ml sodium cyanoborohydride is added. The reaction is incubated at 4° overnight prior to extensive dialysis against saline. To determine the extent of binding of the $(A2'p)_2A$ to the BSA, samples are precipitated with 5% trichloroacetic acid and the proportion of the $[^3H](A2'p)_2A$ bound to the precipitated protein is determined. Between 2 and 7 molecules of $(A2'p)_2A$ were rou-

[12] B. F. Erlanger and S. M. Beiser, *Proc. Natl. Acad. Sci. U.S.A.* 52, 68 (1964).
[13] R. M. D'Alisa and B. F. Erlanger, *Biochemistry* 13, 3575 (1974).

tinely bound per molecule of BSA. There was no apparent difference in the ability of conjugates in this range to elicit anti-"core" antibodies in rabbits, as described below.

Preparation of Antiserum to (A2′p)₂A

Materials

Incomplete and complete Freund's adjuvant (Difco)
(A2′p)₂A-BSA: prepared as described above

Method

(A2′p)₂A-BSA in saline (1 ml of 100 to 800 μg/ml) is homogenized in an equal volume of complete Freund's adjuvant with a blender-type, MSE homogenizer, at 10,000 rpm for about 10 min. The homogenized conjugate (2 ml) is injected at four sites on each thigh (a total of 8×250 μl) of each rabbit. Two weeks later this is repeated with incomplete adjuvant. Serum is obtained by bleeding from the ear vein 7 days later and at monthly intervals at 7–9 days following booster injections in incomplete adjuvant administered as above. Immune sera are stored at $-20°$ after the addition of an equal volume of glycerol. Screening for anti-"core" antibody is done as described below for the RI assay. Dilutions of sera are incubated with [^{32}P](A2′p)₂ApCp to determine the maximum dilution that will bind 50% of this probe.

To date, antisera have been prepared in four different New Zealand white rabbits each injected with different amounts of (A2′p)₂A-BSA within the above range (the amount injected being constant for any given rabbit throughout the course of injections). All the rabbits developed maximum titers sufficient to to give 50% binding of the radioactive probe in the assay below at dilutions of between 3000 and 8000 after the third (first booster) injection. Further bleeding after the monthly booster injections yielded sera of similar titer.

Radioimmune Assay

Materials

[^{32}P](A2′p)₂ApCp prepared as described above
Anti-(A2′p)₂A-BSA sera prepared as described above
Nitrocellulose disks: 2.5 cm, 0.45 μm pore size (Sartorius)
TBS: saline buffered with 35 mM Tris · HCl, pH 7.6
TBS-gelatin: TBS with 0.1% gelatin

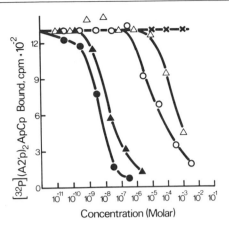

FIG. 2. Radioimmune assay of (2'-5')-oligo(A) (2-5A) "core" and related molecules. Assays were performed as described in the text: (A2'p)$_2$A or (A2'p)$_3$A (●); A2'p5' A (▲); (A3'p)$_2$A (○); adenosine (△); and ATP, CTP, 2'-, 3'-, or 5'-CMP, cytidine, and pCp are all represented by X.

(A2'p)$_2$A and its 5'-mono, di-, and triphosphates were synthesized chemically as described previously.[9,14] Other 2-5A components were synthesized enzymatically as described by Hovanessian *et al.* (this volume [24]).

(A3'p)$_2$A and (A3'p)$_3$A (Boehringer-Mannheim)

pCp (P-L Biochemicals)

All other nucleosides and mono- and dinucleotides were from Sigma.

Method

The radioimmune assays are performed in a final volume of 55 μl; 2000–3000 cpm of [^{32}P](A2'p)$_2$ApCp at 1 to 3 $\times$ 10^6 Ci/mol in 15 μl of TBS-gelatin is added to 15 μl of the material being assayed, i.e., either a standard or unknown. Anti-2-5A "core" serum (25 μl) at a dilution of between 1:1500 and 1:4000 in TBS-gelatin is added. The amounts of probe and antibody are adjusted to give between 50 and 60% of the [^{32}P](A2'p)$_2$ApCp probe bound in the absence of competing material. Incubation is at 30° for 1 hr or overnight at 4°. Samples (40 μl) are applied to nitrocellulose filters that can be washed together, unbound ^{32}P-labeled probe being removed by gentle swirling in three changes of TBS (about 5 ml per filter). The filters are dried and counted in 4 ml of a toluene-based scintillant.

Typical results are shown in Fig. 2. The antisera are relatively specific for the 2'-5' linkage. Fifty percent displacement of the probe is observed

[14] S. S. Jones and C. B. Reese, *J. Am. Chem. Soc.* **101**, 7399 (1979).

with 4 nM (A2'p)$_2$A or (A2'p)$_3$A (in AMP equivalents, i.e., 1.3 nM trimer and 1 nM tetramer; ●, Fig. 2) and 10 nM A2'p5'A (▲, Fig. 2). Much higher concentrations of 3'-5'-linked ApApA (30 μM; ○; Fig. 2) or adenosine (500 μM; △, Fig. 2) are required. ATP, CTP, 2'-, 3'-, or 5'-CMP, pCp, and cytidine were without effect even at 1 mM (X, Fig. 2). The following compounds also displaced 50% of the probe at the concentrations indicated: the 5'-mono-, di-, and triphosphates of (A2'p)$_2$A at 15 nM, 300 nM, and 1.5 μM, respectively; 2'-, 3'-, and 5'-AMP at 5–10 μM and 3'-5'-linked (Ap)$_3$A and ApA at 40 and 100 μM, respectively. Accordingly, the assay will detect higher concentrations of the 5'-phosphorylated components of 2-5A, and there is significant cross-reaction with higher concentrations ($\geq$5 μM) of 2'-, 3'-, or 5'-AMP and $\geq$30 μM (AMP equivalents) of 3'-5'-linked (Ap)$_n$A (n = 1–3). The significance of this in relation to the combined use of the RI and RB assays in the analysis of crude extracts is discussed below.

Essentially identical results have been obtained with [^{3}H](A2'p)$_2$A as probe except that the assay is approximately 1/20th as sensitive, reflecting the lower specific activity of this material.[15]

Radiobinding Assay of 2-5A

The highly specific interaction between 2-5A and the 2-5A-dependent RNase is the basis of a sensitive and convenient assay in which [^{32}P]2-5A-pCp competes with unlabeled 2-5A for RNase binding sites. A major advantage of this assay is that the binding is sufficiently specific that it can be performed with crude cell extracts as the source of the 2-5A-dependent RNase. The results presented below were obtained with rabbit reticulocyte lysates. Essentially identical results have been obtained with the partially purified 2-5A-dependent RNase from the same source and with postmitochondrial supernatant fractions from Ehrlich ascites tumor cells prepared as for use in cell-free protein-synthesizing systems.[16] Less extensive studies with similar extracts from HeLa and mouse L cells suggest that they too can be used in this type of assay.

Materials

Nitrocellulose disks: 2.5 cm, 0.45 μm pore size (Sartorius)
[^{32}P]ppp(A2'p)$_3$ApCp synthesized as described above
Rabbit reticulocyte lysates prepared as described[16]
Buffer A: 85 mM KCl; 20 mM Tris · HCl, pH 7.6; 5 mM magnesium acetate; 1 mM ATP; and 5% glycerol

[15] M. Knight and I. M. Kerr, unpublished work, 1980.
[16] L. Villa-Komaroff, M. McDowell, D. Baltimore, and H. F. Lodish, this series, Vol. 30, p. 709.

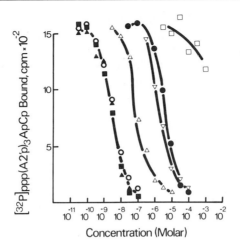

Fig. 3. Radiobinding assay of (2'-5')-oligo(A) (2-5A) and related molecules. Assays were performed as described in the text. ppp(A2'p)$_2$A (▲); pp(A2'p)$_2$A (○); ppp(A2'p)$_3$A (■); p(A2'p)$_2$A (△); (A2'p)$_2$A (▽); pppA2'p5'A (●); and (A3'p)$_2$A (□).

Buffer B: buffer A without ATP
Oligonucleotides (including 2-5A), nucleotides, and nucleosides were obtained as indicated above.

Method

The assays are performed in a final volume of 50 μl. The probe, [^{32}P]ppp(A2'p)$_3$ApCp (3000–6000 cpm at 1 to 3 × 10^6 Ci/mol), is mixed with the standard or unknown sample to be assayed, and buffer A is added to 30 μl followed by 20 μl of reticulocyte lysate. Incubation is performed at 4° for 90 min. The reaction mixtures are then filtered (in 15–30 secs) through nitrocellulose disks presoaked with buffer B, dried, and counted in 4 ml of a toluene-based scintillant. The relative amounts of ^{32}P-labeled probe and lysate are adjusted to obtain approximately 30–50% binding of the probe in the absence of unlabeled competitor.

Typical results are shown in Fig. 3. The 5'-di- and triphosphorylated trimer and 5'-triphosphorylated tetramer components of 2-5A (○, ▲, ■, respectively, Fig. 3) all displaced 50% of the bound probe at concentrations of 3 nM (AMP equivalents). Significantly higher concentrations of the trimer 5'-monophosphate [p(A2'p)$_2$A, 100 nM, △, Fig. 3], "core" [(A2'p)$_2$A, 3 μM, ▽, Fig. 3] or dimer triphosphate [pppA2'p5'A, 3 μM, ●, Fig. 3] are required while mM 3'-5'-linked ApApA (□, Fig. 3), A, ADP, and ATP (of which 0.5 mM is routinely included in the assay) have only a minimal effect. AMP, 2'-, 3'- or 5'-CMP, and pCp at this concentration were inactive.

The rabbit reticulocyte is unusual in our experience in that the trimer di- and triphosphate components of 2-5A are relatively poor activators of the 2-5A-dependent nuclease, the tetramer or higher oligomers being required for efficient activation in this system.[3] It was somewhat surprising, therefore, to find that unlabeled trimer 5'-di- and triphosphates were just as efficient as the 5'-triphosphorylated tetramer in the RB assays (○, ▲, and ■, respectively, Fig. 3). In accord with this, essentially identical results are obtained when the trimer pCp derivative $[^{32}P]ppp(A2'p)_2ApCp$ is used as probe instead of the tetramer pCp derivative $[^{32}P]ppp(A2'p)_3ApCp$. This raises the possibility that the binding observed here is to a 2-5A-specific protein other than the nuclease. The binding activity, however, copurifies with the 2-5A-dependent nuclease,[17] and it seems more likely that the trimer components bind with equal efficiency but only poorly activate the nuclease in this particular system.

Comments

The 3' addition of $[^{32}P]pCp$ to 2-5A and "core" with the T4 RNA ligase provides probes that allow the detection of nanomolar concentrations of 2-5A and "core" with RB and RI assays, respectively. The RI assay is more sensitive for the nonphosphorylated "core" (Fig. 2) whereas the 5'-mono-, di-, and triphosphorylated trimers compete to a decreasing extent in that order. In contrast, the RB assay shows the highest sensitivity to the 5'-tri- and diphosphorylated 2-5A components, but the monophosphates and "core" are detected at higher concentrations (Fig. 3). The two assays are, therefore, complementary in their abilities to detect "core" and 5' phosphorylated 2-5A. The 3'-5'-linked oligoadenylic acids are relatively inactive in both assays as are the individual nucleosides and nucleotides contained in 2-5A and 2-5A-pCp.

The RB and RI assays can, therefore, be used to detect 2-5A and "core" made chemically or with the 2-5A synthetase. These assays are much more convenient than the previously available "biological" assays, and in the 2-5A synthetase assay they provide a much more sensitive means of detecting the product 2-5A than does the incorporation of radioactive ATP. Accordingly, they are now the method of choice for the assay of 2-5A and "core."

Assay of low levels of preformed 2-5A or "core" in crude cell extracts presents the problem of a more complex range of cross-reacting materials. Combination of the RI and RB assays with HPLC analysis (Brown et al., this volume [27]) of such extracts has shown that at low dilutions nonspecific (salt) effects and cross-reacting material (e.g., ≥ 1 mM

---

[17] D. H. Wreschner, C. S. Gilbert, and I. M. Kerr, unpublished work, 1980.

adenosine in the RI assay) can be a problem. Also, although in our experience[18] these problems have not significantly affected the assay of extracts containing higher levels of 2-5A or "core" (i.e., $\geq$150 nM sufficient to allow their assay at $\geq$50-fold dilutions), there is clearly a sufficient range of potentially cross-reacting materials (Figs. 2 and 3) for some further assurance of the identity of the 2-5A or "core" to be required. This is best obtained by analysis in combination with HPLC. Alternative possible approaches, however, include combined T2 RNase and BAP digestion to convert all but 2'-5'-linked oligonucleotides to uncharged (and hence readily removable nucleosides) followed by RI assay or confirmation of positive results in screening RB assays by direct assays of the 2-5A-dependent RNase (Williams *et al.*, this volume [26]).

An interesting aspect of the results obtained in our combined HPLC, RI, and RB analyses of extracts from interferon-treated and control cells with or without virus infection[18] is the natural occurrence of "core" (A2'-p5'A2'p5'A). This does not appear to be simply a breakdown product of 2-5A and may be of significance per se (e.g., in the control of DNA synthesis or cell growth). Whether or not this turns out to be the case, it is clear that by assaying exclusively for 2-5A one is omitting a significant component of the 2-5A system, i.e., "core." Accordingly, in any analysis of the natural occurrence of 2-5A the combined use of HPLC and the RI and RB assays is recommended.

Acknowledgments

We are indebted to Simon Jones and Colin Reese for the gift of chemically synthesized 2-5A trimer 5'-mono-, di-, and triphosphates and are very grateful to Bernard Erlanger for advice and the gift of antisera to 3'-5'-linked oligoadenylic acids used in initial studies not reported here. M. K. was the recipient of a grant from the government of Ghana, and D. H. W. of an EMBO fellowship. P. J. Cayley and R. E. Brown developed the HPLC methods employed (this volume [27]).

[18] M. Knight, P. J. Cayley, R. H. Silverman, D. H. Wreschner, C. S. Gilbert, R. E. Brown, and I. M. Kerr, *Nature (London)* **288**, 189 (1980).

[29] Assay of (2'-5')-Oligo(A) Synthetase with 2',5'-ADP-Sepharose

By Margaret I. Johnston, Robert M. Friedman, and Paul F. Torrence

Research into the cellular changes that result from interferon treatment of cells has led to the discovery of a new enzyme, 2',5'-oligoadenylate synthetase.[1] This enzyme polymerizes ATP into 2',5'-ppp(Ap)$_n$A [(2'-5')-oligo(A)], which is a potent inhibitor of protein synthesis *in vitro*[1-3] and in intact cultured cells.[4] These novel molecules activate a "latent" endonuclease and lead to the degradation of mRNA and of polysomes.[5-7] Extracts from interferon-treated or from control cells contain an enzyme that cleaves the phosphodiester bond of (2'-5')-oligo(A).[3,8] The inhibitor may also be rendered inactive by removal of the terminal 5'-triphosphate moiety.[9] This oligoadenylate system has been implicated not only in the mechanism by which interferon induces an antiviral state, but also in the control of protein synthesis in other cells that are not interferon pretreated.[10]

An unusual and significant feature of this system is that the synthetase requires activation by double-stranded RNA. The enzyme binds tightly to, and retains activity when mixed with, poly(I) · poly(C) attached to an immobile support, such as Sepharose or paper.[10,11] Use of poly(I) · poly(C)-support-bound synthetase results in higher yields of (2'-5')-oligo(A) than a direct solution assay when some cell extracts, such as those from interferon-pretreated mouse L cells, are employed.[12,13] Other

[1] I. M. Kerr and R. E. Brown, *Proc. Natl. Acad. Sci. U.S.A.* **75**, 256 (1978).

[2] L. A. Ball and C. N. White, *Proc. Natl. Acad. Sci. U.S.A.* **75**, 1167 (1978).

[3] A. Schmidt, A. Zilberstein, L. Shulman, P. Federman, H. Berissi, and M. Revel, *FEBS Lett.* **95**, 257 (1978).

[4] B. R. G. Williams and I. M. Kerr, *Nature (London)* **276**, 88 (1978).

[5] B. R. G. Williams, R. R. Golgher, and I. M. Kerr, *FEBS Lett.* **105**, 47 (1979).

[6] A. G. Hovanessian, J. Wood, E. Meurs, and L. Montagnier, *Proc. Natl. Acad. Sci. U.S.A.* **76**, 3261 (1979).

[7] M. J. Clemens and B. R. G. Williams, *Cell* **13**, 566 (1978).

[8] B. R. G. Williams, I. M. Kerr, C. S. Gilbert, C. S. White, and L. A. Ball, *Eur. J. Biochem.* **92**, 455 (1978).

[9] E. M. Martin, N. J. M. Birdsall, R. E. Brown, and I. M. Kerr, *Eur. J. Biochem.* **95**, 295 (1979).

[10] G. R. Stark, W. J. Dower, R. T. Schimke, R. E. Brown, and I. M. Kerr, *Nature (London)* **278**, 471 (1979).

[11] A. G. Hovanessian, R. S. Brown, and I. M. Kerr, *Nature (London)* **268**, 537 (1977).

[12] A. Zilberstein, A. Kimchi, A. Schmidt, and M. Revel, *Proc. Natl. Acad. Sci. U.S.A.* **75**, 4734 (1978).

[13] M. I. Johnston, R. M. Friedman, and P. F. Torrence, *Biochemistry* **19**, 5580 (1980).

METHODS IN ENZYMOLOGY, VOL. 79

crude extracts, such as those from interferon-pretreated HeLa cells, contain significant synthetase activity in a direct solution assay.[14]

We have developed an assay for 2',5'-oligoadenylate synthetase that (a) provides reproducibly high yields of (2'-5')-oligo(A) when crude cell extracts are employed, and (b) provides an opportunity for variation of nucleic acid activator.

Assay Method

Principle. This method is based on the binding of (2'-5')-oligo(A) synthetase from crude extracts to 2',5'-ADP-Sepharose at 4°. After binding, the resin is washed to remove unbound molecules, and the resin-bound synthetase is incubated at 30° in the presence of ATP and a nucleic acid activator. The (2'-5')-oligo(A) in the reaction supernatant is then assayed by its ability to inhibit cell free protein synthesis.

Reagents

   2',5'-ADP-Sepharose (Pharmacia, Uppsala, Sweden), or 2',5'-ADP-agarose (Sigma, St. Louis, Missouri) (linkage via the N-6)
   Poly(I) · poly(C) or other nucleic acid inducer in 0.02 M KCl, 0.01 M HEPES, pH 7.5
   HGI buffer: 20 mM HEPES, pH 7.5, 50 mM KCl, 1.5 mM Mg(OAc)$_2$, 7mM β-mercaptoethanol, 20% (v/v) glycerol
   HGII buffer: 20 mM HEPES, pH 7.5, 90 mM KCl, 10 mM Mg(OAc)$_2$, 7 mM β-mercaptoethanol, 20% (v/v) glycerol
   HGII-ATP: as HGII with 5 mM ATP
   Cell extract containing synthetase activity; S$_{100}$ from interferon-treated mouse L cells prepared as described previously.[15] Briefly, after incubation overnight in the presence of mouse interferon (30 units/ml; 2 × 10^7 NIH units/mg), the cells are harvested and washed with ice-cold buffered saline (0.14 M NaCl, 0.035 M Tris · HCl, pH 7.5) and then lysed either by (a) Dounce homogenization with 1.4 volumes of lysis buffer [0.01 M Tris · HCl, pH 7.5, 0.01 M KCl, 1.5 mM Mg(OAc)$_2$, 7 mM β-mercaptoethanol]; or by (b) suspension and vortexing with 1.4 volumes of lysis buffer containing 0.5% (v/v) NP40. Then Tris · HCl, pH 7.5, KCl, and Mg(OAc)$_2$ are added to final concentrations of 0.03, 0.09, and 0.003 M, respectively. The S$_{100}$ supernatants are prepared first by centrifugation of the cell lysate at 10,000 g for 10 min to yield 10,000 g supernatant (S$_{10}$), and then further centrifugation of the S$_{10}$ supernatant at 100,000 g for 2 hr. The S$_{100}$ supernatants are stored in small aliquots (20–200 μl) in liquid nitrogen and not thawed until just prior to use.

[14] M. A. Minks, S. Benvin, P. A. Maroney, and C. Baglioni, *J. Biol. Chem.* **254**, 5058 (1979).
[15] P. F. Torrence and R. M. Friedman, *J. Biol. Chem.* **254**, 1259 (1979).

Mouse L cell-free system programmed with encephalomyocarditis (EMC) viral RNA. The EMC virus is grown in Krebs II ascites cells[16] and the RNA is extracted and purified as described previously.[17] S_{10} extracts from control mouse L cells are prepared as described above. The S_{10} extracts are treated with 2–10 μg of micrococcal nuclease (Worthington, Millipore) per milliliter at 20° for 5–10 min under the conditions of Pelham and Jackson[18] to reduce endogenous protein synthesis. Extracts are stored in 100–200-μl aliquots in liquid nitrogen and thawed only once at time of use. Protein synthesis is performed as described previously.[15] Other mRNA-dependent protein synthesis systems may be equally satisfactory,[19] e.g., globin-directed synthesis in rabbit reticulocyte lysates.[18,20] The sensitivity of various cell-free systems to inhibition by (2'-5')-oligo(A) may depend on the procedure employed to prepare the extract, particularly the method used to reduce endogenous synthetic activity, which is not sensitive to (2'-5')-oligo(A)-mediated inhibition.[8,13]

Procedure

Step 1. Binding of the Synthetase to the Resin. The 2',5'-ADP-Sepharose (100 μl of packed resin) is washed three times with 15 volumes of HGI buffer by gentle suspension of the resin, followed by centrifugation at 500 rpm for about 2 min, and careful removal of the supernatant wash. After all the supernatant from the last wash is removed, 100 μl of cell extract are added to the resin and the mixture is kept at 4° for 6–18 hr with occasional shaking. The resin then is washed three times with 15 volumes of HGII buffer to remove unbound proteins. Next, the resin-bound synthetase is incubated at 30° for 2–6 hr with 100 μl of HGII-ATP buffer containing 10^{-4} M poly(I) · poly(C) (or other nucleic acid activator). After this incubation, the mixture is centrifuged at 1000 rpm for 2–5 min. The supernatant is removed, heated at 95° for 5 min, and then centrifuged at 10,000 g for 1 min to remove any denatured protein. This heated supernatant is then assayed for its content of (2'-5')-oligo(A).

Step 2. Assay of (2'-5')-oligo(A). The (2'-5')-oligo(A)-containing super-

[16] P. Faulkner, E. Martin, S. Sved, R. C. Valentine, and T. S. Work, *Biochem. J.* **80**, 597 (1961).

[17] I. M. Kerr and E. M. Martin, *J. Virol.* **9**, 559 (1972).

[18] H. R. B. Pelham and R. J. Jackson, *Eur. J. Biochem.* **67**, 247 (1976).

[19] L. Villa-Komaroff, M. McDowell, D. Baltimore, and H. F. Lodish, this series, Vol. 30, p. 709.

[20] L. A. Weber, E. D. Hickey, P. A. Maroney, and C. Baglioni, *J. Biol. Chem.* **252**, 4007 (1977).

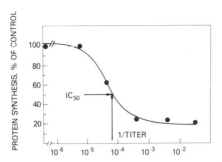

FINAL DILUTION INTO PROTEIN SYNTHESIS REACTION

FIG. 1. Calculation of titers. The supernatant from the synthetase reaction mixture (step 1, standard conditions) is diluted into the protein synthesis reaction mixture. The percentage of inhibition of protein synthesis relative to a water control is determined for each dilution and is plotted here against the final dilution of the inhibitor solution in the protein synthesis reaction mixture. The inverse of the dilution required to inhibit protein synthesis by 50% is termed the titer of the solution.

natant is diluted threefold with sterile distilled water in V-shaped disposable microtiter trays (Linbro Scientific Co., Hamden, Connecticut). Then, 2.5 μl of each dilution, or water for control samples, are diluted 5- or 10-fold into 12.5 or 25 μl of the cell-free protein synthesis system in U-shaped disposable microtiter trays (Linbro). After incubation at 30° for 2 hr, the radioactivity in the trichloroacetic acid-precipitable protein is determined in the samples and in the controls as previously described.[15] The percentage of inhibition observed in the samples relative to the water control is plotted against the final dilution of the reaction mixture supernatants (Fig. 1).

By comparison with curves obtained simultaneously with dilutions of a known concentration of (2′-5′)-oligo(A), the amount of (2′-5′)-oligo(A) in the sample reaction mixtures can be estimated. For instance, tritiated inhibitor may be prepared from [³H]ATP of known specific activity and purified according to published procedures.[14] Alternatively, various (2′-5′)-oligo(A)-containing solutions can be compared through a comparison of their titers, defined as the inverse of the final dilution that is required to inhibit protein synthesis by 50%. These titers can be ascertained from the curves of percentage of inhibition as a function of dilution described above and illustrated in Fig. 1.

In addition, various nucleic acid activators can be compared through calculation of their M_{act} values, defined as the molar concentration of nucleic acid required to produce sufficient (2′-5′)-oligo(A) that, when diluted 10,000-fold, it will inhibit protein synthesis by 33%. These values more precisely reflect the potencies of various nucleic acids that have

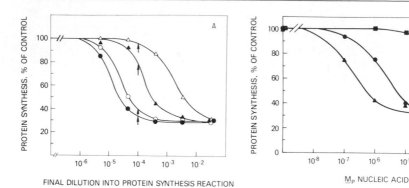

FIG. 2. Calculation of the M_{act} values. (A) Protein synthesis inhibition results are expressed as in Fig. 1 for inhibitor solutions obtained with various concentrations of poly(I) · poly(C): ●, $10^{-4} M$; ○, $10^{-5} M$; ▲, $10^{-6} M$; △, $10^{-7} M$. Arrows point to the percentage of control protein synthesis at a 10^{-4} final dilution of reaction mixture. (B) Results from the titration of *Penicillium chrysogenum* dsRNA (▲); poly(I) · poly(C) (●); QβRNA (■). From graphs, such as panel A, the percentage of control protein synthesis (at a 10^{-4} final dilution of reaction mixture) is determined for synthetase reaction mixtures obtained using several concentrations of nucleic acid activator. As shown here, these percentages of control values are plotted against the concentration of activator in the initial synthetase reaction mixture. The M_{act} values are determined from these figures at 66% of the control.

titers of 1000 or higher when tested under standard conditions at a concentration of 10^{-4} M_p. First, various concentrations (usually 10^{-4} to 10^{-9} M_p) of activator are used in the (2′-5′)-oligo(A) synthesis step 1. Protein synthesis as a function of the dilution of each synthetase reaction mixture is then determined. Figure 2A illustrates the titration of various inhibitor solutions synthesized with various concentrations of poly(I) · poly(C). The percentage of control protein synthesis at a 10^{-4} final dilution of reaction mixture is determined from each curve (arrows, Fig. 2A). This percentage of control value is then plotted against the concentration of nucleic acid activator in the initial synthetase reaction mixture. The M_{act} value is read from the curves at the 66% of control point. Figure 2B shows sample M_{act} calculations for poly(I) · poly(C), *Penicillium chrysogenum* dsRNA, and QβRNA.

Properties of the 2′,5′-ADP-Sepharose-Bound Enzyme

When extracts from interferon-pretreated mouse L cells are employed under the conditions described here, synthesis of (2′-5′)-oligo(A) is linear for about 5 hr, after which time the apparent rate of synthesis declines. The enzyme is relatively stable when bound to 2′,5′-ADP-Sepharose; little activity is lost during incubation at 30°. Although our experiments to date do not discount a small loss of enzyme from the resin, if the resin is washed and reincubated with fresh buffer, ATP and nucleic acid activator, synthesis of inhibitor continues. The levels of (2'-5')-oligo(A) attainable by this assay are comparable to those obtained by the poly(I) · poly(C) Sepharose method and are, at least with mouse L cell extracts, higher than those obtained in a direct solution assay. Our experiments suggest that binding of the extract to the resin helps to eliminate (2'-5')-oligo(A) inactivating enzymes.[13]

Synthesis of (2'-5')-oligo(A) by the bound mouse L cell enzyme is optimal at K^+ concentrations of 0–0.1 M and at Mg^{2+} concentrations of 10–50 mM. Inhibitor synthesis is proportional to the amount of S_{100} from interferon-pretreated cells added to the resin, although linearity is observed only at resin-to-S_{100} ratios greater than five. ATP is essential for the reaction, and the amount of (2'-5')-oligo(A) formed is directly proportional to the concentration of ATP up to about 5 mM. This assay is also very sensitive to the amount and nature of the added nucleic acid activator. Figure 2b illustrates three nucleic acids whose potencies in activating the synthetase vary over 10,000-fold. *Penicillium chrysogenum* dsRNA ($M_{act} = 1.5 \times 10^{-7}$) is a better activator than poly(I) · poly(C) ($M_{act} = 2 \times 10^{-6}$). Q$\beta$RNA, the poorest activator illustrated ($M_{act} = 4 \times 10^{-4}$), may activate the synthetase by virtue of contaminating segments of dsRNA.

In summary, this method is a convenient and quantitative alternative to the assay of (2'-5')-oligo(A) synthetase from crude extracts on poly(I) · poly(C)-Sepharose. In extracts that contain high levels of inactivating enzymes, both methods may be superior to a solution assay in that both resin assays afford a high degree of purification in the binding step. The 2',5'-ADP-Sepharose binding assay offers the additional advantage that the nucleic acid activator can be varied.

[30] An Efficient Chemical Synthesis of Adenylyl(2' → 5')Adenylyl(2' → 5')Adenosine [(2'-5')-Oligo(A)]

By JIRO IMAI and PAUL F. TORRENCE

The unique oligoribonucleotide, 5'-*O*-triphosphoryladenylyl (2' → 5')-adenylyl(2' → 5')adenosine (2-5A) is synthesized by the enzyme 2',5'-oligoadenylate synthetase, which is induced in several cell lines upon

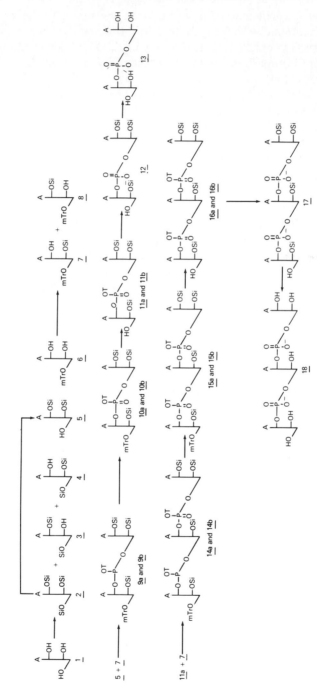

Fig. 1. Synthesis of adenyl (2′ → 5′)-adenyl(2′ → 5′)-adenosine. T, trichloroethyl; mTr, monethoxytrityl; Si, *tert*-butyldimethylsilyl; **a** and **b**, the two diastereomers arising from the chiral phosphotriester center.

interferon treatment.[1-3] The 5'-dephosphorylated "core" (adenylyl(2' →
5')adenylyl(2' → 5') adenosine) (Fig. 1, 18), in contrast to the 5'-triphos-
phate, is not active as a protein synthesis inhibitor in cell-free systems.[4]
This "core" oligomer is active, however, as a protein synthesis inhibitor
in intact cells.[5,6]

Adenylyl(2' → 5')adenylyl(2' → 5')adenosine or its 5'-triphosphate
has been prepared by several different synthetic approaches,[4,6-12] some of
which are ambiguous. The yield of these various procedures varies considerably. We describe here an unambiguous eight-step synthesis that
provides a 37% overall yield based on starting protected nucleosides. This
triester approach using *tert*-butyldimethylsilyl protecting groups also has
been reported by Ogilvie and Theriault.[8]

Since this is a phosphotriester synthesis, a new chiral center is
generated upon formation of the internucleotide bond. The diastereomers
thus formed are referred to as **a** and **b** for each numbered compound in the
synthetic sequence. In this synthesis, the diastereomers are separated
and synthetic sequences are carried out on each separately.

This procedure uses zinc-copper coupled with acetylacetone in dimethylformamide to deblock the trichloroethyl groups of the intermediate
trimer [(16a) and (16b)]. This procedure results in a higher yield and
cleaner reaction than with previously reported methods for removal of the
trichloroethyl group. Furthermore, the use of Chelex in the procedures
considerably improves purification.

Reagents and Equipment

Thin-layer chromatography (TLC) is performed with silica gel GF
plates (Analtech) with the following three solvent systems: (A) ether/
chloroform/methanol (5:4:1); (B) chloroform/methanol (20:1); (C)
n-butanol/ethanol/water/ammonia (60:20:20:1).

[1] I. M. Kerr and R. E. Brown, *Proc. Natl. Acad. Sci. U.S.A.* 75, 256 (1978).
[2] L. A. Ball and C. N. White, *Proc. Natl. Acad. Sci. U.S.A.* 75, 1167 (1978).
[3] A. Schmidt, A. Zilberstein, L. Shulman, P. Federman, H. Berissi, and M. Revel, *FEBS Lett.* 95, 257 (1978).
[4] E. M. Martin, N. J. M. Bridsall, R. E. Brown, and I. M. Kerr, *Eur. J. Biochem.* 95, 245 (1979).
[5] B. R. G. Williams and I. M. Kerr, *Nature (London)* 276, 88 (1978).
[6] A. F. Markam, R. A. Porter, M. J. Gait, R. C. Sheppard, and I. M. Kerr, *Nucleic Acids Res.* 6, 2569 (1979).
[7] M. Ikehara, K. Oshie, and E. Ohtsuka, *Tetrahedron Lett.* 1979, 3677 (1979).
[8] K. K. Ogilvie and N. Y. Theriault, Tetrahedron Lett. 2111 (1979).
[9] S. S. Jones and C. B. Reese, *J. Am. Chem. Soc.* 101, 7399 (1979).
[10] J. Engels and U. Krahmer, *Angew. Chem. Int. Ed. Engl.* 18, 942 (1979).
[11] H. Sawai, T. Shibata, and M. Ohno, *Tetrahedron Lett.*, in press.
[12] H. L. Sleeper and L. E. Orgel, *J. Mol. Evol.* 12, 357 (1979).

^{31}P NMR is determined with at 109 MHz with Varian instrument using chloroform as solvent except for compound (18), which was dissolved in triethylammonium bicarbonate buffer, pH 8.2 (see the table). The external standard is 0.85% phosphoric acid.

Procedure

2′,3′,5′-Tri-O-(tert-butyldimethylsilyl)adenosine (2)

Imidazole (5.98 g, 88 mmol) and *tert*-butyldimethylsilyl chloride (6.65 g, 44.1 mmol) are added to a solution of adenosine [(1), 2.68 g, 10 mmol] in dry dimethylformamide (10 ml). After 20 hr at room temperature, the solvent is removed under vacuum. The residue is dissolved in ether (150 ml) and extracted twice with water (50 ml × 2). The ether layer is dried over anhydrous sodium sulfate and concentrated under vacuum. The residue is subjected to silica gel column chromatography (100 g). The initial eluate with ether–hexane (2 : 1) affords trisilyladenosine (2), mp 143–144° (lit.[13] 142–144°) (5.43 g, 8.92 mmol, 89%). The column then is eluted with ether to give 2′,5′-disilyladenosine (3), mp 170–172.5° (lit.[13] mp 174–177°) (223 mg, 0.45 mmol, 4.5%). The column is further eluted with ethyl acetate to yield 3′,5′-disilyladenosine (4), mp 200–201°, (lit.[13] mp 203–204°) (245 mg, 0.495 mmol, 5.0%).

2′,3′-Di-O-(tert-butyldimethylsilyl)adenosine (5)

Trisilyladenosine [(2), 5.69 g, 9.33 mmol] is dissolved in 80% acetic acid (10 ml) and heated in an oil bath at 100° for 3 hr. The solution is cooled and concentrated under vacuum and then coevaporated twice with toluene (6 ml × 2). The residue is chromatographed on a Florisil column (120 g) with ether. The initial eluate (70 ml) gives unreacted starting material (2) (385 mg, 0.632 mmol, 6.9%). The second eluate (340 ml) affords 2′,3′-disilyladenosine (5), mp 250–253° (lit.[13] mp 250–254°) (3.41 g, 6.89 mmol, 73.8%).

5′-O-Monomethoxytrityl-3′-O-(tert-butyldimethylsilyl)adenosine (7)

5′-O-Monomethoxytrityladenosine (6) (5.39 g, 10.0 mmol) is dried *in vacuo* at 80° for 2 hr; it is then dissolved in dry dimethylformamide and treated with *tert*-butyldimethylsilyl chloride (2.28 g, 15.1 mmol) in the presence of imidazole (2.10 g, 30.8 mmol) at room temperature. The reaction mixture is stirred for 1.5 hr at ambient temperature, and the solvent is

---

[13] K. K. Ogilvie, S. L. Beaucage, A. L. Schifman, N. Y. Theriault, and K. L. Sadana, *Can. J. Chem.* **56**, 2768 (1978).

removed under vacuum. The residue is dissolved in ethyl acetate (100 ml), and this solution is washed twice with water (50 ml × 2). The organic layer is dried over anhydrous sodium sulfate and concentrated under vacuum. The residue is applied to a silica gel column (210 g) and eluted with ethyl acetate. The first eluate (70 ml) provides 5'-monomethoxytrityl-2'-O-silyladenosine (8) (2.16 g, 330 mmol, 33%), which is recrystallized from ether/hexane to give colorless crystals, mp 163–166° (lit.[13] mp 164–167°). The second compound eluted with ethyl acetate (120 ml) is 5'-O-mono-methoxytrityl-3'-O-silyladenosine (7) (2.48 g, 3.80 mmol, 38%), which may be recrystallized from ether/hexane to yield colorless crystals, mp 176–178° (lit.[13] mp 176–178°). Finally the column is washed with methanol to recover unreacted starting material (6) (1.08 g, 2.00 mmol, 20%).

Preparation of 2,2,2-Trichloroethylphosphorodichloridite[14]

2,2,2-Trichloroethanol (14.5 ml, 0.15 mol) is added dropwise to phosphorus trichloride (19.6 ml, 0.225 mol) over a period of 10 min at −10° under dry argon atmosphere. After being kept at room temperature under reduced pressure (15 mm Hg) for 2 hr, the reaction mixture is distilled to yield 2,2,2-trichloroethylphosphorodichloridite (11.8 g, 0.047 mol, 31.5%, bp 52°/0.65 mm Hg).

Condensation of (7) with (5)

To a mixture of collidine (408 mg, 446 μl, 3.37 mmol), trichloroethyl-phosphorodichloridite (0.92 mmol, 136 μl), and dry tetrahydrofuran is added 5'-O-monomethoxytrityl-3'-O-siladenosine (7) (500 mg, 0.765 mmol) dissolved in dry tetrahydrofuran at −78°. The reaction mixture is stirred for 30 min at −78°, and 2',3'-di-O-silyladenosine (5) (360 mg, 0.725 mmol), dissolved in dry tetrahydrofuran (2.50 ml), is added to it over a period of 3 min. The reaction mixture is stirred at −78° for 1 hr followed by warming to room temperature. After the solvent is completely removed under vacuum, the residue is directly applied to a short silica gel column (40 g). The column is washed thoroughly with methylene chloride to remove collidine completely and then eluted with chloroform–methanol (20 : 1) to yield a mixture of two diastereomers (960 mg, 0.717 mmol, 99%). A part of this mixture is applied to preparative silica gel TLC plates which are developed twice in system B. The two diastereomers thus separated are eluted with a mixture of ethyl acetate and methanol (1 : 1) to afford (9a) (R_f, 0.39; system B), mp 116–118°, ^{31}P NMR δ 140.7; and (9b) (R_f 0.19, system B), mp 132–136°, ^{31}P NMR δ 139.3.

[14] W. Gerrard, W. J. Green, and R. J. Phillips, *J. Chem. Soc.* 1954, 1148 (1954).

Iodine Oxidation of (9)

The mixture of diastereomers (970 mg, 0.724 mmol) (9a) and (9b) are dissolved in water-saturated ether (100 ml), and 0.5 M iodine solution in ether is carefully added to this reaction mixture until the reddish brown color of iodine persists. The reaction mixture is examined by TLC in system B to ensure that the oxidation is completed, and then the ether solution is washed twice with a saturated $NaHCO_3$ solution and once with saturated NaCl solution, dried over anhydrous sodium sulfate, and evaporated under vacuum. The residue is applied to a silica gel column (200 g), which is eluted with chloroform–methanol (50 : 1). Diastereomer (10a) (400 mg, 0.295 mmol, 40.8%; R_f 0.25, system B), mp 126–128° (lit.[8] mp 112–117°), ^{31}P NMR δ −2.14, is eluted first followed by diastereomer (10b) (454 mg, 335 mmol, 43.4%; R_f 0.15, system B), mp 175–180° (lit.[8] mp 142–148°), ^{31}P NMR δ −2.20.

Removal of Monomethoxytrityl Group from (10a) *and* (10b)

Compound (10a) (320 mg, 0.236 mmol) is dissolved in 80% acetic acid (8 ml) and heated in an oil bath at 90° for 15 min. The reaction mixture is then evaporated under vacuum and the residual syrup dissolved in chloroform (10 ml). The chloroform solution is washed twice with saturated $NaHCO_3$ solution and once with saturated NaCl solution, dried over anhydrous sodium sulfate, and concentrated under vacuum. The residue is dissolved in methanol (1 ml) and applied to a stepped column (4 steps) packed with Sephadex LH-20 resin (bed volume in methanol is 110 ml). Elution with methanol affords compound (11a) (247 mg, 0.228 mmol, 97%; R_f 0.27, system A), mp 142–144° (lit.[8] mp 134–138°). Compound (10b) can be unblocked in exactly the same way to yield compound (11b) (R_f 0.17, system A), mp 116–119° (lit.[8] mp 108–112°).

Removal of 2,2,2-Trichloroethyl (TCE) Group from Compound (11a) *or* (11b)

To a solution of compound (11a) (35.4 mg, 32.7 μmol, 975 OD_{260} units) in dry dimethylformamide (654 μl) is added active Zn-Cu couple[15] (23.1 mg) and acetylacetone (16.3 μl, 163 μmol). The reaction mixture is stirred and heated in an oil bath at 55° for 1 hr. The reaction may be monitored by TLC in the following way: 1 μl of the reaction mixture is sampled every 15 min and added to 0.1 M EDTA solution (20 μl). A drop of methanol is added to the solution, and 2 μl of this solution is spotted on a silica gel

[15] E. Le Goff, *J. Org. Chem.* 29, 2048 (1964).

plate, which is developed in system A. As the reaction proceeds, the starting material at R_f 0.27 is transformed to a product that remains at the origin. After cooling, the cloudy green reaction mixture is diluted with methanol (20 μl) to give a clear green solution. Chelex [4 ml settled volume, washed with 0.1 M NH_4HCO_3 (3 × 10 ml) and water (3 × 10 ml)] is added to this colored solution, and the mixture is stirred until the solution becomes colorless (1 hr). After removal of Chelex by filtration, the filtrate is evaporated under vacuum. The residue is dissolved again in methanol (1 ml) to determine the yield (963 OD_{260} units. 98.8%) and to establish homogeneity by TLC (R_f 0.80, system C). The solution is evaporated, and the residue is dried in a vacuum oven at 60° for 12 hr for the next step.

The trichloroethyl group can be removed from compound (11b) in the same manner to give compound (12).

Removal of the tert-Butyldimethylsilyl Group from Compound (12)

Compound (12) (5.4 mg, 5.8 μmol, 173 OD_{260} units) prepared above is treated with 0.7 M tetrabutylammonium fluoride[16] solution in tetrahydrofuran (160 μl) at room temperature for 40 min until TLC (system C) shows that the starting material with R_f 0.8 is completely transformed into a single product with R_f 0.46, identical with authentic adenyl(2' → 5')adenosine. The reaction mixture is evaporated and the residue extracted with ether (1 ml × 2) to remove nonpolar substances. The residue is evaporated again and then dissolved in 0.02 M triethylammonium bicarbonate (TEAB) (0.8 ml) solution. The pH of this solution is adjusted to 8.0 with 10% ammonia, and the solution is applied to a DEAE-Sephadex A25 column (1.2 × 14 cm), equilibrated with 0.02 M TEAB. The column is eluted with a linear gradient (200 ml/200 ml) of 0.02 M to 0.1 M TEAB. The eluate is monitored by UV absorbance at 260 nm, and fractions of 3.2 ml are collected. Only one peak (fractions 17–30) is obtained. The pooled fractions are evaporated under vacuum and then repeatedly evaporated with water until the TEAB is completely decomposed. The resulting residue is dissolvedin methanol (250 μl) to determine the OD value (141 OD_{260} units, 4.7 μmol, 81.6%). Sodium iodide in acetone (1.0 M, 250 ml) and then acetone (2.0 ml) are added to the methanol solution. The white precipitate is collected by centrifugation, washed with acetone (2 × 2 ml), and finally dried in a desiccator. The homogeneous product (13) thus obtained shows the same R_f value as standard adenyl(2' → 5')adenosine on PEI cellulose TLC plates developed by three different solvent systems: R_f 0.48 (0.1 M NH_4HCO_3), R_f 0.50 (0.5 M $Mg(OAc)_2$), and R_f 0.51 (1.0 M LiCl).

[16] I. Kuwajima, T. Murofushi, and E. Nakamura, *Synthesis* 9, 602 (1976).

Condensation of Compound (11a) and (7)

A solution of compound (7) (200 mg, 0.306 mmol) in dry tetrahydrofuran (424 μl) is added at −78° to a mixture of collidine (178.4 μl, 1.35 mmol), 2,2,2-trichloroethylphosphorodichloridite (0.36 mmol, 54.4 μl) and dry tetrahydrofuran (412 μl). The reaction mixture is stirred for 30 min at −78°, and then a solution of compound (11a) (265 mg, 0.245 mmol) in dry tetrahydrofuran (660 μl) is added to it over a period of 3 min. The reaction mixture is stirred at −78° for 1 hr and then warmed to room temperature. After the solvent is completely removed under vacuum, the residue is directly applied to a short silica gel column (20 g). The column is washed thoroughly with chloroform to remove collidine completely followed by elution with chloroform–methanol (10:1) to yield a mixture of two diastereomers (377 mg, 0.198 mmol, 81%). A part of this mixture is applied to preparative silica gel TLC plates that are developed twice in system A. Each diastereomer thus separated is eluted with methanol to yield (14a) (R_f 0.27, system A), mp 123–126°, ^{31}P NMR δ −2.37, 139.7., and (14b) (R_f 0.14, system A), mp 139–142°, ^{31}P NMR δ −2.48, 139.3.

Iodine Oxidation of Compounds (14a) and (14b)

The mixture of two diastereomers (14a and 14b) (310 mg, 0.163 mmol) is dissolved in water-saturated ether (3.0 ml), and a solution of iodine in ether (0.5 M) is carefully added to this reaction mixture until the reddish brown color of iodine persists (about 1 ml of the solution is required). The ether solution is washed twice with a saturated $NaHCO_3$ solution and once with a saturated NaCl solution, dried over anhydrous sodium sulfate and evaporated under vacuum to give a residue (298 mg, 0.156 mmol, 96%). A part of the residue is applied to preparative silica gel TLC plates which are developed twice in system A. Two diastereomers thus separated are eluted with methanol to give (15a) (R_f 0.21, system A), mp 139–142° (lit.[8], mp 129–135°), ^{31}P NMR δ −2.25, −2.80, and (15b) (R_f 0.12, system A), mp 139–142° (lit.[8] mp 175–180°), ^{31}P NMR δ −2.18, −2.41.

Removal of the Monomethoxytrityl Group from (15a) and (15b)

The mixture of diastereomers (15a) and (15b) (74.4 mg, 38.9 μmol) is dissolved in 80% acetic acid (0.7 ml) and heated in an oil bath at 90° for 15 min. The reaction mixture was concentrated under vacuum. The residue is then evaporated twice with toluene (2 ml) and then dissolved in methanol (1 ml) to be applied to a stepped column (4 steps) packed with Sephadex LH-20 (bed volume in methanol, 110 ml). The first eluate with methanol affords a mixture of (16a) and (16b) on concentration (62.1 mg,

37.8 μmol, 97%). A part of this mixture is applied to preparative silica gel TLC plates, which are developed twice with system A. Two diastereomers thus separated are eluted with ethyl acetate-methanol (1 : 1) to afford (16a) (R_f 0.08, system A) and (16b) (R_f 0.04, system A).

Removal of Trichloroethyl Group from the Diastereomeric Mixture of (16a) to (16b)

The mixture of (16a) and (16b) (28.5 mg, 10.7 μmol, 477 OD_{260} units) is dissolved in dry dimethylformamide (214 μl). To this solution is added active Zn-Cu couple[15] (16.5 mg) and acetylacetone (12 mg, 12 μl, 120 μmol), and the resulting suspension is heated in an oil bath at 55° for 3 hr with stirring. After cooling, the green, cloudy reaction mixture is diluted with methanol (10 ml) to give a clear green solution. Chelex (4.0 ml packed volume, treated as described above) is added to this solution, and the entire mixture is stirred for 1 hr or until the solution is decolorized. The Chelex is removed by filtration, and the filtrate is carefully evaporated under vacuum. The residue thus obtained (17) (398 OD_{260} units, 8.90 μmol, 83%) runs as a single spot on silica gel TLC (R_f 0.60, system C).

Removal of tert-Butyldimethylsilyl Group, Synthesis of the Trimer (18)

Compound (17) (398 OD_{260} units, 8.90 μmol) is treated with 0.7 M solution of tetrabutylammonium fluoride[16] in dry tetrahydrofuran (1.0 ml) at room temperature for 3 hr. After removal of tetrahydrofuran under vacuum, the residue was extracted with ether (1 ml × 2) to eliminate nonpolar substances followed by evaporation and dilution with 0.075 M TEAB (1.0 ml). The pH of this solution is adjusted to 8.0 with 10% ammonia, and the solution is applied to DEAE-Sephadex A-25 column (1.2 cm × 25 cm) equilibrated with 0.075 M TEAB. The column was eluted with a linear gradient (250 ml/250 ml) of 0.075 M to 0.150 M TEAB. Fractions of 3.8 ml were collected and monitored by UV absorbance at 260 nm. The symmetrical peak that appeared between fractions 52 and 69 is concentrated under vacuum and evaporated with water several times until TEAB is completely decomposed. The pure product (18) thus obtained (291 OD_{260} units, 6.51 μmol, 73%) shows the same R_f value with (18) prepared by an alternative procedure[11] (R_f 0.23, system C; R_f 0.45, PEI cellulose TLC, 0.1 M ammonium bicarbonate).

On both DEAE-column chromatography and HPLC,[15] synthetic (18) comigrates with the bacterial alkaline phosphatase-produced "core" of biosynthetic trimer labeled with [3H]ATP. Digestion of synthetic (18) with a mixture of bacterial alkaline phosphatase and snake venom phos-

NUCLEAR MAGNETIC RESONANCE CHARACTERIZATION OF SYNTHETIC INTERMEDIATES[a]

Structure	C-1'H	C-2'H	C-3'H	C-5'H$_A$	C-5'H$_B$	C-2H	C-8H	$(CH_3)_3$C-Si	CH_3-Si	CH_3OAr
(5)	5.77 (1H)d 8 Hz	5.03 (1H)dd 4 Hz, 8 Hz	4.33 (1H)d 6 Hz	3.94 (1H)d 13 Hz	3.66 (1H)d 13 Hz	7.80 (1H)s	8.29 (1H)s	0.96 (9H)s 0.75 (OH)s	0.13 (6H)s -0.12 (3H)s -0.58 (3H)s	—
(7)	5.99 (1H)d 5 Hz	~4.6 (1H)m	~4.6 (1H)m	3.50 (1H)dd 4 Hz, 10 Hz	3.22 (1H)dd 4 Hz, 10 Hz	8.01 (1H)s	8.23 (1H)s	0.89 (9H)s	0.08 (3H)s 0.00 (3H)s	3.74 (3H)s
(8)	6.03 (1H)d	4.99 (1H)t	~4.30 (1H)m	3.56 (1H)dd 3 Hz, 11 Hz	3.38 dd 3 Hz, 11 Hz	8.01 (1H)s	8.23 (1H)s	0.85 (9H)s	0.02 (3H)s -0.09 (3H)s	3.78 (3H)s
(9a)	6.15 (1H)d 5 Hz 5.87 (1H)d 6 Hz	5.26 (1H)m 4.89 (1H)t 4 Hz	4.63 (1H)t 5 Hz	3.54 (1H)dd 4 Hz, 11 Hz	3.23 (1H)dd 4 Hz, 11 Hz	8.02 (1H)s 7.93 (1H)s	8.23 (1H)s 8.21 (1H)s	0.92 (9H)s 0.80 (9H)s 0.79 (9H)s	0.09 (6H)s 0.06 (3H)s -0.04 (3H)s -0.05 (3H)s -0.26 (3H)s	3.74 (3H)s
(9b)	6.04 (1H)d 2 Hz 5.83 (1H)d 5 Hz	5.38 (1H)m 4.90 (1H)dd 5 Hz, 7 Hz	4.62 (1H)t 4 Hz	3.46 (1H)dd 4 Hz, 11 Hz	3.10 (1H)dd 4 Hz, 11 Hz	7.85 (1H)s 7.82 (1H)s	8.22 (1H)s 8.16 (1H)s	0.90 (9H)s 0.82 (18H)s	0.08 (3H)s 0.05 (6H)s -0.02 (3H)s -0.06 (3H)s -0.16 (3H)s	3.72 (3H)s
(10a)	6.23 (1H)d 4 Hz 5.86 (1H)d 5 Hz	5.60 (1H)m 4.92 (1H)t 4 Hz	4.75 (1H)t 4 Hz	3.56 (1H)dd 4 Hz, 11 Hz	3.26 (1H)dd 4 Hz, 11 Hz	8.05 (1H)s 7.94 (1H)s	8.24 (1H)s 8.19 (1H)s	0.94 (9H)s 0.81 (9H)s 0.80 (9H)s	0.09 (6H)s 0.06 (6H)s -0.06 (3H)s -0.24 (3H)s	3.77 (3H)s
(10b)	6.18 (1H)d 3 Hz 5.84 (1H)d 4 Hz	5.76 (1H)m 4.94 (1H)m		3.48 (1H)dd 2 Hz, 11 Hz	3.16 (1H)dd 4 Hz, 11 Hz	7.88 (1H)s 7.84 (1H)s	8.26 (1H)s 8.08 (1H)s	0.94 (9H)s 0.86 (18H)s	0.14 (3H)s 0.10 (3H)s 0.06 (3H)s -0.02 (6H)s -0.12 (3H)s	3.74 (3H)s
(11a)	6.08 (1H)d 6 Hz 5.88 (1H)d 5 Hz	5.60 (1H)m 4.84 (1H)m	4.67 (1H)d 4 Hz	3.95 (1H)dd 1 Hz, 12 Hz	3.68 (1H)dd 1 Hz, 12 Hz	8.01 (1H)s 7.97 (1H)s	8.15 (1H)s 8.01 (1H)s	0.92 (18H)s 0.82 (9H)s	0.18 (3H)s 0.16 (3H)s 0.08 (6H)s -0.02 (3H)s	—
(11b)	5.85 (2H)d 6 Hz	5.70 (1H)m 4.52 (1H)t 6 Hz		3.80 (1H)br. d 13 Hz	3.55 (1H)br. d 13 Hz	7.94 (1H)s 7.62 (1H)s	8.14 (1H)s 8.03 (1H)s	0.84 (9H)s 0.77 (18H)s	-0.17 (3H)s 0.07 (3H)s 0.05 (3H)s -0.02 (3H)s -0.04 (3H)s -0.09 (6H)s	—

(14a)	6.17 (2H)d 4 Hz 5.88 (1H)d 5 Hz	5.55 (1H)m 5.32 (1H)m 4.94 (1H)t 5 Hz	4.67 (2H)m	3.53 (1H)dd 5 Hz, 12 Hz	3.25 (1H)dd 5 Hz, 12 Hz	8.06 (1H)s 7.99 (1H)s 7.97 (1H)s	8.26 (1H)s 8.23 (1H)s 8.21 (1H)s	0.91 (18H)s 0.82 (18H)s	0.15 (3H)s 0.12 (3H)s 0.09 (6H)s 0.05 (3H)s −0.03 (6H)s −0.23 (3H)s −0.23 (3H)s	3.75 (3H)s
(14b)	6.11 (1H)d 4 Hz 6.07 (1H)d 1 Hz 5.89 (1H)d 5 Hz	5.33 (1H)m 5.23 (1H)m 5.00 (1H)m	4.97 (1H)m	3.45 (1H)br.d 9 Hz	3.10 (1H)br.d 9 Hz	8.05 (1H)s 7.95 (1H)s 7.77 (1H)s	8.28 (1H)s 8.19 (1H)s 8.11 (1H)s	0.92 (18H)s 0.84 (18H)s	0.14 (3H)s 3.13 (3H)s 3.10 (3H)s 0.09 (3H)s 0.08 (3H)s 0.00 (6H)s −0.15 (3H)s	3.77 (3H)s
(15a)	6.20 (1H)d 4 Hz 6.15 (1H)d 4 Hz 5.84 (1H)d 4 Hz	5.61 (1H)m 5.52 (1H)m 4.93 (1H)t 4 Hz	4.74 (1H)t 4 Hz 4.68 (1H)t 4 Hz	3.54 (1H)dd 4 Hz, 11 Hz	3.25 (1H)dd 4 Hz, 11 Hz	8.05 (1H)s 7.97 (2H)s	8.25 (1H)s 8.19 (1H)s 8.16 (1H)s	0.87 (9H)s 0.86 (9H)s 0.80 (18H)s	0.10 (3H)s 0.09 (6H)s 0.06 (6H)s −0.01 (3H)s −0.04 (3H)s −0.20 (3H)s	3.74 (3H)s
(15b)	6.25 (1H)d 3 Hz 6.15 (1H)br.s 5.91 (1H)d 4 Hz	5.70 (1H)m 5.25 (1H)m 5.05 (1H)t 4 Hz	4.95 (1H)m	3.49 (1H)dd 4 Hz, 11 Hz	3.19 (1H)dd 4 Hz, 11 Hz	8.05m (1H)s 7.99 (1H)s 7.82 (1H)s	8.28 (1H)s 8.13 (1H)s 8.09 (1H)s	0.91 (oH)s 0.87 (9H)s 0.85 (9H)s 0.83 (9H)s	0.13 (3H)s 0.11 (6H)s 0.09 (3H)s 0.07 (3H)s 0.03 (6H)s −0.11 (3H)s	3.71 (3H)s
(16a)	6.16 (1H)d 4 Hz 6.03 (1H)d 7 Hz 5.83 (1H)d 4 Hz	5.57 (1H)m 5.41 (1H)m 4.90 (1H)t 4 Hz	4.61 (1H)m	3.93 (1H)d	3.68 (1H)dd 4 Hz, 11 Hz	8.00 (1H)s 7.97 (2H)s	8.23 (1H)s 8.19 8.16 (1H)s	0.89 (9H)s 0.86 (9H)s 0.85 (9H)s 0.81 (9H)s	0.13 (3H)s 0.11 (3H)s 0.04 (12H)s −0.04 (3H)s −0.16 (3H)s	—

"PMR spectra were determined either at 100 MHZ (5–11) or at 220 MHZ (14–16) in chloroform with tetramethylsilane as the internal standard. Only resonances that more clearly resolved are presented in this table. The data are presented in the following order: chemical shift (δ, ppm), number of protons, multiplicity, and, where appropriate, coupling constant. The multiplicity is represented as: singlet, s; doublet, d; triplet, t; doublet of doublets, dd; multiplet, m. Broad is abbreviated as br.

phodiesterase gives adenosine as the only discernible product on TLC. Under conditions where T2 ribonuclease completely degrades adenylyl(3' → 5')adenylyl (3' → 5')adenosine to adenosine and 3'-AMP, the product (18) is completely stable as determined by TLC. The ^{31}P NMR of synthetic (18) is identical with (18) prepared by an alternative route[11] and shows two singlets at chemical shifts of −0.34 and −0.70 ppm (see the table).

Acknowledgment

The authors are deeply indebted to Dr. H. Shindo of this Institute for determination of ^{31}P NMRs.

[17] P. J. Bridgen, J. Imai, M. I. Johnston, and P. F. Torrence, unpublished observations.

[31] Chromatographic Assay on DEAE-Cellulose for (2'-5')-Oligo(A) Synthesis in Cell Extracts

By Michael A. Minks and Corrado Baglioni

Treatment of animal cells with interferon results in the induction of synthesis of new proteins.[1,2] One of the induced proteins is an enzyme that, in the presence of double-stranded RNA (dsRNA), polymerizes ATP into a series of nucleotides linked by the unusual 2',5'-phosphodiester bond.[3] We have designated these nucleotides (2'-5')-oligo(A) to reflect both the unusual phosphodiester linkage and the oligomeric nature of the series.[4] Correspondingly, the enzymic activity that synthesizes the oligonucleotides is designated (2'-5')-oligo(A) polymerase.

The (2'-5')-oligo(A) inhibits *in vitro* protein synthesis at subnanomolar concentrations via the activation of a latent endonuclease.[5,6] The (2'-5')-oligo(A) inhibits protein synthesis and causes RNA degradation also when introduced in intact cells.[6,7] Formation of (2'-5')-oligo(A) may inhibit the

[1] S. L. Gupta, B. Y. Rubin, and S. L. Holmes, *Proc. Natl. Acad. Sci. U.S.A.* 76, 4817 (1979).
[2] E. Knight, Jr., and B. D. Korant, *Proc. Natl. Acad. Sci. U.S.A.* 76, 1824 (1979).
[3] C. Baglioni, *Cell* 17, 255 (1979).
[4] M. A. Minks, S. Benvin, P. A. Maroney, and C. Baglioni, *Nucleic Acids Res.* 6, 767 (1979).
[5] C. Baglioni, M. A. Minks, and P. A. Maroney, *Nature (London)* 273, 684 (1978).
[6] M. J. Clemens and B. R. G. Williams, *Cell* 13, 565 (1978).
[7] B. R. G. Williams and I. M. Kerr, *Nature (London)* 276, 88 (1978).

replication of RNA viruses in interferon-treated cells[5,8] and detectable levels of these oligonucleotides have indeed been observed in interferon-treated L cells infected with encephalomyocarditis virus.[9]

To measure the levels of (2'-5')-oligo(A) polymerase in extracts of animal cells and to study the activation of this enzyme by dsRNA, we have developed an assay for (2'-5')-oligo(A) synthesis. This assay is based on the conversion of [³H]ATP into labeled (2'-5')-oligo(A) that is separated from less charged nucleotides by DEAE-cellulose chromatography. The enzymic activity can be measured from the rate of conversion of ATP into (2'-5')-oligo(A).

Preparation of Cell Extracts

Reagents

> Eagle's saline
> Buffer A: 10 mM KCl, 1.5 mM Mg(OAc)$_2$, 0.5 mM dithiothreitol, 20 mM HEPES, adjusted to pH 7.4 with KOH
> Triton X-100

Procedure

We have measured (2'-5')-oligo(A) synthesis in extracts prepared from cells in suspension or in monolayer culture. The assay has been used successfully with HeLa cells, primary human foreskin fibroblast lines, lymphoblastoid cells (Daudi), and mouse L cells.

Cells are pelleted by centrifugation at 800 g for 5 min and washed thrice with ice-cold saline. All further manipulation is at 0° unless otherwise indicated. The cells are swollen by the addition of 2 volumes of buffer A (relative to the packed cell volume), and after standing for 10 min the cells are broken by homogenization. A Dounce homogenizer is used with a tight-fitting plunger, and cell breakage is monitored by phase contrast microscopy. Homogenization is stopped at about 95% cell breakage. The number of passes of the plunger depends on cell type and is in the range of 8–20 strokes.

The postmitochondrial supernatant is prepared by centrifuging the extract at 30,000 g in a refrigerated centrifuge. The supernatant (S$_{30}$) is aspirated from the centrifuge tubes and stored in small aliquots under liquid nitrogen. When stored in this manner, samples containing about 10 mg of protein per milliliter retain activity for at least a year. The loss of activity on freezing and thawing is minimal.

[8] A. G. Hovanessian, J. Wood, E. Meurs, and L. Montagnier, *Proc. Natl. Acad. Sci. U.S.A.* **76**, 3261 (1979).

[9] T. W. Nilsen and C. Baglioni, *Proc. Natl. Acad. Sci. U.S.A.* **76**, 2600 (1979).

When handling small cultures of cells (about 5×10^6) it has been found advantageous to substitute disruption with detergent for the homogenization step, since losses of material are minimized. Cells are pelleted and washed as described above, but are swollen in hypotonic medium containing the nonionic detergent Triton X-100. In order to produce workable volumes we have used 1 ml of buffer A per 7×10^7 cells to obtain extracts approximately one-fourth the concentration of those prepared by homogenization. After swelling, the cells are disrupted by agitation on a vortex mixer. Cell breakage is monitored by phase contrast microscopy and further agitation is given to produce the desired degree of disruption. The amount of detergent added differs with cell type and is determined experimentally. For example, HeLa cells are homogenized with 0.1% (v/v) Triton X-100 in buffer A whereas 0.02% Triton X-100 is sufficient to disrupt lymphoblastoid cells (Daudi) but leave the nuclei intact. The extract should be stabilized by the addition of 0.2 volume of 50% (v/v) glycerol in buffer A and further processed as described above. Without glycerol these postmitochondrial supernatants lose activity on freezing and thawing. The presence of glycerol has no effect on the $(2'-5')$-oligo(A) assay described below.

Assay for $(2'-5')$-Oligo(A) Synthesis

The $(2'-5')$-oligo(A) polymerase is present in the cell extract in an inactive form and can be activated by incubation with high molecular weight synthetic dsRNA.[10] The assay described monitors the conversion of labeled ATP into $(2'-5')$-oligo(A) with added poly(I) · poly(C).

Reagents

Master mix: $1 M$ K(OAc), $0.25 M$ Mg(OAc)$_2$, $0.1 M$ [^{3}H]ATP, $10 mM$ dithiothreitol, $40 mM$ fructose 1,6-diphosphate, $0.2 M$ HEPES, adjusted to pH 7.4 with KOH
Poly(I) · poly(C): 0.2 mg/ml in $0.1 M$ K (OAc), $20 mM$ HEPES, adjusted to pH 7.4 with KOH
Buffer B: $90 mM$ KCl, $20 mM$ Tris · HCl, pH 7.6
Buffer C: $350 mM$ KCl, $20 mM$ Tris · HCl, pH 7.6

Procedure

Incubations are prepared to contain 0.1 volume of master mix, 0.1 volume of poly(I) · poly(C) solution, and an aliquot of cell extract contain-

[10] B. R. G. Williams, R. R. Golgher, R. E. Brown, C. S. Gilbert, and I. M. Kerr, *Nature* (*London*) **282**, 582 (1979).

ing approximately 50 μg of protein per 25-μl incubation. The concentration of [³H]ATP is adjusted by dilution with cold ATP to have about 0.1 μCi or 200,000 cpm per 25-μl incubation. After 60 min at 30° the reaction is terminated by heating to 95° for 3 min to denature proteins that degrade 2'-5'-oligo(A).² The samples are then diluted with 1 ml of buffer B and may be kept at room temperature until applied to the DEAE-cellulose columns.

The (2'-5')-oligo(A) is separated from the less charged ATP by stepwise elution from DEAE-cellulose columns. It is advisable to deaerate DEAE-cellulose and buffer B prior to use to prevent the formation of gas bubbles in the columns. Whatman DE-52 cellulose is equilibrated with buffer B according to the manufacturer's instructions, and 0.5-ml columns are prepared in 0.7 × 4 cm polypropylene chromatography columns (Bio-Rad). These are washed with 5 ml of buffer B immediately prior to sample application; the heat-treated samples are then passed over the columns three times to allow complete retention of (2'-5')-oligo(A). The columns are then washed with 25 ml of buffer B. This step removes ATP and other less charged nucleotides. The (2'-5')-oligo(A) is then eluted directly into scintillation vials by application of 2 × 1 ml buffer C; 18 ml of scintillant (Scintiverse; Fisher) are added to the vials, and the radioactivity is determined by liquid scintillation counting. The amount of (2'-5')-oligo(A) formed is calculated from the known concentration and radioactivity of the starting [³H]ATP.

The chromatographic separation provides an accurate determination of (2'-5')-oligo(A), since more than 99% of the ATP is removed during the wash with buffer B and at least 99% of the material eluting with 0.35 M KCl is retained upon rechromatography on a second DEAE-cellulose column after dilution to 90 mM KCl; elution of the retained nucleotides with

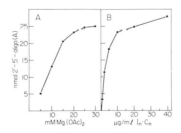

FIG. 1. Effect of magnesium (A) and dsRNA concentration (B) on (2'-5')-oligo(A) synthesis. Reaction conditions are as described in the text except that in (A) the Mg(OAc)₂ concentration was varied and in (B) the Mg(OAc)₂ concentration was 25 mM and the dsRNA concentration was varied. Nanomoles of ATP converted into (2'-5')-oligo(A) from an input of 125 nmol are indicated on the ordinate.

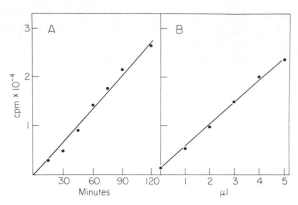

FIG. 2. Rate of (2'-5')-oligo(A) synthesis with respect to time (A) and amount of extract (B). Conditions are as described in the text with 5 μl of extract per incubation (50 μg of protein) in (A) and the indicated amount of extract incubated for 2 hr in (B). Data were taken from Minks *et al.*[11]

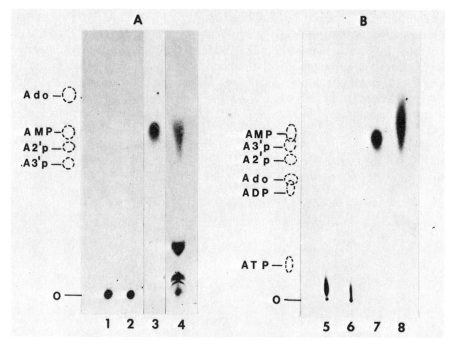

FIG. 3. Chromatography of (2'-5')-oligo(A) and its digestion products on poly-ethyleneimine cellulose. Solvents were 1 *M* acetic acid (*A*) and 1 *M* LiCl (*B*). Lanes 1 and 5, (2'-5')-oligo(A); 2 and 6, ribonuclease T2 digest; 3 and 7, phosphodiesterase digest; 4 and 8, alkaline phosphatase digest. The position of unlabeled markers is shown. From Minks *et al.*[11]

0.35 M KCl is quantitative. A control incubation without added poly(I) · poly(C) and a time zero control are included among the assays. These incubations provide, respectively, a measurement of the incorporation of ATP into oligonucleotides not dependent on (2'-5')-oligo(A) polymerase activity and of impurities in the [³H]ATP that may elute with (2'-5')-oligo(A). We have consistently found less than 500 cpm eluted with buffer C in these control incubations.

The reaction conditions described have been optimized for the HeLa polymerase with respect to magnesium acetate concentration (Fig. 1A) and dsRNA concentration (Fig. 1B). Furthermore, the reaction proceeds linearly with respect to time (Fig. 2A) and the rate of reaction is proportional to the amount of cell extract added (Fig. 2B).[11] These are ideal conditions under which to determine either the amount of polymerase in various extracts, or the requirements of the polymerase for various ions or different activators.[12]

Despite its obvious simplicity, only (2'-5')-oligo(A) formation is monitored by this assay, since there is no increase above background in the 0.35 M KCl fraction of samples incubated without dsRNA or treated with dsRNA but not incubated. The recovered radioactivity represents the dsRNA-dependent incorporation of ATP into oligomeric material. These oligomers have chromatographic behavior and nuclease sensitivities in keeping with a general formula pppA(2'p5'A)$_n$ (Fig. 3). This method is fast, reliable, and suited to the processing of multiple samples. It can therefore be recommended for its simplicity.

[11] M. A. Minks, S. Benvin, P. A. Maroney, and C. Baglioni, J. Biol. Chem. 254, 5058 (1979).
[12] M. A. Minks, D. K. West, S. Benvin, and C. Baglioni, J. Biol. Chem. 254, 10180 (1979).

[32] Assay for (2'-5')-Oligo(A)-Dependent Endonuclease Activity

By Corrado Baglioni, Patricia A. Maroney, and Timothy W. Nilsen

The observation of Kerr and co-workers[1] that double-stranded RNA (dsRNA) is a potent inhibitor of protein synthesis in extracts of interferon-treated cells led to the discovery of oligoadenylates linked by 2'-5'-phosphodiester bonds.[2] These oligonucleotides were designated (2'-5')-

[1] I. M. Kerr, R. M. Friedman, R. E. Brown, L. A. Ball, and J. C. Brown, J. Virol. 13, 9 (1974).
[2] I. M. Kerr and R. E. Brown, Proc. Natl. Acad. Sci. U.S.A. 75, 256 (1978).

METHODS IN ENZYMOLOGY, VOL. 79

oligo(A) and were shown to activate an endonuclease present in extracts of both control and interferon-treated cells.[3,4] This finding explained the previous observation of Lengyel and co-workers[5] that incubation of the latter extracts with dsRNA and ATP activates an endonuclease; this activation occurs upon formation of (2'-5')-oligo(A). The enzyme that synthesizes these oligonucleotides is designated (2'-5')-oligo(A) polymerase, and an assay for this enzymic activity in cell extracts is described in this volume.[6] This procedure can be scaled up for the preparation of large quantities of (2'-5')-oligo(A) simply by increasing the reaction volume. With an extract of interferon-treated HeLa cells the rate of conversion of ATP into (2'-5')-oligo(A) is extremely high, since as much as 50% of the input ATP is converted into (2'-5')-oligo(A).[7] These oligonucleotides, therefore, can be obtained in millimolar concentration after chromatography on DEAE-cellulose.[7] The (2'-5')-oligo(A) can also be chemically synthesized[8] and has become commercially available (P-L Biochemicals). The availability of these compounds will undoubtedly favor studies of their action on cell metabolism and of their role in the antiviral state induced by interferon.

Although an effect of (2'-5')-oligo(A) on protein synthesis and RNA breakdown can be shown in intact cells,[9,10] the endonuclease activation test in cell extracts represents the method of choice for the identification of (2'-5')-oligo(A) and for its quantitation at concentrations in the nanomolar range. The endonuclease can be assayed either indirectly by inhibition of protein synthesis[4] or directly by measuring mRNA breakdown.[3] The latter assay will be described in this chapter. Other assays for RNA cleavage described in the literature are essentially qualitative, since the labeled RNA is examined after incubation with cell extract or partially purified endonuclease by gel electrophoresis.[11] Cleavage is detected by the appearance of degraded RNA in the autoradiographs of the gels. This method is convenient to examine multiple samples either for the presence of endonuclease, as for example during purification of this enzyme, or for

---

[3] C. Baglioni, M. A. Minks, and P. A. Maroney, *Nature (London)* 273, 684 (1978).

[4] M. J. Clemens, and B. R. G. Williams, *Cell* 13, 565 (1978).

[5] G. C. Sen, B. Lebleu, G. E. Brown, M. Kawakita, E. Slattery, and P. Lengyel, *Nature (London)* 264, 370 (1976).

[6] M. A. Minks and C. Baglioni, this volume [31].

[7] M. A. Minks, S. Benvin, P. A. Maroney, and C. Baglioni, *J. Biol. Chem.* 254, 5058 (1979).

[8] A. F. Markham, R. A. Porter, M. J. Gait, R. C. Sheppard, and I. M. Kerr, *Nucleic Acids Res.* 6, 2569 (1979).

[9] B. R. G. Williams and I. M. Kerr, *Nature (London)* 276, 88 (1978).

[10] A. G. Hovanessian, J. Wood, E. Meurs, and L. Montagnier, *Proc. Natl. Acad. Sci. U.S.A.* 76, 3261 (1979).

[11] L. Ratner, R. C. Wiegand, P. J. Farrell, G. C. Sen, B. Cabrer, and P. Lengyel, *Biochem. Biophys. Res. Commun.* 81, 947 (1978).

the presence of (2'-5')-oligo(A). Kerr and collaborators[12] showed that (2'-5')-oligo(A) is present in interferon-treated cells infected with encephalomyocarditis virus in this way. Oligonucleotides were extracted from the cells and resolved by high-pressure liquid chromatography; each fraction was tested for endonuclease activation.

The measurement of endonuclease activity is based on an assay previously developed for the detection of nuclease activity in cell-free protein synthesizing systems.[13] The assay is based on the labeling with [³H]-uridine of the endogenous mRNA of cell-free systems prepared from HeLa cells. The poly(A)-containing RNA was measured before and after incubation by its retention on oligo(dT)-cellulose at high ionic strength and its elution with buffers at low ionic strength.[14] Since a random cleavage of each mRNA molecule results in a loss of half the mRNA annealed to oligo(dT), this assay is well suited for the detection of few endonucleolytic breaks.

In subsequent applications of this method we found it convenient first to synthesize labeled mRNA and to add isolated poly(A)-containing RNA to assays for nuclease activity. Since the only labeled RNA species is that measured by oligo(dT)-cellulose chromatography, quantitation of the results is simplified. It should be pointed out, however, that this assay can be successfully carried out with cell extracts only when the nuclease activity in the absence of added (2'-5')-oligo(A) is low, because extensive degradation of the mRNA does not allow detection of the (2'-5')-oligo(A)-activated endonuclease. In addition, after the cleavages catalyzed by this endonuclease, further degradation of the mRNA fragments produced may result from exonuclease activities. These exonucleases may not attack intact mRNA because of the protection afforded by the "cap" at the 5' terminus and by the poly(A) at the 3' terminus. In our assay, however, exonucleases do not extensively digest either intact mRNA or its fragments, possibly because they are present in very low amounts or because the mRNA is protected by its association with cytoplasmic proteins.

Preparation of Cell Extracts

Reagents

Eagle's saline
Buffer A: 10 mM KCl, 1.5 mM Mg(OAc)$_2$, 0.5 mM dithiothreitol, 20 mM HEPES, adjusted to pH 7.4 with KOH

[12] B. R. G. Williams, R. R. Golgher, R. E. Brown, C. S. Gilbert, and I. M. Kerr, *Nature (London)* 282, 582 (1979).
[13] E. D. Hickey, L. A. Weber, and C. Baglioni, *Biochem. Biophys. Res. Commun.* 80, 377 (1978).
[14] R. E. Pemberton, P. Liberti, and C. Baglioni, *Anal. Biochem.* 66, 18 (1975).

Procedure

We have measured $(2'\text{-}5')$-oligo(A)-dependent endonuclease activity only in extracts prepared from suspension cultures of HeLa cells by homogenizing the cells without detergents. Extracts prepared from other cell lines have given variable backgrounds of nuclease activity, and addition of detergents to HeLa cells resulted in extracts with high nuclease activity. To prepare extracts, the cells are collected in log phase, washed three times with cold saline, and pelleted by centrifugation. The pellet is resuspended in twice its volume of buffer A; after 10 min on ice the cells are broken with less than 10 strokes of a tight-fitting Dounce homogenizer. Cell breakage is monitored by phase microscopy, and homogenization is stopped when about 90% of the cells are broken. Further homogenization may break lysosomes and produce extracts with a high background of nuclease activity. An S_{30} fraction is prepared by 5 min of centrifugation at $30,000\,g$, and the supernatant is removed and stored in small fractions at $-70°$. Samples stored in this way retain activity for several months and can be frozen and thawed a few times.

Radioactive mRNA

Any radioactive RNA can be used in the endonuclease assay. In previous experiments we measured the cleavage of HeLa cell poly(A)-containing mRNA, both free and polysome-bound,[3] but to run several assays it is convenient to have an abundant and inexpensive source of labeled mRNA. Viral mRNAs are suitable for this purpose since they are made in large quantities by infected cells or can be synthesized *in vitro* from isolated virions. The mRNA of some viruses is also made up of few molecular species; e.g., picornavirus mRNA is a 35 S molecule, and vesicular stomatitis virus (VSV) mRNA is made up of five species. The VSV mRNA is particularly suitable for the endonuclease assay because it is both ''capped'' and possesses a $3'$-terminal poly(A).

We have described in detail the preparation of VSV mRNA from infected HeLa cells.[15] It is more convenient, however, to prepare this mRNA by *in vitro* transcription either from permeabilyzed virions[16] or from viral replicative complexes prepared from infected cells. We have followed the latter procedure to prepare VSV mRNA of high specific activity as described by Toneguzzo and Gosh.[17] The RNA synthesized by either procedure is purified by hybridization to oligo(dT)-cellulose and elution with low ionic strength buffer. Transcription reactions containing relatively small amounts of protein can be directly adjusted to $0.5\,M$ LiCl,

[15] L. A. Weber, M. Simili, and C. Baglioni, this series, Vol. 60, p. 351.
[16] D. Baltimore, A. Huang, and M. Stampfer, *Proc. Natl. Acad. Sci. U.S.A.* **66,** 572 (1970).
[17] F. Toneguzzo and H. P. Gosh, *J. Virol.* **17,** 477 (1976).

0.5% sodium dodecyl sulfate, and 10 mM Tris · HCl, pH 7.5. The samples are applied to small columns of oligo(dT)-cellulose and eluted as previously described.[15] The poly(A)-containing RNA is precipitated with ethanol after addition of 0.2 M NaCl and 25 μg of carrier tRNA per milliliter. The RNA is repeatedly washed with ethanol and redissolved in water as described.[15]

Endonuclease Assay

Reagents

Buffer B: 0.6 M K(OAc)$_2$, 7.5 mM Mg(OAc)$_2$, 20 mM fructose 1,6-diphosphate

Radioactive poly(A)-containing mRNA: about 1000 cpm/μl

(2'-5')-Oligo(A): 0.01 mM in water and appropriate dilutions

Buffer C: 0.5 M LiCl, 1 mM EDTA, 0.5% sodium dodecyl sulfate, 10 mM Tris · HCl, pH 7.5

Buffer D: 1 mM EDTA, 0.5% sodium dodecyl sulfate, 10 mM Tris · HCl, pH 7.5

Oligo(dT)-cellulose: type T-3, Collaborative Research, Inc., Waltham, Massachusetts

Procedure

The assays are assembled on ice. Each assay contains in 0.05 ml volume: 30 μl of cell extract, 10 μl of buffer B, 5 μl of radioactive mRNA, 5 μl of 2'-5'-oligo(A) to obtain a final concentration between 1 and 100 nM. The reactions are incubated at 30° for up to 90 min and terminated by the addition of 1 ml of buffer C. Each sample is applied to a 0.4 × 0.6 cm column of oligo(dT)-cellulose equilibrated with buffer C and passed three times over the same column to obtain complete annealing of the mRNA. The column is then washed with 10 ml of buffer C and finally eluted with two 1-ml aliquots of buffer D. These aliquots are combined and counted directly after addition of 18 ml of Scintiverse (Fisher). A reaction not incubated and one incubated without added (2'-5')-oligo(A) are analyzed in parallel. The unincubated sample measures the total poly(A)-containing mRNA and that incubated without (2'-5')-oligo(A) measures the loss of mRNA due to nuclease activity independent of (2'-5')-oligo(A) addition. Most of the mRNA applied should be eluted with buffer D in the unincubated sample, although after prolonged storage of highly radioactive mRNA some losses may occur because of fragmentation of mRNA. About 70–80% of the input mRNA should be eluted with buffer D in the sample incubated with cell extract only.

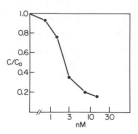

FIG. 1. Dependency of endonuclease activity on (2'-5')-oligo(A) concentration. Reactions prepared as indicated in the text were incubated for 60 min with the concentration of ppp5'A2'p5'A2'p5'A indicated on the abscissa. This trinucleotide was chemically synthesized.[8] The endonuclease activity is expressed as the ratio C/C_0, where C is the amount of poly(A)-containing mRNA recovered from incubations with added (2'-5')-oligo(A) and C_0 is the amount of this RNA in a control incubation. Viral vesicular stomatitis virus mRNA was used in this and the following experiments.

This assay can also be applied to follow the cleavage of mRNA associated with polysomes. In this case, the cell extract is prepared from HeLa cells incubated with [³H]uridine and 0.04 µg of actinomycin D per milliliter to inhibit synthesis of ribosomal RNA.[3] A great proportion of the labeled RNA in these extracts is mRNA associated with polysomes. Inhibitors of polypeptide chain elongation, like sparsomycin or cycloheximide, are added before (2'-5')-oligo(A) to prevent ribosome movement along mRNA. Endonuclease-dependent cleavage can take place only between ribosomes in the mRNA regions that are not protected by their association with ribosomes. The cell extract is analyzed as described above by determining the content of poly(A)-containing mRNA in incubations with added (2'-5')-oligo(A). The endonuclease activity can also be directly shown by examining the polysome pattern before and after incubation with (2'-5')-oligo(A). Polysome breakdown can result only from endonuclease activity.[4,7]

Calculations and Results

The (2'-5')-oligo(A)-dependent endonucleolytic cleavages can be expressed as the ratio of cpm[+(2'-5')-oligo(A)]/cpm[−(2'-5')-oligo(A)]. This value is calculated for different concentrations of (2'-5')-oligo(A) as shown in Fig. 1. The range of concentrations that show dose dependency for endonuclease activation is relatively narrow, and it is also variable for different cell extracts. The reasons for this variability are not fully understood, but are probably related to the presence in cell extracts of a phosphodiesterase activity that degrades (2'-5')-oligo(A).[18-20] Even a short preincubation of these oligonucleotides with cell extract may prevent sub-

[18] M. A. Minks, S. Benvin, P. A. Maroney, and C. Baglioni, *Nucleic Acids Res.* **6**, 767 (1979).

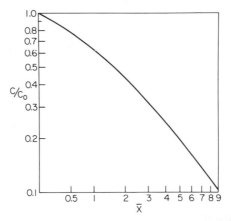

FIG. 2. Relationship between the ratio C/C_0 of poly(A)-containing RNA and the average number $\bar{X}$ of endonucleolytic cleavages per mRNA molecule. See text for explanations.

sequent activation of the endonuclease; this is best shown by adding labeled mRNA after the preincubation.[18] Inactivation of (2′-5′)-oligo(A) is favored by low ATP concentrations in the assay, and for this reason the ATP-regenerating sugar fructose 1,6-diphosphate is included in the reactions. Also ATP at 5 mM concentration may be added to prevent (2′-5′)-oligo(A) degradation.[7] The endonuclease activation may be lost upon removal of the β and γ terminal phosphates, since the presence of at least two terminal phosphates is required for activation of the endonuclease.[19] Moreover, only trimer and larger oligonucleotides are active in most cell extracts, although tetramer is required for the endonuclease activation in reticulocyte lysates.[12]

The endonuclease activity calculated as described above can be expressed as the ratio C/C_0, where $C = $ cpm$[+(2′-5′)$-oligo(A)$]$ and $C_0 = $ cpm$[-(2′-5′)$-oligo(A)$]$. This ratio can be converted to average number of cleavages per mRNA molecule ($\bar{X}$) by deriving from $C/C_0 = 1/(1 + \bar{X})$ the relationship $\bar{X} = (C_0/C) - 1$. This calculation is valid for values of $\bar{X} \geq 10$, but for smaller values of $\bar{X}$ the probability that an RNA molecule is cleaved X times according to a Poisson distribution should be taken into account. Values of $\bar{X}$ can then be calculated from $C/C_0 = (1 - e^X)/X$ (Fig. 2).[21] This method of calculating average cleavages per mRNA molecule

[19] B. R. G. Williams, I. M. Kerr, C. S. Gilbert, C. N. White, and L. A. Ball, *Eur. J. Biochem.* **92**, 455 (1978).

[20] A. Schmidt, A. Zilberstein, L. Shulman, P. Federman, H. Berissi, and M. Revel, *FEBS Lett.* **95**, 257 (1978).

[21] C. Baglioni, S. Benvin, P. A. Maroney, M. A. Minks, T. W. Nilsen, and D. K. West, *Ann. N. Y. Acad. Sci.*, **350**, 497 (1980).

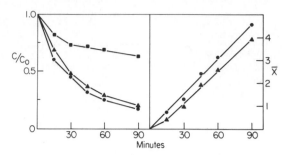

FIG. 3. Time course of mRNA degradation in extracts of control (●——●) and interferon-treated (▲——▲) cells with 280 nM (2'-5')-oligo(A), or in extract of control cells without added (2'-5')-oligo(A) (■——■). In the left panel, the loss of poly(A)-containing mRNA is shown by the decrease of the C/C_0 ratio (see Fig. 1). In the panel at right, the average number of cleavages per mRNA molecule $\bar{X}$ is calculated as described in Fig. 2.

can be applied to mixtures of mRNA molecules of different length with satisfactory results. The theoretical value for $\bar{X}$ corresponds to an experimentally determined value only if (a) the mRNAs are uniformly labeled; (b) the 3'-terminal poly(A) is unlabeled and is not cleaved by the endonuclease; and (c) the RNA fragments produced by endonucleolytic cleavage are not degraded by a 5'-exonuclease. The first two conditions are obtained, since the mRNA is uniformly labeled with [3H]uridine and the (2'-5')-oligo(A)-activated endonuclease does not cleave poly(A).[21] The third condition is difficult to verify, although in several experiments we have observed little degradation in HeLa cell extracts of mRNA fragments containing an unblocked 5'-terminal end.[21]

The time course of mRNA cleavage by the endonuclease can be determined by sampling aliquots from an incubation and analyzing each sample as described above. The kinetics of C/C_0 decrease slows down with time because the probability that a segment of RNA attached to poly(A) is cleaved by endonuclease decreases as the segment becomes smaller and the loss of the RNA is less with each successive cut. When the average cuts per mRNA are calculated as described above, however, linear kinetics of mRNA cleavage are obtained (Fig. 3). The surprising finding is that relatively few cleavages are introduced in a 90-min incubation. This is possibly explained by the presence of relatively large amounts of substrate RNA in the cell extract assayed for (2'-5')-oligo(A)-activated endonuclease. Until purified preparations of endonuclease are tested, it is difficult to establish whether cleavages occur preferentially in specific mRNA regions.

The assay described is relatively simple and provides a direct measurement of endonuclease activity. This assay has been employed in studies of analogs of (2'-5')-oligo(A) modified in either the adenosine moiety[22] or the ribose (unpublished observations) with excellent results. The assay is inexpensive, since very little labeled mRNA is used in the oligo(dT)-cellulose columns can be reused after elution with buffer D. The results obtained with each cell extract are highly reproducible, but new preparations of cell extract should be first tested for nuclease activity and titrated with a (2'-5')-oligo(A) solution of known concentration to obtain accurate results.

[22] M. A. Minks, S. Benvin, and C. Baglioni, *J. Biol. Chem.* **255**, 5031 (1980).

[33] Enzymic Synthesis of (2'-5')ppp3'dA(p3'dA)$_n$, the Cordycepin Analog of (2'-5')-Oligoadenylic Acid

By ROBERT J. SUHADOLNIK, PAUL DOETSCH, JOSEPH M. WU, YOSUKE SAWADA, JOSEPH D. MOSCA, and NANCY L. REICHENBACH

(2'-5')(A)$_n$ synthetase is induced in interferon-treated mammalian and avian cells. The enzyme is activated in cell extracts by double-stranded RNA and converts ATP into (2'-5')-oligonucleotides.[1] Extracts from other interferon-treated cells and extracts from rabbit reticulocytes have high levels of this synthetase.[2-7] Viral and cellular RNA are degraded in the presence of (2'-5')pppA(pA)$_n$.[8] The degradation of the RNA occurs following the activation of a latent endoribonuclease by the (2'-5')pppA(pA)$_n$. The half-life of (2'-5')pppA(pA)$_n$ in HeLa cell extracts is about 2.5 min.[9] We report here the enzymic conversion of the ATP analog, 3'-dATP (cordycepin 5'-triphosphate), to the (2'-5')pppA(pA)$_n$ analog, (2'-5')ppp3'dA(p3'dA)$_n$, by extracts of rabbit reticulocytes. This (2'-5')pppA(pA)$_n$ analog is unique in that it is a more potent inhibitor of protein synthesis in extracts from lysed rabbit reticulocytes and is not hydrolyzed by HeLa cell extracts as is (2'-5')pppA(pA)$_n$. The enzymic

[1] I. M. Kerr and R. E. Brown, *Proc. Natl. Acad. Sci. U.S.A.* **75**, 256 (1978).
[2] M. J. Clemens and B. R. G. Williams, *Cell* **13**, 566 (1978).
[3] L. A. Ball and C. N. White, *Proc. Natl. Acad. Sci. U.S.A.* **75**, 1167 (1978).
[4] M. A. Minks, S. Benvin, P. A. Maroney, and C. Baglioni, *J. Biol. Chem.* **254**, 5058 (1979).
[5] A. Schmidt, A. Zilberstein, L. Shulman, P. Federman, H. Berissi, and M. Revel, *FEBS Lett.* **95**, 257 (1978).
[6] G. R. Stark, W. J. Dower, R. T. Schimke, R. E. Brown, and I. M. Kerr, *Nature (London)* **278**, 471 (1979).
[7] B. R. G. Williams, C. S. Gilbert, and I. M. Kerr, *Nucleic Acids Res.* **6**, 1335 (1979).
[8] A. G. Hovanessian, R. E. Brown, and I. M. Kerr, *Nature (London)* **268**, 537 (1977).
[9] M. A. Minks, S. Benvin, P. A. Maroney, and C. Baglioni, *Nucleic Acids Res.* **6**, 767 (1979).

synthesis of the 2′,5′-phosphodiester bond with 3′-dATP by $(2′-5′)(A)_n$ synthetase is an extension of an earlier report by Suhadolnik and co-workers[10] on the formation of a 2′,5′-phosphodiester bond following incorporation of 3′-dAMP into the RNA of human epithelial #1 cells. Preliminary communication of this work has appeared.[11]

Reagents

Extracts of rabbit reticulocytes (Clinical Convenience, Madison, Wisconsin): Stable at $-70°$ for 6 months. Hemin (25 μM) is added immediately upon thawing.

Human fibroblast interferon (obtained from Dr. A. S. Rabson, Director, Cancer Biology and Drug Program, NIH): 1.0×10^6 units/mg protein

Poly(rI) · poly(rC)-agarose (P-L Biochemicals Inc.): 0.5×1.5 cm columns (0.3 ml) in buffer A.

DEAE-cellulose (DE-52), chromatography paper (3 MM, 46×57 cm), and GF/A glass fiber filters (2.3 cm) (Whatman)

Sephadex G-10 (Pharmacia Fine Chemicals)

Amberlite XAD-4, 200–400 mesh (Pierce Chemical Co.)

3′-dATP (prepared according to the procedure of Suhadolnik et al.[12])

ATP, GTP, 3′-dATP, creatine phosphokinase and phosphocreatine (Sigma Chemical Co.)

[G-³H]3′-dATP, 640 mCi/mmol (prepared according to the procedure of Suhadolnik et al.[12]).

[α-³²P]3′-dATP, 3000 Ci/mmol (Amersham/Searle)

[8-³H]ATP, 22 Ci/mmol (Amersham/Searle)

[¹⁴C]Leucine, 335 mCi/mmol (New England Nuclear)

$(2′-5′)pppA(pA)_2$ and $(2′-5′)A(pA)_2$ (P-L Biochemicals Inc.)

$(2′-5′)3′dA(p3′dA)_2$ (chemically synthesized according to the procedure of Charubala and Pfleiderer[13])

Buffer A: 0.02 M HEPES (pH 7.5), 0.1 M KCl, 0.002 M Mg(OAc)$_2$, 0.002 M dithiothreitol, 10% glycerol; buffer prepared at 25°

Buffer B: 0.02 M HEPES (pH 7.5), 0.09 M KCl; buffer prepared at 25°

Buffer C: 0.02 M HEPES (pH 7.5), 0.35 M KCl; buffer prepared at 25°

Amino acid mixture: 0.50 mM of the following amino acids: alanine, arginine, asparagine, aspartic acid, cysteine, glycine, glutamine,

[10] J. G. Cory, R. J. Suhadolnik, B. Resnick, and M. A. Rich, *Biochim. Biophys. Acta* **103**, 646 (1965).

[11] P. Doetsch, J. M. Wu, Y. Sawada, and R. J. Suhadolnik, *Nature (London)* **291**, 355 (1981).

[12] R. J. Suhadolnik, R. Baur, D. M. Lichtenwalner, T. Uemsatsu, J. H. Roberts, S. Sudhakar, and M. Smulson, *J. Biol. Chem.* **252**, 4134 (1977).

[13] R. Charubala and W. Pfleiderer, *Tetrahedron Lett.* **21**, 4077 (1980).

glutamic acid, histidine, isoleucine, lysine, methionine, phenylalanine, proline, serine, threonine, tryptophan, tyrosine, and valine

Master mixture: amino acid mixture, 81.6 μl; ATP (100 mM, pH 7.0) plus GTP (20 mM, pH 7.0), 8.16 μl; creatine phosphokinase (900 units/ml) plus phosphocreatine (200 mM), 40.8 μl; hemin (0.90 mM), 11.5 μl; KCl (2 M, pH 7.5) plus MgCl$_2$ (36 mM, pH 7.5), 34 μl; [^{14}C]leucine (335 mCi/mmol), 41.46 μl

Sodium hydroxide, 0.2 N

Trichloroacetic acid, 5% (w/v)

Hydrogen peroxide, 30%

Bacterial alkaline phosphatase (25 units/mg protein) and T2 ribonuclease (1060 units/mg protein (Sigma Chemical Co.))

Snake venom phosphodiesterase I (15 units/mg protein) (Worthington Biochemical Corp.)

Sodium metaperiodate: 1.25 mM solution in water

Thin-layer chromatograms (No. 13254 with fluorescent indicator) (Eastman Organic Chemicals)

Solvents for chromatography: solvent A: isobutyric acid/conc. ammonia/water, 66 : 1 : 33, v/v/v; solvent B: 1-propanol/conc. ammonia/water, 60 : 30 : 10, v/v/v

Scintillation fluid: Formula 949 (New England Nuclear), ACS (Amersham).

Enzymic Synthesis of [^{3}H]- or [^{32}P]($2'$-$5'$)ppp3'dA(p3'dA)$_n$ in Reticulocyte Lysates

Poly(rI) · poly(rC)-agarose columns bound with ($2'$-$5'$)(A)$_n$ synthetase are prepared by adding 1.2 ml of reticulocyte lysate to the column and washing with 25 ml of buffer A. The columns are incubated at 30° for 17 hr with either [8-^{3}H]3'-dATP (10 μCi, 640 Ci/mmol) or [α-^{32}P]3'-dATP (25 μCi, 3000 Ci/mmol). To synthesize the ($2'$-$5'$)pppA(pA)$_n$ nucleotides, [8-^{3}H]ATP (10 μCi, 22 Ci/mmol) is added to lysates as described above. At the end of the incubation, the columns are eluted with 2 ml of buffer A. Ninety-five percent of the radioactivity is recovered. The radioactive material eluting with buffer A is applied to DEAE-cellulose columns (0.6 × 2.1 cm) and washed with 42 ml of buffer B. The oligonucleotides are displaced with 5 ml of buffer C. Oligonucleotide formation is determined by the amount of radioactivity displaced from the DEAE-cellulose column with buffer C divided by the total radioactivity recovered. The yield of ($2'$-$5'$)ppp3'dA(p3'dA)$_n$ from 3'-dATP after a 17-hr incubation is 3%; the yield of ($2'$-$5'$)pppA(pA)$_n$ from ATP after a 17-hr incubation is 19%.

Isolation of the (2'-5')-Oligonucleotides Synthesized from ATP or 3'-dATP

The (2'-5')-oligonucleotides synthesized following incubations with either ATP or 3'-dATP that are eluted from the poly(rI) · poly(rC)-agarose columns are diluted to 0.05 M KCl and applied to a DEAE-cellulose column (0.5 × 17 cm). The nucleotides and (2'-5')-oligonucleotides are displaced with a 0.05 to 0.15 M linear gradient of NaCl (40 ml/40 ml), 0.05 M Tris · HCl (pH 8.0) in 7 M urea; 1-ml fractions; flow rate, 4 ml/hr. (2'-5')ppp3'dA(p3'dA)$_2$ is displaced with a peak tube at 36 ml (Fig. 1). Authentic (2'-5')pppA(pA)$_2$ is also displaced with a peak tube at 36 ml.

Proof of Structure of (2'-5')ppp3'dA(p3'dA)$_2$

Four procedures have been used to prove the structure of (2'-5')-ppp3'dA(p3'dA)$_2$.[11] The structural characterization is performed on the oligonucleotide analog obtained from buffer C following DEAE-cellulose chromatography or after desalting by gel filtration (Sephadex G-10, in glass-distilled water, 0.9 × 87 cm column) of the (2'-5')ppp3'dA(p3'dA)$_2$ from the NaCl-gradient elution method (Fig. 1). Bacterial alkaline phosphatase (BAP) digestion verifies the presence of a 5'-terminal triphosphate. Thin-layer chromatography in solvent A of the BAP digest showed a radioactive region on X-ray film that has the same R_f as unlabeled core (2'-5')3'dA(p3'dA)$_2$ (R_f 0.81) and one region with the same R_f as inorganic phosphate (R_f 0.40). Treatment of (2'-5')ppp3'dA(p3'dA)$_2$ with snake venom phosphodiesterase I yielded 3'-dAMP as the only digestion product. An aliquot of the BAP digest of [^{32}P](2'-5')ppp3'dA(p3'dA)$_2$ is applied to an Amberlite XAD-4 column, 200–400 mesh (0.5 g) in 0.05 M triethylammonium bicarbonate (pH 7.4).[14] Inorganic phosphate is displaced with 0.1 M triethylammonium bicarbonate (2–6 ml). Thirty-four percent of the ^{32}P is isolated as inorganic phosphate; the theoretical yield is 33.3%. There was no hydrolysis of (2'-5')ppp3'dA(p3'dA)$_2$ by T2 ribonuclease, which is further evidence for the 2',5'-linkage.

Inhibition of Translation by (2'-5')ppp3'dA(p3'dA)$_2$

Inhibition of translation is determined by measuring the incorporation of [^{14}C]leucine (0.25 μCi, 350 mCi/mmol) into trichloroacetic acid-insoluble polypeptides by the procedure of Wu *et al.*[15] Each tube contains 25 μl of rabbit reticulocyte lysate, 3.45 μl of bovine serum albumin (26 mg/ml), 5 μl of (2'-5')-oligonucleotide (1 μM), 3.5 μl of glass-distilled water, and 13.05 μl of master mixture. Tubes containing all components

[14] T. Uematsu and R. J. Suhadolnik, *J. Chromatogr.* **123**, 347 (1976).
[15] J. M. Wu, C. P. Cheung, A. R. Bruzel, and R. J. Suhadolnik, *Biochem. Biophys. Res. Commun.* **86**, 648 (1979).

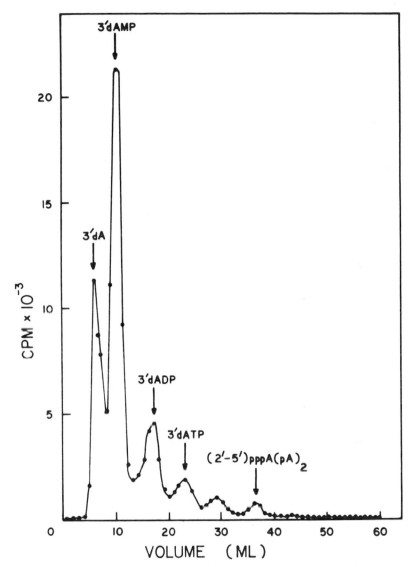

FIG. 1. Chromatography of the analog, $(2'-5')ppp3'dA(p3'dA)_n$, from incubations with 3'dATP on DEAE-Cellulose. Material washed from poly(rI) · poly(rC)-agarose columns from a 17-hr incubation with 3'-dATP is applied to a DEAE-cellulose column (Whatman DE-52, 0.5 × 17 cm). The nucleotides are displaced with a 0.05 to 0.15 M linear gradient of NaCl (40 ml/40 ml), 0.05 M Tris · HCl (pH 8.0) in 7 M urea; 1-ml fractions; flow rate, 4 ml/hr. Reprinted, with permission, from Doetsch et al.[11]

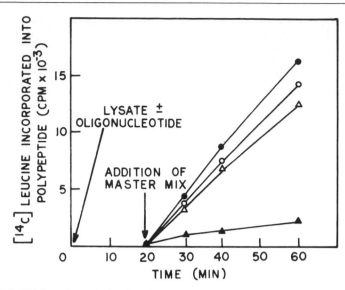

FIG. 2. Inhibition of translation by $(2'-5')pppA(pA)_2$ and $(2'-5')ppp3'dA(p3'dA)_2$. Inhibition of translation with $(2'-5')pppA(pA)_2$ (100 nM) ($\triangle$) and 100 nM plus periodate ($\bigcirc$) is compared to that of $(2'-5')ppp3'dA(p3'dA)_2$ (100 nM) or 100 nM plus periodate ($\blacktriangle$) and control ($\bullet$). $(2'-5')$-Oligonucleotide samples (5 pmol) are treated with 1.25 nmol of sodium metaperiodate[16] in a final volume of 10 μl for 20 min at 25° in the dark. Periodate-treated oligonucleotides are tested in translation assays as described in the text. Periodate in samples containing no oligonucleotide does not inhibit translation. Reprinted, with permission, from Doetsch *et al.*[11]

minus the master mixture are preincubated for 20 min at 30° before initiation of translation (addition of master mixture). At 0, 10, 20, and 40 min after the initiation of translation, 5-μl aliquots are removed and added to tubes containing 200 μl of 0.2 N NaOH. After all assays are stopped in this manner, 50 μl of 30% hydrogen peroxide are added and the tubes are incubated at 30° for 15 min followed by the addition of 5 ml of ice cold 5% trichloroacetic acid. The tubes are held on ice for 30 min followed by filtration on Whatman GF/A glass fiber filters on a Millipore filtration manifold. The filters are washed with an additional 15 ml of 5% trichloroacetic acid, dried under an infrared lamp, and assayed for radioactivity in a liquid scintillation spectrophotometer.

The inhibition of translation by the analog, $(2'-5')ppp3'dA(p3'dA)_2$, is compared with the trimer adenylate, $(2'-5')pppA(pA)_2$ (Fig. 2). The trimer triphosphate analog, $(2'-5')ppp3'dA(p3'dA)_2$, is about four times more potent an inhibitor of translation than is $(2'-5')pppA(pA)_2$. The inhibition of translation by $(2'-5')ppp3'dA(p3'dA)_2$ plus[16] or minus periodate does not change, but there is a small change in the inhibition of translation after

[16] H. Fraenkel-Conrat and A. Steinschneider, this series, Vol. 12B, p. 243.

periodate treatment of (2'-5')pppA(pA)$_2$. These data indicate that there is no ribosyl moiety on the 3' terminus of (2'-5')ppp3'dA(p3'dA)$_2$.

Metabolic Stability of (2'-5')ppp3'dA(p3'dA)$_n$ in HeLa Cell Extracts

The stability of (2'-5')ppp3'dA(p3'dA)$_n$ in HeLa cell extracts has been compared to that of (2'-5')pppA(pA)$_n$. Although there is no hydrolysis of (2'-5')ppp3'dA(p3'dA)$_n$ (12 μM) by HeLa extracts after a 45-min incubation, there is 50% hydrolysis of (2'-5')pppA(pA)$_n$ (5 μM) in about 2 min (Fig. 3). Incubation mixtures (0.125 ml) contain 0.6 part of HeLa cell extract,[17] 2.5 mM Mg(OAc)$_2$, either [^{32}P](2'-5')ppp3'dA(p3'dA)$_n$ (12 μM) or [^{3}H](2'-5')pppA(pA)$_n$ (5 μM), 120 mM KCl, 20 mM HEPES (pH 7.4), and 1 mM dithiothreitol. Incubations are performed at 30°. At the times indicated, 25-μl samples are withdrawn and heated for 3 min at 95°; the amount of undegraded oligonucleotide is determined by DEAE-cellulose chromatography as described.[9] Degradation of oligonucleotides is monitored as a decrease of radioactive material eluting with 0.35 M KCl compared to the zero time point (100% undegraded). The chemically synthesized core (2'-5')3'dA(p3'dA)$_2$ (500 μM) is not hydrolyzed when incubated in HeLa cell extracts for 45 min as determined by cellulose thin-layer chromatography (solvents A and B). Neither cordycepin nor 3'-dAMP are detected.

Concluding Comments

We have shown that (2'-5')ppp3'dA(p3'dA)$_n$, an analog of (2'-5')pppA(pA)$_n$, can be synthesized enzymically from 3'-dATP by the (2'-5')(A)$_n$ synthetase from lysed rabbit reticulocytes, thus indicating that the 3'-hydroxyl on the ribose moiety of ATP is not essential for the synthesis of (2'-5')pppA(pA)$_n$. (2'-5')ppp3'dA(p3'dA)$_2$ is about four times more potent an inhibitor of translation than is (2'-5')pppA(pA)$_2$. Neither the enzymically synthesized analog, (2'-5')ppp3'dA(p3'dA)$_n$, nor the chemically synthesized core analog, (2'-5')3'dA(p3'dA)$_2$, are hydrolyzed by the 2',5'-phosphodiesterase found in HeLa cell extracts. Results from this laboratory have shown that (2'-5')ppp3'dA(p3'dA)$_n$ is also enzymically synthesized by extracts of human fibroblast interferon-treated L cells.[17a]

The synthesis of the trimer triphosphate analog of (2'-5')pppA(pA)$_n$ contrasts with reports from three other laboratories. Lengyel and co-workers[18] reported that (2'-5')(A)$_n$ synthetase from interferon-treated

[17] L. A. Weber, E. R. Feman, and C. Baglioni, *Biochemistry* **14**, 5315 (1975).

[17a] P. W. Doetsch, R. J. Suhadolnik, Y. Sawada, J. D. Mosca, M. B. Flick, N. L. Reichenbach, A. Q. Dang, J. M. Wu, R. Charubala, W. Pfleiderer, and E. E. Henderson, *Proc. Natl. Acad. Sci. U.S.A.*, in press.

[18] H. Samanta, J. P. Dougherty, and P. Lengyel, *J. Biol. Chem.* **255**, 9807 (1980).

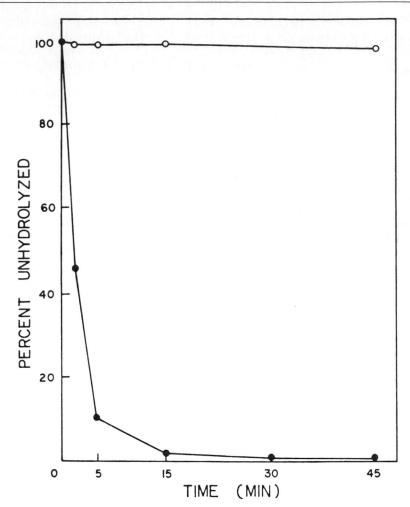

FIG. 3. Stability of $(2'-5')pppA(pA)_n$ (●) and $(2'-5')ppp3'dA(p3'dA)_n$ (○) in HeLa cell extracts. Reprinted, with permission, from Doetsch *et al.*[11]

Ehrlich ascites tumor cells can link adenylate moieties to the $2'$-hydroxyl of $3'$-dATP. Thang and co-workers[19] reported that $3'$-dATP acts as a chain terminator with respect to the addition of one $3'$-deoxyadenylate moiety to either dimer or trimer $(2'-5')$-$pppA(pA)_n$ following 3-hr incubations with rabbit reticulocyte $(2'-5')(A)_n$ synthetase. Baglioni and co-workers[20] reported that $3'$-dATP inhibits the HeLa cell $(2'-5')(A)_n$ synthetase in the

[19] J. Justesen, D. Ferbus, and M. N. Thang, *Proc. Natl. Acad. Sci. U.S.A.* 77, 4618 (1980).
[20] M. A. Minks, S. Benvin, and C. Baglioni, *J. Biol. Chem.* 255, 5031 (1980).

presence of ATP with respect to formation of $(2'-5')pppA(pA)_n$. In view of the reports that $(2'-5')$-oligonucleotides inhibit mitogen-stimulated DNA synthesis in lymphocytes as does interferon,[21] it is possible that analogs of $(2'-5')pppA(pA)_n$ might regulate these activities and could thus be employed as chemotherapeutic agents to replace interferon. For example, we have shown that enzymically and chemically synthesized $(2'-5')$-$3'dA(p3'dA)_2$ inhibits morphological transformation of Epstein–Barr virus-infected human umbilical cord lymphocytes.[17a] Furthermore, the core analog also inhibits HSV-1 infection of lymphocytes. These actions of the core $(2'-5')3'dA(p3'dA)_2$, with no toxicity to the host cell, are strong evidence that the inhibition of transformation and inhibition of viral infection in lymphocytes can be accomplished in the absence of interferon.

Acknowledgments

This work was supported in part by U.S. Public Health Service Grant GM27201, NRSA Grant AM07162, and National Science Foundation Grant PCM-8111752.

[21] A. Kimchi, M. Shure, and M. Revel, *Nature (London)* **282**, 849 (1979).

[34] Assay of Double-Stranded RNA-Dependent Endonuclease Activity

By JOHN A. LEWIS and REBECCA FALCOFF

After treatment with interferon, tissue culture cells possess enhanced levels of various double-stranded RNA (dsRNA)-dependent enzyme activities.[1,2] One such activity is an endonuclease capable of degrading a variety of RNA species, including mRNA and rRNA.[3-5] It has been shown[6-8] that the activation of endonuclease activity is mediated by a low

[1] C. Baglioni, *Cell* 7, 255 (1979).

[2] J. A. Lewis, *in* "Horizons in Biochemistry and Biophysics" (L. A. Kohn and R. M. Friedman, eds.), p. 121. Dekker, New York, 1981.

[3] G. B. Brown, B. Lebleu, M. Kawakita, S. Shaila, G. C. Sen, and P. Lengyel, *Biochem. Biophys. Res. Commun.* 69, 114 (1976).

[4] I. M. Kerr, R. E. Brown, M. J. Clemens, and C. S. Gilbert, *Eur. J. Biochem.* 69, 551 (1976).

[5] L. Ratner, G. C. Sen, G. E. Brown, B. Lebleu, M. Kawakita, B. Cabrer, E. Slattery, and P. Lengyel, *Eur. J. Biochem.* 79, 565 (1977).

[6] M. J. Clemens and B. R. G. Williams, *Cell* 13, 565 (1978).

[7] C. Baglioni, M. A. Minks, and P. A. Maroney, *Nature (London)* 273, 684 (1978).

[8] E. Slattery, M. Ghosh, H. Samanta, and P. Lengyel, *Proc. Natl. Acad. Sci. U.S.A.* 76, 4778 (1979).

molecular weight oligonucleotide, pppA2'p5'A2'p5'A or (2'-5')-oligo(A), which is synthesized in the presence of ATP and dsRNA by an interferon-induced enzyme.[9] The endonuclease is present in all cells, whether treated with interferon or not,[7] but exists in a latent form that is activated by (2'-5')-oligo(A).[6-8]

In order to assay the presence of endonuclease activity in relatively crude cell-free extracts, we have used Mengo virus RNA, which can be prepared easily and is a good substrate for the enzyme.[10] After incubation of extracts with Mengo virus [³H]RNA and an activator [either dsRNA or (2'-5')-oligo(A)], endonuclease activity is detected either by the release of cold acid-soluble radioactivity or sedimentation on sucrose density gradients.

Preparation of Mengo Virus [³H]RNA Stocks

Materials

Mouse L-929 cells
Minimal essential medium (Earle's salts) (Gibco)
Newborn calf serum, antibiotics (Gibco)
Actinomycin D (Sigma)
[³H]Uridine (1 mCi/ml; 20–40 Ci/mmol) (Amersham)

Procedure. Monolayer cultures of mouse L cells are grown in minimal essential medium (Earle's salts) supplemented with 7% newborn calf serum (heat inactivated), penicillin (100 units/ml), streptomycin (50 μg/ml), and gentamycin (40 μg/ml). When the monolayers, in glass Roux bottles or plastic 705-ml flasks, are approximately confluent, the medium is removed and replaced with 5 ml of Mengo virus seed diluted in Dulbecco's phosphate-buffered saline (PBS) to 2.5×10^8 PFU/ml (multiplicity of infection approximately 5 PFU/cell). After a 60-min adsorption period at room temperature, the infecting virus is removed, the monolayers are rinsed with PBS, and then 30 ml of medium supplemented with 2% newborn calf serum and antibiotics plus actinomycin D (5 μg/ml) and [³H]uridine (5 μCi/ml; 25 Ci/mmol). Infection is allowed to proceed at 37° for 16 hr, and the cells are examined for cytopathic effect. The bottles are then frozen at −70°. After two cycles of freezing and rapid thawing, the culture medium is poured into centrifuge bottles and centrifuged at 5000 rpm (4000 g) for 10 min. The supernatants, which usually titer 2 to 5×10^9 PFU/ml, are then stored at either −20° or −70°.

[9] I. M. Kerr and R. E. Brown, *Proc. Natl. Acad. Sci. U.S.A.* 75, 256 (1978).
[10] J. A. Lewis, E. Falcoff, and R. Falcoff, *Eur. J. Biochem.* 86, 497 (1978).

Purification of Mengo Virus [³H]RNA

Before extraction of the viral RNA, the Mengo virus is partially purified by one of the two procedures described below. Great care is taken in these steps, as in the extraction of RNA, to avoid contamination with ribonucleases. All glassware and pipettes are autoclaved and then baked at 200°. Where practicable, the solutions are treated with diethyl pyrocarbonate (0.01% v/v) and then boiled for 20 min or incubated at 37° for 2 hr. Disposable gloves are worn at all times.

Materials

    Diethyl pyrocarbonate (Sigma)
    Polyethylene glycol 6000 (Sigma)
    Cesium chloride (Schwarz-Mann)
    DNase, ribonuclease free (Miles)
    Trypsin, type III (Sigma)
    2-Mercaptoethanol (Sigma)
    Solution 1: 100 mM Tris · HCl, pH 8.2, 20 mM EDTA
    Solution 2: 10 mM Tris · HCl, pH 7.5, 5 mM magnesium acetate
    Solution 3: 10% (w/v) sodium dodecyl sulfate (SDS)
    Solution 4: 1.0 M Na$_2$HPO$_4$, pH 8.7

Procedure 1. Virus stock is thawed and pooled in a large Erlenmeyer flask with a stirring bar, and 2-mercaptoethanol is added to a final concentration of 0.01% (v/v). Solid polyethylene glycol 6000 is added slowly while stirring continuously at 0° to a final concentration of 6% (w/v). After complete dissolution, the mixture is left stirring gently for at least 60 min and centrifuged at 5000 rpm for 20 min at 2°. The supernatant is decanted and discarded; the pellet is resuspended in solution 1 (approximately 10 ml per liter of virus stock) with the aid of a Dounce homogenizer (tight piston). Cesium chloride is added to the virus suspension to give a density of 1.37 g/ml (refractive index = 1.369). The suspension is centrifuged at 50,000 rpm in an angle rotor (Beckman Ti 60 or Ti 50 according to the volume required) for 20 hr at 18°. The virus forms a band at a density of 1.34 g/ml (although a fraction of denser particles exists[11]) and is collected with a Pasteur pipette, diluted with 2 volumes of sterile H$_2$O, and centrifuged over a cushion of 20% (w/v) sucrose in solution 1 at 150,000 g for 1.5 hr at 4°. The final pellet of virus is resuspended in solution 1 (1 ml per liter of starting material) and either stored at −70° or phenol-extracted immediately as described below.

The polyethylene glycol step is somewhat variable, and a rather longer

---

[11] R. Perez-Bercoff, M. Gandler, and E. Preisig, *Virology* **85**, 378 (1978).

period (up to 16 hr) may be preferred to ensure complete recovery of virus. It should be noted that the cesium chloride step is usually accompanied by a large loss in material, although the loss in titratable virus is only 20–50%. If it is difficult to resuspend the virus after CsCl purification, it may be necessary to add Na_2HPO_4 as described for procedure 2.

Procedure 2. This procedure is a modified form of that described by Kerr and Martin.[12] Virus is concentrated from stocks by polyethylene glycol 6000 precipitation as described above. The pelleted virus is resuspended in solution 2 (approximately 2.5 ml per liter of starting material) with the aid of a Dounce homogenizer. DNase (ribonuclease-free) is added from a stock solution (2.5 mg/ml in solution 2) to a final concentration of 25 μg/ml, and the suspension is incubated at 37° for 20 min. Trypsin is added from a stock solution (2.5 mg/ml in solution 2) to a final concentration of 25 μg/ml, and the suspension is incubated at 37° for 10 min. The trypsin treatment is accompanied by significant clarification of the cloudy suspension. The tubes are cooled to room temperature, and 0.1 volume of solution 3 is added. This causes almost complete clarification. After 3 min, 0.25 volume of 30% (w/v) polyethylene glycol 6000 is added, and the suspension is mixed gently at room temperature for 60 min. Virus is pelleted by centrifugation at 10,000 rpm for 20 min at 20°. The pellet is resuspended in solution 2 (approximately 10 ml per liter of starting material) with the aid of a Dounce homogenizer. The opaque suspension is clarified by dropwise addition of solution 4 up to a maximum concentration of 0.2 M Na_2HPO_4. (If the concentration of virus is too high, it is difficult to clarify.) The clear solution is rapidly adjusted to pH 7.5 by the addition of several drops of 1 M acetic acid and layered over 4.0 ml of 30% (w/v) sucrose in solution 2. The gradients are centrifuged in a Beckman SW 41 rotor at 35,000 rpm for 4 hr at 2°. The pelleted virus is resuspended in solution 2 (1 ml per liter of starting material). It is best not to use Na_2HPO_4 to clarify the suspension at this stage, as the phosphate will be precipitated in the subsequent ethanol precipitation step.

Extraction of Mengo Virus [³H]RNA

Materials

Phenol, redistilled (Fisher)
Chloroform (Fisher)
Isoamyl alcohol (Fisher)
tRNA, bakers' yeast (Sigma)
Ribosomal RNA, phenol extracted from L cell polysomes
Solution 5: 2.0 M sodium acetate, pH 5.5

---

[12] I. M. Kerr and E. M. Martin, *J. Virol.* **9**, 559 (1972).

Procedure. Virus purified by either of the above methods is adjusted to 1% SDS and incubated at 45° for 5 min in glass centrifuge tubes to disrupt the virions. Phenol saturated with solution 1 is added, and the mixture is incubated at 45° for 10 min with frequent shaking. One volume of chloroform–isoamyl alcohol (24 : 1) is added, mixed, and incubated at 45° for 5 min. The organic phase is separated by centrifugation at 2000 rpm for 5 min at room temperature, and the aqueous layer is removed with a Pasteur pipette. The organic phase is extracted sequentially with 1 volume and 0.5 volume of solution 1 at 45° for 5 min, and all the aqueous phases are combined. An equal volume of phenol (saturated with solution 1) and of chloroform–isoamyl alcohol (24 : 1) is added to the aqueous phase and, after extraction by shaking at 45° for 10 min, the organic phase is removed. This extraction is repeated once more. The final aqueous phase is adjusted to 0.2 M sodium acetate with 0.1 volume of solution 5. The RNA is precipitated by addition of two volumes of absolute ethanol at −20° for 16 hr. After centrifugation at 10,000 g for 20 min (preferably in a swinging-bucket rotor) at 2°, the pellet is dried under vacuum, dissolved in sterile H_2O, and adjusted to 0.2 M sodium acetate, pH 5.5, with solution 5; the RNA is reprecipitated with 2 volumes of absolute ethanol and left at −20° for 16 hr. RNA is recovered by centrifugation as above, dried under vacuum, and finally dissolved in sterile H_2O (or 5 mM HEPES, adjusted to pH 7.5 with KOH) at a concentration of 1 mg/ml, assuming an extinction coefficient of $E_{260}^{1 \text{ mg/ml}} = 25$.[13] The ratio $A_{260} : A_{280}$ should be greater than 1.7. The yield of RNA is usually approximately 1 mg for each liter of virus stock.

Assay of [³H]RNA

Materials

> Solution 6: 50 mM Tris · HCl, pH 7.5, 500 mM NaCl, 50 mM EDTA, 1% SDS
> Solution 7: 10 mM Tris · HCl, pH 7.5, 100 mM NaCl, 10 mM EDTA, 0.2% SDS

Translation. The integrity of the extracted RNA can be tested by translation in an L cell extract (S_{10})[10] or in an mRNA-dependent reticulocyte lysate[14] supplemented with exogenous tRNA at 40 μg/ml and analysis of the polypeptide products by SDS–PAGE as described elsewhere.[10]

Sucrose Density Gradient Centrifugation. The size of the extracted RNA can be checked by centrifugation in a sucrose density gradient. An aliquot

[13] I. M. Kerr, N. Cohen, and T. S. Work, *Biochem. J.* **98**, 826 (1966).
[14] H. R. B. Pelham and R. J. Jackson, *Eur. J. Biochem.* **67**, 247 (1976).

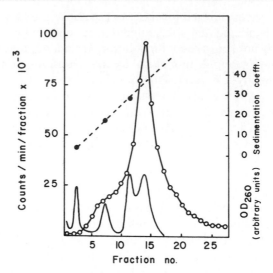

FIG. 1. Integrity of Mengo virus [³H]RNA. Purified viral [³H]RNA (5 μg) was analyzed by sucrose density gradient centrifugation as described in the text. Optical density markers (——) show the sedimentation of 4 S, 18 S, and 28 S RNA. The fastest sedimenting peak (36.5 S) corresponds to the viral RNA and coincides with a peak of radioactivity (O——O). Filled circles (●) represent sedimentation coefficients.

(e.g., 5 μg) of Mengo virus [³H]RNA is mixed with 25 μg of ribosomal RNA and 10 μg of tRNA to serve as marker RNA species, and 0.2 volume of solution 6 is added, together with 2 μl of diethyl pyrocarbonate. After incubation at 37° for 3 min, the mixture is layered on a linear 15 to 30% (w/v) sucrose gradient (in buffer 7) and centrifuged at 45,000 rpm in an SW50.1 rotor for 120 min at 20°. The gradients are analyzed by passage through a UV monitor (Isco), and individual fractions are assayed for radioactivity with Bray's scintillation fluid or Aquasol (New England Nuclear). A profile such as that shown in Fig. 1 is obtained with a peak sedimenting at approximately 36 S. With an optical path length of approximately 1 mm and a full-scale deflection scale of 0.2 A, a sample of 5 μg can be detected on the optical density profile.

Endonuclease Assay

Materials

Poly(I) · poly(C) (Miles)
Solution 8: 20 mM HEPES [4-(2-hydroxyethyl)-1-piperazine ethanesulfonic acid], adjusted to pH 7.5 with KOH, 120 mM KCl, 3.35 mM magnesium acetate, 7 mM 2-mercaptoethanol, 0.8 mM spermidine, 10% (v/v) glycerol

Procedure. In order to test for endonuclease activity in cell-free extracts, it is first necessary to activate the latent nuclease either by addition of exogenous (2′-5′)-oligo(A) or, in the case of interferon-treated cell extracts, by incubation in the presence of dsRNA and ATP to permit the formation of this substance. Cell-free extracts (S_{10} or S_{30} preparations) are prepared from L cells as described in detail elsewhere,[10] passed over Sephadex G-25 equilibrated with solution 8, and stored under liquid nitrogen. Assays are usually carried out in a final volume of 100 μl containing: 60 μl of S_{10} extract, 20 mM Hepes adjusted to pH 7.5 with KOH, 100 mM KCl, 2.0 mM magnesium acetate, 0.48 mM spermidine, 7 mM 2-mercaptoethanol, 6% (v/v) glycerol, 1 mM ATP, 0.25 mM GTP, 5 mM creatine phosphate, 180 μg of creatine phosphokinase per milliliter, and 62.5 μM amino acids. Pactamycin is included at a final concentration of 5 μM to block protein synthesis initiation in order that differences in the rate of mRNA degradation should not be partially attributable to differential protection of polysomal mRNA as a result of different rates of translation by the individual extracts. Poly(I) · poly(C) is then added at a concentration of 100 ng/ml, and the mixture is incubated at 32° for 30 min. At the end of this period, Mengo virus [³H]RNA is added at a concentration of 40 μg/ml, and the incubation is continued. At various times thereafter, aliquots are removed to monitor the release of trichloroacetic acid-soluble radioactivity and the integrity of the RNA.

Release of Acid-Soluble Radioactivity. Aliquots (usually 10 μl) are added to 0.1 ml of cold 10% (w/v) trichloroacetic followed by 10 μl of a stock solution (2 mg/ml) of bovine serum albumin. After 30 min at 0°, the mixtures are centrifuged in a microfuge (10,000 g for 10 min), and radioactivity in the supernatant is determined with Bray's scintillation fluid.

Integrity of Mengo Virus [³H]RNA. An aliquot (25 μl) is added to 2 μl of diethyl pyrocarbonate (to inhibit ribonuclease activities) followed by 200 μl of solution 7. Marker RNA species (25 μg ribosomal RNA and 10 μg tRNA) are added, and the mixtures are incubated at 32° for 3 min before layering over linear 15 to 30% (w/v) sucrose (in solution 7) density gradients for analysis as described above. The results of an experiment of this kind are shown in Fig. 2. Degradation of viral [³H]RNA is not affected by the addition of dsRNA to an extract prepared from control cells. In the absence of added dsRNA, the extent of degradation of mRNA by interferon-treated cell extracts is the same as for control extracts, but addition of dsRNA leads to the almost complete destruction of Mengo virus [³H]RNA within 20–30 min.

Comments. Some degradation of the viral RNA is observed in all cell extracts, whether treated with interferon or not, and this is probably due to a variety of endogenous ribonucleases. The addition of dsRNA to extracts of interferon-treated cells markedly stimulates the extent of this

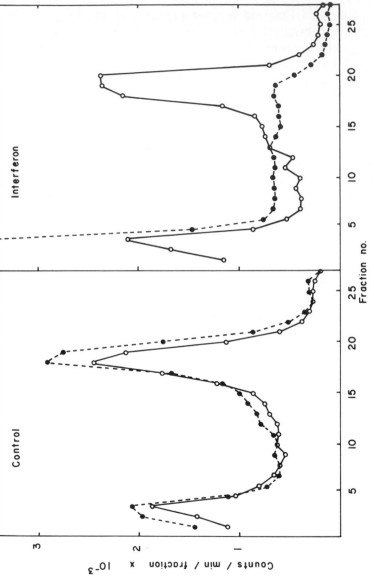

FIG. 2. Degradation of [³H]RNA by extracts of interferon-treated cells. Incubations of extracts with dsRNA and Mengo virus [³H]RNA and analysis by sucrose density gradient centrifugation were performed as described in the text. Incubation with [³H]RNA was for 30 min. In the absence of dsRNA (O———O), the rate and extent of degradation were the same for control and interferon. No effect of dsRNA [100 ng/ml of poly(I) · poly(C), ●---●] on control extracts is detectable, but the viral RNA was almost completely degraded by interferon extracts under these conditions.

degradation, and analysis of the size of the RNA remaining usually shows that within 20 min virtually no large RNA species remain. This extensive degradation is probably due to the action of both the dsRNA activated endonuclease and other cellular enzymes. Initial cleavage by the former enzyme is likely to promote further degradation by other nucleases.

Acknowledgments

We thank Ms. Josianne Sanceau for expert technical assistance. J. A. L. was supported by an EMBO Long-Term Fellowship and is a Sinsheimer Scholar.

[35] Procedures for the Assay of (2'-5')-Oligoadenylic Acids and Their Derivatives in Intact Cells

By E. M. MARTIN, D. M. REISINGER, A. G. HOVANESSIAN, and B. R. G. WILLIAMS

A major effect of interferon on cells is to promote the synthesis of the enzyme (2'-5')-oligoadenylic acid synthetase.[1-5] When extracts from interferon-treated cells, or rabbit reticulocyte lysates (which normally contain high levels of the synthetase[6,7]) are incubated with double-stranded RNA and ATP, a series of 2'-5'-linked oligoadenylic acid triphosphates $[ppp5'A(2'p5'A)_n$, where $n = 1$ to 5 or 6$]^{3,8}$ are produced; the trimer and higher oligomers are extremely potent in inhibiting the translation of host and viral messenger RNAs. This inhibition is brought about by the reversible activation of a ribonuclease.[7,9-12]

[1] A. G. Hovanessian, R. E. Brown, and I. M. Kerr, *Nature (London)* **268**, 537 (1977).

[2] I. M. Kerr, R. E. Brown, and A. G. Hovanessian, *Nature (London)* **268**, 540 (1977).

[3] I. M. Kerr and R. E. Brown, *Proc. Natl. Acad. Sci. U.S.A.* **75**, 256 (1978).

[4] L. A. Ball and C. N. White, *Proc. Natl. Acad. Sci. U.S.A.* **75**, 1167 (1978).

[5] A. Zilberstein, A. Kimchi, A. Schmidt, and M. Revel, *Proc. Natl. Acad. Sci. U.S.A.* **75**, 4734 (1978).

[6] A. G. Hovanessian and I. M. Kerr, *Eur. J. Biochem.* **84**, 149 (1978).

[7] M. J. Clemens and B. R. G. Williams, *Cell* **13**, 565 (1978).

[8] E. M. Martin, N. J. M. Birdsall, R. E. Brown, and I. M. Kerr, *Eur. J. Biochem.* **95**, 295 (1979).

[9] B. R. G. Williams, I. M. Kerr, C. S. Gilbert, C. N. White, and L. A. Ball, *Eur. J. Biochem.* **92**, 455 (1978).

[10] A. Schmidt, A. Zilberstein, L. Schulman, P. Federman, H. Berissi, and M. Revel, *FEBS Lett.* **95**, 257 (1978).

[11] M. A. Minks, S. Benvin, P. A. Maroney, and C. Baglioni, *Nucleic Acids Res.* **6**, 767 (1979).

[12] E. Slattery, N. Ghosh, H. Samanta, and P. Lengyel, *Proc. Natl. Acad. Sci. U.S.A.* **76**, 4778 (1979).

There is growing evidence that these $(2'-5')$-oligoadenylic acids may play a more general role in the regulation of cellular activity other than their specific role in mediating interferon's action.[13] An understanding of this role will come from a much broader survey of the effects of $2'-5'(A)_n$ on different cells and tissues. However, most studies on the biological activities of these compounds have been carried out in cell-free protein-synthetic systems programmed with various cellular and viral messenger RNAs. This has restricted investigations on the effects of $2'-5'(A)_n$ to those cells suitable for the preparation of protein-synthetic assays and has confined observations to effects on protein synthesis or RNA degradation. The major problem in investigating the effects of $ppp5'A$ $(2'p5'A)_n$ on intact cells has been the obvious difficulty posed by the highly charged triphosphate group, which prevents the drug from penetrating the cell membrane. This chapter describes two methods by which penetration of the cell membrane by the oligonucleotide triphosphates may be achieved. An alternative approach is to remove the triphosphate groups, which renders the oligonucleotides inactive in inhibiting protein synthesis in cell-free systems, but allows them to penetrate cells, and be processed by the cell to some active form. A procedure for the treatment of intact cells with the non-triphosphorylated $2'-5'(A)_n$ is therefore also described.

Inhibition of Protein Synthesis by $ppp5'A(2'p5'A)_n$ in Cells Permeabilized by Hypertonic Medium

Principle

Castellot *et al.*[14] reported that BHK-21 cells could be rendered reversibly permeable to a variety of small charged molecules by treatment with hypertonic NaCl solutions, and they used the method to demonstrate the inhibition of DNA synthesis by cytosine arabinoside triphosphate. Cells made permeable in this way retained a high degree of structural and physiological integrity and, after appropriate "resealing" under isotonic conditions, such cells exhibited normal physiological activities. Williams and Kerr[15] and Williams *et al.*[16] used this method of hypertonic permeabilization to demonstrate the ability of both $ppp5'A(2'p5'A)_2$ and $ppp5'A(2'p5'A)_3$ to inhibit protein synthesis, reversibly activate a nuclease, and inhibit virus growth in intact BHK cells.

The method described below is based on these latter reports.

[13] G. R. Stark, W. J. Dower, R. T. Schimke, R. E. Brown, and I. M. Kerr, *Nature (London)* **278**, 471 (1979).

[14] J. J. Castellot, Jr., M. R. Miller, and A. B. Pardee, *Proc. Natl. Acad. Sci. U.S.A.* **75**, 351 (1978).

[15] B. R. G. Williams and I. M. Kerr, *Nature (London)* **276**, 88 (1978).

[16] B. R. G. Williams, R. R. Golgher, and I. M. Kerr, *FEBS Lett.* **105**, 47 (1979).

Materials

BHK-21 Cells. Cells are grown at 37° as monolayer cultures in plastic tissue culture flasks or dishes in Dulbecco's modified Eagle's minimum essential medium (MEM) supplemented with 10% calf serum (using a humidified 5% CO_2 atmosphere in the case of dishes). Dishes are seeded with 3 to 5 × 10^5 cells per 40 cm^2 and allowed to grow for 20–24 hr. Monolayers should be just subconfluent before use.

Hypertonic Medium. Serum-free Dulbecco's modified Eagle's MEM containing 4.2 g of NaCl per 100 ml (0.87 M) is used to render cells permeable. The appropriate concentration of the drug to be tested is added during permeabilization.

Resealing Medium. To restore the permeability of the cell membrane to normal, the permeabilized cells are incubated in Dulbecco's modified Eagle's MEM, minus the amino acid to be used for the protein synthesis assay (methionine), plus 10% calf serum; 0.1 M ATP, 0.1 mM GTP, and 0.1 M creatine phosphate may also be added[14] (see Comments). The effects of oligoadenylic acids on DNA synthesis may be assayed by using a resealing medium consisting of 35 mM HEPES buffer (pH 7.4), 50 mM sucrose, 80 mM KCl, 4 mM MgCl$_2$, 7.5 mM potassium phosphate (pH 7.4), 0.75 mM CaCl$_2$, 10 mM phosphoenolpyruvate, 1.25 mM ATP, 0.1 mM CTP, 0.1 mM GTP, 0.1 mM UTP, and 0.25 mM each of dATP, dCTP, dGTP, and [^{3}H]dTTP (0.4 Ci/mmol).[14] Effects on RNA synthesis may be observed by using a resealing medium consisting of 35 mM HEPES buffer (pH 7.4), 80 mM KCl, 4 mM MgCl$_2$, 7.5 mM potassium phosphate (pII 7.4), 0.75 mM CaCl$_2$, 50 mM sucrose, 0.5 mM MnCl$_2$, 4 mM ATP, 0.25 mM CTP and GTP, 0.01 mM [^{3}H]UTP (4 Ci/mmol), and 200 μg of tRNA.[14]

ppp5'A(2'p5'A)$_n$ Solutions. (2'-5')-Oligonucleotide triphosphates may be synthesized in bulk by reticulocyte 2'-5'(A) synthetase bound to poly(I) · poly(C)-Sepharose and incubated with ATP.[17] The dimer (pppApA), trimer (pppApApA), and tetramer (pppApApApA) thus produced may be separated and purified as described in Kerr and Brown[9] or Martin *et al.*[8] Solutions of the drugs at a concentration of 1 mM may be prepared by dissolving them in phosphate-buffered saline[18] and adjusting the absorbance at 260 nm (1 cm) to 30 (for the dimer), 45 (for the trimer), or 60 (for the tetramer). The solutions may be sterilized by membrane filtration and are stable at room temperature for some days or at $-20°$ indefinitely.

Note. Concentrations of (2'-5')-oligonucleotides are frequently reported in the literature as concentrations with respect to the adenosine

---

[17] A. G. Hovanessian, R. E. Brown, E. M. Martin, W. K. Roberts, K. Knight, and I. M. Kerr, this volume [24].

[18] R. Dulbecco, *Proc. Natl. Acad. Sci. U.S.A.* **38**, 747 (1952).

content, i.e., the absorbance at 260 nm of 1 mM solutions of all oligomers being 15. This convention is not used here.

Assay Procedure

Sterile precautions should be observed at all stages until addition of trichloroacetic acid. Flasks or dishes of monolayer cultures of BHK-21 cells (approximately 40 cm^2) are made permeable by first washing four times with phosphate-buffered saline and then incubating at 37° with 1.0 ml of hypertonic medium containing the drug to be tested. The progress of permeabilization is monitored periodically by observing the uptake of trypan blue from a solution freshly prepared by mixing four parts of 0.2% w/v aqueous solution of trypan blue with one part of NaCl (42.5 g/liter). When about 95% of the cells are rendered permeable (40–50 min), the hypertonic medium is removed and the cells are washed twice with warm (37°) phosphate-buffered saline. Resealing medium (1.0 ml) containing 1.0 μCi of the desired labeled amino acid per milliliter (e.g., [35S]methionine or a mixture of [14C]leucine, valine, and phenylalanine), and previously warmed to 37° is added, and incubation at 37° is continued for 1 hr. The cells are washed twice with ice-cold phosphate-buffered saline, then 1.0 ml of ice-cold 5% trichloroacetic acid is added and the cells are allowed to stand for 30 min at 4°. The cells are then washed four times with 5% trichloroacetic acid and digested in 1 ml of 0.1 N NaOH for 30 min at 37°. Then 0.1 ml of 1 N HCl is added, and the neutralized solution is transferred to a scintillation vial; 10 ml of a scintillant solution suitable for aqueous samples (e.g., 3 g of PPO, 257 ml of Triton X-100, 106 ml of ethanol, 37 ml of ethylene glycol, and 600 ml of xylene[19]) are added. The radioactivity is measured, and the inhibition of protein synthesis is expressed as a percentage of the incorporation of radioactivity obtained with an untreated control.

Comments

Kinetics of Incorporation. The permeabilization procedure works best on subconfluent monolayers of cells. Although much lower doses of drug are effective in producing inhibition when the resealing medium contains 0.1 M ATP, cells are frequently sensitive to the toxic effects of the ATP. More consistent results are obtained if the ATP regenerating system is omitted. Effective resealing, as shown by the loss of trypan blue staining, is not affected by this omission and takes less than 30 min. Thereafter, the rate of amino acid incorporation should be linear for some hours and equal to the rate of protein synthesis in nonpermeabilized cells.[15] Cells not

[19] U. Fricke, *Anal. Biochem.* **63**, 555 (1975).

TABLE I

EFFECT OF ppp5'A(2'p5'A)$_2$ ON PROTEIN
SYNTHESIS BY BHK CELLS PERMEABILIZED BY
HYPERTONIC TREATMENT[a]

Concentration of pppApApA (M)	Inhibition (%)
3×10^{-11}	1.3
3×10^{-10}	7.2
3×10^{-9}	56.7
3×10^{-8}	76
3×10^{-7}	90

[a] From Williams and Kerr.[15]

treated with hypertonic medium are unaffected by the presence of high concentrations (10 μM) of pppApApA. The inhibition by the drug is transitory, and it is gradually reversed with incubation beyond 2 hr.[16]

Dose Responses. Table I shows a dose-response curve for ppp5'A (2'p5'A)$_2$ with 0.1 M ATP included in the resealing medium. A 50% inhibition is obtained with a concentration of about 2 nM, which compares favorably with the dose required to produce a 50% inhibition of protein synthesis in cell-free systems (0.1–0.3 nM). The tetramer [ppp5'-A(2'p5'A)$_3$] produces a similar level of inhibition, but neither the (2'-5') dimer nor the (3'-5') trimer inhibit at concentrations up to 10 μM. The dephosphorylated trimer (A2'p5'A2'p5'A) causes a 50% inhibition in the concentration range of 5–25 μM only if 0.1 M ATP is included in the resealing medium.

Effects on Nucleic Acid Synthesis. The effects of oligoadenylic acids on RNA or DNA synthesis may be assayed by using modified resealing media as indicated above. Levels of (2'-5')trimer or tetramer that produced 50% inhibition of protein synthesis had no effect on DNA or RNA synthesis during the first 24 hr of treatment. Degradation of polyadenylated RNA in treated cells may be monitored by prelabeling the cells in medium containing 2% calf serum and 50 μCi/ml of [^3H]uridine for 30 min before washing in serum-free medium and permeabilizing in the presence of the (2'-5')-oligonucleotide. After incubating for 40 min in resealing medium, the cells are solubilized in 0.1 M NaCl, 0.01 M Tris · HCl (pH 8.8), 2 mM EDTA, 1% (v/v) 2-mercaptoethanol, 1% (v/v) Triton N101, and 0.5% (w/v) sodium deoxycholate. The lysate is centrifuged at 20,000 g for 10 min, the supernatant is adjusted to 0.4 M with respect to NaCl, sodium dodecyl sulfate is added to 0.5% (w/v), and the lysate is chromatographed on oligo(dT)-cellulose as described by Williams et al.[16]

Use of Other Cells. The method was developed for use with BHK-21 cells, but is probably satisfactory with minor modifications for use with other common continuously passaged cell lines. However, it is not satisfactory for human lymphoblastoid cells, which fail to reverse trypan blue permeability when treated with isotonic media.

Rendering Cells Permeable to ppp5'A(2'p5'A)$_n$ by Calcium Phosphate Coprecipitation

Principle

Graham and Van der Eb[20] had observed that the infectivity of adenoviral DNA could be improved if the DNA was previously adsorbed onto a calcium phosphate suspension before adding to cells. They suggested that the precipitate of calcium phosphate, with its adsorbed DNA sedimented onto cells, became adsorbed to the cell membrane and promoted uptake of the DNA through a calcium-requiring process. The procedure developed by them was used by Hovanessian *et al.*[21] to take up a mixture of (2'-5')-oligonucleotide triphosphates into a variety of cells and show that they inhibited protein synthesis, promoted increased ribonuclease activity, reduced nucleic acid synthesis, and inhibited the growth of VSV. The assay method described below is based on these reports.

Materials

MRC5 Cells. Human embryonic fibroblast (MRC5) cells are grown as monolayer cultures in plastic multiwell dishes (24 × 16 mm in diameter, Linbro Scientific Inc., Hamden, Connecticut), seeded with 1×10^5 cells per well in growth medium (Eagle's basal medium containing Earle's salts and 10% fetal calf serum), and incubated for 36–44 hr at 37° in a humidified 5% CO_2 atmosphere. For the most reproducible results, the monolayers should be only 60–90% confluent before use. *Note.* This assay may be performed using Vero, HeLa, or L-929 cells; all take up ppp5'A(2'p5'A)$_n$ satisfactorily using this method, but the dose-response curves vary slightly from each other.[21]

(2'-5')-Oligoadenylic Acid Solutions. (2'-5')-Oligonucleotide triphosphates may be prepared, purified, and sterilized as described above.

Protein Synthesis Assay Medium. Protein synthesis is assayed by incubation with medium minus the amino acids to be used for labeling (e.g., [^{35}S]methionine or [^{14}C]leucine, valine, and phenylalanine) plus 10% calf serum, and containing 1 μCi/ml of the appropriate labeled amino acid.

[20] F. L. Graham and A. J. Van der Eb, *Virology* **52**, 456 (1973).
[21] A. G. Hovanessian, J. N. Wood, E. Meurs, and L. Montagnier, *Proc. Natl. Acad. Sci. U.S.A.* **76**, 3261 (1979).

Assay Procedure

Sterile precautions should be maintained throughout the assay until the addition of trichloroacetic acid. Growth medium is aspirated and HEPES-buffered saline (pH 7.05) (8.0 g of NaCl, 0.37 g of KCl, 0.125 g of $Na_2HPO_4 \cdot 2H_2O$, 1.0 g of glucose and 5.0 g of HEPES per liter) containing different concentrations of (2'-5')A is added to the cell monolayers. Then a 2.0 M $CaCl_2$ solution (prepared in H_2O, sterilized by filtration and stored at $-20°$) is added at a final concentration of 125 mM (final volume in each well is 0.5 ml). Adsorption is carried out at room temperature for 45 min, then 0.5 ml/well of serum-free growth medium is added. The cells are then incubated in a humidified 5% CO_2 atmosphere at 37° for 90 min. The medium is then aspirated and replaced with 0.5 ml of protein synthesis assay medium, prewarmed at 37°, per well. The cells are incubated at 37° for 1 hr. They are washed twice with ice-cold phosphate-buffered saline, then 1 ml of 5% trichloroacetic acid is added per well. Thereafter, the procedure is as described above for the hypertonic treatment assay.

Comments

Kinetics of Incorporation. Incorporation of amino acids should be linear for at least 3 hr in the untreated control and should occur at a rate almost identical to that of cells that have not been treated with $CaCl_2$. However, RNA synthesis may show a lag of up to 6 hr before linear synthesis occurs, presumably due to a requirement to recover from the lack of serum in the medium.[21] The inhibitory effects of low concentrations (<20 nM) of the drug on protein and RNA synthesis are transitory, reversal to normal synthetic rates occurring 6 hr after removal of the drugs. At higher concentrations the inhibitory effect may persist for up to 48 hr.[22]

Dose Responses. Table II shows the dose-response curves obtained using MRC5 and Vero cells. These results were obtained with a mixture of oligonucleotide triphosphates containing the trimer and higher oligomers [ppp5'A(2'p5'A)$_n$]. The inhibitory response is comparable to that obtained in hypertonic treatment assay, 50% inhibition being obtained with concentrations of ppp5'A(2'p5'A)$_n$ of the order of 1–3 nM.

Influence of Incubation Parameters. The efficiency with which charged molecules are taken up by cells is exquisitely sensitive to pH. Below pH 6.9, no precipitation of calcium-phosphate occurs; above pH 7.2, the precipitate forms rapidly, is heavier and coarser, and is not effective in promoting uptake.[20] It is advisable to determine the optimum concentration of $CaCl_2$ for different cell types. The length of exposure of the cells to the drug–calcium phosphate complex is also important; the times indicated in the assay procedure have been found to be optimum for the cells used.

[22] A. G. Hovanessian and J. N. Wood, *Virology* **101**, 81 (1980).

TABLE II
EFFECT OF ppp5'A(2'p5'A)$_n$ ON PROTEIN SYNTHESIS BY
MRC5 AND VERO CELLS IN THE CALCIUM PHOSPHATE
COPRECIPITATION ASSAY[a]

Concentration of (2'-5')A (M)	Inhibition (%) of amino acid incorporation in	
	MRC5 cells	Vero cells
3×10^{-10}	20	39
10^{-9}	35	65
3×10^{-9}	71	80
10^{-8}	84	86
3×10^{-8}	90	92
10^{-7}	95	93

[a] From Hovanessian et al.[21]

Application to Other Cells. A number of cell lines have been shown to take up charged molecules by this method and to have suffered only minimal toxic effects of the calcium treatment. Such lines include Vero, BHK-21, HeLa, L-929, and KB cells. There may, however, be up to 100-fold differences in sensitivity to (2'-5')-oligonucleotides between different cell lines. Lymphoblastoid cells (e.g., Daudi) are not satisfactory as up to 80% are killed by the calcium treatment,[23] and the method may be satisfactory only for monolayers of cells.

Effect of Oligonucleotide Triphosphate on DNA Synthesis. The assay may be modified to study effects of oligonucleotide triphosphates on DNA synthesis by replacing the protein synthesis assay medium with 0.5 ml of growth medium containing 10 μCi/ml of [^{3}H]thymidine. Both growth of L-929 cells and their DNA synthesis are inhibited by concentrations of ppp5'A(2'p5'A)$_n$ comparable to those causing inhibition of protein synthesis, but unlike the latter effect, which is seen almost immediately, inhibition of DNA synthesis requires 24 hr to develop maximally. At low concentrations of the drug (<20 nM) both effects are transitory.[22]

Effect of Oligonucleotide Triphosphate on RNA Synthesis. RNA synthesis may be monitored by including [^{3}H]uridine (2 μCi/ml) in the protein synthesis assay medium and measuring incorporation for 4 hr after addition of the incubation medium. A 50% inhibition of RNA synthesis occurs with concentrations of ppp5'A(2'p5'A)$_n$ in the range of 3–10 nM.[21] This is presumably due to the activation of a ribonuclease by the drug, as it promotes degradation of viral RNA in infected cells.[22]

[23] D. M. Reisinger and E. M. Martin, unpublished results.

Comparison with the Hypertonic Treatment Assay. Both methods for inducing the uptake of oligonucleotide triphosphates into intact cells have a similar sensitivity, with 50% inhibition of protein synthesis in the 3–9 nM range. The calcium phosphate coprecipitation method produces less damage to cells and is therefore more reproducible and can be applied to a wider range of cells; it is probably now the method of choice.

The Use of Non-Triphosphorylated Oligonucleotides

Principle

Removal of the 5'-triphosphate group from ppp5'A(2'p5'A)$_n$ radically reduces its polarity, thus permitting its penetration of the cell membrane. Loss of the triphosphate abolishes its ability to activate ribonuclease, and hence (2'-5')ApApA is inactive in the cell-free protein–synthetic assay system.[2] However, intact cells probably contain kinases capable of replacing the triphosphate group, as (2'-5')ApApA is active in inhibiting protein synthesis in intact cells, albeit at higher concentrations than those that are effective in cell-free systems.[15,23] Moreover, it is capable of inhibiting the growth of at least one type of cell, the Burkitt's lymphoma-derived human lymphoblastoid cell line Daudi.[23] (2'-5')ApApA causes a marked inhibition of [³H]thymidine incorporation into DNA of intact Daudi cells, and this inhibition is more pronounced than its effect on either protein or RNA synthesis. The mechanism of its inhibitory effect on DNA synthesis is not known. Nevertheless, this action may be used as a basis for assaying the effects of non-triphosphorylated oligonucleotides on intact cells, as described below.

Materials

Daudi Cells.[24] These cells are grown as stationary suspension cultures at 37° in a growth medium of RPMI-1640 medium (adjusted to pH 7.4 with 1 N NaOH rather than with NaHCO$_3$) supplemented with 10% fetal bovine serum (batch-tested for ability to support growth of Daudi cells), 32 μg/ml of gentamicin, 60 μg/ml of penicillin, and 20.3 mM HEPES. They may be grown in any suitable closed vessel, and subcultured when the cell concentration reaches about 1.5×10^6 cells/ml by adding growth medium to adjust the concentration to 2 to 3×10^5 cells/ml. The cultures should not be agitated during growth. It is preferable to use cells from frozen stocks, as their sensitivity to drugs may alter with continuous subculturing for periods longer than 3 months. The growth of this cell line

[24] E. Klein, G. Klein, J. S. Nadkarni, J. J. Nadkarni, H. Wigzell, and P. Clifford, *Cancer Res.* 28, 1300 (1968).

is sensitive to interferon, being 50% inhibited by concentrations of 2–5 international units of interferon per milliliter.

(2'-5')ApApA Solutions. The non-triphosphorylated oligonucleotides may be prepared from ppp5'A(2'p5'A)$_n$ by bacterial alkaline phosphatase treatment and repurification or by synthesis from 2',3'-AMP by Michelson's[25] method as described by Martin *et al.*[8] Bulk solutions are prepared at a concentration of 2 mM (i.e., an absorbance at 260 nm of 90) in phosphate-buffered saline and sterilized by membrane filtration. The solutions are stored at $-20°$. Dilutions of the drug for the assay of inhibitory activity are made in phosphate-buffered saline from these concentrated stocks.

Procedure

Daudi cells are subcultured the day before the assay by adjusting their concentration to 5×10^5 cells/ml and incubating. On the day of the assay, fresh growth medium is added to bring the cell concentration back to 5×10^5/ml. Portions (2.0 ml) of the cell suspension are dispensed into 16×125 mm plastic screw-top tubes (Lux Scientific Corpn., Newbury Park, California). Appropriate dilutions of the oligonucleotide (0.2 ml) are added, the tubes are placed in a suitable rack on their side at a slight angle to provide stationary culture conditions, and the racks are incubated at 37° for 2 hr. [³H]Thymidine (0.1 ml of 5 μCi/ml, specific activity approximately 5 Ci/mmol) is then added, and the incubation is continued for 30 min. Incorporation is stopped by the addition of 10 ml of ice-cold phosphate-buffered saline containing 2 mM thymidine. The cells are then collected by filtration through a 2.5 cm glass fiber filter disk (e.g., Whatman GF/C filter) and washed with a further 10 ml of saline. Ice-cold 5% trichloroacetic acid (10 ml) is added to the cells on the filter and sucked through. This procedure is repeated four times to wash the cells free of acid-soluble material. The cells are then washed with 5 ml of methanol and dried at 60° for 30 min. The filter papers are transferred to a glass vial, covered with a toluene scintillant (e.g., 3 g of PPO plus 0.3 g of dimethyl POPOP per liter of toluene), and counted in a liquid scintillation counter. Percentage inhibition of [³H]thymidine incorporation is calculated in the usual manner.

Comments

Kinetics of Inhibition. (2'-5')ApApA (5 μM) causes a measurable inhibition of [³H]thymidine incorporation within 30 min, 50% of the maximum inhibition being observed within the first 3 hr. The inhibition then in-

[25] A. M. Michelson, *J. Chem. Soc.* p. 1371 (1959).

creases progressively, but slowly, reaching a maximum at about 18 hr after the addition of the drug. It is not transitory and can be maintained for up to 4 days. At any stage, however, it may be reversed by washing out the drug. This is in contrast to the effect of interferon, which does not begin to cause inhibition of [³H]thymidine incorporation until about 15 hr after treatment, although maximum inhibition is seen at 22 hr.[23]

Cause of the Inhibition. (2'-5')ApApA inhibits the growth of Daudi cells, and so it can be assumed that the inhibition of [³H]thymidine incorporation measured in this assay does reflect, at least in part, inhibition of DNA synthesis. However, its direct effect on DNA synthesis has yet to be demonstrated. The inhibition is not produced by adenosine, nor the 2'-5' dimer, nor by the 3'-5' trimer oligoadenylate, and hence is not a nonspecific effect of adenosine.

Dose Responses. Table III shows a typical response of Daudi cells to the presence of various concentrations of (2'-5')ApApA. At concentrations of about 7 μM, 50% inhibition occurs, indicating that this assay is about one-thousandth as sensitive as the assays for the effect of ppp5'-A(2'p5')A$_n$ on amino acid incorporation in intact cells described above. This large difference in sensitivity may reflect a low efficiency of uptake into the cell (for which evidence exists), an inefficient phosphorylating mechanism in the cell, or the possibility that the effect of the drug on DNA synthesis is unrelated to its effects on protein synthesis.

Application to Other Cells. The effect of (2'-5')ApApA on [³H]thymidine incorporation has been observed only in transformed lymphoblastoid cell lines whose growth is sensitive to the action of interferon. It is not observed with lines of L-929 cells on BHK-21 cells whose growth is not affected by interferon. Its effects on whole animals have not been investigated.

TABLE III
EFFECT OF A2'p5'A2'p5'A ON [³H]THYMIDINE INCORPORATION BY DAUDI CELLS[a]

Concentration of (2'-5')ApApA (M)	% Inhibition of [³H]thymidine incorporation
5×10^{-7}	6
10^{-6}	14
5×10^{-6}	32
10^{-5}	64
2.5×10^{-5}	77

[a] Reisinger and Martin.[23]

Effects of ApApA on Protein and RNA Synthesis. When the labeled precursor in the assay system is replaced by ^{14}C-labeled amino acids or ^{3}H]uridine to study the effects of the drug on protein and RNA synthesis, it is found that neither is affected to the extent that ^{3}H]thymidine incorporation is inhibited. For example, 1 μM (2'-5')ApApA did not affect the synthesis of either macromolecule, whereas 10 μM caused a 20% inhibition of protein synthesis and a 35% stimulation of RNA synthesis.[23]

Conclusions

(2'-5')-Oligoadenylic acids are extremely potent inhibitors of protein synthesis, being effective at the subnanomolar level in cell-free systems. However, the setting-up of cell-free system assays, involving the preparation and testing of cell extracts and messenger RNAs, is not to be undertaken lightly. Methods using intact cells such as those outlined above have distinct advantages. They require less skill, are less costly, and are probably as reproducible, although this may depend on the relative experience of the operator with cell extracts versus whole cells. They have the disadvantage of some loss of sensitivity, requiring about 10- to 100-fold higher concentrations of the oligonucleotide triphosphates. On the other hand, these intact-cell systems offer the opportunity to investigate the possible manifold effects of the drugs within the cell, as instanced by the studies referred to above or the effects of (2'-5')ApApA on DNA synthesis.

Acknowledgments

The authors are very grateful to Dr. I. M. Kerr for initiating their interest in (2'-5')-oligonucleotides, for providing materials for their research, and for valuable comments on this manuscript.

[36] Assay of Eukaryotic Initiation Factor eIF-2 Activity with Extracts of Interferon-Treated Cells

By JOHN A. LEWIS and REBECCA FALCOFF

Treatment of tissue-culture cells with interferon elicits a variety of responses,[1] among which are the appearance of several double-stranded RNA (dsRNA)-dependent enzyme activities.[1,2] In analogy with results

[1] J. A. Lewis, *in* "Horizons in Biochemistry and Biophysics" (L. A. Kohn and R. M. Friedman, eds.), p. 121. Dekker, New York, 1981.
[2] C. Baglioni, *Cell* 17, 255 (1979).

obtained in the rabbit reticulocyte lysate system,[3-5] dsRNA has been shown to enhance markedly the phosphorylation of at least two proteins in extracts of interferon-treated cells.[6-9] One of these proteins migrates on sodium dodecyl sulfate (SDS)–polyacrylamide gel electrophoresis in the same position as eIF-2α (MW 38,000), the smallest subunit of initiation factor eIF-2 that mediates the binding of methionyl-tRNA$_i^{met}$ to 40 S ribosomal subunits. The other protein (MW 67,000) is thought to be the kinase that is responsible for phosphorylation of eIF-2α.[3] Interestingly, a protein of MW 67,000 is one of several proteins that are found to be synthesized in greatly increased amounts after cells are treated with interferon.[10-12]

In the reticulocyte lysate, increased phosphorylation of eIF-2 is accompanied by reduced binding of Met-tRNA$_i$ to 40 S preinitiation complexes.[13] We have shown that incubation of interferon-treated cell extracts with dsRNA leads to a reduction in their capacity to initiate protein synthesis.[9] Thus, the incorporation of [^{35}S]methionine in the NH$_2$-terminal positions of peptides synthesized in vitro was analyzed by a modification[9] of the Edman degradation procedure[14] and found to be reduced by dsRNA. Furthermore, the incorporation of [^{35}S]methionine into a peak sedimenting at approximately 40 S was measured by sedimentation on sucrose density gradients and also shown to be sensitive to dsRNA.

Formation of Methionyl-tRNA · 40 S Ribosome Complexes

Materials

[^{35}S]Methionine, 5 mCi/ml, 500–1200 Ci/mmol (Amersham)
Cetyltrimethylammonium bromide (Aldrich)
Yeast RNA (Sigma)
Poly(I) · poly(C) (Miles)

[3] P. J. Farrell, K. Balkow, T. Hunt, R. J. Jackson, and H. Trachsel, Cell 11, 187 (1977).
[4] D. Levin and I. M. London, Proc. Natl. Acad. Sci. U.S.A. 75, 1121 (1978).
[5] G. Kramer, J. M. Cimadevilla, and B. Hardesty, Proc. Natl. Acad. Sci. U.S.A. 73, 3078 (1976).
[6] B. Lebleu, G. C. Sen, S. Shaila, B. Cabrer, and P. Lengyel, Proc. Natl. Acad. Sci. U.S.A. 73, 3107 (1976).
[7] W. K. Roberts, A. Hovanessian, R. E. Brown, M. J. Clemens, and I. M. Kerr, Nature (London) 264, 477 (1976).
[8] A. Zilberstein, P. Federman, L. Shulman, and M. Revel, FEBS Lett. 68, 119 (1976).
[9] J. A. Lewis, E. Falcoff, and R. Falcoff, Eur. J. Biochem. 86, 497 (1978).
[10] L. A. Ball, Virology 94, 282 (1979).
[11] E. Knight and B. D. Korant, Proc. Natl. Acad. Sci. U.S.A. 76, 1824 (1979).
[12] S. L. Gupta, B. Y. Rubin, and S. L. Holmes, Proc. Natl. Acad. Sci. U.S.A. 76, 4817 (1979).
[13] C. H. Darnbrough, S. Legon, T. Hunt, and R. J. Jackson, J. Mol. Biol. 76, 379 (1973).
[14] W. R. Gray, this series, Vol. 25, p. 333.

Solution 1: 20 mM K$^+$ HEPES, pH 7.5, 100 mM KCl, 2.5 mM magnesium acetate, 7 mM 2-mercaptoethanol

Solution 2: 0.1 mM sodium cacodylate, pH 5.5, 100 mM KCl, 2.5 mM magnesium acetate, 7 mM 2-mercaptoethanol

Procedure. Cell-free extracts (S$_{10}$) are first incubated with or without dsRNA in order to activate the interferon-induced inhibitory activities. Formation of methionyl-tRNA · 40 S ribosomal subunit complexes is then assayed by brief incubation with [^{35}S]methionine and subsequent analysis by sucrose density gradient centrifugation.[13]

Incubation mixtures are prepared containing in a final volume of 150 μl: 90 μl of S$_{10}$ extracts preincubated and passed through Sephadex (prepared as described elsewhere[9]), 20 mM K$^+$ HEPES, pH 7.5, 100 mM KCl, 2.0 mM magnesium acetate, 0.45 mM spermidine, 7 mM 2-mercaptoethanol, 6% (v/v) glycerol, 1 mM ATP, 0.25 mM GTP, 5 mM creatine phosphate, 180 μg of creatine phosphokinase per milliliter, and 62.5 μM cold amino acids with the exception of methionine. Duplicate tubes are prepared for each extract of control or interferon-treated cells and dsRNA [poly(I) · poly(C)] is added to one of each pair at a final concentration of 100 ng/ml. After incubation at 32° for 30 min, 5 μl of [^{35}S]methionine (3.3 mCi/ml; 224 Ci/mmol) are added and the incubation is continued for a further 10 min. Alternatively, 2 μg of [^{35}S]Met-tRNA$_i^{Met}$ can be added and incubation continued for 2 min.[9] The reaction mixtures are rapidly cooled by immersion in an ethanol–ice mixture ($-10°$), and 0.2 ml of solution 1 is added. The mixtures are layered over linear 15 to 30% (w/v) sucrose (in solution 1) density gradients and centrifuged at 45,000 rpm for 150 min in an SW50.1 (Beckman) rotor at 2°. Gradients are analyzed by passage through a UV monitor (Isco), and 0.2-ml fractions are collected. Amino acyl-tRNA and peptidyl-tRNA are precipitated by the addition of 1.0 ml of 0.5 M sodium acetate, pH 5.5, and 1.0 ml of 2% cetyltrimethylammonium bromide[13] and incubation at 32° for 10 min. Precipitates are filtered on Whatman GF/C glass-fiber disks and rinsed several times with H$_2$O before drying under an infrared lamp and assaying for radioactivity in a toluene-based scintillation fluid.

The Met-tRNA$_i^{Met}$, whether added exogenously or formed *in situ* by the addition of radioactive methionine, is rapidly degraded in cell-free extracts, and it is important to work rapidly. Usually the total counts per minute in the 40 S peak are lower when Met-tRNA$_i^{Met}$ is used. We have modified the procedure[15] by substituting a low pH buffer (pH 5.5) for solution 1 as described elsewhere.[16] Thus, after completion of the incuba-

[15] V. M. Pain, J. A. Lewis, P. Huvos, E. C. Henshaw, and M. J. Clemens, *J. Biol. Chem.* 255, 1486 (1980).

[16] R. G. Crystal, N. A. Elson, and W. F. Anderson, this series, Vol. 30, p. 101.

tion, 0.2 ml of solution 2 are added and the mixtures are analyzed on 15 to 30% (w/v) sucrose solution 2 density gradients. This results in considerably increased amounts of radioactivity sedimenting in the 40 S region. It is important to use the low concentration cacodylate buffer because ribosomes will readily be precipitated at this pH by many other buffers, such as 20 mM sodium acetate, pH 5.5.

The results of such an experiment are given in Fig. 1, which shows that preincubation of an interferon extract with dsRNA markedly inhibits uptake of [35S]methionine into the 40 S region. The effect of dsRNA can be overcome by addition of excess initiation factors in the form of a crude salt wash fraction of rabbit reticulocyte ribosomes[16] (Table I). Extracts of control cells do not normally exhibit dsRNA-dependent inhibition of preinitiation complex formation (results not shown), at least with the concentrations of dsRNA used (100–400 ng/ml). Occasionally, however, extracts of control cells that had been stored for long periods of time or had been preincubated after storage in liquid nitrogen did show some dsRNA sensitivity.

Although binding of methionyl-tRNA to the 40 S subunit is a GTP-dependent step, we have observed rather variable results when GTP was omitted from the reaction mixtures. We attribute this to the presence of endogenous GTP even in extracts passed through Sephadex. The GTP is probably present complexed to eIF-2 and other GTP-binding enzymes. Although we have usually utilized preincubated extracts that have essen-

TABLE I

EFFECT OF DOUBLE-STRANDED RNA AND INITIATION
FACTORS ON METHIONYL-tRNA · 40 S RIBOSOMAL
SUBUNIT COMPLEX FORMATION[a]

Additions	[35S]Methionine in 40 S peak (fmol)	
	− dsRNA	+ dsRNA
None	24.3	3.9 (16)
Initiation factors	23.2	17.8 (77)

[a] Conditions for measuring uptake are described in the text. Double-stranded RNA [poly(I) · poly(C)] was added at a final concentration of 100 ng/ml, and incubation with 2 μg of [35S]methionyl-tRNA (113,830 cpm/μg) was for 10 min at 32°. Numbers in parentheses are percentages of the appropriate control. One femtomole of methionine is equivalent to 34 cpm.

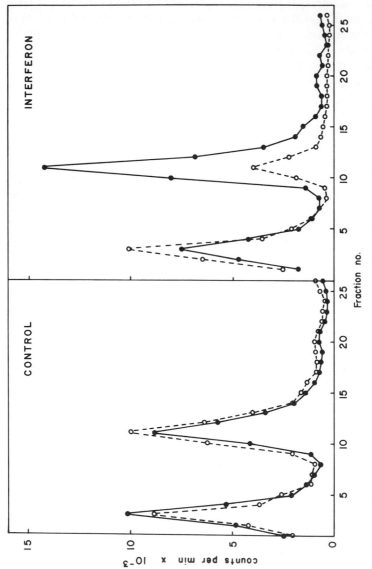

FIG. 1. Effect of dsRNA on Met-tRNA · 40 S ribosomal subunit complex formation by control and interferon-treated cell extracts. Incubations were performed as described in the text with [^{35}S]methionine (110 μCi/ml; 224 Ci/mmol), and reaction mixtures were analyzed by centrifugation on 15 to 30% sucrose density gradients at 45,000 rpm for 150 min in an SW50.1 rotor. Direction of centrifugation is from left to right. Cetyltrimethylammonium bromide-precipitable counts are shown. ——●, incubations without dsRNA; O----O, incubations with dsRNA [100 ng/ml of poly(I) · poly(C)].

tially no protein synthetic activity, comparable results have been obtained in extracts treated with micrococcal nuclease (10 units/ml) in the presence of 1 mM CaCl$_2$. After incubation at 20° for 10 min, the nuclease is inactivated by the addition of ethylene glycol–bis(β-aminoethyl ether) N,N'-tetraacetic acid (EGTA) to a concentration of 2 mM. The formation of Met-tRNA$_i^{Met}$ · 40 S ribosomal subunit complexes can also be monitored in extracts possessing endogenous translational activity. Incorporation of radioactivity into the polysome region of gradients sometimes obscures the 40 S peak slightly, but this may be overcome by the addition of cycloheximide (50 μg/ml). Since this may result in a disruption of the steady state normally observed, interpretation of the results may be complicated. Finally, we have observed that in some cell-free systems (e.g., Ehrlich ascites tumor S$_{10}$) initiation may be impeded at low methionine concentrations (less than 1 μM), and it may be wise to test a range of methionine concentrations before selecting the final conditions for further experiments.

Assay of [^{35}S]Methionine Uptake into NH$_2$-Terminal Positions of Peptides

Materials

Phenylisothiocyanate (Fisher)
Pyridine(Fisher)
Phosphorus pentoxide (Fisher)
Trifluoroacetic acid (Fisher)
Ethylene dichloride (Fisher)
Pactamycin (UpJohn Co.)

CAUTION: Several of these reagents are noxious, and extreme care should be exercised to limit exposure to them. Ethylene dichloride vapor will dissolve several plastics, and careful selection of laboratory ware is suggested.

Procedure. Preincubated S$_{10}$ extracts (passed through Sephadex) of control and interferon-treated cells are prepared as described previously.[9] Incubation mixtures (100 μl) are prepared in duplicate containing 60 μl of S$_{10}$ extract and the same concentrations of reagents as listed in the preceding section plus [^{35}S]methionine (245 μCi/ml; 50 Ci/mmol). To one of the duplicates dsRNA [poly(I) · poly(C)] is added at a final concentration of 100 ng/ml. In addition, control incubations are performed containing 5 μM pactamycin to block protein synthesis initiation. These mixtures are incubated at 32° for 30 min to activate the dsRNA-dependent enzymic activities. At this point, unlabeled Mengo virus RNA, prepared as described in detail in this volume,[17] is added (40 μg/ml) and the incubation is con-

[17] J. A. Lewis and R. Falcoff, this volume [34].

tinued for 10 min. (When a large number of incubations are performed, it is advisable to stagger the start of each incubation by 30 sec). In some cases, initiation factors (high salt-wash fraction of rabbit reticulocyte ribosomes[16]) are added with the mRNA at a final protein concentration of 125 μg/ml. At the end of this incubation period, 40 μl of each reaction mixture is pipetted in duplicate onto Whatman No. 3 MM filter paper disks (2.4 cm in diameter). The disks are plunged into cold 10% trichloracetic acid and left for 30 min. The disks are then heated at 95° for 10 min in 5% trichloroacetic acid, washed several times in trichloroacetic acid and then dried by several washes in ethanol, ethanol–ether (1 : 1), and ether. After drying under an infrared lamp, the disks are transferred to a dry Erlenmeyer flask (500 ml) to which 20 ml of 50% aqueous pyridine are added, followed by 10 ml of 10% (v/v) phenylisothiocyanate in pyridine. The flask is shaken and carefully gassed with nitrogen, sealed with two layers of Parafilm, and incubated at 45° for 60 min. Excess reagent is removed, and the disks are transferred to a sheet of aluminum foil, which is placed in a desiccator over phosphorus pentoxide and solid potassium hydroxide. After evacuation, the desiccator is placed in an oven at 60° for 90 min. The dried disks are then transferred to individual, predried silica glass test tubes. (It is suggested that metal test tube racks be used and that tubes be kept in a fixed sequence, as some reagents will attack plastics and obliterate markings.) To each test tube is added 0.6 ml of trifluoroacetic acid (caution: use a fume hood) such that the disks are just saturated with liquid. After gassing with nitrogen, the tubes are sealed with a silicone rubber bung or Parafilm and heated at 45° for 45 min. The test tubes are unsealed and placed in a desiccator under vacuum overnight. The dried disks are extracted sequentially with 2.0 ml, 1.5 ml, and 1.0 ml of ethylene dichloride for 30 min at room temperature, and the extracts are pooled in glass or plastic scintillation vials and dried under vacuum for several hours. (Use a vapor trap if a house vacuum line or pump is used.) The residue is dissolved directly in a toluene-based scintillation fluid and counted in a liquid scintillation counter. From time to time, control reactions are performed in which either phenylisothiocyanate or trifluoroacetic acid are omitted. In these control reactions, less than 10% of the normal radioactivity is present in the final extract. When the entire Edman degradation is performed on incubations containing 5 μM pactamycin or lacking Mengo virus RNA, between 5 and 15% of the radioactivity in complete incubations is obtained.

The results of an analysis of this type are shown in Table II. The inclusion of dsRNA had no effect upon the incorporation of methionine into the NH$_2$-terminal position of peptides by control extracts, but strongly inhibited the uptake by interferon-treated cell extracts. This in-

TABLE II

EFFECT OF DOUBLE-STRANDED RNA AND INITIATION FACTORS ON
INCORPORATION OF [^{35}S]METHIONINE INTO NH$_2$-TERMINAL
POSITIONS OF POLYPEPTIDES[a]

Additions	[^{35}S]Methionine in NH$_2$-terminal position (fmol)	
	Control	Interferon
None	9.5	8.3
dsRNA	9.5 (100)	2.0 (24)
Initiation factors	11.2	6.7
dsRNA + initiation factors	9.5 (85)	3.2 (48)

[a] Incorporation was assayed as described in the text with 60 μl of
S$_{10}$, [^{35}S]methionine (245 μCi/ml; 50 Ci/mmol) and plus or
minus poly(I) · poly(C) (100 ng/ml) and rabbit reticulocyte initiation factors (125 μg/ml). Each figure is the mean of duplicate
assays, and a blank representing incorporation in the presence
of 5 μM pactamycin has been subtracted. Numbers in parentheses are percentages of the values obtained in the absence
of dsRNA. One femtomole is equivalent to 50 cpm.

hibition was somewhat reduced when the extract was supplemented with
reticulocyte initiation factors, although the extent of this reversion is usually less than is observed for the formation of [^{35}S]Met-tRNA · 40 S subunit complexes.

Comments

Analysis of Met-tRNA$_i^{Met}$ · 40 S ribosomal subunit complex formation
is a convenient measure of eIF-2 activity in crude extracts and has the
advantage, in the present application, of being independent of mRNA.
Consequently, the parallel activation of an endonuclease activity by
dsRNA should not affect the measurement of initiation complex formation. Since other steps in the protein synthetic pathway could be rate
limiting, we have attempted to compare the relative rates of initiation by
direct measurement of the production of NH$_2$-terminal methionine residues and have utilized both preinitiation complex formation and the
Edman degradation procedure to give an overall view of the initiation process. The Edman procedure, however, does have certain drawbacks,
since activation of endonuclease activity causes degradation of mRNA,
which probably results in fewer intact 5' termini being available for initia-

tion; this probably explains the limited ability of initiation factors to reverse the inhibition. Nonetheless, the extent of dsRNA-induced inhibition of initiation activity observed by the two procedures is in good agreement, and the results show that the dsRNA-dependent phosphorylation of peptides of MW 38,000 and 67,000 is accompanied by a reduction in the rate of initiation, probably as a result of a reduced Met-tRNA$_i^{Met} \cdot$ 40 S subunit complex formation.

Acknowledgment

J. A. L. was supported by an EMBO long-term fellowship and is presently a Sinsheimer Scholar.

[37] Procedures for Isolation and Assay of Translation-Inhibitory Ribosome-Associated Proteins[1]

By CHARLES E. SAMUEL

Interferons inhibit the production of infectious progeny virions of a wide range of unrelated DNA and RNA viruses.[2,3] In many animal virus–host cell systems, the level of virus genome expression inhibited as a result of interferon treatment is the translation of viral mRNA into protein.[3,4] Cell-free extracts prepared from interferon-treated cells catalyze the translation *in vitro* of a variety of exogenously added viral mRNAs much less efficiently than extracts prepared from untreated cells.[5-8] The inhibitor(s) of protein synthesis observed in extracts prepared from interferon-treated murine, human, and simian cells are associated with the ribosomal system and can be separated from ribosomes either by washing

[1] This work was supported in part by research grants from the National Institute of Allergy and Infectious Diseases (AI-08909 and AI-12520) and the American Cancer Society (MV-16), and by a Research Career Development Award from the U.S. Public Health Service.
[2] N. B. Finter, ed., "Interferons and Interferon Inducers." North-Holland Publ., Amsterdam, 1973.
[3] W. E. Stewart, II, "The Interferon System." Springer-Verlag, Vienna and New York, 1979.
[4] C. E. Samuel, S. M. Kingsman, R. W. Melamed, D. A. Farris, M. D. Smith, N. G. Miyamato, S. R. Lasky, and G. S. Knutson, *Ann. N. Y. Acad. Sci.* **350**, 473 (1980).
[5] E. Falcoff, R. Falcoff, B. Lebleu, and M. Revel, *J. Virol.* **12**, 421 (1973).
[6] R. M. Friedman, D. H. Metz, R. M. Esteban, D. R. Tovell, L. A. Ball, and I. M. Kerr, *J. Virol.* **10**, 1184 (1972).
[7] S. L. Gupta, M. L. Sopori, and P. Lengyel, *Biochem. Biophys. Res. Commun.* **54**, 777 (1973).
[8] C. E. Samuel and W. K. Joklik, *Virology* **58**, 476 (1974).

isolated ribosomes with buffers containing concentrated salt[5,8–10] or by incubation of crude extracts prior to ribosome isolation.[7,11]

The inhibition of translation in interferon-treated cell-free systems occurs at both the level of initiation and the level of elongation of polypeptide chain biosynthesis. The inhibition at the level of initiation, characterized by a reduction in the association of virus mRNA to ribosomes[12] and a reduction in the formation of virus-specific methionyl-X initiation peptides,[13] is accompanied by a reduction in protein synthesis initiation factor eIF-2 activity[14–16] and the phosphorylation of two ribosome-associated proteins, the 38,000-dalton α-subunit of eIF-2 and a 64,000–69,000-dalton protein designated P_1.[17–19] The phosphorylation of eIF-2α and P_1 correlates with the appearance of the ribosome-associated inhibitor of translation,[9,18] is dependent upon the duration of interferon treatment,[9,18,20] and shows a dose dependence similar to that for the development of the antiviral state.[20] The inhibition at the apparent level of elongation appears to be due, in part, either to the nucleolytic degradation of the viral mRNA[21,22] or to a deficiency of certain minor tRNA species[23,24] or to both.

Procedures are described below for the isolation of translation-inhibitory ribosome-associated proteins from interferon-treated cells and for the assay of these proteins in cell-free protein synthesizing systems prepared from Krebs ascites tumor cells.

Isolation of Interferon-Mediated Inhibitors of Translation

Growth, Interferon Treatment, and Harvest of Cells. Clone L-929 mouse fibroblast cells are routinely grown in roller culture (1330-cm^2 growth area

[9] C. E. Samuel, D. A. Farris, and D. A. Epstein, *Virology* 83, 56 (1977).
[10] C. E. Samuel and D. A. Farris, *Virology* 77, 556 (1977).
[11] D. A. Epstein, T. C. Peterson, and C. E. Samuel, *Virology* 98, 9 (1979).
[12] W. K. Joklik and T. C. Merigan, *Proc. Natl. Acad. Sci. U.S.A.* 56, 558 (1966).
[13] C. E. Samuel, *Virology* 75, 166 (1976).
[14] K. Ohtsuki, F. Dianzani, and S. Baron, *Nature (London)* 269, 536 (1977).
[15] J. A. Cooper and P. J. Farrell, *Biochem. Biophys. Res. Commun.* 77, 124 (1977).
[16] R. Kaempfer, R. Israeli, H. Rosen, S. Knoller, A. Zilberstein, A. Schmidt, and M. Revel, *Virology* 99, 170 (1979).
[17] C. E. Samuel, *Proc. Natl. Acad. Sci. U.S.A.* 76, 600 (1979).
[18] C. E. Samuel, *Virology* 93, 281 (1979).
[19] C. E. Samuel, this volume [22].
[20] A. Kimchi, L. Shulman, A. Schmidt, Y. Chernavorsky, A. Fradin, and M. Revel, *Proc. Natl. Acad. Sci. U.S.A.* 76, 3208 (1979).
[21] G. E. Brown, B. Lebleu, M. Karvgkita, S. Sharla, G. C. Sen, and P. Lengyel, *Biochem. Biophys. Res. Commun.* 69, 114 (1976).
[22] D. A. Epstein and C. E. Samuel, *Virology* 89, 240 (1978).
[23] A. Zilberstein, B. Dudock, H. Berissi, and M. Revel, *J. Mol. Biol.* 108, 43 (1976).
[24] R. Falcoff, E. Falcoff, J. Sanceau, and J. A. Lewis, *Virology* 86, 507 (1978).

per glass bottle) in Eagle's MEM containing 7% fetal calf serum or, more recently, 5.6% newborn calf serum and 1.4% fetal calf serum. Confluent cells are treated at 37° with 300 units of interferon per milliliter and harvested without trypsinization by scraping into cold (0–4°) isotonic buffer (35 mM Tris · HCl, pH 6.8 at 25°; 146 mM NaCl; 11 mM glucose).[19] Cells are then washed three times in isotonic buffer and, after the last wash, are suspended in a small volume of isotonic buffer, transferred to a graduated conical centrifuge tube, and pelleted at 800 g for 10 min in an IEC CRU-5000 refrigerated centrifuge. The volume of the cell pellet is recorded, and then as much buffer as possible is carefully removed with a pipette.

Krebs ascites tumor cells are maintained *in vivo* by injection of 0.15 ml of ascitic fluid (about 1 to 2 × 10^7 cells) into the peritoneal cavity of female albino mice; the cells are passed every 7–9 days. Krebs cells in the ascitic form are treated with 1 × 10^4 units of interferon by injection into the peritoneal cavity 6–7 days after cell passage. About 8 hr later, an identical second dose of 1 × 10^4 units of interferon is administered, and cells are collected 12–18 hr later. Control mice are treated with a volume of phosphate buffer equal to that of the interferon administered. Liquid tumors are collected from the mice, filtered through cheesecloth into cold isotonic buffer, and washed five times at 4° by differential centrifugation at 65 g for 5 min in isotonic buffer. The packed-cell volume is then estimated as described above for L-929 cells.

Preparation of Ribosome and Cell Sap Fractions. Ribosome and cell sap fractions are prepared from nonpreincubated S$_{10}$ cell-free extracts as described in this volume.[19] The cell sap fractions are passed through a column of Sephadex G-25 before storage at −90°; ribosomal salt-wash fractions are obtained by washing ribosomes with buffer containing KCl concentrations as indicated in the figure legends and text and are dialyzed before storage at −90°; preincubated, salt-washed ribosomes are obtained by washing crude ribosomes with buffer containing 1 M KCl, preincubation, and then recovery by centrifugation.[8]

Preparation of Cell-Free Protein Synthesizing System

Cell-free protein synthesizing extracts of Krebs cells are prepared by a method based on that described by Mathews and Korner.[25] All operations, unless otherwise stated, are carried out at 4°. Cell-free S$_{10}$ extracts are prepared as described in detail elsewhere,[19] and then preincubated at 37° and passed through a column of Sephadex G-25. For the preincubation step, the S$_{10}$ extract is supplemented with ATP (Sigma or P-L Biochemicals) to a final concentration of 1 mM, GTP (P-L Biochemicals) to 0.2

[25] M. B. Mathews and A. Korner, *Eur. J. Biochem.* 17, 328 (1970).

mM, creatine phosphate (Sigma) to 10 mM, and creatine phosphokinase (Sigma) to 200 μg/ml, incubated at 37° for 40 min, and then centrifuged at 10,000 g for 10 min in a Sorvall RC-5 centrifuge and SS34 rotor. The resulting supernatant solution is passed through a column of Sephadex G-25 (Pharmacia, medium mesh, 2.5 × 20 cm column for 25 ml extract) equilibrated with 20 mM HEPES buffer, pH 7.5, containing 120 mM KCl, 5 mM Mg(OAc)$_2$, and 1 mM DTT; the most concentrated fractions are pooled and stored in 0.2- to 0.5-ml aliquots at −90°. The protein concentrations of the preincubated, Sephadex G-25-treated S_{10} extracts routinely range from 8 to 12 mg/ml.

Preparation of Messenger RNA

Reovirus. The Dearing strain of reovirus type 3 is grown in mouse L fibroblast cells in suspension culture and purified by the procedure of Smith *et al.*[26] modified to include final dialysis of the purified virus preparation against 0.05 M Tris · HCl, pH 8.0, and 0.1 M KCl.[27] Reovirus mRNA is then prepared as the *in vitro* transcription product of the core-associated double-stranded RNA-dependent single-stranded RNA polymerase by the method of Skehel and Joklik[28] except that the reaction mixture was modified to contain 10 μM S-adenosyl L-methionine. Messenger RNA synthesized *in vitro* is purified from the reaction mixture as described by Samuel and Joklik.[8]

Vaccinia. The WR strain of vaccinia virus is grown as described by Becker and Joklik[29] and purified as described by Joklik.[30] The preparation of vaccinia virus cores and the synthesis *in vitro* of vaccinia virus mRNA are as described by Kates and Beeson[31]; mRNA synthesized *in vitro* is purified from the reaction mixture as described for reovirus mRNA.[8]

Assay of Interferon-Mediated Inhibitors of Translation

Interferon-mediated inhibitors of translation are measured by determining the ability of subcellular fractions prepared from interferon-treated as compared to untreated cells to affect the translation of viral mRNA catalyzed by the mouse ascites cell-free protein synthesizing system prepared from untreated Krebs cells.

[26] R. E. Smith, H. J. Zweerink, and W. K. Joklik, *Virology* 39, 719 (1969).
[27] C. E. Samuel and W. K. Joklik, *Virology* 74, 403 (1976).
[28] J. J. Skehel and W. K. Joklik, *Virology* 39, 822 (1969).
[29] Y. Becker and W. K. Joklik, *Proc. Natl. Acad. Sci. U.S.A.* 51, 577 (1964).
[30] W. K. Joklik, *Biochim. Biophys. Acta* 61, 290 (1962).
[31] J. Kates and J. Beeson, *J. Mol. Biol.* 50, 1 (1976).

Reagents. All chemicals used are of reagent grade; solutions are prepared with deionized, glass-distilled water.

N-2-Hydroxyethylpiperazine-N'-2-ethanesulfonic acid (HEPES), 0.5
 M, pH 7.5
KCl, 3.0 *M*
Mg(OAc)$_2$, 0.1 *M*
Dithiothreitol (DTT), 0.1 *M* (Calbiochem or Sigma)
ATP, 0.1 *M* neutralized with KOH (Sigma or P-L Biochemicals)
GTP, 0.02 *M* neutralized with KOH (P-L Biochemicals)
CTP, 0.06 *M* neutralized with KOH (Sigma)
Creatine phosphate, 0.5 *M* neutralized with KOH (Sigma)
Creatine phosphokinase (Sigma)
L-Amino acids, 19 except L-leucine, 2 m*M* each neutralized with
 KOH (Sigma)
L-[4,5-^{3}H$_2$]Leucine, 0.5 mCi/ml, 50–60 Ci/mmol (Schwarz/Mann or
 New England Nuclear)
Trichloroacetic acid, 50% w/v
Ethanol
Ether

Reaction Mixture. In vitro protein synthesis is measured by determining the incorporation of [^{3}H]leucine into hot trichloroacetic acid-insoluble material. The standard reaction mixture (50 μl) contains 30 m*M* HEPES (pH 7.5), 120 m*M* KCl, 3 m*M* Mg(OAc)$_2$, 1 m*M* ATP, 0.2 m*M* GTP, 0.6 m*M* CTP, 1 m*M* DTT, 10 m*M* creatine phosphate, 0.3 mg creatine phosphokinase per milliliter, 50 μ*M* each of the unlabeled L-amino acids, 50 μCi of L-[^{3}H]leucine per milliliter, 0.3 ml of ascites extract per milliliter of reaction mixture, and, as indicated, either no exogenously added mRNA or 100 μg of reovirus mRNA or 50 μg of vaccinia mRNA per milliliter and varying amounts of ribosomal salt-wash fraction. The reaction mixture is incubated at 25° for the amount of time indicated. Aliquots (usually 10 μl) are then removed and adsorbed into filter paper discs (25 mm from Schleicher and Schuell), which are dropped into cold 10% trichloroacetic acid and subsequently treated with 5% trichloroacetic acid at 90° for 15 min followed by two washes in 5% trichloroacetic acid at 4°, one wash in ethanol : ether (1 : 1) and one wash in ether.[32] The dried filter disks are then placed into toluene containing 2,5-diphenyloxazole (0.4%) and 1,4-bis[2-(5-phenyloxazolyl)benzene] (0.01%), and the radioactivity on them is measured in a Beckman LS-230 liquid scintillation spectrometer for 10 min or for periods of time sufficient to give standard errors of counting rates of less than 1.0%.

---

[32] H. Noll, *in* "Techniques in Protein Biosynthesis" (P. N. Campbell and J. R. Sargent, eds.), Vol. 2, p. 171. Academic Press, New York, 1969.

EFFECT OF INTERFERON TREATMENT OF THE ABILITY OF SALT-WASHED RIBOSOMES, RIBOSOMAL SALT-WASH, AND CELL SAP FRACTIONS TO TRANSLATE REOVIRUS AND VACCINIA VIRUS mRNAs[a,b]

| Origin of | | | [3H]Leucine incorporated in response to | | | |
| Salt-washed ribosomes (cells) | Riboscmal salt-wash fraction (cells) | Cell sap fraction (cells) | Reovirus mRNA | | Vaccinia mRNA | |
			cpm	%	cpm	%
Untreated	Untreated	Untreated	19,346	100	10,178	100
Interferon-treated	Untreated	Untreated	16,841	87	7,649	75
Untreated	Interferon-treated	Untreated	5,126	27	1,704	17
Untreated	Untreated	Interferon-treated	17,765	92	6,489	64
Interferon-treated	Interferon-treated	Interferon-treated	5,857	30	1,073	11
Interferon-treated	Interferon-treated	Untreated	13,139	68	5,170	51

[a] From Samuel and Joklik.[8]

[b] Reaction mixtures contained in a total volume cf 100 μl: 3.6 A_{260} units of salt-washed preincubated ribosomes, 50 μg of 0.12–1.0 M KCl ribosomal salt-wash protein, and 200 μg of cell sap protein. The concentrations of reovirus and vaccinia mRNA were 65 and 35 μg/ml, respectively.

General Comments. The optimum concentrations of KCl and Mg(OAc)$_2$ vary slightly between extracts and mRNAs and must therefore be determined for each preparation. Preincubation of ascites S$_{10}$ fractions at 37° for 40 min reduces the endogenous activity by more than 90%. [^{3}H]Leucine incorporation in the standard preincubated ascites cell-free protein-synthesizing system is routinely stimulated 7- to 10-fold by exogenously added reovirus mRNA above the endogenous level. Endogenous activity can also be reduced by micrococcal nuclease treatment; however, EGTA and pTp-treated micrococcal nuclease-treated cell-free systems appear to be responsive to interferon-mediated inhibition of translation only at the level of protein phosphorylation, not at the level of (2-5)-oligoadenylate-mediated mRNA degradation.[33]

Concluding Comments

Treatment of uninfected mouse ascites tumor cells or L-929 fibroblast cells with interferon greatly decreases the ability of S$_{10}$ fractions prepared from them to translate viral message, but has no significant effect on their ability to translate endogenous message or synthetic messages such as poly(U). As shown in the table the inability of extracts prepared from interferon-treated cells to translate viral mRNA may be traced to the ribosomal salt-wash fraction. When S$_{10}$ fractions of interferon-treated and control cells are fractionated into salt-washed ribosomes, factors dissociated from ribosomes by washing with 1 M KCl, and cell-sap fractions, the capacity of heterologous combinations of these fractions to translate reovirus and vaccinia virus mRNA depends predominantly on whether the ribosomal salt-wash source is from untreated or interferon-treated cells.

The inability of ribosomal salt-wash fractions from interferon-treated cells to support the translation of viral mRNA is not due to the absence of factor(s) that are normally present in untreated cells, but rather to the presence of additional factor(s) in interferon-treated cells that are inhibitory. The presence of these inhibitory factors depends upon the duration and dose of interferon treatment. The effect on the translation of reovirus mRNA catalyzed by a cell-free protein synthesizing system prepared from untreated ascites cells of varying amounts of ribosomal salt-wash fractions prepared from ribosomes isolated from untreated L-929 cells and L-929 cells treated with interferon for periods of 2, 6, 12, and 18 hr is shown in Fig. 1. Ribosomal salt-wash fractions prepared from cells treated for 12 or 18 hr significantly inhibit translation; translation is slightly stimulated by ribosomal salt-wash fractions prepared from untreated cells and

[33] N. G. Miyamoto and C. E. Samuel, *Virology* **107,** 461 (1980).

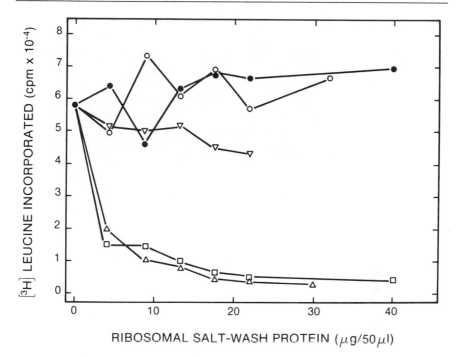

FIG. 1. Effect of duration of interferon treatment on the expression of the interferon-mediated translation-inhibitory ribosome-associated proteins. Mouse L-929 cells were treated for varying periods of time as indicated, and 0.3–0.8 M KCl ribosomal salt-wash fractions were prepared and assayed for their ability to affect reovirus mRNA translation catalyzed by the mouse ascites cell-free protein-synthesizing system prepared from untreated Krebs cells. Ribosomal salt-wash from untreated cells (●——●) and from cells treated with 300 units of interferon per milliliter for 2 hr (○——○), 6 hr (▽——▽), 12 hr (△——△), and 18 hr (□——□). From Samuel et al.[9]

cells treated for 2 hr, and slightly inhibited by salt-wash fractions prepared from cells treated for 6 hr.

The polypeptide complement ascertained by electrophoresis in SDS-polyacrylamide gels of ribosomal salt-wash fractions obtained by washing ribosomes from untreated and interferon-treated cells successively in buffers containing 0.05, 0.07, 0.17, 0.30, 0.60, and 1.0 M KCl is shown in Fig. 2; each ribosomal wash fraction was also assayed for its ability to inhibit the translation of reovirus mRNA. A polypeptide with a mobility between that of reovirion μ and σ and marker polypeptides is clearly detectable in the 0.3–0.6 M KCl and the 0.6–0.8 M KCl salt-wash fractions from interferon-treated cells; this polypeptide is not discernible in the anlogous salt-washes of ribosomes from untreated cells (arrow). Salt-wash fractions

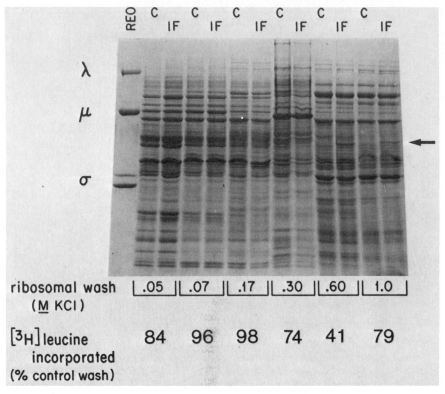

FIG. 2. Sodium dodecyl sulfate (SDS)–polyacrylamide gel patterns of ribosomal salt-wash fractions prepared from ribosomes of untreated (C) and interferon-treated (IF) ascites cells. Ribosomes were washed with buffer that contained increasing concentrations of KCl, ranging from $0.05\,M$ to $1.0\,M$, as indicated; the ribosomes were washed successively at the indicated salt concentrations for 30 min with stirring at 4°, then isolated by centrifuging for 1.5 hr at $150,000\,g$; each supernatant was then recentrifuged at $150,000\,g$ for 4.5 hr before dialysis. The arrow indicates the position of the ribosome-associate polypeptide, which is greatly enhanced by interferon treatment. REO, reovirion (λ, μ, σ) marker proteins.

The effect of the ribosomal salt-wash fractions from C and IF cells on the incorporation of [³H]leucine in response to reovirus mRNA (100 μg/ml) was determined in reaction mixtures that contained in a total volume of 100 μl: 3.6 A_{260} units of salt-washed preincubated ribosomes, 50 μg of 1 M KCl ribosomal salt-wash, and 200 μg of cell sap fraction, all prepared from untreated cells. In addition, comparable amounts of each of the respective wash fractions from untreated or interferon-treated cells were added (these were samples from the same ribosomal salt-wash fraction as were electrophoresed). Results are expressed as the percentage of incorporation obtained with control wash for each pair (C, IF) of washes. Adapted from Samuel and Joklik.[8]

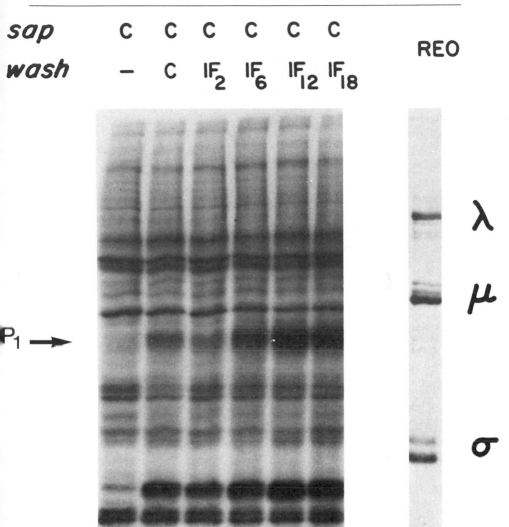

FIG. 3. Effect of duration of interferon treatment on the interferon-mediated phosphorylation of ribosome-associated proteins. Phosphorylation reaction conditions were as described elsewhere in this volume[19] and contained, where indicated, cell sap fraction (sap) (69 μg of protein) prepared from untreated cells (C) and 0.3–0.8 M ribosomal salt-wash fraction (wash) (8.8 μg of protein) prepared from untreated cells (C) or from cells treated with interferon (IF) for 2, 6, 12, or 18 hr. Reovirus dsRNA was present in all phosphorylation reactions. ^{32}P-labeled products were analyzed by sodium dodecyl sulfate (SDS)–polyacrylamide gel electrophoresis and autoradiography. The arrow indicates the position of the interferon-dependent P_1 protein phosphorylation that correlates with the appearance of the ribosome-associated inhibitor of translation. REO, ^{14}C-labeled reovirion (λ, μ, σ) marker proteins. From Samuel et al.[9]

from interferon-treated ribosomes that contain the largest amounts of the additional polypeptide component as judged by SDS–polyacrylamide gel electrophoresis (the 0.3–0.6 M KCl wash fractions) also possess the greatest ability to inhibit the translation of reovirus mRNA (Fig. 2).

Phosphorylation of ribosome-associated protein P_1 is one of the most sensitive biochemical markers of interferon action in murine and human cells.[19] The effect of duration of interferon treatment on the phosphorylation of P_1 present in ribosomal salt-wash fractions prepared from untreated L-929 cells and L-929 cells treated with interferon for 2, 6, 12, and 18 hr is shown in Fig. 3. Protein P_1, with a mobility between that of reovirion μ and σ marker polypeptides, is maximally enhanced after 12 hr of interferon treatment. The correlation between the duration of interferon treatment required for maximal expression of the ribosome-associated inhibitor of translation (Fig. 1) and for phosphorylation of the ribosome-associated protein P_1 (Fig. 3) is possibly indicative of a relationship between phosphorylation and inhibition of translation.

[38] Assay of Glycosyltransferase Activities in Microsomal Preparation from Cells Treated with Interferon

By R. K. Maheshwari, I. K. Vijay, Kenneth Olden, and Robert M. Friedman

There is a 30- to 200-fold reduction[1] in the production of infectious vesicular stomatitis virus (VSV) released from L cells treated with 10–30 reference units/ml of interferon (IF); however, in these cultures virus particle production, as measured by VSV particle-associated viral RNA, virus nucleocapsid protein, and viral transcriptase was inhibited by less than 10-fold. In addition, there was a significant reduction in glycoprotein (G) and membrane protein (M) of VSV particles released from IF-treated cells.[2] Evidence supporting the deficiency of G protein in VSV released from IF-treated cells was derived from electron microscopic studies. Under conditions where glycoprotein spikes were clearly detectable on the surface of VSV released from cells not treated with IF, very few spikes were observed on VSV released from IF-treated cells.[3] All these results suggested that IF-treated cells produced VSV particles with low

[1] R. K. Maheshwari and R. M. Friedman, J. Gen. Virol. 44, 261 (1979).

[2] R. K. Maheshwari, F. T. Jay, and R. M. Friedman, Science 207, 540 (1980).

[3] R. K. Maheshwari, A. E. Demsey, S. B. Mahanty, and R. M. Friedman, Proc. Natl. Acad. Sci. U.S.A. 77, 2284 (1980).

infectivity and that this low infectivity may be releated to the reduced amount of G and M protein incorporated into such particles.

It is known that G protein of VSV contains two "complex" type asparagine-linked oligosaccharide chains.[4] The role of polyisoprenoid glycolipid intermediates in the biosynthesis of N-glycosidically bound oligosaccharides is well established.[5-8] Tunicamycin (TM) is a potent inhibitor of protein glycosylation produced by *Streptomyces lysosuperficus*.[9] *In vitro* studies[10] from a number of laboratories clearly demonstrated that TM blocked only the first step in the lipid-linked pathway of eukaryotes, the synthesis of *N*-acetylglucosaminylpyrophosphoryldolichol from UDP-*N*-acetylglucosamine and dolichylphosphate. Subsequent steps in the pathway and the final transfer of an oligosaccharide to protein were not affected. *In vitro* studies with different systems have documented the fact that TM prevented the glycosylation of proteins containing N-glycosidically linked carbohydrate. That TM blocked glycosylation *in vivo* by a mechanism identical to that demonstrated *in vitro* has been provided by studies with hen oviduct tissue slices.[11] The antiviral activity of TM has also long been known; its effects seem to depend on the virus strain and cell type used. In BHK cells, TM caused a marked decrease in the yield of infectious VSV,[12] but this effect was not due to lowered specific infectivity of virions containing unglycosylated viral glycoproteins; rather it was related to a block in virion assembly.[13] In a different system, however, Nakamura and Compans[14] have shown that TM had relatively little effect on the production of influenza virus particles; however, the virions produced in the presence of TM lacked the spiked appearance of normal virus particles. In a related study, TM prevented the formation of Semliki Forest and fowl plague viral particles.[15] It has further been shown in influenza and VSV that polarity in the maturation sites was maintained in Madin–Darby canine kidney (MDCK) cell monolayers treated with TM.[16] Such results taken together indicated that glycosyla-

[4] J. R. Etchison, J. S. Robertson, and D. F. Summers, *Virology* **78**, 375 (1977).

[5] J. J. Lucas and C. J. Waechter, *Mol. Cell. Biochem.* **11**, 67 (1976).

[6] C. J. Waechter and W. J. Lennarz, *Annu. Rev. Biochem.* **45**, 95 (1976).

[7] F. W. Hemming, *Biochem. Soc. Trans.* **5**, 1223 (1977).

[8] A. J. Parodi and L. F. Leloir, *Biochim. Biophys. Acta* **559**, 1 (1979).

[9] A. Takatsuki, K. Arima, and G. Tamura, *J. Antibiot.* **24**, 215 (1971).

[10] D. Struck and W. Lennarz, *in* "The Biochemistry of Glycoproteins and Proteoglycans (W. J. Lennarz, ed.), pp. 35–73. Plenum, New York, 1980.

[11] D. K. Struck and W. J. Lennarz, *J. Biol. Chem.* **252**, 1007 (1977).

[12] R. Leavitt, S. Schlesinger, and S. Kornfeld, *J. Virol.* **21**, 375 (1977).

[13] I. Tabas, S. Schlesinger, and S. Kornfeld, *J. Biol. Chem.* **253**, 716 (1978).

[14] K. Nakamura and R. W. Compans, *Virology* **84**, 303 (1978).

[15] R. T. Schwarz, J. M. Rohrschneider, and M. F. G. Schmidt, *J. Virol.* **19**, 782 (1976).

[16] M. G. Roth, J. P. Fitzpatrick, and R. W. Compans, *Proc. Natl. Acad. Sci. U.S.A.* **76**, 6430 (1979).

tion of viral glycoprotein was not required for the production of these enveloped virus particles, but that deficient glycosylation appeared to result in noninfectious particles lacking in glycoprotein.

In vitro studies have shown that the synthesis of *N*-acetylglucosaminyl lipids with properties of dolichol derivatives that serve as precursors for the dolichol-bound oligosaccharide lipid intermediates in protein glycosylation was decreased in membrane preparations from IF-treated mouse L or human foreskin cells compared to similar membrane preparations from untreated cells. This effect could be partially responsible for the results that we observed with VSV released from L cells treated with IF.[1-3] Such virions were selectively deficient in G and M proteins. It might also explain observations on Moloney murine leukemia virus (MMLV) produced by IF-treated TB cells. This system yielded MMTV with markedly decreased infectivity.[17] The MMLV virions were also deficient in the viral envelope glycoproteins gp 69/71. Therefore, the decreased amount of glycoprotein in the virions may be directly related to a reduced rate of synthesis of lipid-linked oligosaccharides at an early step during biosynthesis of asparagine-linked glycoprotein in IF-treated cells. Such effects on glycosylation in IF-treated cells may also be pertinent to the varied activities of interferons on cell growth and on the immune system.

Experimental Procedures

Materials. UDP-*N*-Acetyl-[3H]glucosamine (sp. act. 6.6 mCi/mol), GDP-[14C]mannose (sp. act. 210 μCi/μmol) and UDP-[3H]glucose (sp. act. 3 mCi/μmol) were purchased from New England Nuclear. Unlabeled sugar-nucleotides and dithiothreitol (DTT) were the products of Sigma. All other reagents were from commercial sources and were of the highest purity available. The radioactive sugar-nucleotides were diluted with unlabeled substrates to yield the desired specific radioactivity in the incubations.

Preparation of Cell Membranes

Mouse L cell or human foreskin fibroblast cells were grown in 150-cm^2 flasks to confluency (1 $\times$ 10^7 cells) and were treated with mouse and human interferon, respectively, for 18 hr. All monolayers were washed twice with saline, and cells were scraped and centrifuged at 800 g for 10 min. The cell pellets were suspended in four volumes of ice-cold buffer consisting of 0.1 M Tris · HCl, pH 7.0, 0.25 M sucrose, and 1 mM EDTA. The cells were homogenized by 12 strokes of a Dounce homogenizer (glass) with pestle B (large clearance) followed by 12 strokes with pestle A

[17] R. M. Friedman, R. K. Maheshwari, F. T. Jay, and C. Czarniecki, *Ann. N. Y. Acad. Sci.* **350**, 533 (1980).

(small clearance). This procedure was followed by 8 strokes with a Ten Broeck homogenizer (Corning).

The crude homogenates were centrifuged at 9000 g for 15 min. The resulting supernatant solutions were centrifuged at 39,000 g for 30 min. The final crude particulate fractions were suspended in the homogenization buffer and immediately frozen in Dry Ice and kept frozen at $-70°$ until used (usually overnight). Freezing did not appear significantly to affect the glycosyltransferase activities.

The frozen membranes were thawed at room temperature and dispersed in about 100 volumes of ice-cold wash buffer (50 mM Tris · HCl, 5 mM 2-merceptoethanol, and 0.5 mM EDTA, pH 7.6). After centrifugation at 48,000 g for 30 min, the particulate preparations were resuspended in the same buffer to a protein concentration of approximately 5 mg/ml. The objective of washing the membranes was to minimize the levels of endogenous sugar-nucleotides, which would otherwise reduce the specific activity of the substrate and lower the level of radioactivity incorporated into products. The protein was determined by the Lowry procedure[18] with bovine serum albumin as standard.

Assay of Glycosyltransferase Activities

To study the incorporation of N-acetylglucosamine into glycolipids,[19] incubations contained the following components in a total volume of 0.1 ml: 50 mM Tris · HCl, pH 7.6, 10 mM DTT, 10 mM MgCl$_2$, 5 μM UDP-N-acetyl-[^{3}H]glucosamine, specific activity 1–6 mCi/μmol, and approximately 0.2–0.3 mg of membrane protein. After incubation at 37° for 10 min, incubations were quenched with 3 ml of CHCl$_3$/CH$_3$OH (2 : 1) and vortexed immediately; 50 μl (approximately 1 mg of protein) of a microsomal preparation from the lactating bovine mammary tissue[20] was added to serve as carrier membrane, and the glycolipids were isolated by the multiple extraction procedure. With a washed membrane preparation, N-acetylglucosamine is primarily incorporated into glycolipids soluble in CHCl$_2$/CH$_3$OH (2 : 1), and there is little, if any, incorporation of the label into the CHCl$_3$/CH$_3$OH/H$_2$O (10 : 10 : 3) soluble lipid-linked oligosaccharides. As most of the asparagine-linked glycoproteins of both the high mannose and the complex type[21] have a number of mannose residues on the nonreducing end of the chitobiosyl unit in their saccharide structures, it would appear that most of the endogenous GDP-mannose is removed by washing.

[18] O. H. Lowry, N. J. Rosebrough, A. L. Farr, and R. J. Randall, *J. Biol. Chem.* **193**, 265 (1951).

[19] W. W. Chen and W. J. Lennarz, *J. Biol. Chem.* **252**, 3473 (1977).

[20] I. K. Vijay, and S. R. Fram, *J. Supramol. Struct.* **7**, 251 (1977).

[21] R. Kornfeld and S. Kornfeld, *in* "The Biochemistry of Glycoproteins and Proteoglycans" (W. J. Lennarz, ed.), pp. 1–34, Plenum, New York, 1980.

The transfer of mannose from GDP-[^{14}C]mannose into glycolipids soluble in $CHCl_3/CH_3OH$ (2:1) and $CHCl_3/CH_3OH/H_2O$ (10:10:3) was studied in standard incubations containing 50 mM Tris · HCl, pH 7.2, 10 mM DTT, 10 mM MnCl$_2$, 0.05 mM UDP-N-acetylglucosamine, 5 μM GDP-[^{14}C]mannose (sp. act. 210 μCi/μmol), and approximately 0.2–0.3 mg of membrane protein, in a total volume of 0.1 ml.[22] After incubation at 37° for 14 min, incubations were quenched with 3 ml of $CHCl_3/CH_3OH$ (2:1) and processed as above for the transfer of N-acetylglucosamine into different glycolipids.

For the biosynthesis of glucosyl lipids, incubations contained 50 mM Tris · HCl (pH 7.4), 10 mM DTT, 10 mM MgCl$_2$, 0.05 mM each of UDP-N-acetylglucosamine and GDP-mannose, and 5 μM UDP-[^{3}H]-glucose (sp. act. 1 mCi/μmol) and 0.2–0.3 mg of the membrane protein in a total volume of 0.1 ml.[23] After incubation at 37° for 7.5 min, incubations were quenched with 3 ml of $CHCl_3/CH_3OH$ (2:1) and processed as given above for the incorporation of N-acetylglucosamine and mannose into different glycolipids.

For both products soluble in $CHCl_3/CH_3OH$ (2:1) and $CHCl_3/CH_3OH/H_2O$ (10:10:3), one-tenth of the extracted glycolipid fraction was evaporated under N_2 and counted in 3 ml of ScintiVerse (Fisher Scientific Company). After evaporation under N_2, the remainder of the extract was subjected to mild acid hydrolysis with 0.1 N HCl at 90° for 15 min. This treatment selectively cleaves the carbohydrate-lipid linkage to release neutral oligosaccharides. The hydrolyzates were freeze-dried and subsequently partitioned between 2 ml each of chloroform and water. The radioactivity was quantitatively recovered in the aqueous phase. The lower chloroform layer was extracted one more time with 2 ml of water. The combined aqueous phase was passed through a short (0.5 × 3 cm) column of mixed-bed resin, AG 501, and eluted with 4 ml of water. The column eluate was freeze-dried and subjected to paper chromatography on Whatman No. 1 paper, developed in ethyl acetate/pyridine/acetic acid/water, 5:5:1:3, for 24 hr to separate the saccharides obtained from glycolipids soluble in $CHCl_3/CH_3OH$ (2:1). Under these conditions, mono-, di-, tri-, tetra-, and pentasaccharides are well resolved. To separate the higher saccharides obtained from glycolipids soluble in $CHCl_3/CH_3OH/H_2O$ (10:10:3), paper chromatograms were developed in butanol/pyridine/water (4:3:4), for 4 days. One-centimeter segments of the paper chromatogram were cut out, soaked in 0.3 ml of water in scintillation vials, and counted after the addition of 3 ml of ScintiVerse.

[22] J. J. Lucas, C. J. Waechter, and W. J. Lennarz, *J. Biol. Chem.* **250**, 1992 (1975).
[23] P. W. Robbins, S. S. Krag, and T. Tiu, *J. Biol. Chem.* **252**, 1780 (1977).

[39] Procedures for Studying Transcription and Translation of Viral and Host Nucleic Acids in Interferon-Treated Cells

By WOLFGANG K. JOKLIK

The purpose of this chapter is to describe how to study the effect of treating cells with interferon on the rate at which virus- and host-coded RNA species are transcribed and translated within them. The basic experimental design involves pulse-labeling of cells, infected or not infected, untreated or treated with interferon, with RNA and protein precursors, and then measurement of the relative amounts of RNA transcripts or proteins that are formed as a function of either interferon concentration, time after infection, or whatever other variable is under investigation. The problem thus resolves itself into a description of the techniques used to quantitate, in relative terms, specific species of RNA transcripts and proteins.

A. Quantitation of Viral RNA Transcripts

There are two basic techniques for quantitating viral RNA transcripts in infected cells. The first, and most specific, involves measurement of the amount of RNA that can be hybridized to a large excess of complementary DNA or RNA. The second, and simpler, takes advantage of the fact that conditions can sometimes be arranged so that the newly formed viral RNA is the only newly formed RNA of its size that is present in the cytoplasm of infected cells; in such cases one merely prepares cytoplasmic cell fractions, subjects them to sucrose or glycerol density gradient centrifugation, and measures the amount of radioactive label in bands known to contain only virus-specified RNA. The particular strategy that is followed and the nature of the hybridization probe that one uses differ from virus to virus. I will therefore describe separately the techniques applicable to members of the more commonly investigated virus families. I will not address variables such as choice of cell, multiplicity of infection, nature of interferon used, interferon concentration and duration of interferon treatment, or when, after infection, the measurement is made, since none of these variables affects how the viral transcripts or proteins are best measured.

1. Isolation of Viral RNA Transcripts from Infected Cells

Viral RNA transcripts may be isolated from isolated whole cells, or from isolated nuclei, or from the cytoplasmic cell fraction. RNA isolated from nuclei represents mainly primary transcripts of viral genomes that replicate in the nucleus; RNA isolated from the cytoplasm represents mainly messenger RNA (mRNA) associated with polyribosomes. Thus the former represents unprocessed transcripts; the latter, processed transcripts. Quantitation of the former is therefore a measure of the kinetics of transcription, whereas comparison of the relative rates at which primary transcripts are formed in the nucleus and at which mRNA appears in the cytoplasm provides a measure of transcript processing and migration of processed transcripts from the nucleus to the cytoplasm. Finally, quantitation of virus-specified RNA extracted from whole cells provides an overall estimate of total viral transcripts formed in infected cells.

a. *Isolation of Viral RNA Transcripts from Whole Cells.* The cells will have been grown either in monolayer or in spinner culture; the method to be described refers to the cells in one confluent 100-mm in diameter petri dish (10^7 cells) or to the cells in 10–20 ml of spinner culture (also 10^7 cells). The cells may be labeled or unlabeled; the labeling, unless specified otherwise, will have been with 50–200 μCi of [^{3}H]uridine (about 20 Ci/mmol), for 15–30 min. The cells in monolayer culture are then rinsed with ice-cold solution A[1] containing 2 mM Ca^{2+}; 2 ml of ice-cold extraction buffer are then added, and the plates are swirled until the cells have lysed.[1a] The lysate is vortexed and an equal volume of phenol (88% Mallinckrodt, no preservative added), prewarmed to 70° and saturated with extraction buffer lacking sodium dodecyl sulfate (SDS), is added; the mixture is vortexed again and stored at 0° until all samples have been collected. As for spinner culture cells, after the labeling period cells are added to an equal volume of crushed frozen solution A containing 2 mM Ca^{2+} and are centrifuged; the cell pellet is washed with solution A and suspended in 4 ml of distilled water, and 0.2 volume of 5× concentrated extraction buffer is added. The cells are then vortexed, an equal volume of prewarmed phenol saturated with extraction buffer but lacking SDS is added, and the mixture is vortexed. The lysed cells are stored at 0° until all samples have been collected.

The RNA is then phenol-extracted at 55° by vigorous shaking in a heated water bath for 5 min. The aqueous and phenol phases are separated by centrifugation, and the phenol phase and interface are reextracted by vortexing with 0.5 volume of ice-cold extraction buffer; this second aqueous phase is combined with the first. The combined aqueous phases are then extracted with 0.5 volume of phenol–chloroform–isoamyl alcohol (49.5 : 49.5 : 1), and the final aqueous phase is precipitated with 2.5 volume of 95% ethanol containing 0.15 M sodium acetate. The precipitated RNA is stored overnight at $-20°$.

[1] The composition of buffers and solutions is given in Section D at the end of this chapter.
[1a] F. Wolf and D. Schlessinger, *Biochemistry* **16**, 2783 (1977).

The RNA samples are then resuspended in 1 ml of extraction buffer, and phenol is extracted as described above, except that all steps are done at 0°. These additional extractions after the first EtOH precipitation are not essential but should be performed if it is necessary to remove traces of protein, which may cause aggregation of the RNA; this in turn may interfere with subsequent analysis. It should also be noted that the use of 0.5% SDS in the extraction buffer is optimal for inhibiting nucleases; higher levels of detergent lead to extraction of much more DNA. The final aqueous phase is again precipitated with 2.5 volume of EtOH containing 0.15 M sodium acetate and stored at $-20°$.

b. Isolation of Viral RNA Transcripts from Nuclei. Nuclei are isolated as follows[2]: Cells are collected by centrifugation or by scraping of monolayers, as the case may be, washed once with cold Eagle's minimal essential medium, and suspended at the rate of 10^6 cells/ml in cold isotonic lysis buffer. After 15 min with occasional gentle agitation by drawing through a short-tipped Pasteur pipette, the mixture is centrifuged at 900 g for 5 min. The pelleted nuclei are resuspended at the same rate in cold lysis buffer containing 1% Triton N-101 and 5% sodium deoxycholate, and gently agitated with a Pasteur pipette. This treatment removes the cell membrane and most visible cytoplasmic tags. The suspension is then layered over 1.5 ml of 0.3 M sucrose in 2 mM MgCl$_2$ and 10 mM Tris · HCl (pH 7.6) in a 15-ml tube, which is centrifuged at 2500 rpm for 5 min. The pellet consists of detergent-cleaned nuclei. RNA is extracted from these nuclei[3] by resuspending them in 1 ml of 0.05 M sodium acetate (pH 5.1)–10 mM EDTA–0.5% SDS, vortexing vigorously, and adding an equal volume of phenol prewarmed to 70° and saturated with the sodium acetate–ethylenediaminetetraacetic acid (EDTA) buffer lacking SDS. The mixture is shaken vigorously in a water bath at 55°. The aqueous and phenol phases are then separated, and the phenol phase and interface are reextracted as described in Section A,1,a, with phenol saturated with the acetate-EDTA buffer. The further isolation and purification of RNA is carried out as described above.

c. Isolation of Viral RNA from Cytoplasm. Cells are collected as described above. The cell pellet is washed once with buffer E and resuspended in 1 ml of buffer E; Nonidet P-40 (NP40) (5% in buffer E, 0.1 ml) is then added, and the mixture is stirred on ice for 20 min. The lysed cells are centrifuged at 700 g for 10 min and to the supernatant, which contains the cytoplasmic fraction, is added 1 ml of 0.1 M sodium acetate (pH 5.0), 0.2 ml of 10% SDS, and 2 ml of phenol (as above) saturated with buffer E.[4]

[2] B. P. Brandhorst and E. H. McConkey, *J. Mol. Biol.* **85**, 451 (1974).
[3] This is a modification of the method of R. Soiero and J. E. Darnell, *J. Mol. Biol.* **44**, 551 (1969).
[4] R. D. Palmiter, *Biochemistry* **13**, 3606 (1974).

The sample is shaken for a few seconds, allowed to stand for 5 min, and shaken again. Then 2 ml of chloroform are added, and the mixture is shaken again for several seconds. It is then centrifuged at 700 g for 10 min; the lower (chloroform–phenol) phase is removed, and an additional 2 ml of chloroform are added to the interphase and the aqueous phase; the mixture is again shaken and centrifuged as described above. The chloroform phase is again removed, and the procedure is repeated until the white precipitate at the interface either disappears or no longer changes in appearance. Usually two more chloroform extractions are needed. Following the last chloroform extraction, the upper (aqueous) phase is withdrawn and the RNA in it is precipitated with 2 volumes of EtOH and stored at $-20°$.

2. Isolation of Poly(A)-Containing RNA Transcripts

It is frequently desirable to separate virus-specified transcripts, both unprocessed and processed, from ribosomal RNA and transfer RNA, which generally account for over 90% of cellular RNA. This is accomplished best by chromatography on oligo(dT)-cellulose, which is carried out as follows.[5]

The ethanol-precipitated RNA isolated by SDS–phenol–chloroform extraction is collected by centrifuging at 10,000 g for 20 min. The pellets are washed twice with 80% EtOH and dried under vacuum. The RNA pellets are dissolved in 10 mM Tris · HCl (pH 7.4) and 2 M KCl is added to a final concentration of 0.5 M. The RNA sample is then loaded onto a 2 ml bed volume column of oligo(dT)-cellulose (Collaborative Research Inc, Waltham, Massachusetts, type 3, which binds 2–4 A_{260} units of poly(A) per 100 mg of dry cellulose), which had been equilibrated with column buffer. After 15 min to allow for hybridization, nonadsorbed material is removed by three 5-ml washes with column buffer, and poly(A)-containing RNA is eluted with successive 1-ml washes of 10 mM Tris · HCl (pH 7.4), at a flow rate of about 0.5 ml/min. The fractions containing eluted RNA are combined and made 0.2 M with respect to NaCl; the RNA is then precipitated with 2 volumes of EtOH and stored at $-20°$. In order to achieve complete separation of poly(A)-containing and ribosomal RNA, a second passage through oligo(dT) cellulose is usually required.

3. Isolation of Hybridization Probes

Viral RNA transcripts are quantitated by hybridizing them to an excess of probe. Two modifications of this method will be described here.

[5] H. Aviv and P. Leder, *Proc. Natl. Acad. Sci. U.S.A.* **69**, 1408 (1972).

The first employs labeled transcripts, which are hybridized to a very large excess of unlabeled probe; the second employs unlabeled transcripts, which are hybridized to an excess of labeled probe. I will describe first the preparation of unlabeled double-stranded (ds) reovirus RNA and of the dsDNAs of poxviruses, herpesviruses, adenoviruses, papovaviruses, and adeno-associated viruses; and then the preparation of single-stranded (ss) parvovirus DNA and retrovirus cDNA, and of the negative-stranded RNAs of rhabdoviruses, paramyxoviruses, orthomyxoviruses, and bunyaviruses, as well as the techniques used to label them radioactively.

   a. Isolation of ds Reovirus RNA. Reovirus (usually the Dearing strain of reovirus type 3) is generally grown in suspension cultures of HeLa or L cells.[6] Cells are concentrated by centrifugation to 10^7 cells/ml in Eagle's medium containing 20 mM Mg^{2+} and 1% fetal calf serum. Virus inoculum [10–20 plaque-forming units (PFU)/cell] is added and allowed to adsorb for 20 min at 37°; the cells are then diluted to 10^6 cells/ml in growth medium (Joklik's modified Eagle's medium containing 4% calf serum and 1% fetal calf serum), and incubated at 34° for 48 hr. They are then harvested by centrifugation and washed once with Earle's saline. All subsequent steps are carried out at 4°. The cells are resuspended in 10 mM Tris · HCl (pH 8)–0.25 M NaCl–10 mM 2-mercaptoethanol (2-ME) at a concentration of 10^8 cells/ml, and homogenized in a VirTis 23 homogenizer for 2 min at half speed. Freon 113, 0.25 volume, is then added, and homogenization is repeated. The aqueous and organic phases are separated by centrifugation at 3000 rpm for 15 min, and the organic phase is reextracted with an equal volume of homogenization medium. The combined aqueous phases are then reextracted with 0.25 volume of Freon 113 as above. Virus in the resultant aqueous phase is pelleted by centrifugation for 60 min in a Spinco L2-65B ultracentrifuge at 25,000 rpm in a SW27 rotor. The virus pellet is resuspended overnight in SSC and layered onto a 20 to 40% (w/w) sucrose density gradient containing 1 mM phosphate buffer (pH 7.2). After centrifugation in the SW27 rotor for 60 min at 24,000 rpm, the virus band, which is near the middle of the gradient, is collected and diluted with 3 volumes of SSC; the virus is again pelleted as above. The pellet is then resuspended in 1 ml of SSC and layered onto a preformed CsCl gradient, density 1.20–1.40 g/ml, and centrifuged for 2 hr at 24,000 rpm in a SW27 rotor. Fractions containing virus are pooled and dialyzed overnight against several liters of 0.05 M Tris · HCl (pH 8).

   The amount of reovirus is estimated by the empirically determined relation[6]: 5.42 optical density units at 260 nm are equivalent to 1 mg of reovirus, which is equivalent to 1.13×10^{13} reovirus particles. Virus yields are usually between 10 and 20 mg of virus per 10^9 cells.

[6] R. E. Smith, H. J. Zweerink, and W. K. Joklik, *Virology* 39, 791 (1969).

Double-stranded RNA is extracted from reovirus particles as follows. Purified reovirus is suspended in a 0.05 M Tris · HCl (pH 8) at a concentration of 1 mg/ml, then SDS is added to a concentration of 1%; the mixture is incubated at 65° for 60 min and extracted three times at room temperature with phenol saturated with STE buffer. The aqueous phase is extracted three times with chloroform–isoamyl alcohol (24 : 1), made 0.15 M with respect to NaCl, and the RNA is then precipitated with 2.5 volume of EtOH at −20°. The RNA is then dissolved in STE buffer and passed through a Sephadex G-100 column to remove oligonucleotides.[7] Double-stranded RNA emerges with the flowthrough fraction: fractions containing it are pooled, made 0.15 M with respect to NaCl, and precipitated with 2.5 volume of EtOH at −20°.

b. Isolation of Vaccinia Virus DNA. Highly purified vaccinia virus (20 mg) (prepared as described by Joklik[8]) is pelleted by centrifuging at 10,000 g for 40 min and resuspended with brief (3 sec) sonication with a microtip sonicator probe in 1 ml of SSC, and 30 μl of 2-ME are added.[9] After 1 hr at room temperature 0.3 ml of proteinase K (2 mg/ml, heated at 100° for 10 min to inactivate any deoxyribonuclease) is added, and the mixture is incubated for 2 hr at 37°. Then SDS is added to a final concentration of 0.5%, and incubation at 37° is continued for 60 min with gentle agitation. The SDS concentration is then increased to 1%, and NaCl is added to a concentration of 0.1 M; one equal volume of chloroform-saturated phenol is added, and the mixture is vortexed and centrifuged at 1000 g for 5 min. The aqueous phase is reextracted twice more with fresh chloroform-saturated phenol; then 2 volumes of EtOH are added to it, and the precipitated DNA is stored at −20°.

c. Isolation of Herpes Simplex Virus (HSV) DNA. HeLa cells grown in plastic roller bottles (10^8 cells/bottle) are infected at a multiplicity of 10 PFU/cell with HSV type 1 strain KOS and incubated in medium 199 containing 5% fetal calf serum. At 20 hr after infection virus from the cell overlay medium and from the cytoplasmic cell fraction is purified as follows.[10] The overlay medium is poured off and stored on ice; the cells are scraped into ice-cold 0.15 M NaCl and pelleted by centrifuging at 1000 g for 5 min. The cell pellet is suspended in RSB containing 0.5% NP40, and the suspension is kept at 0° for 10 min. The saline supernatant is added to the overlay medium, and the mixture is centrifuged at 6000 g for 10 min to pellet the floating cells, which are pooled with the bulk of the cells, which are then disrupted with 10 strokes of a tight-fitting glass Dounce homogenizer, followed by centrifugation at 8000 g for 8 min. The HSV in

[7] J. L. Nichols, A. R. Bellamy, and W. K. Joklik, *Virology* **49,** 562 (1972).
[8] W. K. Joklik, *Virology* **18,** 9 (1962).
[9] K. Oda and W. K. Joklik, *J. Mol. Biol.* **27,** 395 (1967).
[10] L. E. Holland, K. P. Anderson, J. R. Stringer, and E. K. Wagner, *J. Virol.* **31,** 447 (1979).

the supernatant is pooled with the medium supernatant and pelleted by centrifugation at 27,000 g for 20 min. All centrifugations are performed at 0° to 4°. The virus-containing pellet is resuspended in a small volume of 0.1 M NaCl–10 mM N-2-hydroxyethylpiperazine-N'-2-ethanesulfonic acid (HEPES)–20 mM EDTA (pH 7.4), gently lysed by the addition of 0.1 volume of 20% Sarkosyl and 0.05 volume of 10% SDS, and digested with 300 μg of proteinase K per milliliter (which had been heated at 100° for 10 min to inactivate any deoxyribonuclease) for 3 hr at 45°. Viral DNA is purified by a single cycle of isopycnic centrifugation in 3 M CsCl–0.01 M EDTA (pH 8) (30 hr at 20° at 50,000 rpm).[11] The DNA that bands at 1.726 g/ml is collected and precipitated with 2 volumes of EtOH at −20°.

  d. Isolation of Adenovirus DNA. There are numerous serotypes of adenovirus, those most commonly studied being serotypes 2, 5, and 12. Adenovirus 2 is grown and purified, and its DNA extracted, as follows.[12]

  Four liters of KB cells growing in suspension culture are infected at a multiplicity of about 20 PFU/cell. At 36–48 hr after infection the cells are collected by centrifugation at 300 g for 20 min and resuspended in 40 ml of 0.01 M Tris · HCl (pH 7.5). The cells are homogenized with an equal volume of Freon 113; after centrifugation to separate the phases, the aqueous phase is extracted again. To the milliliters (n) of the resulting aqueous phase, $n/2$ g of CsCl are then added; the suspension is centrifuged to equilibrium (7 hr at 50,000 rpm). Adenovirus type 2 bands at 1.334 g/ml. The virus-containing fractions are collected and dialyzed against 0.01 M Tris · HCl (pH 7.2–7.5). Upon storage at 4° for several days, the virus particles precipitate out of suspension and become susceptible to digestion with Pronase B (heated to 100° to inactivate DNase) for 30 min at 37°. The mixture is extracted three times with an equal volume of phenol saturated with 1 M Tris · HCl (pH 7.2), and the phenol is removed by dialysis for 24 hr against several changes of SSC. The DNA is precipitated by the addition of 2 volumes of EtOH and stored at −20°.

  e. The Isolation of Parvovirus DNA. There are two groups of mammalian parvoviruses: the members of the genus parvovirus, and the members of the genus adeno-associated virus (AAV). The genome of both is ssDNA: but while the genome of the former is negative-stranded DNA (that is, it possesses the opposite polarity to mRNA), populations of virions belonging to the latter genus consist of equal numbers of particles in which plus- and negative-stranded DNA is encapsidated. The two types of hybridization probes for parvovirus mRNA, the preparation of which will be described here, are the dsDNA of adeno-associated viruses, and the negative-stranded ssDNA of viruses belonging to the genus parvovirus. These probes are prepared as follows.

[11] E. K. Wagner, K. K. Tewari, R. Kolodner, and R. C. Warner, *Virology* 57, 436 (1974).
[12] W. Doerfler, *Virology* 38, 587 (1969).

i. DOUBLE-STRANDED AAV DNA. When AAV particles are extracted at high temperature or in high salt, the plus and minus strands that they contain anneal to yield dsDNA. The preparation of this DNA will be described for AAV 2.

AAV 2 is grown in suspension cultures of KB cells with adenovirus type 2 as helper.[13] At 48 hr after infection the cells are harvested by centrifuging, suspended in PBS at a rate of 2 to 4 × 10⁷ cells/ml, and disrupted by sonication. Deoxycholate and trypsin are added to final concentrations of 2 and 0.2%, respectively, and the mixtures are incubated at 37° for 30 min. Debris is removed by centrifuging at 2000 g for 10 min, and the virus is pelleted by centrifuging at 22,000 rpm for 3 hr at 4°. The pellet is resuspended in 10 mM Tris · HCl (pH 7.9) (at the rate of 4 ml per liter of suspension culture) with mild sonication, and is then purified on sucrose–CsCl step gradients consisting of 5 ml of a 1.41 g/ml CsCl solution in 50 mM Tris · HCl (pH 8.9)–1 mM EDTA–0.15% Sarkosyl–0.1% Brij 58–0.1 M NaCl underneath 2 ml 30% sucrose in the same buffer but containing 0.2% instead of 0.15% Sarkosyl.[15] After centrifuging at 39,000 rpm at 4° for 24 hr, the virus bands are visualized with vertical light against a black background, and the "full" virus band between 1.41 and 1.46 g/ml is collected with a Pasteur pipette and dialyzed against 10 mM Tris · HCl (pH 7.5).

DNA is extracted from AAV 2 by centrifugation in alkaline sucrose density gradients (5 to 20% (w/w) sucrose in 0.3 N NaOH–0.7 M NaCl–1 mM EDTA–0.15% Sarkosyl; 3 hr at 42,000 rpm, 20°). Fractions containing DNA, which has a sedimentation coefficient of about 18 S, are pooled, and dialyzed against SSC. The DNA is then allowed to anneal to duplex molecules by heating at 66°.[14] The DNA is dissolved in 0.03 M sodium acetate (pH 4.5)–10 mM ZnSO₄–0.3 M NaCl and treated with 0.2 unit of S1 nuclease per milligram of DNA for 30 min at 37° in order to remove residual ssDNA. The mixture is then deproteinized by shaking with two lots of 1 volume each of phenol to remove the S1 nuclease, and the DNA is precipitated from the final aqueous phase with 2 volume of EtOH at −20°.

ii. NEGATIVE-STRANDED DNA OF MEMBERS OF THE GENUS PARVO-VIRUS. The two best studied viruses in this group are the minute virus of mice (MVM) and the hamster oncolytic virus H1. There follows a description of how to prepare negative-stranded H1 DNA.[15] MVM DNA may be prepared as described by Astell et al.[16] and Tattersall.[17]

[13] K. I. Berns and J. A. Rose, J. Virol. 5, 693 (1970).
[14] L. M. de la Maza and B. J. Carter, Virology 82, 409 (1977).
[15] M. R. Green, R. M. Lebovitz, and R. G. Roeder, Cell 17, 967 (1979).
[16] C. R. Astell, M. Smith, M. B. Chow, and D. C. Ward, Cell 17, 691 (1979).
[17] G. J. Bourguignon, P. J. Tattersall, and D. C. Ward, J. Virol. 20, 290 (1976).

H1 is grown in RL5E cells, a rat cell line that is transformed by murine sarcoma virus. Suspension cultures of these cells are infected at a multiplicity of 5 PFU/cell and harvested by centrifugation 40–48 hr later. After washing in PBS, the cells are suspended at a concentration of 5×10^7 cells/ml in 50 mM Tris · HCl (pH 8.9)–1 mM EDTA–0.1% Brij 58 (Sigma) and are disrupted by freeze-thawing three times followed by sonication. The pH is maintained at about 8.9 throughout this procedure by the addition of 1 N NaOH. Cellular debris is removed by centrifuging at 3500 rpm for 15 min, and mitochondria are removed by centrifuging at 10,000 rpm for 10 min. The supernatant is made 10 mM with respect to Mg^{2+}, and 50 μg/ml of deoxyribonuclease 1 (Sigma) and 50 μg/ml of pancreatic ribonuclease (Sigma, type XI-A) are added; the mixture is incubated at 37° for 30 min. Trypsin (Worthington) and chymotrypsin (Worthington) are then added to 50 μg/ml, and incubation is continued at 37° for 30 min. After centrifugation at 10,000 rpm for 10 min, the supernatant is layered onto a sucrose–CsCl step gradient as described above for AAV.

In order to extract its DNA, the virus is treated with 50 μg of deoxyribonuclease I (Worthington) per milliliter in 10 mM Tris · HCl (pH 7.5)–5 mM Mg^{2+}-0.15 M NaCl at 37° for 90 min and precipitated by the addition of 0.1 volume of 3 M sodium acetate (pH 5.5) and 2.5 volume of EtOH at $-20°$. The precipitate is dissolved in 100 μl of 30 mM NaOH–2 mM EDTA, and the solution is layered onto a 5 to 20% alkaline sucrose density gradient (5 to 20%, w/w, sucrose in 0.2–0.8 M NaOH and 0.8–0.2 M NaCl, all containing 2 mM EDTA). After centrifuging at 55,000 rpm at 4° for 4 hr, the gradient is collected from the bottom and the fractions containing H1 virion DNA are pooled and dialyzed against SSC and precipitated with 2 volumes of EtOH at 20°. The yield is about 100 μg of H1 DNA per 1×10^9 infected cells.

f. Preparation and Isolation of Retrovirus cDNA. There are numerous groups of retroviruses; among those that are studied most intensively are avian, murine, feline, and primate retroviruses. The procedures for quantitating retrovirus RNA are here specified for avian retroviruses; procedures for other retrovirus RNAs are similar in principle.

Avian retroviruses, such as RSV strain Prague C or B77, are harvested from the supernatant fluid of infected cells by centrifuging for 30 min at 80,000 g after first removing cell debris by centrifuging at 8000 g for 15 min. The viral pellet is resuspended in 5 mM Tris · HCl (pH 8.6)–1 mM EDTA at a concentration of about 3 mg/ml; although the virus is still crude, it is satisfactory for synthesizing cDNA.[18] The reaction mixture contains, in a volume of 1 ml: 0.1 M Tris · HCl (pH 8.3), 10 mM Mg^{2+}, 5 mM dithiothreitol, 100 μg of actinomycin D, 0.1% Triton X-100, 0.1 mM each of dATP, dCTP, and dGTP, 50 μCi (about 3 μM) of [^{3}H]dTTP (20

[18] A. L. Schincariol and W. K. Joklik, *Virology* **56**, 532 (1973).

Ci/mmol) and 300 μg of crude virus. If the concentration of Triton X-100 is raised above 0.01%, the yield of DNA decreases markedly. Actinomycin D, at a concentration of 100 (or 200) μg/ml reduces the yield of DNA by 75%; but the transcripts formed in its presence contain a far higher proportion of single-stranded species and far fewer reiterated sequences than those formed in its absence. The reaction mixture is incubated for 8–10 hr under N_2 at 37°. It is then made 10 mM with respect to EDTA, 0.1 M with respect to NaCl, and 0.5% with respect to SDS; 50 μg of calf thymus DNA and 500 μg of Pronase (heated at 100° for 10 min to inactivate any DNase) are added, and the mixture is incubated at 37° for 60 min. The DNA is deproteinized three times with buffer-saturated phenol and once with chloroform, and precipitated with 2 volumes of EtOH at −20°. The precipitate is dissolved in 10 mM phosphate buffer (pH 6.8), and single and double strands are separated by hydroxyapatite column chromatography.[19] Single strands (85–90% of the total) elute at 0.16 M phosphate; they are pooled and passed through Sephadex G-50 to remove phosphate, precipitated with EtOH as above, and dissolved in 5 mM EDTA (pH 7.4). The solution is then made 0.3 M with respect to NaOH and kept at 100° for 5 min to hydrolyze RNA; it is then neutralized with HCl after the addition of 0.1 volume of 1 M Tris · HCl (pH 7.4) and chromatographed on a Sephadex G-50 column equilibrated with 0.1 M NaCl–10 mM Tris · HCl (pH 7.4)–1 mM EDTA. The labeled DNA from the excluded region of the column is precipitated with ethanol as above. The specific activity of cDNA prepared in this manner is between 3 and 10 × 10^3 cpm/pmol (1 to 3 × 10^7 cpm/μg).

g. *Isolation of Negative-Stranded Rhabdovirus RNA*. The most intensively studied rhabdovirus is vesicular stomatitis virus (VSV), which will be used to illustrate the procedure used for isolating rhabdovirus RNA.[20]

The VSV is grown in monolayers of BHK cells infected at a multiplicity of 3. After 15 hr of infection, the medium is removed and clarified by low speed centrifugation; the virus in it is recovered by pelleting at 20,000 rpm for 2 hr in the SW27 rotor. The pellet containing the virus is resuspended in ET buffer and isopycnically banded in a 7 to 52% (w/w) sucrose (in ET buffer) density gradient at 20,000 rpm for 15 hr in a SW41 rotor. Fractions containing the virus are pooled and diluted threefold with ET buffer; virus particles are recovered by pelleting as above.

In order to isolate the negative-stranded RNA from VSV particles, virus is suspended in 1 ml of SET buffer. The suspension is made 1% with respect to SDS and extracted twice with phenol, which is removed from the final aqueous phase with ether. Viral RNA is recovered by precipita-

[19] G. Bernardi, *Biochim. Biophys. Acta* **174**, 435 (1969).
[20] G. Stamminger and R. A. Lazzarini, *Cell* **3**, 85 (1974).

tion with 2.5 volumes of EtOH in 0.4 M NaCl at $-20°$. It may be further purified by centrifugation into a 10 to 30% (w/w) sucrose density gradient containing SET–0.1% SDS. VSV RNA sediments with a sedimentation coefficient of 42 S. Fractions containing it are pooled and the RNA is again precipitated with EtOH as above and stored at $-20°$.

h. Isolation of Negative-Stranded Paramyxovirus RNA. Among the most intensively studied paramyxoviruses are Newcastle disease virus (NDV), Sendai virus (parainfluenza virus type 1), and SV5. The procedures used to quantitate mRNAs specified by them will be illustrated by the example of Sendai virus. Procedures applicable to the other paramyxoviruses are analogous.

Negative-stranded Sendai virus RNA is isolated as follows.[21] Three-day-old confluent cultures of bovine kidney cells (MDBK) are infected with 10 mean egg-infective doses per cell of Sendai virus (Harris strain). At 36 hr after infection the medium is removed and centrifuged for 10 min at 5000 g to remove cellular debris, and the virus is pelleted through a cushion of 25% glycerol in TNE for 2 hr at 36,000 rpm at 8°. The virus pellet is dissolved in TNE containing 0.5% SDS and 0.2 mg of proteinase K (Merck) per milliliter at an approximate concentration of 2 mg/ml and incubated for 15 min at 25°, and 150 μl amounts are then centrifuged on 5 to 23% (w/w) sucrose (in 0.1 M LiCl–10 mM Tris · HCl (pH 7.4)–4 mM EDTA–0.1% SDS) density gradients for 105 min at 50,000 rpm at 7° in Spinco SW56 rotors. Fractions containing the 46–50 S RNA are pooled, and the RNA in them is precipitated with 2.5 volumes of EtOH in 0.4 M NaCl at $-20°$.

i. Isolation of Influenza Virus Negative-Stranded RNA. Numerous strains of influenza virus are used in research. The procedures described here use the influenza A strain WSN(HON1).

Virus is grown in the allantoic cavity of 11-day-old chick embryos and the allantoic fluid is concentrated 20- to 40-fold either by adsorption to and elution from chick erythrocytes or by ultrafiltration using an Amicon CH-4 concentrator.[22] The virus concentrate is clarified by centrifugation at 10,000 g for 10 min and centrifuged to equilibrium (52,000 g, SW27 rotor, 16 hr) on 20-ml density gradients of 25–70% (w/w) sucrose in STE buffer. Fractions containing the visible virus band are removed and collected, pooled, diluted threefold with STE, and centrifuged at 30,000 rpm for 60 min to pellet the virus, which is then recentrifuged in a 25 to 70% (w/w) sucrose density gradient for 2.5 hr at 80,000 g. The final virus band is pelleted, resuspended in STE, lysed with 1% SDS, and extracted re-

[21] D. Kolakofsky, E. Boy de la Tour, and H. Delius, *J. Virol.* **13**, 261 (1974).
[22] W. J. Bean, Jr., G. Spiriam, and R. G. Webster, *Anal. Biochem.* **102**, 228 (1980).

peatedly with STE-saturated phenol–chloroform (1 : 1). Viral RNA is precipitated with 2 volumes of EtOH at −20°.

Alternatively influenza virus strain WSN may be grown in MDBK cells infected at a multiplicity of 0.1.[23] After an adsorption period of 1 hr at 37°, Eagle's medium supplemented with 2% calf serum is added, and the cells are incubated at 37° for 40 hr. The medium is then poured off, debris and cells are removed by centrifuging it at 2000 g for 10 min, and virus is pelleted by centrifuging at 30,000 rpm for 60 min. The crude virus is resuspended in STE and isolated in density gradients as described above.

j. Isolation of Negative-Stranded Bunyavirus RNA. Several bunyaviruses are under active investigation; the one that is used for purposes of describing the techniques used for quantitating bunyavirus mRNA is snowshoe hare virus (SSHV).[24]

SSHV is grown in monolayers of BHK cells infected in a multiplicity of about 0.01. After incubation at 33° for 48 hr the virus is purified as follows: the tissue culture fluid is clarified by low speed centrifugation (10,000 g for 20 min in a Beckman J-21 centrifuge) to remove cell debris, and the virus is precipitated by the addition of polyethylene glycol 6000 (Carbowax 6000, 70 g/liter) and NaCl (23 g/liter).[25] The mixture is stirred for 1 hr at 4°, and the virus precipitate is collected by centrifugation at 10,000 g for 20 min, resuspended in TSE buffer, clarified by centrifugation, and loaded onto a combination equilibrium–viscosity gradient of glycerol and potassium tartrate (16 ml of 30%, w/w, glycerol in TE buffer in the inside feeder chamber of the density gradient maker, and 14 ml of 50%, w/w, potassium tartrate in the outer mixing chamber). After centrifugation for 3 hr at 4° (40,000 rpm in a Spinco SW41 rotor), the visible virus band is collected with a pipette and dialyzed overnight at 4° against TSE buffer. Generally no further purification is required. If it is desired to purify the virus further, it is loaded onto a density gradient of 20 to 70% (w/v) sucrose in 1 M NaCl–10 mM Tris · HCl (pH 7.4)–2 mM EDTA, which is centrifuged at 4° and 35,000 rpm for 3 hr. The virus band is harvested as above, diluted fourfold with TSE buffer, and pelleted through a 1-ml cushion of 30% (w/v) sucrose in TSE buffer by centrifugation for 3 hr at 35,000 rpm. The final virus pellet is resuspended in TSE buffer.

In order to extract RNA, the virus is lysed with 1% SDS, and the RNA is extracted three times with an equal volume of a phenol mixture (100 g of phenol, 14 g of m-cresol, 0.1% 8-hydroxyquinoline).[26] The RNA is precipi-

[23] R. M. Krug, *Virology* **44**, 125 (1971).
[24] A. C. Vezza, P. M. Repik, P. Cash, and D. H. L. Bishop, *J. Virol.* **31**, 426 (1979).
[25] J. F. Obijeski, D. H. L. Bishop, F. A. Murphy, and E. L. Palmer, *J. Virol.* **19**, 985 (1976).
[26] J. F. Obijeski, D. H. L. Bishop, E. L. Palmer, and F. A. Murphy, *J. Virol.* **20**, 664 (1976).

tated from the aqueous phase with 2 volumes of EtOH at $-20°$ after the addition of LiCl to 2 M.

4. Labeling of the Single-Stranded Nucleic Acid Hybridization Probes

The single-stranded nucleic acid hybridization probes may be labeled in two ways. First they may be labeled *in vivo*; that is, they are extracted from labeled virus. The amount of label that is used depends on the degree of labeling that is desired. For making reasonably large amounts of virus, originating from 1–4 liters of cells or medium, the amounts of label used will vary from 1 to 10 μCi/ml of [³H]uridine or [³H]thymidine; when making smaller amounts of virus the amount of radioactive label may be increased to 20 μCi/ml or more. The duration of labeling will generally be from 3 to 5 hr after infection until the rate of virus multiplication diminishes below, say, 25% of the maximum rate. Infected cells may be treated with actinomycin D (1 μg/ml) to diminish labeling of host cell RNA species; but actinomycin D cannot be used in cells infected with H1 virus or influenza virus.

The second, and usually more convenient, method of labeling the probes is by iodinating them *in vitro*.[22] The following stock reagents are used for iodination: 2× heparin buffer (0.2 M sodium acetate, pH 5.0–200 μg/ml sodium heparin), 0.5 mM KI, ¹²⁵I (Amersham-Searle, 100 mCi/ml, carrier-free, pH 8–9), and TlCl₃ (6 mg/ml). The TlCi₃ solution is stable at this concentration and may be stored frozen.

The iodination reactions are prepared by mixing 4 μg of viral RNA (or DNA) (cthanol precipitates washed three times with 70% EtOII) in 8 μl of water with 15 μl of 2× heparin buffer, 5 μl of KI, 2 μl of ¹²⁵I, and 1 μl of TlCl₃, and incubating the mixture at 45° for 15 min. After chilling in ice, the reaction is stopped by the addition of 1 μl of 2-ME, and the mixture is diluted with 6 ml of 0.1 M sodium acetate (pH 5.0) containing 10% glycerol and 10 μg of yeast RNA. Free iodine is removed from the mixture by passage through a 9-ml column of Sephadex G-25. Fractions containing RNA (or DNA) are located with a Geiger counter, pooled, extracted once with phenol–chloroform, and precipitated with EtOH as above.

5. Quantitative Hybridization Assay of Viral RNA Transcripts

Viral RNA transcripts are isolated as poly(A)-containing RNA as described in Section A,2 and are quantitated by measuring the extent to which they are capable of hybridizing to the probes the isolation of which is described in Section A,3. These probes are of four types: dsRNA, ds- or ssDNA (which are used in the same manner), cDNA, and negative-stranded RNA. The methods of using these probes are as follows:

    a. Double-Stranded RNA. The principle is to hybridize pulse-labeled

cytoplasmic RNA from infected cells to an excess of dsRNA and to measure the amount of radioactively labeled RNA that is rendered resistant to
RNase.[27]

Cytoplasmic RNA from infected cells and dsRNA are dissolved in 1
mM EDTA (pH 7.4) and mixed to provide 25-μl mixtures that contain 10
μg dsRNA. They are denatured by adding 9 volumes of dimethyl sulfoxide
(DMSO) and incubating at 45° for 30 min; the salt concentration is
then adjusted to 0.03 M NaCl–10 mM Tris · HCl (pH 7.4)–2 mM EDTA,
the DMSO concentration is reduced to 67%, and the samples are incubated at 37° for 24 hr. The RNA is then precipitated with 2 volumes of
EtOH–0.1 M NaCl at −20°; the precipitates are washed with 80% EtOH,
dissolved in 2× SSC, and incubated with 10 μg of RNase per milliliter for
15 min at 37°. The RNA is again precipitated with 2 volumes of EtOH, and
the radioactivity that is resistant to RNase is measured. Since under these
experimental conditions the excess of reovirus dsRNA over ssDNA is
likely to be 100-fold or more, the efficiency with which reovirus ssRNA is
measured is close to 100%.

b. *Double-Stranded and Single-Stranded DNA.* The probes that fall into
this group are the dsDNAs of poxviruses, herpesviruses, adenoviruses,
papovaviruses, and certain parvoviruses, and the negative-stranded DNA
of other parvoviruses (see above). The ethanol-precipitated DNAs are
washed three times with 70% EtOH, dissolved in SSC, denatured in 0.1 M
NaOH at 100° for 5 min, diluted with about 20 volumes of 2 M NaCl to
bring the pH to 11.4, and loaded onto 25 mm in diameter Millipore HA
nitrocellulose filters at the rate of 5 μg of DNA per filter.[28,29] The filters are
then washed with 10 ml of 6× SSC, dried at room temperature for 4 hr and
at 80° for an additional 2 hr in a vacuum oven. This mode of applying DNA
to nitrocellulose filters ensures the virtual absence of paired snapback
sequences in the probe.

The RNA samples that are to be quantitated are pelleted from the
EtOH in which they are stored, the pellets are washed three times with
70% ethanol, resuspended in 10 mM 2× TESS, and incubated with DNA
filters at 65° for 40 hr.[30] Following hybridization, the filters are washed for
30 min in 2 liters of 2× TESS and treated for 60 min with 10 μg of RNase
per milliliter in 2× TESS at 37°; they are then washed with 2× SSC, dried,
and counted. The amount of radioactivity that is bound to the filters is a
measure of the amount of virus-specified RNA in the original sample of
poly(A)-containing RNA.

[27] Y. Ito and W. K. Joklik, *Virology* **50**, 189 (1972).
[28] D. Gillespie and S. Spiegelman, *J. Mol. Biol.* **12**, 829 (1965).
[29] M. Mellie, E. Ginelli, G. Corneo, and R. di Lernia, *J. Mol. Biol.* **93**, 23 (1975).
[30] M. M. Harpold, R. M. Evans, M. Salditt-Georgieff, and J. E. Darnell, *Cell* **17**, 1025 (1979).

c. Retrovirus cDNA. The principle is to anneal about 0.1 μg of cDNA containing about 10^5 cpm with RNA isolated from infected cells and to measure the amount of probe that is rendered resistant to S1 nuclease,[18] a nuclease that is specific for ssDNA. The fraction of the probe that is rendered S1- resistant is kept small (below 15%) to obviate the difficulty of sequences that are transcribed frequently from portions of the provirus from saturating the probe.

Ethanol precipitates of both poly(A)-containing viral RNA transcripts and cDNA probe are washed three times with 70% EtOH, dissolved in HB, and mixed so as to provide 25-μl reaction mixtures that contain about 1 μg of cDNA (10^5 cpm). The reaction mixtures are sealed in capillaries and heated at 65° for 48 hr. The mixtures are then expelled into 0.03 M sodium acetate (pH 4.5)–10 mM $ZnSO_4$–0.3 M NaCl and are incubated at 37° for 30 min without and with sufficient S1 nuclease (determined empirically) to degrade the amount of cDNA added. One drop of carrier yeast tRNA (2 mg/ml) is added, and mixtures are precipitated with 5% trichloroacetic acid. The precipitates are collected on GF/C glass fiber filters, dried and counted in 10 ml of Omnifluor-toluene. Since the specific activity of the cDNA is known, the amount of RNA present in each infected cell sample is readily calculated from the fraction of cDNA that is rendered resistant to S1 nuclease.

d. Negative-Stranded RNA. The probes that fall into this group are the RNAs of rhabdoviruses, ortho- and paramyxoviruses, and bunyaviruses. The principle of the hybridization procedure is to anneal samples of the RNA from the interferon-treated cells with fixed amounts of labeled viral RNA (labeled either *in vivo* or with ^{125}I after extraction) and to measure the fraction of viral RNA that is rendered resistant to RNase digestion.[24] It is important to keep this fraction small (preferably below 12.5%) so that transcripts of a small portion of the viral genome do not saturate the genome sequences that are available.

The specific activity of the viral RNA should be 10^5 to 10^6 cpm/μg. The cellular RNA sample and viral RNA probe are both dissolved in 0.4 M NT and mixed to provide 25-μl reaction mixtures, with all mixtures containing 10^4 cpm of RNA probe. The reaction mixtures are sealed in capillaries and incubated at 60° for 48 hr. The samples are then expelled into 200 μl of NT; two 100-μl samples are then incubated with and without, respectively, 2 μg of RNase A for 30 min at 37°, then precipitated with trichloroacetic acid; radioactivity is measured in the precipitate. A curve is first constructed with varying amounts of the RNA sample from cells *not* treated with interferon; and that amount of all RNA samples (on an $OD_{260 nm}$ basis, and therefore on the basis of cell number) is then chosen which results in no more than 12.5% of the probe being rendered RNase-resistant.

6. Quantitation of Viral RNA Transcripts in the Cytoplasm by Density Gradient Centrifugation

The rate of transcription of viral genomes can sometimes be measured without recourse to hybridization probes if transcription occurs in the cytoplasm. Since it takes 10–15 min for newly synthesized host messenger and ribosomal RNA to migrate from the nucleus to the cytoplasm, the only labeled RNA species in the cytoplasm of cells that are infected with such viruses and that are pulse-labeled for less than 15 min is viral mRNA. The only exception is a relatively small amount of 4 S tRNA, which does enter the cytoplasm within 10–15 min; but this tRNA is readily separated from the viral mRNAs by density gradient centrifugation.

a. Labeling of Infected Cells. Cells (2×10^7) are pulse-labeled for 10 min with 10–20 μCi of [^{3}H]uridine. If cell monolayers are used, they are then washed with several 10-volume lots of ice-cold Earle's saline and then scraped into 3–5 ml of fresh ice-cold Earle's saline; if spinner culture cells are used, they are pipetted onto 10–20 volumes of crushed, frozen Earle's saline.

b. Breaking Open Cells and Isolation of Cytoplasmic Fractions. The cells are collected by centrifugation at 700 g for 5 min, washed once with Earle's saline, and resuspended in 1 ml of RSB. They are allowed to swell for 5–10 min at 0° and then broken in a Dounce homogenizer calibrated for the number of strokes necessary to break over 90% of cells. The homogenate is then centrifuged at 800 g for 2–3 min, and the supernatant fraction is used as the cytoplasmic fraction.

c. Sucrose Density Gradient Centrifugation. Sodium dodecyl sulfate is added to the cytoplasmic fraction to a concentration of 1%, and the mixture is layered onto 11-ml sucrose density gradients (15–30%, w/w, sucrose in 5 mM Tris · HCl, pH 7.3–0.1 M NaCl–0.5% SDS), which are centrifuged at 20° for about 7 hr or until 28 S ribosomal RNA has moved about two-thirds of the distance to the bottom of the tube. The density gradients are collected in 0.3–0.4-ml fractions after passage through a spectrophotometer capable of automatically recording optical density at 260 nm. The optical density marker is provided by the 28 and 18 S ribosomal RNA present in these cytoplasmic fractions. After the addition of 50 μg of RNA or protein as carrier, material insoluble in 15% trichloroacetic acid is collected on nitrocellulose filters, and radioactivity in it is measured in a liquid scintillation spectrometer.

This method is applicable, with modifications, to the quantitation of transcripts and mRNAs of reoviruses, poxviruses, rhabdoviruses, paramyxoviruses, bunyaviruses, togaviruses, and picornaviruses. For some viruses, such as picornaviruses (poliovirus), it may be necessary to increase the amount of [^{3}H]uridine used for pulse-labeling in order to obtain a sizable amount of radioactivity in the viral RNA band. For some

RNA viruses (but not orthomyxoviruses) it may be desirable to add actinomycin D to a final concentration of 2 μg/ml 1 hr before pulse-labeling to inhibit the formation of host cell mRNA and ribosomal RNA (if such has not already been switched off by the viral infection); actinomycin D should not be added too long before pulse-labeling to avoid interference with the antiviral state induced by interferon. It is always necessary to run controls with uninfected cells to make certain that no radioactive label is present in those regions of the density gradients where the viral RNA band is located; that is, the background in such locations should be very low.

B. Quantitation of Host-Coded RNA Transcripts

There are three classes of host-coded RNA transcripts: those that give rise to mRNAs, those that give rise to ribosomal RNAs, and those that give rise to transfer RNAs. There follows a description of how each of these classes of transcripts is quantitated.

1. Messenger RNAs and Their Precursors

Within the context of the present studies, host-coded "messenger"-type RNA transcripts are best quantitated as poly(A)-containing RNAs. The reason for this is primarily the fact that focusing on possession of poly(A) as the distinguishing feature of this class of transcripts provides a convenient and highly efficient means of isolating them. There is no evidence that interferon interferes with polyadenylation; on the contrary, there is positive evidence that it has no effect on the polyadenylation of vaccinia virus mRNA.[31] Although not all transcripts of the "messenger" type may be polyadenylated, measuring those that are polyadenylated nevertheless provides a measure of all.

In order to investigate the effect of interferon on host mRNA transcription, infected cells are therefore pulse-labeled briefly (about 15 min) with [³H]uridine as described in Section A,6,a, RNA is isolated either from whole cells or from their nuclear or cytoplasmic fractions, as described in Section A,1, and polyadenylated RNA species are isolated as described in Section A,2. The amount of radioactivity in polyadenylated RNA is then measured.

2. Ribosomal RNAs and Their Precursors

Cells may be used in monolayer or in spinner culture.[1]

a. Monolayer Cultures. Cells in plates or flasks, untreated or treated with interferon, are washed with Joklik's modification of Eagle's medium (Gibco) lacking methionine and supplemented not only with dialyzed

---

[31] N. Klesel, W. Wolf, G. Hiller, and C. Jungwirth, *Virology* 57, 570 (1974).

serum (of the type used for the particular cell being used), but also with 10 mM formate and 20 μM adenosine and guanosine (to prevent incorporation of the label that is to be used (see below) into nucleic acid bases). All media used subsequently must be supplemented in this manner.

After a period of 1–2 hr in this medium for equilibration, the cells are labeled with [methyl-³H]methionine (4 Ci/mmol, Amersham/Searle) at the rate of 20 μCi/ml. After the desired labeling period, the cells are rinsed with ice-cold solution A containing 2 mM Ca^{2+}. All further procedures are those described in Section A,1,a.

b. Spinner Cultures. Cells that have been treated with interferon in spinner culture are collected by centrifuging at 2000 g for 10 min and are resuspended in the same medium as specified above. They can be labeled 30 min later. At the end of the labeling period portions of the culture are added to an equal volume of crushed frozen solution A containing 2 mM Ca^{2+}, after which they are treated as described in Section A,1,a.

c. RNA Extraction. RNA is extracted from these cell samples as described in Section A,1,a.

After deproteinization, the ethanol-precipitated RNA is washed twice with 70% EtOH and resuspended in 0.5 ml of extraction buffer. Two portions from each sample are then taken. One is diluted into 0.1 N NaOH, and the absorbance at 260 nm is measured; the other is counted for radioactivity. The measured absorbance and counts per minute are used to calculate the specific activity of the RNA in each sample. The samples are again precipitated with 2.5 volumes of EtOH containing 0.15 M sodium acetate and stored at −20°.

In preparation for polyacrylamide gel electrophoresis, the samples are resuspended in 50–100 μl of 10 mM sodium acetate–1 mM EDTA–0.5% SDS in 90% formamide, and bromophenol blue (0.1%), and marker ¹⁴C-labeled HeLa cell ribosomal RNA are added. The samples are warmed to 55° for 5 min, chilled rapidly to 0°, and then applied to polyacrylamide gels.

d. Polyacrylamide Gel Electrophoresis. The technique used for polyacrylamide gel electrophoresis is essentially that of Weinberg and Penman[32] with the following modifications: gels 13.5 cm long are used, containing 2.2% acrylamide, 0.25% diacrylate (v/v), and 10% glycerol. The gels are made the day before use and are run at 4° at 5 mA/gel constant current for 8 hr. After electrophoresis, the gels are frozen and sliced into 1.8-mm slices, and the slices are dissolved overnight at 37° in capped vials containing 1 ml of 10% Protosol (New England Corp.) dissolved in toluene scintillator. Four milliliters more of toluene scintillator are then added to each vial, and the samples are counted with settings to permit simultaneous

[32] R. A. Weinberg and S. Penman, *J. Mol. Biol.* **38**, 289 (1970).

determination of ^{3}H and ^{14}C. The counts per minutes in the various RNA bands are then summed for each gel.

With a labeling of 5 min under these conditions, and with HeLa cells, only 45 S ribosomal RNA precursor is labeled; with a labeling period of 60 min there are roughly equal amounts of label in 45 S, 32 S, 28 S, and 18 S RNA. With longer labeling periods, increasing amounts of label are present in the mature 28 S and 18 S ribosomal RNA species. Therefore this method permits study of the effect of interferon on both ribosomal RNA precursor transcription, as well as on the processing of the primary transcripts.

3. Transfer RNAs

Cells are labeled with [^{3}H]uridine (5 μCi/ml) for 60 min, and the cytoplasmic cell fraction is prepared as described in Section A,1,c.[33] It is dialyzed against 20 mM Tris · HCl (pH 7.5) and centrifuged at 20,000 g for 15 min. It is then mixed with an equal volume of phenol (88%, AR, Mallinckrodt Chemical Works), and the mixture is stirred at room temperature for 1 hr and centrifuged at 20,000 g for 15 min. The phenol layer is reextracted twice with 0.5 volume of 20 mM Tris · HCl (pH 7.5). To the combined aqueous layers is added 0.1 volume of 20% potassium acetate followed by 2 volumes of EtOH. The suspension is kept overnight at −20° and then centrifuged at 25,000 g for 20 min. The precipitate is extracted twice with one-tenth of the original volume of 1 M NaCl, the combined NaCl extracts are mixed with an equal volume of 1 M Tris · HCl (pH 9), and the mixture is incubated at 37° for 45 min. The solution is then dialyzed against 1 liter of deionized water at 4° for 24 hr, with two changes of water. The dialyzed tRNA preparation is then applied to a DEAE-cellulose (DE-11, Whatman) column (0.5 × 10 cm) equilibrated with 10 mM Tris · HCl (pH 7.5). The column is washed thoroughly with 10 mM Tris · HCl (pH 7.5) and 0.1 M NaCl. The tRNA is then eluted with 1 M NaCl; fractions containing material absorbing at 260 nm are pooled, and the radioactivity present in them is determined.

C. The Quantitation of Translation Products

1. The Quantitation of Virus-Coded Translation Products

The principle of the technique for quantitating virus-specified translation products is to pulse-label infected cells with a radioactively labeled amino acid, to precipitate virus-coded proteins in the cytoplasmic cell fraction with antiviral antiserum, and to collect the antigen–antibody

[33] K. Bose, M. K. Chatterjee, and N. K. Gupta, this series, Vol. 29, p. 522.

complexes on activated *Staphylococcus* A (Staph A) immunoadsorbent. At that stage either the amount of radioactive label adsorbed to the bacteria is measured directly, or the specific antigens are liberated from the complexes and analyzed in SDS–polyacrylamide gels, which are then subjected to fluorography. The amount of radioactive label present in each desired protein band is then measured.

a. *Antiserum*. Antiviral antiserum is generally made in rabbits. Usually female New Zealand white rabbits [Cesarian obtained–barrier sustained (Charles River)], which are pathogen-free, are preferred. Antiserum can often be made under two sets of circumstances. The first obtains when the virus in question can multiply in rabbits. In that case one inoculates intracutaneously and/or intramuscularly with the maximum amount of virus (optimally, purified virus suspended in SSC) that can be injected without killing the animals (best determined by inquiry of scientists with experience in this area). Virus then multiplies in the animals, and antibodies are made to all virus-specified proteins, both structural and nonstructural. In the second set of conditions, the virus is one that cannot grow in rabbits; in that case 1 mg of virus is mixed with Freund's complete adjuvant (Difco) and is injected subcutaneously and intramuscularly (in equal amounts). It is possible to combine these two techniques, that is, to inject a virus that will multiply in rabbits at the same time both in the active state and in Freund's adjuvant. After 2–4 weeks the animals are bled by cardiac puncture, and the serum is collected and heated at 50° for 30 min to destroy complement. Its titer is best determined by mixing a small amount of virus, for example 300 PFU, with serial dilutions of it, incubating the mixtures at 0° for 4 hr, and plating them on susceptible cells. The titer of the antiserum is the reciprocal of the dilution that reduces the number of plaques by 50%.

Before being used for the specific immunoprecipitations that are described below, the antiserum should be preabsorbed with a normal cell extract; the extract is prepared by suspending normal cells in NET-1.0 buffer containing 1 mM phenylmethylsulfonyl fluoride (PMSF) at a concentration of 5×10^6 cells/ml, incubating the mixture at 4° for 15 min with gentle mixing, and removing the nuclei by centrifuging at 3000 g for 5 min. Equal volumes of antiserum and of this supernatant are then mixed and incubated at room temperature for 30 min, when an equal volume of activated Staph A immunoadsorbent (see below) is added, and incubation is continued for another 15 min at room temperature. The bacteria are then removed by centrifuging at 10,000 g for 1 min in a Microfuge, and the serum is stored at −70°.

b. *Activated Staph A Immunoadsorbent*. Activated Staph A immunoadsorbent can be bought from several sources; however, it is easily prepared,[34] and it is often advantageous to make a large batch of it so as to have on hand material of uniform quality.

An inoculum of the Cowan I strain of *Staphylococcus aureus* is first grown up in 8 lots of 5 ml of Bacto-tryptic soy broth (using screw-capped vials and incubating at 37° for 6–7 hr with occasional shaking). The contents of these vials are then added to flasks containing 2 liters of Bacto-tryptic soy broth, and the flasks are incubated for 12–14 hr with stirring. The bacteria are pelleted (4500 rpm, 20 min, 4°), the pellets are resuspended in PBS-azide buffer, washed twice in the same buffer, and pelleted again in a centrifuge tube of known weight; the pellet is resuspended in PBS-azide buffer to yield a 10% (w/v) suspension. Formalin is then added to a final concentration of 1.5% (v/v); the mixture is stirred for 90 min at room temperature, the bacteria are pelleted, and the pellet is washed once with PBS-azide buffer and resuspended in PBS-azide buffer to yield a 10% (w/v) suspension, which is then heated at 80% for 5 min and chilled. The bacteria are pelleted, washed twice with PBS-azide buffer, resuspended in PBS-azide buffer to yield a 10% (w/v) suspension, dispensed into 1-dram glass vials, 1–2 ml/vial, and stored at −70°.

This formalin-fixed heat-killed Staph A is then activated for use as an immunoadsorbent as follows.[34] The bacteria in 1 ml of a 10% suspension are pelleted and resuspended in NET-0.5 buffer. The mixture is allowed to stand at room temperature for 15 min; the activated Staph A is then pelleted, the pellet is washed three times with NET-0.05 buffer, and the pellet is resuspended in sufficient NET-0.05 buffer to yield a 10% (v/v) suspension. The activated Staph A suspension is stored at 4°, and should be used within 6 hr of activation.

   c. *Labeling of Cells and Preparation of Cell Lysates.* The essential feature of the technique is to label protein to a very high specific activity so that the amount of protein that has to be loaded onto gels (see below) is as small as possible.

Cell monolayers (or suspension cultures) (3×10^6 cells) are starved for 1 hr in growth medium lacking methionine (using dialyzed serum) and labeled for 60 min in such medium supplemented with 100 μCi of [^{35}S]-methionine per milliliter. The cells are then incubated for another 30 min in complete medium containing no label, harvested into Earle's saline, washed and resuspended in 0.3 ml of 0.8 M KCl–10 mM Tris · HCl (pH 7.8)–1 mM PMSF–1% Triton X-100. Subsequent manipulations are carried out on 100-μl samples of these lysates. To each sample is added 400 μl of 10 mM Tris · HCl (pH 7.8)–1 mM PMSF–1% Triton X-100. The lysates are centrifuged for 1 hr at 45,000 rpm, and the supernatants are used for the subsequent immunoprecipitation.

[34] S. W. Kessler, *J. Immunol.* 115, 1617 (1975).

d. *Immunoprecipitation of Viral Proteins.* Cell lysates are first preabsorbed. To 50 μl of supernatants prepared as described above, 5 μl of preimmune serum (serum obtained before rabbits are immunized) is added, and the mixture is incubated for 30 min at room temperature; 50 μl of activated Staph A immunoadsorbent is then added, and after 15 min at room temperature the bacteria are pelleted as described above.

To 25 μl of the preabsorbed supernatant, 25 μl of preabsorbed antiserum are then added. After 1 hr at room temperature, 25 μl of activated Staph A immunoadsorbent is added; the incubation is continued for 15 min, and the bacteria are then pelleted.

One may proceed as follows. If the object of the experiment is merely to quantitate the total amount of virus-coded protein, the pelleted bacteria are washed in NET-0.05 buffer and the amount of radioactivity in the pellet is determined. If it is desired to examine the distribution of radioactive label among a variety of virus-coded proteins, the viral antigens are eluted from the Staph A–antibody–antigen complex and analyzed by analytical SDS–polyacrylamide gel electrophoresis. In that case the pelleted bacteria are washed three times with NET-0.05 buffer, and adsorbed labeled viral proteins are recovered by resuspending the pellets in 70 μl of "solubilizing mixture," which contains 50 mM Tris · HCl (pH 6.8)–2% SDS–5% 2-ME–10% glycerol–0.002% BPB; the suspension is heated at 100° for 5 min. The bacteria are then pelleted, and the supernatants are applied to SDS–polyacrylamide gels.

e. *Polyacrylamide Gel Electrophoresis.* The SDS–polyacrylamide gel electrophoresis is performed according to any of the numerous published and currently used techniques. It is advisable to refer to recent publications dealing with the proteins of the particular virus that is under investigation and to model the analytical electrophoresis system according to the one that gives particularly clear patterns for the particular virus that is used. Fluorograms are prepared as described by Laskey and Mills.[35] The films are then scanned with a densitometer, and the amount of radioactivity present in each viral protein band is calculated.

The amount of cells, as well as the amount of label and the size of the sample that should be used at each stage of the analysis, may not be optimal for all cell–virus systems: clearly the amount of viral protein that is synthesized will depend on the permissiveness of the particular cell for the particular virus that is used. The figures quoted above apply to reovirus growing in mouse L fibroblasts. Adjustments in these figures may have to be made for cell–virus systems in which less virus-specified proteins are synthesized. However, the basic principle remains, namely, viral

[35] D. A. Laskey and A. D. Mills, *Eur. J. Biochem.* **56,** 335 (1975).

proteins should be labeled as extensively as possible to avoid overloading the gels.

2. The Quantitation of Host-Cell Proteins

If the object is merely to quantitate the total amount of protein that is synthesized, it is only necessary to pulse-label cells with a radioactively labeled amino acid for a given period of time, such as 90 min, to chase for 30 min in unlabeled medium, to pellet and wash the cells (in Earle's saline), and to measure the amount of radioactivity that is insoluble in 5% trichloroacetic acid. If, on the other hand, the object is to measure the rate of synthesis of some specific protein, then it is necessary to use some technique such as measurement of a function (such as an enzyme activity) or it is necessary to prepare a specific antiserum. In that case, the technique to be used is identical to that described in Section C,1.

D. Composition of Buffers and Solutions

RSB: 10 mM NaCl, 10 mM Tris · HCl (pH 7.4), 1.5 mM Mg^{2+}

Acetate–EDTA buffer: 0.05 M sodium acetate (pH 5.1), 10 mM EDTA

Isotonic lysis buffer: 0.15 M NaCl, 2 mM Mg^{2+}, 10 mM Tris · HCl (pH 7.6), 0.05% Triton N-101 (Sigma)

Buffer E: 25 mM Tris · HCl (pH 7.5), 25 mM NaCl, 5 mM Mg^{2+}

Column buffer: 10 mM Tris · HCl (pH 7.4), 0.5 M KCl

SSC: 0.15 M NaCl, 0.015 M sodium citrate (pH 7.4)

ET buffer: 1 mM Tris · HCl (pH 7.4), 1 mM EDTA

SET buffer: 0.01 M NaCl, 0.05 M Tris · HCl (pH 7.4), 1 mM EDTA

TNE buffer: 0.05 M Tris · HCl (pH 7.5), 0.025 M NaCl, 2 mM EDTA

STE buffer: 0.1 M NaCl, 10 mM Tris · HCl (pH 7.4), 1 mM EDTA

TSE buffer: 0.15 M NaCl, 10 mM Tris · HCl (pH 7.4), 2 mM EDTA

TE buffer: 2 mM EDTA, 2 mM Tris · HCl (pH 7.4)

TESS: 0.01 M [N-tris(hydroxymethyl)methyl-2-aminoethanesulfonic acid (pH 7.4)], 0.3 M NaCl, 10 mM EDTA, 0.2% SDS

NT: 0.4 M NaCl, 10 mM Tris · HCl (pH 7.2)

HB: 0.3 M NaCl, 10 mM Tris · HCl (pH 7.2), 1 mM EDTA, 1% SDS

PBS: 0.136 M NaCl, 10 mM sodium phosphate (pH 7.2)

Solution A: 0.27 M NaCl, 5 mM KCl, 0.06 M Na$_2$HPO$_4$–2 mM KH$_2$PO$_4$ (pH 7.4)

Extraction buffer: 0.1 M Tris · acetate (pH 5.4), 0.01 M EDTA, 0.5% SDS

NET-1.0 buffer: 0.15 M NaCl, 5 mM EDTA, 50 mM Tris · HCl (pH 7.4), 0.02% NaN$_3$, 1.0% (v/v) Triton X-100

NET-0.5 buffer: 0.15 M NaCl, 5 mM EDTA, 50 mM Tris · HCl (pH 7.4), 0.02% NaN$_3$, 0.5% (v/v) Triton X-100

NET-0.05 buffer: 0.15 M NaCl, 5 mM EDTA, 50 mM Tris · HCl (pH 7.4), 0.02% NaN$_3$, 0.05% (v/v) Triton X-100

PBS-azide buffer: 150 mM NaCl, 40 mM NaH$_2$PO$_4$ (pH 7.2), 0.02% NaN$_3$

Earle's saline: 0.115 M NaCl, 5.3 mM KCl, 5.6 mM glucose, 26 mM NaHCO$_3$, 1.14 mM Mg^{2+}, 10 mM Na$_2$HPO$_4$ (pH 7.2)

[40] Induction of Interferon by Nonmultiplying Virus

By WOLFGANG K. JOKLIK

Despite intensive work, little is known concerning the mechanisms by which infection with viruses induces interferon. Several efforts have been made with temperature-sensitive virus mutants to determine whether some specific reaction during the viral multiplication cycle is responsible for inducing interferon. At first it was thought that the intracellular formation of double-stranded RNA (dsRNA) (in the form of replicative forms or intermediates) may correlate with ability to induce interferon (at least by RNA-containing viruses), but this was found not to be the case: studies with *ts* mutants of Semliki Forest virus[1] and Sindbis virus[2,3] have shown that, whereas some RNA$^-$ mutants do not induce interferon at nonpermissive temperatures, others do so. Furthermore, mutants of measles virus have been isolated that are indistinguishable from wt virus in all respects except inability to induce interferon[4,5]; and work with a set of reovirus ts mutants showed that neither early nor late mutants were able to induce interferon at nonpermissive temperatures, which suggested that the process that led to interferon induction was one of the very latest in virus morphogenesis.[6]

Viruses unable to replicate, such as viruses in nonpermissive cells and inactivated viruses, can also induce interferon. Examples of the former are Newcastle disease virus (NDV) in L cells[7] and reovirus in fish or chick cells[8,9]; examples of the latter are viruses inactivated by ultraviolet (UV)

[1] D. Lomniczi and D. C. Burke, *J. Gen. Virol.* **8**, 55 (1970).

[2] G. J. Atkins, M. D. Johnston, L. N. Westmacott, and D. C. Burke, *J. Gen. Virol.* **25**, 381 (1974).

[3] G. J. Atkins and C. L. Lancashire, *J. Gen. Virol.* **30**, 157 (1976).

[4] J. McKimm and F. Rapp, *Proc. Natl. Acad. Sci. U.S.A.* **74**, 3056 (1977).

[5] J. McKimm and F. Rapp, *Virology* **76**, 409 (1977).

[6] M. T. Lai and W. K. Joklik, *Virology* **51**, 191 (1973).

[7] F. Dianzani, S. Gagnoni, C. E. Buckler, and S. Baron, *Proc. Soc. Exp. Biol. Med.* **133**, 324 (1970).

[8] H. K. Oie and P. C. Loh, *Proc. Soc. Exp. Biol. Med.* **136**, 369 (1971).

[9] W. F. Long and D. C. Burke, *J. Gen. Virol.* **12**, 1 (1971).

irradiation, such as UV-irradiated NDV,[10-13] Colorado tick fever virus,[14] and reovirus.[6,9,15] A difficulty here, as with other inducers, is that there is considerable variation in response depending on the nature of the cell and of the interferon inducer. Thus unirradiated NDV induces interferon in mouse cells, but not in chick cells; UV-irradiated NDV induces interferon in chick cells, but not in L cells; and UV-irradiated reovirus induces interferon in mouse cells, but not in chick cells. Most probably these differences reflect the fact that the expression of eukaryotic genes is not controlled by simple "on-off" switches, but by complicated multicomponent mechanisms that are subject to feedback at several levels.

The purpose of this chapter is to describe several systems in which viruses unable to multiply can induce interferon. Some of the systems that have been examined are listed above. The three that will be described here in some detail are (a) the induction of interferon in avian cells by reovirus; (b) the induction of interferon in murine cells by UV-irradiated reovirus; and (c) the induction of interferon in human cells by bluetongue virus.

Induction of Interferon in Avian Cells by Reovirus

The cells chosen for this segment are duck cells, more specifically duck embryo fibroblasts, which are totally nonpermissive for reovirus. The reasons why duck cells are chosen over the more commonly used chick cells are that (a) systems for preparing chick interferon have already been described e.g., 9; (b) we have recently purified duck interferon to a specific activity that is at least one order of magnitude higher than what has been reported for chick interferon[16] (it is at least 1%, and possibly well over 10% pure); and (c) duck cells have the advantage over chick cells in harboring far less endogenous retrovirus genetic information, so that they are superior to chick cells at least for studies on the effect of interferon on retrovirus multiplication. It should be noted that chick interferon fails to protect duck cells, and vice versa.

Preparation of Duck Embryo Fibroblasts (DEF). Primary DEF are prepared from 12- 13-day-old Pekin duck embryos by modification of the method of Vogt.[17] Briefly, after sterile removal of the shell and outer

[10] D. C. Burke and A. Isaacs, *Br. J. Exp. Pathol.* 39, 452 (1958).
[11] M. Ho and M. K. Breinig, *Virology* 25, 331 (1965).
[12] J. S. Youngner, A. W. Scott, J. V. Hallum, and W. R. Stinebring, *J. Bacteriol.* 92, 862 (1966).
[13] S. S. Gandhi, D. C. Burke, and C. Scholtissek, *J. Gen. Virol.* 9, 97 (1970).
[14] E. J. Dubovi and T. G. Akers, *Proc. Soc. Exp. Biol. Med.* 139, 123 (1972).
[15] D. R. Henderson and W. K. Joklik, *Virology* 91, 389 (1978).
[16] R. F. Ziegler and W. K. Joklik, *J. Int. Res.* (in press).
[17] P. K. Vogt, *in* "Fundamental Techniques in Virology" (K. Habel and N. P. Salzman, eds.), p. 198. Academic Press, New York, 1969.

membrane, embryos are removed from the eggs through the use of a sterile rectal hook and are placed in a 150-mm plastic tissue culture dish containing sterile PBS1. The embryos are decapitated and eviscerated with sterile forceps, rinsed in a second tissue culture dish containing PBS1,[18] and placed in a third dish containing fresh PBS1. This is repeated for 15–20 embryos per preparation. A 0.25% trypsin solution, 10 ml per embryo, is added to a 300-ml trypsinizing flask containing a magnetic stirring bar. The embryos are placed into a sterile, 10-ml glass syringe fitted with a 1.5 inch, 16-gauge needle and are then passed through the needle into the trypsinizing flask. Ten embryos are trypsinized in each 300-ml flask.

The trypsin–embryo suspension is stirred for 15 min at 37°, the flask is tilted to allow the larger debris to settle, and the supernatant containing the trypsinized cells is poured sterilely into centrifuge bottles containing 20 ml of cold PGM per embryo. Trypsin is again added to the flask, and the trypsinization steps are repeated. If the extent of digestion of the embryos is then still insufficient, the trypsinization steps are repeated yet again. The cells are pelleted (1500 rpm, 15 min, 4°), the supernatant is discarded, and the cell pellet is resuspended in warm PGM (10 ml per embryo). The cells are counted after diluting 0.2 ml of the cell suspension into 10 ml of PBS1 with a Coulter counter Model F_N with an attenuation setting of 1, an aperture setting of 128, and a threshold setting of 25. The cells are seeded into 150-mm plastic tissue culture dishes in 25 ml of PGM (1.2×10^7 cells/dish) or into 1330 cm^2 glass roller culture bottles in 200 ml of PGM (2.5×10^8 cells/roller bottle). Tissue culture dishes are incubated at 37° in a water-jacketed incubator in a 5% CO_2–95% air atmosphere; roller culture bottles are placed on a Bellco roller bottle apparatus in a 37° warm room and are rotated at 6 rph.

Purification of Reovirus. The Dearing strain of reovirus serotype 3 is purified as described in this volume (Section A,3,a of Chapter 39).

Induction of Duck Embryo Fibroblast Interferon. The optimal conditions for inducing interferon in confluent monolayers of DEF are as follows[16]: To a 150-mm tissue culture dish containing 2×10^7 DEF growing in monolayer culture is added an inoculum containing 2×10^{12} reovirus particles (added multiplicity, 100,000 virus particles/cell) in 1 ml of Puck's saline A plus 1 ml of GM containing only 0.1% calf serum. After 2 hr at 37°, 23 ml of GM containing 0.1% calf serum are added, and the cells are incubated at 37° for 48 hr. Reovirus does not grow in DEF. The medium is then harvested, clarified by centrifugation, acidified to pH 2, and stored at 4°. This induction process, which has been optimized for reovirus multiplicity, effect of UV irradiation on reovirus (no advantage), duration of

[18] The composition of media, buffers, and solutions is given at the end of the chapter.

period of incubation, and medium serum concentration, yields 12,500 microtiter units (MTU) of interferon per 10^7 DEF.

If DEF growing in 1330 cm^2 glass roller culture bottles are used (2.5 × 10^8 cells per bottle), the inoculum consists of $1.25 × 10^{13}$ reovirus particles (added multiplicity 50,000 virus particles/cell) in 10 ml of a 1 : 1 mixture of Puck's saline A and GM containing only 0.5% calf serum. After 2 hr at 37°, 40 ml of GM containing 0.5% calf serum are added; the cells are incubated for another 58 hr, when the medium is harvested as described above. This induction process, which has been optimized for medium volume, length of incubation period, and multiplicity, yields $2.4 × 10^4$ MTU of interferon per 10^7 DEF.

Microtiter Assay of Duck Interferon. Duck interferon is assayed either by a cytopathic effect (CPE) reduction assay combining the techniques of Henderson and Joklik[15] and Tilles and Finland[19] (the "microtiter assay"), or by a plaque reduction assay.

For the microtiter assay, 100 µl of GM are added to each well of a Linbro microtiter 96-well tissue culture plate (8 rows of 12 wells each) using an 8 × 100 µl Titertek manifold. One hundred microliters of each interferon sample to be assayed is added to the first well of each row, and simultaneous serial twofold dilutions are then performed using an 8 × 100 µl manifold, 100 µl being discarded from the last well (final volume, 100 µl per well). To each well $4 × 10^4$ DEF are added in 50 µl of GM with an 8 × 50 µl manifold, and the plates are incubated for 24 hr at 37°, during which time the cells attach to the surface to form monolayers and, if interferon is present, become resistant to virus challenge. The monolayers are then challenged with 200 PFU of VSV per well in 50 µl of GM, the plates are incubated for an additional 36 hr at 37°, and the monolayers are examined under a light microscope for evidence of virus-induced cell destruction. Control monolayers, with no interferon added, show complete cell destruction. In interferon-treated cell monolayers, the change from complete protection to complete destruction usually occurs within a range of three wells. One MTU of interferon is defined as the reciprocal of the dilution that is midway between the highest dilution of interferon that completely protects the cell monolayer and the lowest dilution of interferon that permits complete destruction of the cell monolayer. With this assay 200 interferon samples can readily be titrated per day.

Plaque Reduction Assay of Duck Interferon. The plaque reduction assay for duck interferon is that of Lai and Joklik.[6] In brief, DEF are seeded into 60-mm tissue culture dishes at a density of $2 × 10^6$ cells/dish. Interferon samples are diluted serially in GM. When the DEF monolayers are confluent, the medium is removed and a 3-ml aliquot of each interferon dilu-

[19] J. G. Tilles and M. Finland, *Appl. Microbiol.* **16**, 1706 (1968).

tion is added to duplicate dishes. The dishes are incubated for 24 hr at 37°; the interferon-containing medium is then removed, 85 PFU of VSV are added to each dish in 0.1 ml of GM, and the dishes are incubated at 37° for 1 hr. Each dish is then overlaid with 5 ml of a mixture of 41% (v/v) prewarmed 2% Noble agar in H_2O and 59% (v/v) 2 × GM containing 50 μg of gentamycin per milliliter. The overlay is allowed to harden at room temperature, and the dishes are incubated until plaques are visible on the control plates by light microscopy (about 36 hr). Three milliliters of additional overlay containing 0.3 ml of 1 : 300 neutral red (Gibco) is then added to each dish, the dishes are incubated for an additional 6 hr at 37°, and the plaques are counted on a light box with indirect illumination. One plaque reduction unit of duck interferon is the dilution of interferon that reduces the number of plaques that develop in its presence to 50% of the number of plaques that develop in control monolayers (1 PRD_{50}).

Induction of Murine Interferon by UV-Irradiated Reovirus

Mouse L929 fibroblasts are grown in spinner culture in growth medium [Joklik's modified MEM (Gibco)] supplemented with 2% fetal calf serum and 3% calf serum. Aliquots of 4 × 10^6 cells are taken and placed into 100-mm plastic dishes in 11 ml of medium. The cultures are confluent 24 hr later, and the cells are then aged for another 24 hr at 37°. The medium is then removed and the desired inoculum is added (see below).

Reovirus type 3 (strain Dearing) is grown and purified as described in this volume [39]. For UV irradiation, 5 mg of virus, at a concentration of 1 mg/ml (1.13 × 10^{13} virus particles/ml) in SSC, are placed into a 100-mm plastic dish and UV irradiated for 120 sec, without shaking, with two 15 W GE germicidal lamps located at a distance of 8.8 cm.[15] At this dose each virus particle receives an average of 30 lethal hits (the titer is reduced by more than 10 logs), and the ability to transcribe reovirus dsRNA into ssRNA is inhibited by more than 95%. The ability of such virus to induce interferon in mouse L-929 fibroblasts is 100 times greater than that of nonirradiated virus. The reason for this appears to be labilization of the reovirus capsid shells and introduction of its dsRNA into the cell.[15]

For the actual induction of interferon, cell monolayers prepared as described above are inoculated with 0.2 ml of SSC containing 3 × 10^{10} virus particles UV irradiated as described above (multiplicity, 3000 virus particles per cell). After an adsorption period of 1 hr at 37°, the cells are washed with growth medium (see above), 7 ml of fresh growth medium are added, and the cells are incubated for 24 hr at 38°. The medium is then harvested and clarified by centrifugation; the pH is adjusted to 2 at 4°, which is the form in which the interferon samples may be stored. This induction process, which has been optimized for effect of UV dose, multiplicity, temperature of incubation, and length of incubation period, yields about 1 international interferon unit per cell.[15]

Induction of Human Interferon by Bluetongue Virus

Bluetongue virus (BTV) is a member of the reovirus family. It is a pathogen of ruminants, particularly of cattle and sheep, but cannot multiply in human cells. In spite of that, it is a potent inducer of interferon in several types of human cells, including leukocytes as well as normal and transformed cells of epithelial and fibroblastoid origin. Different cell lines differ greatly in the extent to which BTV induces the formation of interferon in them; further, it seems that the attenuated American BT-8 (vaccine) strain of BTV is by far the best interferon inducer. Although no detailed work has been carried out, it is likely that the mechanism by which BTV induces the formation of interferon is the same as that by which UV-irradiated reovirus does so, namely through breakdown of the virus particle and liberation of the dsRNA that it contains into the interior of the cell. Nothing is known concerning the steps that then ensue and that result in the derepression of the interferon gene.

I will describe here the induction of fibroblast interferon by the human bladder carcinoma cell line HT-1376.[20,21] The growth and purification of BTV is described elsewhere in this volume. HT-1376 cells are grown as monolayers in 32-ounce bottles in RPMI-1640 medium supplemented with 10% fetal calf serum. The cells are used when they reach about 2×10^7 cells/bottle. The growth medium is then poured off, and 1 ml of BTV containing 2×10^7 PFU in maintenance medium (see below) is added. After 30–60 min to allow for adsorption, 15 ml of maintenance medium are added. The maintenance medium consists of MEM supplemented with nonessential amino acids and 2% fetal calf serum. The supernatant fluid is harvested 24 hr later and centrifuged for 10 min at 350 g to remove cell debris; the medium is then adjusted to pH 2 at 4°. It is stored under these conditions until required for use.

The amount of interferon induced by HT-1376 cells induced with BTV usually does not exceed 0.02 interferon unit per cell; this value is considerably less than the value of almost 1 that is achieved by UV-irradiated reovirus in murine L-929 fibroblasts (see below). Irradiating BTV with UV light does not increase its interferon-inducing potential. Although the amount of interferon induced by BTV in human cells is considerably less

[20] P. Jameson and S. E. Grossberg, *Arch. Virol.* **62**, 209 (1979).
[21] P. Jameson and S. E. Grossberg, *in* "Human Interferon" (W. R. Stinebring and P. J. Chapple, eds.), p. 27. Plenum, New York, 1978.

than that induced by UV-irradiated reovirus in murine cells, BTV is nevertheless a valuable interferon inducer for human cells. In particular, as pointed out above, BTV will induce not only the formation of fibroblast interferon in HT-1376 cells, but also the formation of leukocyte interferon in human leukocytes (discussed elsewhere in Volume 78). In human leukocytes BTV is about twice as potent an inducer as NDV, which in turn is about five times as potent as 100 μg of poly(I) · poly(C) per milliliter.[21]

Composition of Media, Buffers, and Solutions

GM: Ham's F10; 10% (v/v) tryptose phosphate broth; penicillin, 50 units/ml; streptomycin, 50 μg/ml; fungizone, 2 μg/ml; 5% calf serum (pH 7.2)

PGM: GM plus 3% extra calf serum (total calf serum concentration equals 8%) and 2% chick serum (pH 7.2)

PBS1: 0.14 M NaCl, 8 mM Na$_2$ HPO$_4$, 1.5 mM KH$_2$PO$_4$, 2.5 mM KCl (pH 7.0)

Puck's saline A: 0.8% NaCl, 0.04% KCl, 0.009% Na$_2$HPO$_4$ · H$_2$O, 0.006% KH$_2$PO$_4$, 0.1% glucose, 0.035% NaHCO$_3$, 0.4% MgCl$_2$ · 6 H$_2$O, 0.002% phenol red (pH 7.2)

SSC: 0.15 M NaCl, 0.015 M sodium citrate (pH 7.4)

[41] Assay of Effects of Interferon on Plasminogen Activator[1]

By EDWARD W. SCHRODER and PAUL H. BLACK

Interest in the effects of interferon on plasminogen activator (PA) stems from two important characteristics of the enzyme. First, the release of PA by cells *in vitro* correlates with the transformed[2-4] or actively growing untransformed[3,4] cell phenotype. Second, intracellular PA is a membrane-associated enzyme[5-7] that is enriched in plasma membrane

[1] Publication No. 27 of the Hubert H. Humphrey Cancer Research Center at Boston University, Boston, Massachusetts

[2] L. Ossowski, J. C. Unkeless, A. Tobias, J. P. Quigley, D. B. Rifkin, and E. Reich, *J. Exp. Med.* **137**, 112 (1973).

[3] I. N. Chou, S. P. O'Donnell, P. H. Black, and R. O. Roblin, *J. Cell. Physiol.* **91**, 31 (1977).

[4] For review, see P. H. Black, *Adv. Cancer Res.* **32**, 75 (1980).

[5] J. Unkeless, K. Dano, G. M. Kellerman, and E. Reich, *J. Biol. Chem.* **249**, 4295 (1974).

[6] J. K. Christman, G. Acs, S. Silagi, and S. C. Silverstein, *in* "Proteases and Biological Control" (E. Reich, D. B. Rifkin, and E. Shaw, eds.), p. 827. Cold Spring Harbor Press, Cold Spring Harbor, New York, 1975.

[7] J. P. Quigley, *J. Cell Biol.* **71**, 472 (1976).

fractions of actively growing cells[7,8] and is released into the medium in a soluble form, presumably by a membrane-shedding mechanism.[4] The association of PA shedding with periods of cellular proliferation, and evidence that interferon can inhibit PA shedding,[9] raise the possibility that growth medium PA levels may serve as a readily quantitated correlate of the antiproliferative activity of interferon. The plasma membrane location of PA suggests that PA may be a useful model for studies both of membrane shedding and of cell surface marker expression. A large number of cell types have been shown to have readily measurable levels of PA[4] including normal and transformed fibroblasts of human and nonhuman origin, macrophages, and polymorphonuclear leukocytes. A method for assaying the effect of interferon on plasminogen activator will be described using, as an example, simian virus 40-transformed Swiss mouse 3T3 (SV3T3) cells treated with mouse L-cell interferon. It is hoped that our experience with this model will be useful to others who may be working with different cell types or interferon sources.

Cell Culture Methods

Reagents

Growth medium: Dulbecco's modification of Eagle's minimal essential medium (DMEM) supplemented to 5% (v/v) with fetal calf serum and with 250 units of penicillin and 250 μg of streptomycin per milliliter

Cells. SV3T3 cells[10] were resuspended twice weekly by trypsinization and used in our laboratory between passages 15 and 25.

Interferon. Mouse L cell interferon, partially purified by antibody affinity chromatography[11] to a specific activity of 3×10^7 reference units per milligram of protein, was stored lyophilized at $-70°$, and reconstituted with 0.5% bovine serum albumin in phosphate-buffered saline, pH 7.4 (PBS) and diluted to 10^5 units/ml. Reconstituted interferon may be dispensed in small aliquots and stored frozen at $-70°$ for several months without loss of activity.

Procedure. SV3T3 cells plated at a density of 5×10^3/cm² in DMEM containing 5% fetal calf serum, and incubated at $37°$ in a humidified atmosphere of 95% air and 5% CO_2, reach a density of approximately 3×10^5 cells/cm² after 4 days. Upon further incubation, the cell sheet tends to detach from the dish. The effects of interferon on PA levels of SV3T3 cells are best demonstrated when the cells have reached a density of 2 to 4 $\times$

[8] S. Jaken and P. H. Black, *Proc. Natl. Acad. Sci. U.S.A.* 76, 246 (1979).
[9] E. W. Schroder, I. N. Chou, S. Jaken, and P. Black, *Nature (London)* 276, 828 (1978).
[10] P. H. Black, *Virology* 28, 760 (1966).
[11] C. A. Ogburn, K. Berg, and K. Paucker, *J. Immunol.* 111, 1206 (1973).

10^4 cells/cm². At this density, both intracellular and shed PA are readily detected with relatively short exposures to serum-free medium, and before cell detachment becomes a problem. The optimal cell density will undoubtedly differ for other cell types and should be determined in preliminary experiments.

In a typical experiment, SV3T3 cells are plated in tissue culture dishes 35 mm in diameter as described above. After 24 hr, the medium is changed to fresh DMEM containing 5% fetal calf serum and supplemented with varying amounts of interferon. The cells are incubated for an additional 24 hr, after which they will have reached a density of 2 to 4×10^4/cm². The medium is removed, and the cells are washed twice with warm, serum-free DMEM and replaced with 1.0 ml/dish of serum-free DMEM containing interferon or bovine serum albumin as a control. Cells are incubated for 10–12 hr, the harvest fluids (HF) are collected, and cell lysates are prepared as described below.

PA Sample Preparation. The serum-free HF from control and interferon-treated cells is collected, chilled on ice, and centrifuged to remove cellular debris. The cell sheet is washed twice with warm PBS, and scraped into 1.0 ml of cold 0.1% Triton X-100 in 0.1 M Tris · HCl, pH 8.1, with a rubber policeman. The resulting cell lysates and the clarified HFs are stored frozen at $-30°$ until assayed for PA activity.

PA Assay

PA activity is conveniently measured using a fibrinolytic assay, essentially as described by Strickland and Beers[12] and as applied with minor modifications to Swiss 3T3 cells by Chou *et al.*[13] This assay is an indirect measure of PA, which is based upon the plasminogen-dependent release of fibrinopeptides from plastic tissue culture dishes coated with ^{125}I-labeled fibrin.

Reagents and Equipment

^{125}I-labeled fibrinogen: Bovine fibrinogen (fraction I, 90% clottable, Pentex) is iodinated by the IC1 method of Helmkamp *et al.*[14] Reaction mixtures containing 1–5 mg of bovine fibrinogen per milliliter and 2–5 mCi of NaI^{125} have typically yielded ^{125}I-labeled fibrinogen with a specific activity of 2 to 3×10^8 cpm/mg protein. The iodinated protein is dialyzed against PBS containing 0.3 N NaCl, and stored frozen at $-70°$.

[12] S. Strickland and W. H. Beers, *J. Biol. Chem.* **251**, 5694 (1976).

[13] I. N. Chou, R. O. Roblin, and P. H. Black, *J. Biol. Chem.* **252**, 6256 (1977).

[14] R. W. Helmkamp, R. L. Goodland, W. F. Bale, I. L. Spar, and L. E. Mutschler, *Cancer Res.* **20**, 1495 (1960).

Plasminogen: Human plasminogen is partially purified from pooled human serum by lysine–Sepharose chromatography,[15] dialyzed against PBS, adjusted to a protein concentration of 1 mg/ml, and stored frozen at $-30°$ in small aliquots.

Unlabeled fibrinogen: 0.1 mg of bovine fibrinogen per milliliter in PBS

DMEM containing 10% fetal calf serum, thrombin source

Tris · HCl, 0.1 M, pH 8.1.

Triton X-100, 0.1% in 0.1 M Tris · HCl, pH 8.1

Tissue culture dishes, 24 wells, 16 mm well diameter

Gamma scintillation spectrometer

Substrate Preparation. Assay plates are prepared by adding 0.25 ml of a solution containing 0.1 mg of unlabeled fibrinogen per well and 8×10^5 cpm of labeled fibrinogen per milliliter in PBS. The plates are dried for 2–3 days at 45° and stored at room temperature. For use, assay plates are activated by incubation for 2 hr at 37° with 1.0 ml of DMEM containing 10% fetal calf serum per well. The average per well loss of substrate during activation is determined by counting several portions of the activation medium in the gamma counter. The plates are then washed twice with 1 ml of 0.1 M Tris · HCl, pH 8.1, per well and used immediately.

Procedure. The PA samples are assayed in a reaction mixture containing 5–10 μg of human plasminogen, up to 0.1 ml of HF or cell lysate sample, and sufficient 0.1% Triton X-100 in 0.1 M Tris · HCl, pH 8.1, to bring the final volume to 1.0 ml. The HF and cell lysate sample volumes are selected by prior assay to yield from 10 to 40% hydrolysis of the remaining substrate (prior to activation, corrected only for radioactive decay), within 4 hr at 37°. One or two volumes of each sample within this linear range of the assay is assayed in duplicate or triplicate, and the mean percentage of hydrolysis is calculated. Although it is difficult rigorously to define a unit of activity for this assay, we have arbitrarily considered the amount of PA that generated enough plasmin to yield 10% hydrolysis of the starting substrate (corrected for decay) in 4 hr under the experimental conditions described, to be 1 unit of PA activity.[13] Appropriate blanks include assays of selected HFs and cell lysates in the absence of plasminogen, and the background hydrolysis due to the plasminogen preparation in the absence of sample. These blanks are run routinely, and the total background is subtracted from experimental values before units of PA activity are calculated.

Although concentrations of HF of up to 10% of the total reaction volume do not significantly inhibit the fibrinolytic reaction as described, larger volumes of HF samples may be inhibitory and should be avoided.

[15] K. C. Robbins and L. Summaria, this series, Vol. 45, p. 261.

EFFECT OF INTERFERON ON PLASMINOGEN
ACTIVATOR (PA) ACTIVITY OF SV3T3 CELLS

| Interferon[a] (units/ml) | Percent of control units PA[b] | |
	HF	Cell lysate
0	100 ± 4	100 ± 6
50	83 ± 7	120 ± 7
100	72 ± 7	135 ± 12
500	42 ± 8	106 ± 6

[a] Interferon was assayed in 3T3 cells by a neutral red dye uptake method patterned after that of N. B. Finter [*J. Gen. Virol.* 5, 419 (1969)] using vesicular stomatitis virus as the challenge virus. Units are expressed as reference units based on NIH reference standard G-002-904-51.

[b] The PA data are derived from triplicate assays of each of two culture dishes that had been treated in parallel and are shown as the arithmetic mean values ± standard error of the mean. PA units are as described in the text.

Direct addition of up to 500 units of interferon per milliliter did not alter assayable levels of PA.

Sample Data. With the procedures described, we have determined that interferon treatment of SV3T3 cells reduces the amount of PA that is shed into the culture medium, perhaps by inhibiting some aspect of the shedding process.[9] The table illustrates the results of a representative experiment. The amount of PA detected in the HF is diminished in a manner that is dependent upon interferon dose, whereas intracellular PA levels are either slightly elevated or unchanged.

Possible Application

As indicated by several chapters in this volume, there has been considerable interest in the antiproliferative activity of interferons against transformed and tumor cells, and some indication that interferons may be useful antitumor agents for certain malignancies. To date, the potency of interferon preparations used for such studies has been determined and expressed in terms of the classical antiviral properties of interferon. Most cells are several orders of magnitude more sensitive to the antiviral activities of interferon than to the antiproliferative effects. Although the mechanisms of interferon-induced antiproliferative effects on cells are

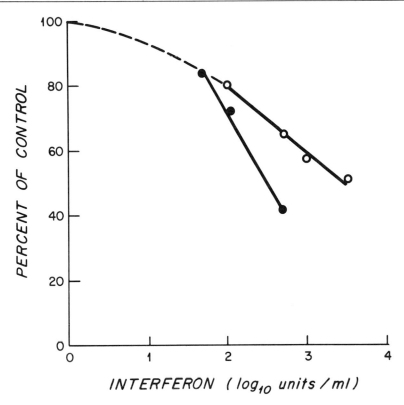

FIG. 1. Responses of SV3T3 cell plasminogen activator (PA) shedding and proliferation to various interferon doses. ●, Percentage of control PA units per 1.0 ml of harvest fluid. ○, percentage of control cells per culture after 3 days of interferon treatment as described in the text.

poorly understood, it is likely that these mechanisms will be found to differ from those responsible for the antiviral functions. Thus, the development of an assay of interferon potency based upon a cellular function that is coupled to the proliferative state of those cells might offer a more appropriate means of standardizing the anticellular potency of interferon than do assays based upon antiviral activities.

Although we have not applied the extensive analyses to the effects of interferon on PA shedding which would be necessary to produce a functional assay for interferon potency, graphic representation of the data presented in the table suggests that the effects of interferon on PA, as described, have some of the characteristics required for such an assay (Fig. 1). The decrease in PA shedding by interferon-treated SV3T3 cells is linear over a limited dose range, as is the antiproliferative response. Although the two curves do not coincide, the sensitivities of the cells to the two effects of interferon are more closely related to each other than to the antiviral dose response (not shown) on which the numbers of interferon units indicated in the abscissa are based.

The assay of interferon effects on PA as described in this chapter shares some of the disadvantages common to many of the existing biological assays for interferon potency. In its present form, the total procedure requires several days, the use of some expensive equipment, and the use of a radiolabeled substrate with the related problems of safe use and disposal. However, synthetic substrates have been described for serine proteases,[16] and it may be possible to develop a less cumbersome procedure in the near future.

[16] P. L. Coleman, H. G. Latham, Jr., and E. N. Shaw, this series, Vol. 45, p. 12.

[42] Assay of Effect of Interferon on Intracellular Enzymes

By T. Sreevalsan, E. Lee, and Robert M. Friedman

Interferons possess a myriad of biological activities other than the antiviral action; for instance, several reports have unequivocally documented the growth inhibitory effects of interferons in normal and malignant cells.[1] In addition to the effect on cell division, interferons display a pleiotropic effect on enzymes in intact cells. Several reports show that interferons may enhance, modify, or inhibit cellular enzymes and functions, depending upon the cell system used.

Apart from the enzyme precursors for protein kinase and (2'-5')-oligoadenylate synthetase expressed during the establishment of the antiviral activity, interferons can cause enhanced production or activity of some enzymes in tissue culture cells. Thus, the induction of arylhydrocarbon hydroxylase by benzanthracene was stimulated by mouse interferon in fetal mouse cells[2]; production of prostaglandin E in human fibroblast cultures was also stimulated by interferon[3]; human leukocyte interferon enhanced the release of histamine from human leukocytes exposed to ragweed antigen E or anti IgE[4]; and interferon treatment of mouse L cells resulted in a significant increase in the total level of tRNA methylase activity.[5]

[1] I. Gresser, *Adv. Cancer Res.* **16**, 97 (1978).

[2] D. W. Nebert and R. M. Friedman, *J. Virol.* **11**, 193 (1973).

[3] M. Yaron, I. Yaron, D. Gurari-Rotman, M. Revel, H. R. Lindner, and U. Zoy, *Nature (London)* **267**, 457 (1977).

[4] S. Ida, J. J. Hooks, R. P. Siraganian, and A. L. Notkins, *J. Exp. Med.* **145**, 892 (1977).

[5] K. R. Rozee, S. Lee, and J. McFarlane, *Can. J. Microbiol.* **15**, 969 (1969).

Examples of inhibition rather than stimulation or alteration of cellular enzymes or antigens by interferons have been reported. Although the effect of interferons in whole cells is less dramatic than that caused by conventional inhibitors of protein synthesis, several instances have been described indicating the selective inhibition of certain proteins or enzymes by interferons. Treatment of cells with mouse or rat interferon resulted in a significant suppression of the dexamethasone-induced tyrosine amino transferase in rat heptoma cells.[6] Similarly hydrocortisone-induced synthesis of glycerol-3-phosphate dehydrogenase in rat glial cells[7] or glutamine synthetase in chick embryonic neural retinal cells[8] was inhibited when cells were treated with rat or chick interferon, respectively. In addition, the induction of ornithine decarboxylase (ODC), a key enzyme in polyamine biosynthesis, in mouse Swiss 3T3 cells can be inhibited by mouse interferon.[9] Recent experiments have shown that S-adenosyl-1-methionine decarboxylase (SAM-decarboxylase), another inducible enzyme in the polyamine biosynthesis pathway, can also be inhibited by interferon in 3T3 cells (Lee and Sreevalsan, unpublished observations). In general there appears to be some selectivity in the types of enzymes inhibited by interferon since, in chick retinal cells, the induction of glutamine synthetase was inhibited whereas the level of enzymes such as acetylcholine esterase or lactic dehydrogenase was unaffected.[8] Although serum induced the synthesis of ODC and glucose transport proteins in quiescent Swiss 3T3 cells, only the former was inhibited by interferon.[9] The selective inhibition exerted by interferons is further exemplified by the observation that in Friend leukemia cells the dimethyl sulfoxide-induced synthesis of globin was inhibited, but there was no effect on gross protein synthesis.[10]

We have chosen for the present discussion the example of the inhibitory effect of mouse interferon on ODC and SAM-decarboxylase induction in quiescent mouse Swiss 3T3 cells. Swiss mouse 3T3 cells, a line of cells originally isolated and characterized by Todaro and Green,[11] provide an excellent experimental cell system to study the effect of interferon on cellular enzymes. These cells can be propagated in serial cultures. Upon

[6] G. Beck, P. Poindron, D. Illinger, J. P. Beck, J. P. Ebel, and R. Falcoff, *FEBS Lett.* **48**, 297 (1974).

[7] D. Illinger, G. Coupin, M. Richards, and P. Poindron, *FEBS Lett.* **64**, 391 (1976).

[8] T. Matsuno, N. Shirasawa, and S. Kohno, *Biochem. Biophys. Res. Commun.* **70**, 310 (1976).

[9] T. Sreevalsan, J. Taylor-Papadimitriou, and E. Rozengurt, *Biochem. Biophys. Res. Commun.* **87**, 679 (1979).

[10] G. B. Rossi, A. Dolei, L. Cioe, A. Benedetto, G. P. Materese, and F. Belardelli, *Proc. Natl. Acad. Sci. U.S.A.* **74**, 2036 (1977).

[11] G. J. Todaro, and H. Green, *J. Cell Biol.* **17**, 299 (1963).

reaching confluency the cells cease dividing and remain viable in the quiescent state for several days.[11] The addition of serum or a mixture of three well-defined polypeptide hormones, epidermal growth factor (EGF), insulin, and vasopressin, to such cells when in the G_0/G_1 phase, stimulates a complex spectrum of biochemical changes that precede the onset of DNA synthesis.[12] A dramatic increase of 50- to 100-fold over quiescent cultures in the level of ODC and SAM-decarboxylase occurs in stimulated cells along with numerous other prereplicative events. Mouse interferon, when added to cultures at the time of stimulation, can suppress these enzymes as well as inhibit the initiation of DNA synthesis.[9,13]

Both ODC and SAM-decarboxylase are examples of inducible enzymes in tissue culture cells. A variety of different cell cultures including Swiss 3T3, when stimulated to proliferate, respond with a large increase in the level of both enzymes. Several agents, including polypeptide hormones, hypoxanthine, asparagine, cyclic AMP, and tumor promoters, such as phorbol esters, are potent inducers of ODC.[14] Indeed, induction of these enzymes is an invariant component of the integrated response of proliferating cells.[15] Ornithine decarboxylase is a cytosol enzyme with an absolute requirement for pyridoxal phosphate[14] and has been purified from rat liver[16] and a virus-transformed line (SV3T3) of 3T3 cells.[17] The enzyme exhibits a short half-life (10–20 min) in both intact tissue and cell cultures.[14] The SAM-decarboxylase has been purified[18] and the enzymic activity can be stimulated *in vitro* by putrescine or inhibited by spermine.[18] Like ODC, SAM-decarboxylase also possesses a short half-life (35–60 min) and rapid inducibility in cell cultures.[15] Both enzymes can be assayed in the laboratory easily by determining the amount of $^{14}CO_2$ released from appropriately labeled ornithine or S-adenosylmethionine.

Preparation of Quiescent Cultures of 3T3 Cells

Swiss mouse 3T3 cells[11] are maintained as monolayer cultures using Dulbecco's modification of Eagle's medium containing 10% calf serum, penicillin (50 units/ml), and streptomycin (50 mg/ml). Cells are grown in a humidified atmosphere of a 90% air–10% CO_2 mixture at 37°. Stocks are seeded with 5×10^4 cells per 100-mm tissue culture plastic plate and subcultured at the third day with 0.25% trypsin to disperse the cells.

[12] E. Rozengurt, A. Legg, and P. Pettican, *Proc. Natl. Acad. Sci. U.S.A.* **76**, 1284 (1979).
[13] F. Balkwill and J. Taylor-Papadimitriou, *Nature (London)* **274**, 798 (1978).
[14] J. Janne, H. Poso, and A. Raina, *Biochim. Biophys. Acta* **473**, 241 (1978).
[15] D. V. Maudsley, *Biochem. Pharmacol.* **28**, 153 (1979).
[16] M. F. Obenrader and W. F. Prouty, *J. Biol. Chem.* **252**, 2860 (1977).
[17] R. J. Boucek and K. J. Lembach, *Arch. Biochem. Biophys.* **184**, 408 (1977).
[18] T. Oka, K. Kano, and J. W. Perry, *Adv. Polyamine Res.* **1**. (1977).

Cultures are tested routinely for contamination with *Mycoplasma*. Quiescent cultures are obtained in the following manner. Tissue-culture dishes (100 mm) are seeded with 3.8×10^5 cells/dish and incubated at 37°. The cultures receive fresh medium after 24 hr of incubation. They become confluent and quiescent by 6 days of incubation. To monitor quiescence, cells can be labeled with [³H]thymidine for 24 hr, and subsequent autoradiography should show that less than 1% of the cell nuclei become radioactively labeled.[12] The quiescent cultures are stimulated by washing twice with warm growth medium containing no serum and then incubating with growth medium containing 10% serum or a mixture of three growth factors (epidermal growth factor, 10 ng/ml; insulin, 1 g/ml; and vasopressin, 10 ng/ml). The onset of DNA synthesis occurs in such cultures by 12 hr after stimulation. A striking elevation in the level of ornithine and SAM-decarboxylases occurs in the stimulated cultures by 6–7 hr after addition of serum or growth factors.

Enzyme Induction in Quiescent Swiss 3T3 Mouse Cells

The cultures (in 100-mm dishes) are washed twice with warm growth medium containing no serum; then each culture receives 8 ml of warm medium containing the mitogen (either 10% serum or a mixture of the three growth factors). In experiments where the inhibitory effect of interferon is to be tested we add interferon to the cultures at this time. Appropriate controls should include (*a*) unstimulated cultures that receive growth medium without the mitogen; and (*b*) stimulated control cultures that receive the growth medium and appropriate mitogens. Duplicate cultures are used for each group. Since the appearance of the enzymes in cells is maximal at 6–7 hr after stimulation, the cultures are harvested at that time. Additionally, to determine the enzymic activity in unstimulated cells some cultures must be harvested at the onset of the incubation. Cultures are harvested at 4° by removing the medium and washing the cells with cold phosphate-buffered saline (PBS). The cells are scraped into cold PBS with a rubber policeman and pelleted at 800 g for 15 min at 4°. The pellet obtained may be stored at least for a week at −70° before performing the enzyme assay.

Assay of Ornithine Decarboxylase Activity

Principle. The enzymic activity is determined by measuring the amount of ¹⁴CO₂ released from [1-¹⁴C]ornithine according to the reaction

$$NH_2(CH_2)_3CH(NH_2)^{14}COOH \rightarrow NH_2(CH_2)_4NH_2 + {}^{14}CO_2$$

The enzyme has an absolute requirement for pyridoxal phosphate and is stimulated by sulfhydryl compounds. The assay procedure as described here is similar to that reported by Hogan[19] with minor modifications.

Reagents

N-2-Hydroxyethylpiperazine-N'-2-ethanesulfonic acid (HEPES)
Dithiothreitol (DTT)
Pyridoxal phosphate
EDTA
Protosol (New England Nuclear Co)
Trichloroacetic acid, 50% w/w
Fluid for scintillation counting: 760 ml of toluene containing 0.45% PPO and 0.008% POPOP + 440 ml of Triton X-100 + 100 ml of 5% trichloroacetic acid)
Stock solution (10×) for the enzyme assay contains 500 mM HEPES (pH 7.1), 50 mM DTT, 5 mM pyridoxal phosphate, and 1.0 mM EDTA. The pH of the solution is adjusted to 7.1; aliquots can be stored at $-20°$ for several weeks.

Assay Procedure. The cell pellet (1 to 1.5×10^7 cells) is suspended in 250 μl of 1 × stock solution, mixed well, and frozen and thawed three times to achieve efficient cell lysis. Then the cell lysate is centrifuged at 17,000 g for 15 min at 4°. The supernatant is recovered and serves as the extract.

The components of the assay are mixed in chilled test tubes (15 mm × 85 mm), kept in ice, and are added in the following order: 1 × of stock solution, 300 μl; cell extract, 100 μl; L-[^{14}C]ornithine in 1 mM ornithine, 100 μl. The final volume of the reaction is 0.5 ml, and it contains 0.2 μCi of L-[1-^{14}C]ornithine (specific radioactivity 45 mCi/mmol), 0.2 mM L-ornithine, 50 mM HEPES (pH 7.1), 0.1 mM EDTA, 0.05 mM pyridoxal phosphate, and 5 mM DTT. The reaction is run in duplicate for each sample. A blank reaction (in duplicate) contains all the components described above except that the 100 μl of enzyme extract are replaced by 100 μl of the cocktail. The tubes are capped with serum stoppers holding a polyethylene well (Kontes). A strip of Whatman No. 1 filter paper (7.5 by 25 mm) containing 20 μl of Protosol is placed in the polyethylene well as described by Kobayashi[20] to adsorb the CO_2 liberated during the reaction. The reaction is initiated by placing the chilled tubes in a 37° shaking water bath. After incubation at 37° for 60 min, the reaction is stopped by the injection of 0.5 ml of 50% trichloroacetic acid. To ensure complete liberation of CO_2, the incubation is continued for an additional 30 min. The filter paper from the polyethylene well is transferred to vials containing 10 ml of

[19] B. L. M. Hogan, *Biochem. Biophys. Res. Commun.* **45**, 301 (1971).
[20] Y. Kobayashi, *Anal. Biochem.* **5**, 284 (1963).

the counting solution and assayed for radioactivity in a scintillation spectrometer. All counts are corrected by subtracting the value obtained where no enzyme (blank) is added. The results are expressed as nanomoles of $^{14}CO_2$ released per milligram of protein in 60 min. Protein is measured by the method of Lowry et al.[21] with crystalline bovine serum albumin as standard. Since DTT present in the extract interferes with the assay, protein in the samples is precipitated with 6 volumes of alcohol or 5% trichloroacetic acid and dissolved in 0.1 M NaOH for the Lowry assay.

Interferon and ODC Induction

In our experience, the addition of serum (10%) to quiescent Swiss 3T3 cells increases ODC activity, which reaches a maximal level (15–20 nmol per milligram of protein in 60 min) by 6 hr and thereafter declines.[9] The rise of enzymic activity in serum-treated cultures is usually 50- to 100-fold over unstimulated cells. The kinetics of induction of the enzyme are similar to those of serum when growth factors are used. However, in cells induced with the growth factor mixture the enzymic activity is 3- to 5-fold higher than with serum. Regardless of the inducer used (serum or growth factors), mouse interferon added to the cultures at the time of induction inhibits the enzyme induction. The degree of inhibition is dependent upon the dose of interferon used; for instance, at 1000 units/ml the inhibition ranges from 70 to 90% of control values. Additionally, mouse interferon preparations of varying specific activity (5×10^6 units/mg protein to 2×10^8 units/mg protein) give similar results, suggesting that the inhibitory effect on enzyme induction is specific. Human lymphoblastoid interferon, even at a concentration of 1×10^4 units/ml, does not inhibit enzyme induction in Swiss 3T3 mouse cells.[9]

Assay of SAM-Decarboxylase

Principle. The assay method consists of measuring the $^{14}CO_2$ released from S-adenosyl[1-^{14}C]methionine. Adenosyl[1-^{14}C]methionine is converted to 5'-deoxy-5'-S-(3-methylthiopropylamine) sulfonium adenosine and $^{14}CO_2$. As reported by Pegg and Williams-Ashman,[22] unlike the *Escherichia coli* enzyme,[23] Mg^{2+} ion is not required for the decarboxylation of adenosylmethionine by extracts from mammalian tissues. The presence of putrescine in the reaction mixture is essential for extensive decarboxylation of the substrate.

[21] O. H. Lowry, N. J. Rosebrough, A. L. Farr, and R. J. Randall, *J. Biol. Chem.* 93, 265 (1951).
[22] A. E. Pegg and H. G. Williams-Ashman, *J. Biol. Chem.* 244, 682 (1969).
[23] C. W. Tabor, this series, Vol. 5, pp. 756 and 761.

Reagents

Sodium phosphate buffer, pH 7.20
DTT
Putrescine
EDTA
S-Adenosyl-L-methionine, S-adenosyl[1-^{14}C]methionine

The rest of the reagents needed are the same as those described under ornithine decarboxylase. The 1× stock solution contains sodium phosphate buffer, 100 mM, pH 7.2; DTT, 5 mM; putrescine, 2.5 mM; and, EDTA, 0.1 mM. This is prepared and stored as a 10× solution at −20°. Since S-adenosylmethionine is unstable at neutral or basic pH,[22] it should be stored at −20° in a solution of sulfuric acid : ethanol (9 : 1, v/v), pH 2.0. Under these conditions the rate of decomposition is less than 1% a year.

Assay Procedure. The procedure for the preparation of cell extract is essentially that described for ornithine decarboxylase except that the 1× stock solution used for suspending the cell pellet contained the components described above. The details of the procedure for enzyme assay are similar to those for ornithine decarboxylase. The final volume of the reaction mixture is 0.5 ml and contains 1.0 μCi of L-[1-^{14}C]-S-adenosyl-L-methionine (60 mCi/mmol), 0.1 mM S-adenosyl-L-methionine as well as the components of the 1× stock solution. The reaction mixture is incubated at 37° for 1 hr. All counts are corrected by subtracting the value obtained when no enzyme (blank) is added. The specific activity of the enzyme is expressed as nanomoles of ^{14}CO$_2$ released per milligram of protein in 60 min.

Interferon and SAM-Decarboxylase Induction

Addition of serum (10%) or a mixture of the three growth factors to quiescent Swiss 3T3 mouse cells results in an increase in enzymic activity of 50- to 100-fold over that in unstimulated cells (Lee and Sreevalsan, unpublished observation). The temporal appearance of the enzyme follows a similar pattern to that of ornithine decarboxylase: after an initial lag of 2 hr the enzymic activity increases until it reaches a maximal value by 7 hr after stimulation, and then declines. As compared to ornithine decarboxylase the maximal specific activity observed with SAM-decarboxylase in Swiss 3T3 cells is low (0.5–1.5 nmol of ^{14}CO$_2$ per milligram of protein per hour). Addition of mouse, but not human, interferon to quiescent cells at the time of stimulation results in a sustained inhibition of enzyme induction, the degree of inhibition being dependent upon the dose of interferon; for instance, mouse interferon at 1000 units/ml caused a 70–90% inhibition of enzyme induction (E. Lee and T. Sreevalsan, unpublished finding). As noted with ornithine decarboxylase induction, interferon preparations of varying specific activity have similar inhibitory effects.

Acknowledgment

The work performed at Georgetown University was supported by American Cancer Society Grant CD-56.

Section IV

Methods to Study Interferon Activity at the Cellular Level

[43] Methods and Procedures for Experiments with Mixed Cell Populations: Transfer of the Antiviral State Induced by Interferon

By J. EDWIN BLALOCK

We have shown that induction of the antiviral state is probably mediated by secondary messenger molecules that are induced at the cell membrane by interferon, signal the nucleus to produce antiviral protein(s), and by cell-to-cell transfer induce viral resistance in adjacent cells.[1,2] These findings resulted from coupling two previous observations. First, many cell types communicate among themselves by gap junctional transfer of metabolites and small control molecules.[3] Second, interferon action shows species preference.[4] We reasoned that, if induction of the antiviral state was mediated by small secondary molecules, these by gap junctional transfer might influence adjacent cells. This was tested by coculturing two different cell species in the presence of interferon to which only one cell species was sensitive and determining whether the other became resistant to virus infection. It was found that under these conditions the cell species not directly sensitive to interferon became resistant to virus infection. The cell-to-cell transfer of viral resistance was initiated by interferon, was rapid, and required ongoing RNA synthesis in the recipient cell.[2] This process represents a major amplification system for interferon action.[5]

Assay of Interferon-Induced Transfer of Viral Resistance

Freshly trypsinized mouse L cells and human amnion (WISH) cells in Eagle's minimal essential medium supplemented with 2% fetal calf serum were cocultured in a 1:1 ratio in Micro Test II tissue culture plates (Falcon Plastics, Oxnard, California). The total number of cells in each well (about 28 mm^2) was 1.5×10^5 in 0.15 ml, which is comprised of 7.5×10^4 cells/well of each of the two cell species. Controls consisted of an equivalent number of either cell species alone. Various concentrations of mouse virus-type interferon in 0.1 ml or an equivalent volume of medium were

[1] J. E. Blalock and S. Baron, *Nature (London)* **269**, 422 (1977).
[2] J. E. Blalock and S. Baron, *J. Gen Virol.* **42**, 363 (1979).
[3] W. R. Lowenstein *in* "Cell Membranes: Biochemistry, Cell Biology and Pathology" (G. Weissmann and R. Claiborne, eds.), p. 105. HP Publ., New York, 1975.
[4] R. Z. Lockhart, *in* "Interferons and Interferon Inducers" (N. B. Finter, ed.), p. 11. North-Holland, Amsterdam, 1973.
[5] J. E. Blalock, *Proc. Soc. Exp. Biol. Med.* **162**, 80 (1979).

METHODS IN ENZYMOLOGY, VOL. 79

TABLE I
TRANSFER OF INTERFERON-INDUCED VIRAL RESISTANCE BETWEEN
CELLS OF VARIOUS SPECIES

Cells[a]		Interferon (300 units/ml)	Log_{10} inhibition of VSV yield from recipient cells[b]
Donor	Recipient		
WISH	BHK-21	Human	0.7
RK-13	WISH	Rabbit	0.8
L	CE	Mouse	0.8
L	Vero	Mouse	0.8
L	WISH	Mouse	1.2
RK-13	Vero	Rabbit	None
WISH	MDCK	Human	None

[a] WISH, human amnion cells; RK-13, rabbit kidney cells; L, mouse fibroblasts; BHK-21, baby hamster kidney cells; CE, secondary chick embryo cells; Vero, African green monkey kidney cells; MDCK, canine kidney cells.

[b] Log_{10} inhibition was calculated according to the formula in the text.

added, and cultures were incubated overnight at 37° in a 4% CO_2 atmosphere. Units of interferon are expressed in terms of the NIH reference mouse interferon. Supernatant fluids were decanted, and each well was infected with 10^3 plaque-forming units (PFU) of poliovirus in 0.1 ml. Virus yields from pooled triplicate cultures were determined approximately 24 hr later by a slightly modified microplaque assay in which 0.5% methylcellulose (1500 centipoise) was substituted for carboxymethyl cellulose.[6]

Results of a typical experiment are shown in Fig. 1. Poliovirus did not replicate in mouse L cells, and mouse interferon did not alter the yield of poliovirus from the human WISH cells. Coculturing mouse L cells and human WISH cells in the absence of interferon did not affect the yield of poliovirus from the WISH cells, whereas mouse interferon caused a dose-dependent reduction in poliovirus yields from human WISH cells in the cocultures. The log_{10} inhibition of virus yields in the cocultures in the presence of mouse interferon is taken as a measure of the transfer of interferon-induced viral resistance.

Choice of Cell Species

The prototype transfer system consists of mouse interferon, mouse L cells, and human WISH cells, but other cell combinations and their respective interferons may be substituted.[7] Table I shows a compilation of

[6] J. B. Campbell, J. Grunberger, M. A. Kochman, and S. L. White, *Can. J. Microbiol.* **21,** 1247 (1975).

[7] T. K. Hughes, J. E. Blalock, and S. Baron, *Arch. Virol.* **58,** 77 (1978).

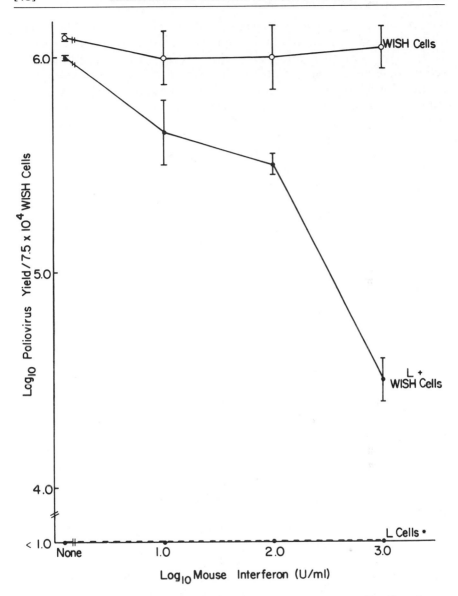

FIG. 1. Interferon-induced transfer of viral resistance from mouse L cells to human amnion (WISH) cells. *Poliovirus does not replicate in mouse cells. Each point represents the mean yield and the 95% confidence limits.

the cell combinations we have tested and the results obtained. The conditions of the assay for these combinations are the same as mentioned above except that vesicular stomatitis virus (VSV) was substituted for poliovirus. The only apparent prerequisites to choosing the species of cells employed are that the donor cell species be sensitive to its respective interferon while the recipient cell species is insensitive to that same interferon. Additionally one must have a recipient cell species that is permissive for an interferon-sensitive virus.

By the appropriate choice of cell types transfer between cells within a species may be also observed. For instance, human amnion WISH and human amnion U cells differ in their kinetic response to human fibroblast interferon. WISH cells respond about 1–2 hr faster than U cells. When U cells and WISH cells are cocultured, the U cells now respond at the same rate as the WISH cells.[8] Also, mouse L cell clones can be obtained that differ 10-fold in their sensitivity and 100-fold in the maximum protection afforded by mouse fibroblast interferon. By cocultivation of a high-responder and a low-responder clone, the response of the low-responder clone approaches that of the high response.[5] These results are interpreted to result from transfer of viral resistance within cells of the same species. In general, transfer within a species is more efficient than between species.[8]

Choice of Virus

Theoretically any interferon-sensitive virus may be employed as the challenge so long as it replicates in the recipient cells. In practice we have used poliovirus, vesicular stomatitis virus, Sindbis virus, and vaccinia virus. All these viruses under the proper conditions can be used to demonstrate transferred resistance. Care must be taken when virus types are used that at higher multiplicities of infection (MOI) can overcome directly induced viral resistance, since transferred resistance is also overwhelmed.[2] For example, at an input MOI of 0.1 PFU/cell vaccinia virus yield was inhibited by mouse interferon (7500 units/ml) in L cells alone (98.5% reduction) as well as in a 1 : 1 L cell and WISH cell mixture (93.3% reduction). At an input MOI of 1 PFU/cell, vaccinia virus was not sensitive to mouse interferon in L cells alone and transfer of resistance to WISH cells was not observed.[1] The MOI effect is one of many characteristics that show that transferred resistance is very similar if not identical to the directly induced antiviral state.

Transferred viral resistance seems to parallel virus sensitivity to the antiviral state, which is directly induced by interferon. In most instances,

[8] J. E. Blalock and G. J. Stanton, *J. Gen. Virol.* **41**, 325 (1978).

one would want to work with the most sensitive virus type possible. Poliovirus is convenient to use since it is relatively sensitive to interferon and has the added advantage that it grows only in human cells. In the mouse interferon, mouse L cell, and human WISH cell system, this allows for the direct demonstration that resistance develops in the human cells. However, we commonly employ VSV as a challenge, since in our experience it is more sensitive to interferon than is poliovirus. When VSV is used as a challenge there is in theory the complication that it grows in both mouse and human cells. In practice, this does not affect the results, since for all intents and purposes VSV does not replicate in mouse L cells at the concentrations of mouse interferon employed. The VSV yield from mouse interferon-treated L cells is usually 2–3 $\log_{10}$ below the WISH cell yield when 10–100 units of mouse interferon are used. When VSV is the challenge virus, the amount of viral resistance that is transferred to WISH cells is based on an expected virus yield. This yield usually gives equivalent results whether it is calculated from the percentage of WISH cells in a L cell–WISH cell mix without interferon or the virus yield from an equivalent number of WISH cells with mouse interferon. Virus yields from interferon-treated or nontreated human WISH cells are very similar to the yield from non-interferon-treated L cells, and the virus yield from mouse interferon-treated L cells is negligible in comparison. The $\log_{10}$ inhibition of VSV yield from a transfer experiment that employed a 1 : 1 ratio of L cells to WISH cells would employ the following formula:

$$- \log_{10} \frac{\text{virus yield from cell mix with interferon}}{1/2 \text{ virus from cell mix without interferon}}$$

A comparison of the results obtained when VSV or poliovirus are used as a challenge in a transfer experiment are shown in Fig. 2. The conditions are as described in the assay procedure section. Interferon-induced transfer of resistance from L cells to WISH cells is demonstrable with both virus types. While a similar degree of protection of WISH cells is eventually achieved with both poliovirus and VSV, viral resistance is detected with less interferon when VSV is the challenge. Therefore, VSV is a more sensitive virus for the detection of transferred viral resistance. As expected, VSV was more sensitive than poliovirus to human interferon treatment of WISH cells alone (data not shown).

Choice of Interferon

In the mouse system we have found that both virus type, and immune type interferon[9] will effect the transfer of viral resistance from mouse L

[9] J. E. Blalock, J. Georgiades, and H. M. Johnson, *J. Immunol.* **122**, 1018 (1979).

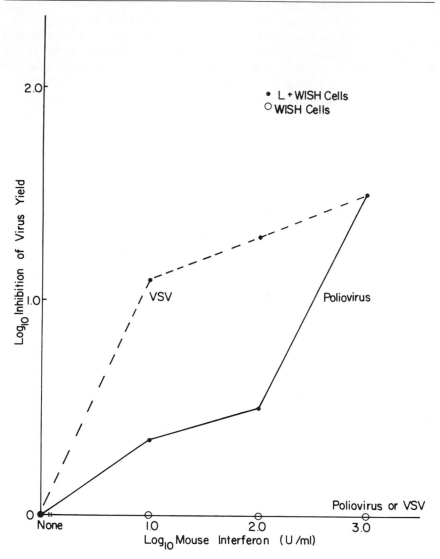

FIG. 2. Interferon-induced transfer of viral resistance: comparison of vesicular stomatitis virus (VSV) and poliovirus yield reduction. Log_{10} inhibition represents reduction in virus yield from WISH cells.

cells to human WISH cells. On a per unit basis, mouse immune-type interferon is about 2.5–10 times more efficient than virus type interferon at causing the transfer of viral resistance (Table II). All interferons used in Table I were virus type. Since both virus-type and immune-type inter-

TABLE II

COMPARISON OF TRANSFER OF ANTIVIRAL RESISTANCE WITH IMMUNE-TYPE AND
VIRUS-TYPE INTERFERONS

Interferon[a] (units/ml)	Ratio[b] of viral inhibition transferred to WISH cells: viral inhibition on L cells		Ratio of interferon transfer activity, immune-type: virus-type
	Immune-type interferon	Virus-type interferon	
3	0.56	0.056	10
10	0.29	0.063	4.6
33	0.10	0.040	2.5
100	0.10	0.025	4

[a] Interferons were produced as described by Blalock et al.[9] and assayed by VSV yield reduction as described in the text.

[b] Ratio = antilog$_{10}$ (log$_{10}$ viral inhibition transferred to WISH cells−log$_{10}$ viral inhibition on L cells). Log$_{10}$ inhibition transferred to WISH cells was calculated according to the formula in the section on choice of virus. Log$_{10}$ inhibition on L cells was calculated by substituting the virus yields from L cells alone into the same formula.

ferons seem to elicit the same or similar sequences of events in the recipient cell during transfer,[9] the choice of interferon type does not seem crucial. We do not, however, have information on human leukocyte-type interferon.

Comments

The transfer of interferon-induced viral resistance is very dependent on the cell density. As expected the cell-to-cell transfer process does not occur when cells are not in contact.[1] Thus it is imperative that transfer experiments be performed at cell confluency or above. In microtiter wells (28 mm^2) confluency occurs at about 1×10^5 cells/well. Cell concentrations above confluency show greater transferred resistance. We usually employ 1.5×10^5 cells/well, but also get very good results at 2.25×10^5 cells/well.

Transfer of viral resistance also depends on the ratio of donor to recipient cells at a given cell density. For transfer from L to WISH cells, the minimum ratio that shows transfer is about 1 L cell to 2 WISH cells. For simplicity we usually employ a ratio of 1 to 1. Greater degrees of transfer, if desired, can be obtained by increasing the ratio of L to WISH cells.

Based on the number of plaques that we usually count in transfer experiments a 0.5 log$_{10}$ reduction in virus yield is significant at a $p < 0.05$.

[44] Procedures to Study the Cytostatic Effects of Interferon on Ehrlich Ascites Cells

By Yasuiti Nagano and Hiroshi Saito

It has not yet been confirmed whether interferon (IF) inhibits the multiplication of tumor cells through its cytocidal[1,2] or its cytostatic[3] action. In order to investigate this question, one must carry out a consecutive count of the number of viable cells and the number of dead cells in a cell population cultured in the presence of IF. One must also look at the percentage of cells in the mitotic stage at each given time. If the effect of IF is hardly exerted while cell division is suppressed by other means, and if the effect of IF is cytocidal, IF should exert its usual effect. An investigation on this point should shed more light on this question.

The experimental procedures for investigating the inhibiting effect of IF derived from mouse L-929 cells on the multiplication of Ehrlich's ascites (EA) cancer cells are as follows.

Cell Count of Viable and Dead Cells

The Ehrlich ascites cells used for this experiment are STD-ddY mouse passage cells. Two weeks after intraperitoneal transplantation, the ascitic fluid, which contains the Ehrlich ascites cells, is collected, diluted with an adequate amount of phosphate-buffered saline (PBS) (Ca^{2+} and Mg^{2+} free) and centrifuged twice at 1000 rpm for 5 min. The sediment is suspended in PBS, and the cells are counted, after which 0.2 ml of the suspension is injected intraperitoneally.

Ehrlich ascites cells (5×10^4) and 5000 units of IF are placed in Eagle's MEM (L-glutamine, 0.03%; sodium hydrogen carbonate, 0.075%; calf serum, 5%), and the total volume of 1.7 ml is poured into plastic dishes (Falcon, 35 mm diameter) and incubated at 37° in a CO_2 (5%) incubator. Two dishes are taken out at various times, and the cells are removed from the container walls by trypsinization. They are stained with 0.5% trypan blue, and the unstained (viable) and stained (dead) cells are counted (Fig. 1).

Mitotic Index

The Ehrlich ascites cell suspension is centrifuged at 1000 rpm for 5 min, and 5 ml of PBS diluted fourfold with water is added to the sediment

[1] M. G. Tovey, D. Brouty-Boyé, and I. Gresser, Proc. Natl. Acad. Sci. U.S.A. 72, 2265 (1975).
[2] M. G. Tovey and D. Brouty-Boyé, Exp. Cell Res. 118, 383 (1979).
[3] Y. Nagano and H. Saito, C. R. Seances Soc. Biol. Ses Fil. 173, 20 (1979).

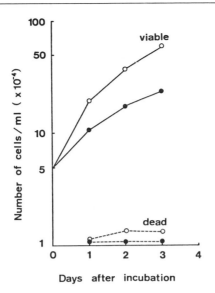

FIG. 1. Effect of interferon (IF) on the multiplication of Ehrlich ascites cells. ○, Mock IF; ●, IF.

(approximately 2×10^5 cells). This is left to stand for 10 min, then centrifuged at 800 rpm for 5 min. An adequate amount of an acetic acid–methanol solution (1 part glacial acetic acid to 3 parts methanol) is added to the sediment for 10 min to permit fixation. The cells are suspended by pipetting, centrifuged at 800 rpm for 5 min, resuspended with an addition of the acetic acid–methanol solution, and centrifuged at 800 rpm for 2 min. The sediment is suspended in an appropriate amount of the acetic acid–methanol solution. One drop of this sediment is then placed on a glass slide, which is tilted to spread the drop evenly over the glass, air-dried, stained for 20 min with Giemsa, and examined microscopically. The cells in the mitotic stage will have lost nuclear membrane, and the chromosomes will have spread. Such cells are then counted.

IF Effect on Cell Populations Whose Division Is Inhibited by Other Means

Inhibition of Division by Cycloheximide. After adding 5000 units of IF to 5×10^4 Ehrlich ascites cells and incubating them for 24 hr in 1.7 ml of MEM, 0.6 μg of cycloheximide in 0.2 ml is added and the incubation is continued. The number of viable and dead cells are counted over the following days.

High-Temperature Culture. After addition of 5000 units of IF to 5×10^4 Ehrlich ascites cells and incubating the cells at 39° in 1.7 ml of MEM, the cells are counted over the following days and compared with those cultured at 37°.

[45] Assay of Growth Inhibition in Lymphoblastoid Cell Cultures

By MARIAN EVINGER and SIDNEY PESTKA

In addition to its antiviral properties, interferon appears to possess the ability to inhibit cell multiplication. Although much controversy has focused on the question of whether growth-inhibitory activity is an intrinsic biological property of interferon, evidence now indicates that the antiviral and growth-inhibitory activities copurify.[1,2] As growth-inhibitory activity is established as a bona fide biological activity of interferon, it is imperative that an accurate, rapid, and sensitive assay be developed for comparison and standardization of growth-inhibition measurements. We describe here an assay for the measurement of the antiproliferative activity of interferon with the human lymphoblastoid Daudi cell line.

Many of the characteristics describing the antiviral activities of interferon are likewise applicable for defining its growth-inhibitory effects, e.g., heat and trypsin sensitivities, acid stabilities, species-specificity, and neutralization by interferon antisera. As reviewed by Stewart,[1] interferon exerts its growth-inhibitory effects on both normal and transformed cells.

Numerous primary and transformed human cell lines have been tested for their growth response to human interferons.[3–5] Of all the cell lines examined, the human lymphoblastoid Daudi cell line has proved to be remarkably sensitive to the growth-inhibitory effects of interferon.[6–8] In addition to this sensitivity, there are several other reasons why the Daudi cell line is an exceptional choice for measurement of antiproliferative effects of interferon. The Daudi cells (originally derived from a patient with Burkitt's lymphoma[3]) are a continuous cell line, are easy to maintain in culture, and grow rapidly. Their existence in suspension cultures facilitates the accurate transfer of cells and quantifying of cell numbers. Comparison of multiple experiments is also possible because the growth

[1] W. E. Stewart, II, in "The Interferon System," p. 238. Springer-Verlag, Vienna and New York, 1979.

[2] M. Evinger, M. Rubinstein, and S. Pestka, Arch. Biochem. Biophys. 210, 319 (1981).

[3] A. Adams, H. Strander, and K. Cantell, J. Gen. Virol. 28, 207 (1975).

[4] J. Hilfenhaus, H. Damm, and R. Johannsen, Arch. Virol. 54, 271 (1977).

[5] J. Hilfenhaus and H. E. Karges, Z. Naturforsch. B 29, 618 (1974).

[6] W. E. Stewart, II, I. Gresser, M. G. Tovey, M.-T. Bandu, and S. Le Goff, Nature (London) 262, 300 (1976).

[7] J. Hilfenhaus and R. Mauler, Ann. N. Y. Acad. Sci. 350, 626 (1980).

[8] J. S. Horoszewicz, S. S. Leong, and W. A. Carter, Science 206, 1091 (1979).

response of Daudi cells to interferon remains relatively constant under controlled conditions. In light of these considerations, the Daudi cell line is an optimal choice for measuring the growth-inhibitory effects of interferon. Consequently, the following methodology will detail the culture conditions for Daudi cells and the techniques for the measurement of growth inhibition in this cell line.

Culture Conditions

Daudi cells can be maintained in suspension culture with RPMI-1640 medium, supplemented with 10% fetal calf serum and 50 $\mu g/ml$ of Gentamicin. Because the Daudi cell line is a producer of Epstein–Barr virus,[3] all culture maintenance and manipulations should be performed under conditions of containment, as specified by the guidelines for human cells harboring oncogenic viruses.[9] The cells are grown in nonagitated sealed T-150 flasks containing 100–125 ml of culture. The growth rate of Daudi cells is optimal at $36 \pm 0.5°$. If the temperature exceeds $37.5°$, the growth rate slows considerably and viability decreases.

Daudi cells are maintained in logarithmic growth by dilution with fresh media to a density of 2 to 4×10^5 cells/ml when the culture attains a density of 1 to 1.5×10^6 cells/ml (early stationary phase). It is possible to maintain 90–95% viability in Daudi cell cultures by passaging every 3 days. Viability of cultures, as determined by exclusion of trypan blue dye, may reach 97–98% of the total cell number. For assay of growth-inhibitory activity, cultures were selected with maximal viability, live cells always constituting >90% of the cell population.

It should be noted that Daudi cell cultures may, on occasion, experience a rather sudden decline in culture viability. For example, culture viability may decrease to 75% within 3 days and to 20% within 5–6 days. Such a demise of the Daudi cultures is characterized initially by a "ragged" irregular appearance of the normally symmetrical cell surface. Subsequently, cellular debris increasingly accumulates in the culture medium. This apparently irreversible process cannot be effectively remedied by changing culture media or growth conditions. The most effective resolution is to initiate new cultures of frozen cell stocks at the first appearance of these signs. Daudi cells respond well to culture after being frozen in liquid N_2 at a concentration of $\sim 10^7$ cells/ml in media containing 10% dimethyl sulfoxide (DMSO). It is necessary to wash the DMSO from the media within a few hours after thawing in order to minimize the cytotoxic effects of DMSO.

[9] "Safety Standards for Research Involving Oncogenic Viruses," National Cancer Institute, Bethesda, Maryland.

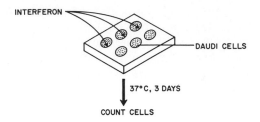

FIG. 1. Schematic illustration of the assay for growth-inhibitory activity.

Growth Assay

Measurement of the growth-inhibitory activity of interferon is straight-forward (Fig. 1). Attention to the following details will ensure a rapid and maximally sensitive response of the Daudi cells to the effects of interferon.

As also reported by Horoszewicz et al.,[8] the sensitivity of the growth assay appears to be highly dependent upon the physiological state of the cell culture. It appears that culture density and viability are the most important parameters in the selection of cultures for this assay, the cultures in the early stationary phase of growth exhibiting the greatest growth response. Early stationary or plateau phase of growth may be defined as cultures (in 10% fetal calf serum) at a density of 1 to 1.2 × 10^6 cells/ml with a population doubling time of ~48 hr. This may be compared to a population experiencing exponential growth between 2 and 9 × 10^5 cells/ml with a doubling time equaling 24–30 hr. When culture densities ranging from 7 × 10^5 to 2 × 10^6 cells/ml were tested, optimal growth response in this assay was consistently achieved with cultures containing 0.9 to 1.2 × 10^6 cells/ml. Although the growth rate decreases in early stationary phase, the culture viability should be at least 90%, preferably >95% for use in the growth-inhibition assay.

The growth-inhibition assay is performed in the following manner. First, Daudi cells are collected by brief centrifugation (6300 g for 5 min), then gently suspended in warmed (36°) media to a density of 1 to 2 × 10^5 cells/ml. When Daudi cells were plated at densities ranging from 5 × 10^4 to 2.5 × 10^6 cells/ml, it was observed that greater viability was maintained and the growth rate remained linear with 1 to 2 × 10^5 cells/ml. Cultures are allowed to equilibrate for 2–3 hr at 36° before dispensing 1-ml volumes into 16-mm wells of Linbro culture plates (No. 76-033-05). The plates are sealed with polyester adhesive sheets (Linbro No. 76-412-05) before incubation for 2 hr at 36°. Interferon diluted in culture media is then added to the wells in small volumes (≤50 μl). The plates are sealed once more with

TABLE I
GROWTH OF DAUDI CELLS IN THE PRESENCE OF
LEUKOCYTE INTERFERON[a]

| | Cell number ($\times 10^{-5}$ cells/ml) | | |
Day	No interferon (control)	Interferon (10 units/ml)	Percent Control
0	1.5	1.5	100
2	4.6	2.5	35
3	7.9	4.0	39
4	11.5	5.0	37

[a] A culture at a density of 1.1×10^6 cells/ml was diluted to 1.5×10^5 cells/ml for use in a growth-inhibition assay. At the designated days, the increase in the number of cells in interferon-treated (10 units/ml) cultures was compared to the increase in untreated (media alone) cultures. Results are expressed as the percentage of control growth.

the adhesive sheets, agitated gently for 4–5 sec to mix the contents, and incubated at 36°. The adhesive covers should completely seal all wells of the plate. Otherwise, slight gaps will alter aeration, thereby resulting in markedly different growth rates of cultures among the individual wells.

Amount of Interferon per Assay

Interferon preparations should be diluted so that 5–30 units of interferon are added to each well. All determinations should be performed in triplicate and should include a control (nontreated media or buffer) well on each plate. Each experiment should also contain dilutions of a standard preparation of interferon with titers from 0 to 50 units/ml for leukocyte interferon and from 0 to 250 units/ml for fibroblast interferon. The presence of a standard curve in each experiment is useful not only for establishing the amount of growth inhibition relative to a predetermined antiviral titer, but also to provide an internal control of cell response from experiment to experiment.

Length of Incubation

The cells, prepared as described above, are incubated at 36.5° for 3 days. As shown in Table I, the difference between the growth of control and interferon-treated cultures is maximal by 2–3 days. The growth rate of the control cultures should remain linear, but cell number should not approach saturation density during the incubation. It is common to observe a lag, i.e., little or no increase in cell number during the first 24 hr of the assay. The cell viability, however, should remain the same in both control and interferon-treated cultures; i.e., the interferon is exerting a cytostatic rather than cytotoxic effect.

Measurement of Growth Inhibition

The inhibitory effects of interferon on proliferation of Daudi cells may be determined in either of two manners: (a) by counting cells; or (b) by measuring incorporation of radioactively labeled precursors into macromolecules, such as [³H]thymidine into DNA.

The simplest method for quantifying the amount of growth inhibition involves counting the number of Daudi cells present after incubation for 3 days. Because cell viability is maintained, the number of cells in each well of the assay plate can be determined by counting either in a hemacytometer or with a Coulter cell counter. The speed and reproducibility of the Coulter counter (Model ZBI) particularly recommend its use with large numbers of samples, e.g., assaying interferon in gradient fractions. The 1-ml assay volume in the wells can be rapidly and conveniently transferred with a Pasteur pipette to a vial containing 9 ml of phosphate-buffered saline (137 mM NaCl, 8 mM Na$_2$HPO$_4$, 1.5 mM KH$_2$PO$_4$, 3 mM KCl, 0.9 mM CaCl$_2$, 0.5 mM MgCl$_2$) for counting. When the Coulter counter is used, viability is also determined for control cultures and for cells from the interferon standard curve.

Results of Growth-Inhibition Assay

The results of a growth-inhibition assay are routinely expressed as the percentage increase in the number of cells in interferon-treated cultures relative to the increase in number of cells in nontreated control cultures; i.e.,

$$\% \text{ control} = \frac{(\text{No. interferon-treated cells})_{\text{day 3}} - (\text{No. interferon-treated cells})_{\text{day 0}}}{(\text{No. control cells})_{\text{day 3}} - (\text{No. control cells})_{\text{day 0}}}$$

Alternatively, as suggested by Horoszewicz et al.,[8] the results may be expressed in terms of the percentage of cell doublings of interferon-treated cells relative to the cell doublings of nontreated control cells.

Maximal Sensitivity of the Growth Assay

As reported previously,[2,6-8] Daudi cells are extremely sensitive to the growth-inhibitory effects of interferon. Although their growth is inhibited by both human leukocyte and fibroblast interferons, this cell line is much more sensitive to the growth-inhibitory effects of leukocyte interferon than to the effects of fibroblast interferon. The growth response of Daudi cells to increasing amounts of both leukocyte and fibroblast interferons is demonstrated in Table II. Under the conditions of the assay, a 50% inhibition of Daudi cell growth was observed with 1–3 units of leukocyte interferon per milliliter. These data correlate well with the report of Stewart et al.[6] that 50% growth inhibition can be achieved with 1 unit of leukocyte interferon per milliliter. Ninety percent growth inhibition is achieved with 60 units/ml of leukocyte interferon. With fibroblast interferon, 50% growth inhibition occurs with 50 units/ml, 90% inhibition with about 200 units/ml.

The maximal amount of growth inhibition that can be attained as a result of treatment of Daudi cells with interferon may vary as a function of the Daudi cell sample. Although there may occasionally be no increase in cell number in cells treated with "maximal amounts" (>100 units/ml of leukocyte or >250 units/ml of fibroblast) of interferon, it is far more common for a portion of the cells to double once after the addition of interferon. Hence, the cell number under conditions of maximal growth inhibition may range from 0 to 50% of the control growth. Whether this reflects a random distribution of cells not "arrested" prior to DNA replication or a proportion of the cell population resistant to the inhibitory effects of interferon has not been determined.

TABLE II
GROWTH RESPONSE OF DAUDI CELLS IN THE PRESENCE OF LEUKOCYTE
AND FIBROBLAST INTERFERONS

Leukocyte interferon		Fibroblast interferon	
Units/ml	Percent control growth	Units/ml	Percent control growth
0	100	0	100
3	47	10	78
6	39	25	69
15	31	50	48
30	17.5	100	35
60	10	200	9

Greater sensitivity to the antiproliferative effects of interferon is observed when the growth-inhibition assay is performed with Daudi cells cultured in media containing lower concentrations of fetal calf serum. Similarly, Daudi cell populations exhibiting a greater degree of maximal growth inhibition may be selected by passaging the cultures in media supplemented with 5% fetal calf serum, rather than 10% fetal calf serum. Possibly, the growth rate in low serum concentrations extends the portion of the cell cycle sensitive to the inhibitory effects of interferon.

[46] The Culture of Human Tumor Cells in Serum-Free Medium

By David Barnes, Jürgen van der Bosch, Hideo Masui, Kaoru Miyazaki, and Gordon Sato

Work from our laboratory in recent years has established that it is possible to grow both primary cultures and established cell lines in culture medium in which the usual serum supplement has been replaced with the proper combination of hormones, nutrients, binding proteins, and attachment factors.[1,2] From the standpoint of interferon research, this technology is valuable, since it allows the design of experiments examining the nature of the biological effects of many substances, including interferon, in model systems that are much more defined and less complicated than those previously available. In addition, serum-free cell culture in principle should allow isolation and purification of secreted substances, such as interferon, in a less tedious and more straightforward manner than is possible with serum-containing medium.

In this chapter we describe methods for the serum-free growth of established cell lines and primary cultures of four kinds of human tumors. We feel that these model systems are particularly valuable for several reasons. First, each may be grown in serum-free medium, offering the advantages mentioned above. Second, each tumor type may be grown as transplantable tumor lines in athymic (nude) mice, allowing both *in vivo* and *in vitro* experiments, and manipulations that are combinations of *in vivo* and *in vitro* experiments. Finally, these tumor types include the three types of cancer that are the major causes of cancer deaths in the United States: lung, colon, and breast carcinoma.

[1] I. Hayashi and G. Sato, *Nature (London)* **259**, 132 (1976).
[2] D. Barnes and G. Sato, *Anal. Biochem.* **102**, 255 (1980).

Human Colon Carcinoma Cell Lines in Serum-Free Medium

Establishment of Human Colon Carcinoma Cell Line HC84S

HC84S cells were established from a transplantable human colon carcinoma (T84) which was established in BALB/c athymic mice.[3] Tumor specimens were obtained from a lung metastasis in a patient with colon carcinoma and injected subcutaneously into nude mice. The tumors produced were successively transplanted into athymic mice.

T84 tumor tissue from athymic mice was minced and plated into culture dishes containing a 1 : 1 mixture of Ham's F12[3a] and Dulbecco's modified Eagle's medium (F12 : DME) supplemented with 2.5% fetal calf serum and 5% horse serum. In a few weeks colonies of epithelial cells grew up. To kill fibroblasts, the cultures were treated with antiserum prepared in rabbits against mouse melanoma cells (B16). Once established, cells were cultivated in F12 : DME supplemented with 2.5% fetal calf serum and 5% horse serum.

Analysis of Hormone and Growth Factor Requirements of HC84S Cells

For growth assays, HC84S cells grown in serum-supplemented medium were harvested by trypsinization, treated with 0.1% soybean trypsin inhibitor, and washed twice with phosphate-buffered saline (PBS) to remove serum components. Cells were plated in F12 : DME supplemented with test materials.

Cell growth was estimated by the amount of protein per plate because HC84S cells adhered to culture dishes and to each other, and it was difficult to separate these cells into single cells by trypsinization or collagenase treatment without a substantial breakdown of cells. Comparison of cell number and protein content were made for HC84S cells grown in serum-supplemented medium and serum-free, hormone-supplemented medium. Cell number varied linearly with protein content in both serum-free and serum-containing medium. The mean was $1.5 + 0.3 \mu g$ of protein per 10^3 cells for those grown in serum-supplemented medium and $1.2 \pm 0.3 \mu g$ of protein per 10^3 cells for those grown in serum-free, hormone-supplemented medium. However, the authors recommend the use of the method of counting nuclei described in detail in this chapter, which was developed after the studies with the HC84S line.

[3] H. Murakami and H. Masui, *Proc. Natl. Acad. Sci. U.S.A.* **77**, 3464 (1980).
[3a] Abbreviations: EDTA, ethylenediaminetetraacetic acid; HEPES, N-2-hydroxyethylpiperazine-N'-2-ethanesulfonic acid; MES, 2(N-morpholino)ethanesulfonic acid; DME, Dulbecco's modification of Eagle's medium (powder); F12, Ham's F12 medium (powder); EGF, epidermal growth factor.

Hormones and growth factors were tested individually for effects on the growth of HC84S cells. Insulin, glucagon, transferrin, epidermal growth factor (EGF), triiodothyronine (T_3), hydrocortisone, selenium, and ascorbic acid stimulated cell growth. The optimal concentration of those substances were determined, and a serum-free hormone-supplemented medium for HC84S cells was developed that consists of F12 : DME supplemented with insulin (2 μg/ml), glucagon (0.2 μg/ml), transferrin (2 μg/ml), EGF (1 μg/ml), hydrocortisone (50 nM), T_3 (0.5 nM), selenium (25 nM) and ascorbic acid (57 μM). We have termed this "HC medium."

The HC84S cells begin to grow exponentially after a lag of 3–5 days in HC medium and thereafter grow with a doubling time of about 5 days, faster than in serum-containing medium. When HC84S cells were plated on collagen gels in HC medium, they grew without a lag. HC84S cells grown in HC medium piled up and produced glandlike structures.[3] This morphology is strikingly different from monolayers formed by the cells grown in serum-supplemented medium.

Establishment of Other Human Colon Cancer Cell Lines in Hormone-Supplemented Medium

We wished to determine whether cells taken from T84 tumors transplanted in athymic mice could be established directly in the defined medium developed for the HC84S cell line. Primary cultures were prepared from T84 tumors on collagen gels and placed in serum-supplemented F12 : DME medium in plates overnight. The next day the medium was replaced with HC medium. The cells grew in this medium at a rate comparable to that in serum-supplemented medium. When each of the supplements was tested for a growth stimulatory effect, transferrin showed the most striking effect with primary cultures, although insulin was the most effective with established HC84S cells. The serum-free, hormone-supplemented medium supported very poorly the growth of fibroblasts in these primary cultures.

Thus far we have established human colon carcinoma cell lines from two transplantable human colon carcinoma lines carried in athymic mice and a carcinoma taken directly from a patient by the use of collagen-coated dishes and HC medium. However, it should be noted that attempts to establish carcinoma cell lines from several other transplantable human colon carcinoma lines have been unsuccessful.

Generally 1–3 months are required before human carcinoma cells in culture grow vigorously and can be considered an established line. During this period of time, most carcinoma cells that existed originally in the primary culture die out. There are two possibilities to explain this observation. One is that during this period a certain subpopulation of the original tumor cells were selected under culture conditions and eventually took over the population. Another possibility is that some cells adapted to the culture conditions and eventually gained the capability to grow in culture. Of course, both factors may be contributing. However, when the colon carcinoma cell lines that were established in culture were injected into athymic mice and tumors produced were cultured under the same conditions, they could be more easily established. This suggests that we selected a population of carcinoma cells from among the whole population.

Primary Cultures of Human Colorectal Carcinomas and Quantitation of Growth in These Cultures

Primary explantations from *in vivo* tissues, performed with the objective of establishing a cell line or strain by serial subcultivation, suffer frequently from one or several of the following disadvantages.

1. Low plating efficiencies of single-cell suspensions (often ranging between 0.01% and 1%) may be an indication of selection of minor subpopulations not representative for the tissue investigated.
2. When seeding cell aggregates of original *in vivo* composition, quantitation of growth *in vitro* is frequently impossible, and the final outcome of such an explanation is equally doubtful.
3. The use of culture media favoring growth of cell types other than the ones looked for frequently leads to overgrowth by undesired cells (fibroblasts) and selection against the cell type of interest.
4. Changes of cellular properties due to genetic or epigenetic alterations favored by the *in vitro* conditions applied lead to the emergence of cell populations that do not occur *in vivo*.[4]

Thus the development of procedures yielding reproducible and representative primary seedings from *in vivo* tissues that can be submitted to quantitative *in vitro* measurements seems highly desirable, first, for the purpose of characterizing the changes accompanying the establishment of cell lines or strains; and second, for the purpose of developing culture conditions that allow the survival and propagation *in vitro* of entire *in vivo* cell populations while preserving their original characteristics. The development of such procedures for epithelial tissues is of special importance since (*a*) such tissues are responsible for most of the specialized functions of organs *in vivo*; and (*b*) it is epithelial cells that give rise to the overwhelming majority of all spontaneous tumors.

Reports on the successful primary explantation of epithelial tissues and characterization of growth in such cultures are rare, especially in the case

[4] G. Todaro and H. Green, *J. Cell Biol.* **17**, 299 (1963).

of human tissues, to which there is limited access for obvious reasons.[5-7] During the past few years the athymic nude mouse has emerged as a suitable experimental host for transplantations of a range of human tumors, allowing a constant source of tissue for repeated *in vitro* experimentation.[8] In the following sections, methods are reported for seeding and measurement of growth of primary cultures of human colonic and rectal tumors, which were serially subcutaneously transplanted in athymic nude mice of BALB/c background. These methods were devised working simultaneously on 11 different tumors, which displayed extremely divergent properties with regard to sensitivity to distingegrating actions of several enzymes and EDTA,[3a] content of necrotic debris (ranging from near 0% in some tumors to up to 90% or more in others), content of mucoid substance (also varying widely among different tumors), and spatial organization. Some tumors consisted of extended massive cellular aggregates, whereas others displayed sparsely distributed small cellular aggregates in a matrix of mucous or necrotic material. Since the methods of seeding and quantitation reported here are applicable to all 11 tumors, it is hoped that they will be useful for the majority of colonic and rectal tumors.

Method for Primary Cultivation of Carcinomas of Human Colon and Rectum

After killing the host animal by neck luxation, the tumor is immediately excised and placed in a petri dish. All work is done with sterile instruments and labware. If not stated differently, all solutions used are prewarmed to 37°. If possible, the collagenous capsule surrounding the tumor is peeled off with the help of two pairs of forceps. After washing the tumor with PBS, 2–4 cm³ are cut with a scalpel into slices of about 1 mm in thickness. The material is then cut into a fine mince with the help of a special tool consisting of a handle with an array of 18 parallel razor blades spaced 0.75 mm apart from each other by washers. The mince, consisting of pieces 100–500 μm in diameter and small debris, is suspended in 50 ml of PBS by pipetting and is centrifuged for 2–3 min in a clinical centrifuge at half-maximal speed. The supernatant is discarded and the top layer of the sediment, consisting of small debris or mucoid material that is devoid of live cell aggregates, is removed by aspiration. Depending on the individual tumor this procedure must be repeated with the remaining sediment once or twice until a top layer of debris or mucoid substance is no

[5] J. Rheinwald and H. Green, *Cell* **6**, 331 (1975).
[6] H. Green, *Cell* **15**, 801 (1978).
[7] J. Yang, J. Richards, Q. Gurman, W. Imagawa, and S. Nandi, *Proc. Natl. Acad. Sci. U.S.A.* **77**, 2088 (1980).
[8] *Proc. Symp. Use Athymic (Nude) Mice Cancer Res. 1977* (1978).

longer detectable after centrifugation, as determined by microscopic examination of the sediment.

In some cases a suspension of the last sediment in culture medium will yield reasonable primary cultures, at least for qualitative purposes. However, in the majority of cases the cell aggregates in the sediment are still associated with large amounts of necrotic and/or mucoid material, both of which gravely impair a uniform and reproducible attachment to the culture dish. Therefore, the following enzymic procedure is applied routinely. The last sediment, not exceeding a volume of 3 cm^3, is suspended in 8 ml of trypsin solution and incubation in a 10-cm tissue culture dish on a warm plate at 37°. Then, consecutively, the following additions are made: 2 ml of collagenase solution, 2 ml of Dispase solution, 5 ml of hyaluronidase solution. For rather sensitive tumors, incubations are allowed to proceed for 2 min with each solution. For more resistant tumors, 5 min are allowed for each step.

The suspension is then centrifuged, and the resulting sediment is freed of debris by aspiration. The sediment is resuspended in 200 ml of F12 : DME, which is stirred magnetically. This suspension is filtered through a nylon screen of 100 μm mesh size. Permeation through this screen must be facilitated by scraping back and forth with a spatula on the screen, otherwise frequently an unnecessarily large amount of valuable material is withheld on the screen. The filtrate is centrifuged, resuspended in 50 ml of culture medium, once more centrifuged, and freed from a top layer of debris.

This last sediment is suspended in an appropriate amount of culture medium at room temperature. Immediately after the addition of 30 μg of acid-solubilized collagen per milliliter, the suspension is plated onto tissue culture dishes (1.5 ml per 35-mm dish) and incubated at 37° in a humidified atmosphere of 5% CO_2 in air. Plating should be performed from a stirred suspension by means of a hand-driven dispensor ("repipetter"). In order to prevent significant sedimentation in the dispensor between the platings of consecutive dishes, attached Teflon tubings should be as short as possible, and their inner diameter should not exceed 2 mm. With the amount of material described, this procedure yields a suspension of cell aggregates containing between 5×10^6 to 2×10^8 cells. The time required for the explantation from the moment of killing the host animal to the moment of transferring the primary cultures to the incubator should not exceed 1.5 hr.

Quantitative attachment of the cell aggregates to the culture plates occurs during the following 10–20 hr (Fig. 1). For this process to take place effectively, the collagen added to the culture medium is essential for some tumors. It is important that the concentration of collagen added not

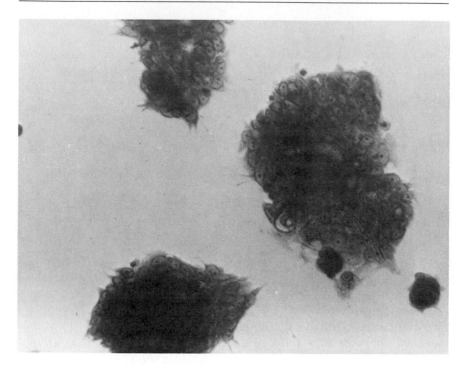

FIG. 1. Primary culture of colon tumor T84, 1 day after seeding.

be high enough to allow the formation of macroscopic gels, since in the presence of gels the counting method described below cannot be applied. Spreading of the cell aggregates on the culture plates occurs within 48 to 72 hr after plating (Fig. 2). Changes of medium are performed every 2 days with culture medium containing 6 μg of collagen per milliliter.

Growth was observed to start in cultures of all 11 tumors at the first day after seeding with population doubling times ranging from 1.5 to 8 days.[9] At the same time abundant cell death can be observed in these cultures, an *in vivo* characteristic of normal intestinal epithelium as well as of tumors originating from this tissue. Owing to the use of serum-free hormone-supplemented medium, growth of fibroblasts is never observed in these cultures. Primary cultures achieved in this way can be kept for several weeks. Secondary cultures can be produced by applying the same enzymic procedures on the primary cultures as described here for the preparation of primary cultures from *in vivo* tissue.[9]

Single-cell suspensions prepared form tumors by prolonged enzymic treatment decay during the first few days of culture and do not give rise to

[9] J. van der Bosch, H. Masui, and G. Sato, *Cancer Res.* 41, 611 (1981).

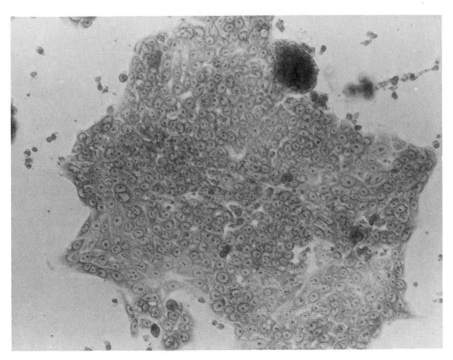

Fig. 2. Primary culture of colon tumor T219, 3 days after seeding.

growth to any extent. In fact, the primary plating of cell aggregates has an advantage over single-cell seedings in that it does not allow selection to occur during the disintegration process and ensures that the natural population is seeded in aggregates of original composition and spatial organization.

Solutions.[3a]

Phosphate-buffered saline, pH 7.3 (PBS): NaCl, 8.0 g; KCl, 0.2 g; KH_2PO_4, 0.2 g; Na_2HPO_4, 1.15 g; D-glucose, 4.0 g; H_2O, to 1000 ml.

Trypsin solution (adjust pH 7.5 with NaOH): trypsin 50 mg/liter; EDTA $2 \times 10^{-4} M$; HEPES $10^{-2} M$; NaCl $1.6 \times 10^{-1} M$; KCl $10^{-2} M$; D-glucose $2.2 \times 10^{-2} M$. Trypsin was obtained with Worthington/Millipore (217 units/mg).

Collagenase solution, pH 7.5: collagenase, 2 g/liter; $CaCl_2$, $1.25 \times 10^{-3} M$; HEPES, NaCl, KCl, D-glucose as given for trypsin solution. Collagenase was obtained from Worthington/Millipore (type CLS III, 125 units/mg).

Dispase solution, pH 7.5: Dispase, 2 g/liter; $CaCl_2$, $2.5 \times 10^{-4} M$; HEPES, NaCl, KCl, D-glucose as given for trypsin solution. Dispase was obtained from Boehringer Mannheim (Grade II).

Hyaluronidase solution, pH 4.0: hyaluronidase 600 mg/liter; EDTA, $2 \times 10^{-4} M$; MES, $3 \times 10^{-2} M$. NaCl, KCl, D-glucose as given for trypsin solution. Hyaluronidase (bovine testes) was obtained from Sigma (type I-S, 270 NF units/mg).

Culture Medium. This is a modification of the medium described by Murakami and Masui.[3]

DME, 6.2 g
F12, 4.9 g
$NaHCO_3$, 1.1 g
HEPES, 3.6 g
Penicillin G, K^+ salt, 120 mg
Streptomycin sulfate, 270 mg
Ampicillin, 25 mg
*Insulin, 2 mg
*Transferrin, 2 mg
*Triiodothyronine (T_3), $2 \times 10^{-10} M$
*EGF, 4 μg
Na_2SeO_3, $5 \times 10^{-9} M$
$NaSiO_3$, $8 \times 10^{-8} M$
$NiSO_4$, $8 \times 10^{-11} M$
$SnCl_2$, $8 \times 10^{-11} M$
$MnSO_4$, $1.7 \times 10^{-10} M$
$(NH_4)_6Mo_7O_{24}$, $1.7 \times 10^{-10} M$
$NaVO_4$, $8 \times 10^{-10} M$
$CdSO_4$, $1.7 \times 10^{-8} M$
H_2O, to 1000 ml

DME and F12 were obtained from Grand Island Biological Company. Purified acid-soluble bovine dermal collagen in pH 3 acetic acid was purchased from Collagen Corporation, Palo Alto, California. Components indicated by an asterisk (*) were added separately immediately before plating as small volumes of concentrated stocks. After suspending cells for plating in this medium at room temperature, add 30 mg of acid-solubilized collagen per liter of medium and plate immediately.

Method for Quantitative Evaluation of Growth in Primary Cultures of Human Tumors of Epithelial Origin

Cell numbers in tissue cultures from tumors of epithelial origin are frequently difficult to evaluate because enzymic treatment of these cultures either gives rise to a polydisperse suspension of cell aggregates,

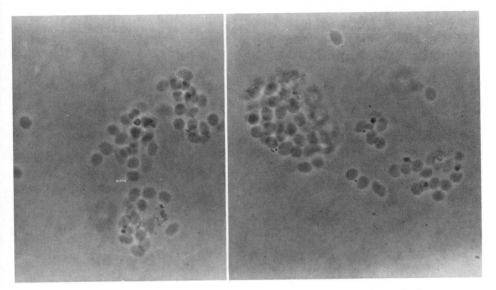

Fig. 3. Aggregates of swollen cell nuclei prepared from a primary culture of colon tumor T84.

which often contain cells of varying size with unknown size frequency distribution, or, under more drastic conditions, leads to single cell suspensions at the expense of loss of an unpredictable percentage of cells. A way out of this dilemma is the procedure outlined below. This procedure gives the details for quantitative preparation and counting of nucleic from such tissue cultures and was adapted for the special needs of these tissues from a method described for microfluorimetric determination of nuclear DNA contents.[10]

Cultures are washed four times with 5 ml of solution A and four times with solution B, the washes following rapidly in sequence after each other. All solutions are applied at room temperature. Volumes given are for tissue cultures on 35-mm plates. Immediately after the last wash, 0.8 ml of detergent solution C is applied per plate. Under this condition cells swell and lyse. In order to facilitate cell lysis and detachment, it is useful to swirl the dishes several times gently. After 5–20 min the swollen nuclei can be observed under the microscope floating singly and in aggregates (Fig. 3). In order to stabilize the nuclei against prolonged action of the detergent and mechanical rupture during pipetting, fixation is effected by the addition of 3.2 ml of formaldehyde solution D. At this stage contents of the culture plate must not be pipetted. Addition of solution D is performed by a single stroke from a dispenser with a wide-bore opening, ensuring gentle and thorough mixing with the contents of the culture plate.

[10] L. Vindelov, *Virchows Arch. B: Cell Pathol.* **24**, 227 (1977).

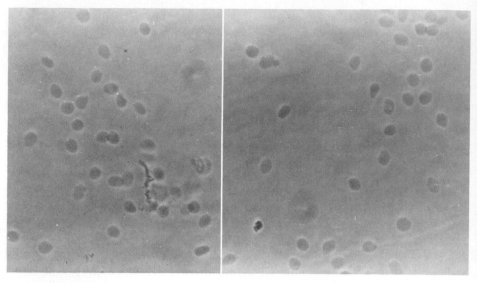

Fig. 4. Suspension of single nuclei prepared by pipetting of formaldehyde-fixed aggregates shown in Fig. 3.

After 10–20 hr the contents (4 ml) of the culture plates are transferred to counting vessels. In order to adjust to an ionic strength appropriate for counting with the Coulter counter, 4 ml of solution E are added. Aggregates of nuclei are dispersed into single nuclei at this stage by gentle pipetting (Fig. 4). For further dilution, solution F is used. With appropriate settings of current, amplification, and threshold, the nuclei in such a suspension can correctly be counted with a Coulter counter and good agreement can be obtained with nuclei counts employing a microscopic counting chamber. The time stability of such nuclear suspensions is satisfactory; counts taken 3 hr after the first measurement do not deviate significantly from the original values (Fig. 5).

This method has given satisfying counting results with primary cultures of 11 different human tumors of the colon and rectum, as well as with a squamous carcinoma of the lung in early culture stages and with adenocarcinoma of the lung at early and late culture stages.[9] Before subjecting cultures to this procedure, they must be checked microscopically for the presence of bi- and multinucleated cells, and the existence of such a population of cells must be taken into account when interpreting the Coulter counter data. Furthermore, it should be mentioned that, if washings with solutions A and B are performed as described above, no dead cells remain on the culture dishes when solution C is added, and thus

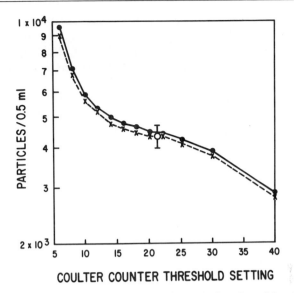

COULTER COUNTER THRESHOLD SETTING

FIG. 5. Particle number counted by Coulter counter as a function of threshold setting. A Coulter counter Model Z_F, equipped with a 100-μm capillary was used at settings $1/C = 1$ and $1/A = 1$. Particle numbers in a preparation of cell nuclei from colon tumor T84 were counted immediately after addition of solution E (●——●) and 3 hr later (×----×). For comparison, the number of nuclei counted with a microscopic counting chamber is shown also (○).

nuclei of the live fraction of such cultures are counted exclusively with this method.

Solutions [3a]

A: 0.9% NaCl, 10 mM HEPES, pH 7.5
B: 2 mM EDTA, 1 mM HEPES, pH 7.5
C: 0.1% Nonidet P-40, 2 mM EDTA, 1 mM HEPES, pH 7.5
D: 3.7% formaldehyde, 1% methanol, 3 mM HEPES, pH 7.5
E: 1.8% NaCl, 20 mM HEPES, pH 7.4
F: 0.9% NaCl, 10 mM HEPES, pH 7.4

Nonidet P-40 was obtained from BDH, England; 37% formaldehyde, 10% methanol solution was obtained from Mallinckrodt.

Growth of a Human Astrocytoma Cell Line (T24) in Serum-Free Medium

Tumor tissue was obtained from an 89-year-old female patient with astrocytoma (grade IV). The tumor tissue from surgery was injected into BALB/c athymic mice and thereafter heterotransplanted in these mice.[11]

[11] H. Masui and T. White, in preparation.

The human astrocytoma cell line (HA24) was established in culture by the following procedure. Tumor tissue from passage 16 in the mice was minced and plated into culture medium (F12 : DME) supplemented with 2.5% fetal calf serum and 5% horse serum. Fibroblasts in primary cultures were killed by anti-mouse serum.

When the cell line was cloned in serum-containing medium by standard procedures and cells from these clones were injected into nude mice (10^6 cells per animal), some of the clones produced tumors in the animal, but many clones did not.[11] One tumorigenic clone (C9) and one non-tumorigenic clone (No. 4) were chosen for analysis of hormone and growth factor requirements in serum-free medium. These cloned lines have been used in studies of the effects of interferon on transformed and nontransformed cells.[12]

Both C9 and clone 4 may be grown in F12 : DME supplemented with grimmel factor[2] ($6\mu g/ml$), insulin (5 $\mu g/ml$), transferrin (25 $\mu g/ml$), hydrocortisone (50 nM), and selenium (30 nM) (HA medium). the growth rates of both cloned lines in this serum-free medium were about 80% of that observed in serum-supplemented medium, while the saturation densities of both cloned lines were about half of that observed in the serum-supplemented medium.

To determine the relative importance of each of the supplements for cell growth in serum-free medium, each supplement was deleted individually from the serum-free, hormone-supplemented medium and the growth rates of both cloned lines in these media were compared. For clone 4, the growth rates in media lacking either insulin, transferrin, selenium, gimmel factor, or hydrocortisone were 103%, 98%, 85%, 23%, and 94%, respectively, of the growth rate in serum-free medium containing all the supplements. For clone C9, the values were 94%, 72%, 54%, 36%, and 89%, respectively. These results indicate that gimmel factor is the most growth stimulatory supplement for both cloned cell lines, but that the tumorigenic clone, C9, is also sensitive to the deletion of transferrin and selenium from the serum-free medium.

In order to grow both cloned lines in serum-free medium, culture dishes must be coated with serum. Culture dishes are incubated with F12 : DME supplemented with 7.5% newborn calf serum for 20–24 hr and then washed with F12 : DME three times before use. If plates are not precoated with serum, most cells do not attach to culture dishes when plated into serum-free media. Cold-insoluble globulin (CIg) can replace the serum coat to a certain extent, and growth is about half as good on CIg-coated dishes compared to that on serum-coated dishes. Coating dishes with polylysine does not support cell growth under these conditions.

[12] S. Slimmer, and H. Masui, and N. Kaplan, this volume [50].

Primary Culture of a Human Lung Epidermoid Carcinoma (T222)

Materials

The following designations are used for the culture media employed:

F12 : DME: a 1-to-1 mixture of Dulbecco's modified Eagle's medium and Ham's F12 medium supplemented with 10 mM HEPES; sodium bicarbonate, 1.1 mg/ml; penicillin, 92 units/ml; streptomycin, 200 μg/ml; and ampicillin, 25 μg/ml

Hormone-supplemented medium for lung epidermoid carcinoma (LE medium): F12 : DME supplemented just before use with 2.5×10^{-8} M selenium (Na$_2$SeO$_3$), insulin, 5 μg/ml; $10^{-9} M$ T$_3$; and glucagon, 0.2 μg/ml

Serum-supplemented medium: F12 : DME supplemented with 5% (v/v) horse serum and 2.5% (v/v) fetal calf serum

Primary Culture

The T222 lung epidermoid carcinoma (squamous cell carcinoma) from a human patient was maintained by successive transplantations in BALB/c nude mice. Epidermoid carcinoma tissue (tissue volume, 0.5–1 cm^3) from a tumor-bearing mouse at passage 23 to 26, from which connective tissue and necrotic areas had been removed, was minced into small pieces approximately 1 mm in diameter in a small volume of serum-supplemented medium in a culture dish. The minced tissues contained a great deal of keratinized cell debris, which were removed by washing the minced tissues by suspension, centrifugation, and resuspension, once with serum containing medium and then twice with F12 : DME. The washed tissues were incubated with 10 ml of F12 : DME containing 1 mg of collagenase per milliliter in a 100-mm diameter culture dish at 37° for 2–4 hr. The length of the incubation was determined by the size of the minced tissue and the extent of cell keratinization. The resulting suspension was roughly fractionated by centrifugation, and cell clumps containing 10–100 cells were collected. The fraction rich in single cells, containing fibroblasts and single epithelial cells that did not grow in culture, was discarded.

The epidermoid carcinoma cells thus obtained did not attach to untreated plastic cell culture dishes in the absence of serum whereas they attached and spread effectively on dishes coated with CIg or serum-supplemented medium. Coated dishes were prepared as follows. A solution of CIg (1 mg/ml in PBS) was rapidly mixed with F12 : DME in culture dishes (1 μg of CIg per square centimeter of growth area), and then incubated at 37° for approximately 60 min. The culture dishes were then washed twice with F12 : DME. In order to coat culture dishes with serum,

F12 : DME containing 10% (v/v) fetal calf serum was incubated in culture dishes at 37° overnight, and then the dishes were washed as above. Newborn calf serum and calf serum also could be used instead of fetal calf serum. However, culture dishes coated with horse serum or collagen gels were not effective for cell attachment. As an alternative method, cells could be plated in serum-supplemented medium, and the medium changed to hormone-supplemented medium LE at 1 or 2 days after plating.

Usually 0.1 ml of packed cells (cell number could not be determined) was suspended in 100 ml of medium, and 0.5 ml of the cell suspension was plated into a culture dish (Falcon, 60 mm diameter, No. 3002) or a culture flask (Falcon, 25 cm² growth area, No. 3013). The plated cells attached and spread 24–48 hr after plating. Culture media were changed every 3 days. If the cells at this stage were harvested by trypsinization and counted with a Coulter counter, the cell number per dish or flask was 1 to 2×10^5.

Growth and Keratinization of Epidermoid Carcinoma Cells

When epidermoid carcinoma cells were incubated in hormone-supplemented medium LE, a 10-fold increase in cell number was seen after 12 days of culture. Although these cells at high or moderate cell densities could grow even in F12 : DME without the supplements of the LE medium, at low cell densities the growth rate decreased markedly as compared with that in the hormone-supplemented medium LE. When incubated in serum-supplemented medium, the cells spread well during the first week, but the increase in cell number was much less than that in the hormone-supplemented medium LE. In addition, the culture dishes with serum-supplemented medium were overgrown with fibroblasts by the second week in culture. Typically, relative growth rates with hormone-supplemented, serum-supplemented and unsupplemented F12 : DME were in the ratio 5 : 1 : 2, respectively.

Of the four substances used for the hormone-supplemented medium LE, selenium and insulin showed the highest relative growth stimulatory activities, though the effect of insulin was not observed in the absence of selenium. T_3 increased cell number, to some extent, by inhibiting cell keratinization described below. Glucagon also showed small additive growth-promoting activities.

When epidermoid carcinoma cells grew to near confluency on the culture dishes containing hormone-supplemented medium LE without T_3 or unsupplemented F12 : DME, the colonies stopped spreading and began producing keratinized cells on basal cell layers. About 30 days after plating, the basal cell layers were completely covered with keratinized cell layers. The keratinization and stratification was more marked in CIg-coated dishes than in serum-coated dishes, and these morphological changes were almost completely inhibited by serum- or T_3-supplemented medium.

Epidermoid carcinoma cells could be subcultured several times in hormone-supplemented medium LE if a sufficient cell density was maintained. However, cell growth decreased upon repeated subculture, and a continuous cell line has not been established in the serum-free, hormone-supplemented medium up to now. The decrease in growth upon subculture seems to be due to marked sensitivity of the cells to protease treatment.

Lung Adenocarcinoma (T291); Established Cell Line and Primary Culture

Materials

Unsupplemented serum-free medium F12 : DME and serum-supplemented medium were the same as those for the lung epidermoid carcinoma (T222). Hormone-supplemented medium (LA medium) contained 10 μg of bovine insulin, and 10 μg of human transferrin, 1 ng of EGF per milliliter; $2.5 \times 10^{-8} M$ selenium (as Na_2SeO_3), $5 \times 10^{-8} M$ hydrocortisone, and 1 mg/ml human fatty acid-free serum albumin.

Primary Culture

Human lung adenocarcinoma was maintained by successive transplantation in BALB/c nude mice. Cells were prepared from adenocarcinoma tissue from tumors in the 12th to 15th passage by essentially the same procedure as described for epidermoid carcinoma. The tumor tissue contained a large quantity of mucus. After the mucus and cell debris were removed, the minced tissue was treated with 1 mg of collagenase per milliliter at 37° for 1–2 hr, and then small cell clumps were collected by centrifugation.

Primary cultures of lung adenocarcinoma, like epidermoid carcinoma, did not attach to plastic dishes in the absence of serum. Although they attached and spread effectively on dishes coated with CIg, they subsequently became rounded and detached from the dishes. This detachment was not seen with dishes coated with serum-supplemented medium. Alternatively, cells could be plated initially in serum-supplemented medium and the medium be replaced with hormone-supplemented medium LA one day after plating.

Approximately 1×10^{-3} ml of packed cells was plated per dish (Falcon 3002) or flask (3013), and 1 to 2×10^5 cells remained attached to dishes 1 day after plating. Because primary adenocarcinoma cells could not be dispersed into single cells by trypsinization, cell number was determined by counting number of nuclei as described for primary culture of colon cancer cells.

If primary cultures of the T291 lung adenocarcinoma were plated, maintained, and subsequently passaged in serum-supplemented medium, a continuous line could be established in culture.[13] In passaging this line, 1.5 to 2.0 × 10^5 cells were plated per dish or flask and 60–70% of the plated cells remained attached 24 hr after plating. Cell number could be determined with a Coulter counter after trypsinization of cultured cells.

Hormone Requirements for Serum-Free Growth

When cells of the established adenocarcinoma line were incubated in hormone-supplemented medium LA, a 20-fold increase in cell number was seen in 2 weeks. Typically, the relative growth rates with the hormone-supplemented (LA), serum-supplemented and unsupplemented serum-free media (F12 : DME) were in the ratio 100 : 70 : 15, respectively. Of the 6 substances that were added to the hormone-supplemented medium LA, insulin was the most stimulatory. Transferrin, EGF, selenium, and fatty acid-free albumin showed small, additive effects. Hydrocortisone seemed to be involved in attachment of these cells; if hydrocortisone was removed from the medium, cells became rounded and easily detached from the dishes. In the hormone-supplemented medium LA, the established adenocarcinoma cell line could be subcultured indefinitely.

The hormone-supplemented medium LA described above was also applicable to primary cultures of lung adenocarcinoma. In this medium cells in primary culture grew at almost the same rate as in serum-supplemented medium. This medium did not support the growth of fibroblasts, whereas the overgrowth of fibroblasts generally was a problem in serum-supplemented medium. The primary adenocarcinoma cells showed essentially the same hormone requirements as the established cell line, although the growth rate was lower than that of the established cell line. Thus far cell lines have not been established from cells cultured exclusively in hormone-supplemented medium because the cells gradually die after several passages. Presumably additional supplements must be found that will improve the ability of the hormone-supplemented medium to support long-term growth of lung adenocarcinoma cells.

Growth of Human Breast Cancer Cells in Serum-Free Medium

The MCF-7 line of human breast tumor cells is a well characterized line which retains in vitro many of the properties one would expect of

[13] H. Masui and K. Miyazaki, in preparation.

human mammary epithelium *in vivo*.[14] These cells may also be grown in nude mice, and the tumors appear to be hormone dependent.[15] We have developed a serum-free, hormone-supplemented medium which will support the long-term, multipassage growth of MCF-7 cells.[16] We have also had some partial success in the use of this medium for the serum-free growth of primary cultures of transplantable human breast tumor lines carried in our laboratory in nude mice and primary cultures of human breast tumors taken directly from patients. This medium, however, is not uniformly applicable to such primary cultures.

MCF-7 in Serum-Free Medium

MCF-7 cells plated in F12:DME containing antibiotics, 1.2 g of sodium bicarbonate per liter, 15 mM HEPES, pH 7.4, and $10^{-8} M$ sodium selenite show little or no growth. However, these cells do grow if this medium is further supplemented with insulin, transferrin, EGF, prostaglandin $F_2 \alpha$ ($PGF_2 \alpha$) and CIg. In most experiments cells in the serum-free medium containing the supplements show a somewhat lower plating efficiency than cells in medium containing 10% fetal calf serum (FCS), resulting in fewer cells at any time in dishes containing serum-free medium with the supplements compared to cell number in dishes with serum-containing medium. However, the growth rate is the same in either medium. Addition of the five factors to cells in medium with 10% FCS does not increase cell number beyond that seen with 10% FCS alone.[16]

A rich basal nutrient medium is critical for serum-free growth of MCF-7. Best growth in mass culture is seen in F12:DME with the five supplements. Cells also grow in F12 with the supplements, but no growth is seen if the supplements are added to the less complex DME. In experiments in which cells are plated at low densities, growth in F12 with the supplements is as good or better than growth in F12:DME with the supplements. This is not unexpected, since F12 was designed for growth of cells at low densities.[17]

Omission of any one of the five supplements from the medium results in reduced growth. Insulin is by far the most important of the supplements; cell number is markedly reduced in its absence. Omission of either transferrin or CIg causes a less severe reduction in cell number. Omission of $PGF_{2\alpha}$ or EGF from the mixture results in small, variable decreases in

[14] H. Soule, J. Vasquez, A. Long, S. Albert, and M. Brennan, *J. Natl. Cancer Inst.* **51**, 1409 (1973).

[15] J. Russo, M. Brennan, and M. Rich, *Proc. Am. Assoc. Cancer Res.* **17**, 116 (1976).

[16] D. Barnes and G. Sato, *Nature (London)* **281**, 388 (1979).

[17] D. Barnes, and G. Sato, *in* "Mammary Cell Biology" (C. M. McGrath, M. J. Brennan, and M. A. Rich, eds.), pp. 277–288. Academic Press, New York, 1981.

cell number. Maximum effect of insulin is seen at a dose of 25–250 ng/ml. Transferrin effects are maximal at about 25 μg/ml, and CIg effects at about 7.5 μg/ml. $PGF_{2\alpha}$ causes small effects over a wide concentration range and is used at 25–100 ng/ml. The optimal dose of EGF varies somewhat with the batch used and is maximal in the range of 10–100 ng/ml.

MCF-7 cells have been grown for over 4 months in serum-free medium with the five supplements. Success in long-term, serum-free growth of MCF-7 cells requires that the cells be treated gently upon passage, since damage from proteases, extremes of pH, and centrifugation becomes more serious in serum-free medium. Occasionally, it is helpful to change the medium the day after passaging to avoid problems with viability, which are presumably due to hydrolases released from cells damaged in the passaging procedure. Long-term, serum-free growth of MCF-7 is easier to maintain in plastic cell culture dishes than in plastic cell culture flasks. The reason for this is unclear, but may relate to quicker equilibration of gases in dishes compared to flasks. F12 : DME is made fresh every 2 weeks and stored frozen in aliquots. All six stimulatory factors are added separately to culture plates as small volumes of sterile concentrated stocks immediately upon plating the cells. Owing to instability of some of the supplements, the complete serum-free medium cannot be made beforehand and stored as one would normally store serum-containing medium.

Although cells in serum-free medium supplemented with the five factors grow as well as those in serum-supplemented medium, the morphology is quite different. Cells in the serum-free medium generally remain rounded, grow as aggregates, and do not spread well on the culture dish. The addition of 5 μg/ml of a spreading promoting factor, which we have termed serum spreading factor, isolated from human serum and first reported by Holmes to exist in human serum results in restoration of the characteristic epithelial shape of MCF-7.[16–18] We have found that this material is active also on other cell types in serum-free medium, including rat glioma (C6), mouse neuroblastoma (N18TG-2), mouse embryonal carcinoma (F9, 1003), rat ovary (RF-1), mouse embryo (3T3, SV40-3T3), mouse pre-adipocyte (1246), and human fetal lung fibroblasts (W138).[19–21] The spreading promoting factor is isolated by passing serum at alkaline pH over glass beads and eluting with sodium and potassium carbonate buffers as described in the following section.

[18] R. Holmes, *J. Cell Biol.* **32**, 297 (1967).
[19] D. Barnes, D. McClure, J. Orly, R. Wolfe, and G. Sato, *J. Supramol. Struct. Suppl.* **4**, 180 (1980).
[20] G. Serrero, M. Darmon, D. Barnes, A. Rizzino, and G. Sato, *In Vitro* **16**, 261 (1980).
[21] D. Barnes, R. Wolfe, G. Serrero, D. McClure, and G. Sato, *J. Supramol. Struct.* **14**, 47 (1980).

CIg is reported to promote adhesion and spreading of some cell types in culture.[22] CIg adheres to the plastic culture dish and presumably exerts its effects by mediating the adhesion of the cells to the culture substrate.[22,23] The serum spreading factor isolated on the glass bead column also adheres to plastic and cannot be removed by several washes of medium or phosphate-buffered saline.[21] This spreading factor seems to be different from CIg, however, because our preparations do not contain any CIg by electrophoretic or immunological analysis.[17,21]

Estradiol has previously been reported by several investigators to be stimulatory to varying degrees for MCF-7 growth.[24] We have seen an increase in cell number upon addition of estradiol under some conditions in serum-free medium, although this effect is generally small and rather variable.[17] Although small effects of estradiol are seen in the presence of each of the six factors separately, particularly in the presence of CIg or the serum spreading factor, no effect is seen if estradiol is added to medium containing all the other factors. It has recently been reported that estradiol may be retained by MCF-7 for long time periods after transfer to serum-free medium, suggesting that the variability and limited extent of the estradiol responses that we observe are related to the effect of residual estrogen.[25] Although we do not include estradiol in our medium, it is conceivable that an alternative serum-free formulation could be developed for MCF-7 that would include estradiol as one of the stimulatory factors.

A serum-free medium also has been developed by Allegra and Lippman for another human mammary tumor cell line, ZR-75-1. This medium contains insulin, transferrin, dexamethasone, fibroblast growth factor, triiodothyronine, and estradiol and uses a basal nutrient medium other than the F12 : DME used for MCF-7.[26]

Serum Spreading Factor

Described in this section are two procedures for the partial purification of serum spreading factor from human serum. Both are modifications of the procedure of Holmes.[18] Procedure 1 results in a preparation in which most of the protein in reduced samples migrates upon sodium dodecyl sulfate (SDS)–polyacrylamide gel electrophoresis in a manner consistent with molecular weights between 60,000 and 90,000.[21] Preliminary experiments indicate that the active factor in these preparations is a glycoprotein

[22] K. Yamada and K. Olden, *Nature (London)* 275, 179 (1979).
[23] J. Orly and G. Sato, *Cell* 17, 295 (1979).
[24] M. Lippman and G. Bolan, *Nature (London)* 256, 592 (1975).
[25] J. Strobl and M. Lippman, *Cancer Res.* 39, 3319 (1979).
[26] J. Allegra and M. Lippman, *Cancer Res.* 38, 3823 (1978).

that migrates in a manner consistent with a molecular weight between 70,000 and 80,000.

Procedure 2 is quicker than procedure 1, and results in higher yields of protein and spreading activity. However, procedure 2 is of limited usefulness in that it results in a preparation that contains an additional major protein band on SDS gels (molecular weight approximately 15,000–25,000) in addition to the bands seen in preparations made by procedure 1. This low molecular weight band in some preparations comprises a high percentage of the total protein in these preparations. Furthermore, preparations made by procedure 2 contain one or more growth-promoting factors in addition to the serum spreading factor. This growth-promoting activity may be seen if cells are plated in F12 : DME in the presence of serum spreading factor preparations made by procedure 2 and the absence of the other growth-stimulatory supplements in the MCF-7 medium. The growth-promoting activity may reside in the low molecular weight band, since this band is absent or greatly reduced in serum spreading factor preparations, which of themselves contain spreading-promoting activity but no growth promoting activity.

Procedure 1. Outdated human plasma is dialyzed overnight against 0.8% sodium chloride, clotted by the addition of 1 mg of calcium chloride per milliliter, and the clot is removed by low speed centrifugation (700 g, 30 min). The resulting serum is adjusted to pH 8.0 with 1 N sodium hydroxide, and 80 ml is put on a 50-cm by 2.5-cm column previously packed with acid-washed glass beads and equilibrated with 0.6 M sodium bicarbonate (pH 8.0). Bed volume of the column should be approximately 250 cm^3. The flow rate of the column should be approximately 2 ml/min. The procedure is carried out at room temperature. The column is eluted sequentially with 250 ml of 0.6 M sodium bicarbonate (pH 8.0), 125 ml of 0.6 M sodium bicarbonate–0.2 M sodium carbonate (pH 9.5), 125 ml of water, 125 ml of 0.15 M potassium bicarbonate–0.5 M potassium carbonate (pH 9.5), and 300 ml of 0.6 M potassium bicarbonate–0.2 M potassium carbonate (pH 9.7). Ten-milliliter fractions are collected.

Peak fractions eluted with 0.6 M potassium bicarbonate–0.2 M potassium carbonate are pooled and concentrated with Amicon CF 25 Centriflo Membrane Cones. The potassium carbonate buffer is exchanged for 10 mM potassium hydroxide by repeated concentration and dilution utilizing the Centriflo Membrane Cones, and the preparation is sterilized by filtration. All tubes used for collection and storage of the serum spreading factor preparations are polypropylene. These preparations may be stored frozen, but our experience is that activity remains relatively stable over several months if the preparations are stored at 4°.

Procedure 2. Preparation of the serum and column are the same as for procedure 1. The serum is adjusted to pH 8.0 with 1 N NaOH. The serum is applied to the glass bead column previously equilibrated with 0.6 M sodium bicarbonate at pH 8.0, and the column is subsequently washed with two bed volumes of 0.6 M sodium bicarbonate at pH 8.0, and one bed volume of H_2O. The adsorbed serum spreading factor is eluted with a solution of 0.6 M potassium bicarbonate and 0.2 M potassium carbonate at pH 9.7. Active fractions are pooled, adjusted to pH 7.4, sterilized by filtration, and frozen. Upon thawing, the active precipitate is dissolved in sterile 10 mM KOH. Some precipitate may form in these preparations upon prolonged storage or repeated freezing and thawing. Suitable glass beads may be obtained from Ferro, Cataphote Division, Jackson, Mississippi (No. 1014, class IV-A).

Summary

In this chapter we describe methods for growth in serum-free medium of several types of human tumors. Our laboratory and others have reported formulations of serum-free medium that will support the growth of other, additional types of human cells in culture that are not described in this chapter but are reviewed elsewhere.[2,27]

The serum-free medium HC described for the human colon tumors was originally devised to support the serum-free growth of a line of human colon tumor cells (HC84S) established in serum-containing medium and derived from a human colon tumor line (T84) transplantable in nude mice. Subsequently, it was found that this medium will support the serum-free growth of several (but not all) other transplantable human colon tumor lines in primary culture as well as primary culture of at least one tumor directly from a patient.

The serum-free medium HA described for the human astrocytoma was devised to support the serum-free growth of a line of human astrocytoma cells (HA24) established in serum-containing medium and derived from a human astrocytoma line (T24) transplantable in nude mice. Both tumorigenic and nontumorigenic clones have been isolated from the HA24 cell line, and both will grow serum-free in the HA medium, although quantitative differences in responses of the different clones to omission of the individual components of the HA medium are seen.

The serum-free medium LA described for the human lung adenocarcinoma was devised to support the serum-free growth of primary cultures of the T291 lung adenocarcinoma transplantable in nude mice. This me-

[27] D. Barnes and H. Masui, *Biomedicine,* in press (1981).

dium will support the continuous serum-free growth of a line of lung adenocarcinoma cells originally established from cultures of the T291 tumor in serum-containing medium. The serum-free medium LE described for the human lung epidermoid carcinoma was devised to support the serum-free growth of primary cultures of the T222 lung epidermoid carcinoma transplantable in nude mice. Neither the LA nor the LE medium will support continuous, multipassage serum-free growth of cell cultures obtained from the T291 or T222 tumors.

MCF-7 is an established cell line, derived in serum-containing medium from a metastatic pleural effusion of a human breast cancer.[14] The medium developed for the continuous serum-free growth of the MCF-7 line will support growth in primary culture to some extent of some human breast tumors, but not all we have tested. In addition, we report in this section procedures for the partial purification of a factor from serum that stimulates the spreading of these cells, as well as others, in serum-free medium.

We also describe in this chapter methods for the isolation of human tumor cells of epithelial origin for primary culture and methods for the quantitation of cells in such cultures. These methods, which are likely to be generally applicable in the isolation of most cell types for primary culture, allow us to obtain cultures with an acceptable plating efficiency while avoiding the ubiquitous problem of fibroblastic overgrowth of the epithelial population. Furthermore, the use of medium in which serum has been replaced by cell type-specific combinations of nutrients, hormones, binding proteins, and attachment factors allows us to reproduce more accurately *in vitro* the nutritional, hormonal and stromal influences to which the cells are subject *in vivo*.[28] In this way we should minimize the genetic or epigenetic alteration and selection of an aberrant population of cells from those of the original culture.

It is interesting to note that in several instances presented in this chapter both the growth and the differentiated properties of the cell types in their respective serum-free, hormone-supplemented media are closer to what we wish to attain in culture than what which we see in cultures of these cells in serum-supplemented medium. Human colon tumor cells, as well as human lung adenocarcinoma and epidermoid carcinoma, grow better in their serum-free media than in serum-containing medium. Furthermore, the keratinization of the epidermoid carcinoma cells is more marked in serum-free medium, as is the formation of villi-like, mucin-secreting structures in serum-free cultures of the colon tumor cells.[3] It is our view that this is further evidence that, by the use of serum-free me-

[28] D. Barnes and G. Sato, *Cell,* **22,** 649 (1980).

dium specifically designed for each cell type, we are more accurately reproducing the *in vivo* environment of the cells.

Acknowledgments

This work was supported by NIH grants N01-CB-74188, 1F32CA06188-01, CA09290-02, and CA-23052. The authors wish to thank Bonnie Wolf, Terry White, and Barbara Miller for technical assistance and many helpful discussions without which this work would not have been possible.

[47] Use of Chemostat Culture for the Study of the Effect of Interferon on Tumor Cell Multiplication

By Michael G. Tovey

In addition to its well known antiviral action, interferon can affect cell division and function.[1] Interferon also exhibits a marked antitumor action in experimental animals[1] and is currently being tested as an antitumor agent in patients.[1-3] It is therefore of considerable theoretical and practical interest to determine the mechanism of inhibition by interferon of tumor cell multiplication. Although *in vitro* cell culture systems have been used extensively to study the effect of interferon on cell multiplication,[1,4-8] such systems are a poor approximation of conditions that exist *in vivo*. It is important to understand the limitations of *in vitro* systems and to find means of overcoming the disadvantages inherent in the use of conventional cell culture. I shall first outline some of the restrictions of conventional cell culture and describe an alternative system, chemostat continuous-flow culture, which overcomes some of the disadvantages of conventional cell culture, and then show how chemostat culture can be used to study the effect of interferon on tumor cell multiplication.

[1] I. Gresser and M. G. Tovey, *Biochim. Biophys. Acta* **516**, 231 (1978).

[2] H. Mellstedt, M. Björkholm, B. Johansson, A. Ahre, G. Holm, and H. Strander, *Lancet* **1**, 245 (1979).

[3] H. Strander, *Blut* **35**, 277 (1977).

[4] I. Gresser, D. Brouty-Boyé, M.-T. Thomas, and A. Macieira-Coelho, *Proc. Natl. Acad. Sci. U.S.A.* **66**, 1052 (1970).

[5] E. Knight, Jr., *Nature (London)* **262**, 302 (1976).

[6] J. De Maeyer-Guignard, M. G. Tovey, I. Gresser, and E. De Maeyer, *Nature (London)* **271**, 622 (1978).

[7] W. E. Stewart, II, *in* "The Interferon System", Springer-Verlag, Vienna and New York, 1979.

[8] L. M. Pfeffer, J. S. Murphy, and I. Tamm, *Exp. Cell Res.* **121**, 111 (1979).

Limitations of Conventional Cell Culture Systems for Study of the Effect of Interferon on Tumor Cell Multiplication

The traditional method of culturing animal cells is in batch culture. Cells are inoculated into a quantity of nutrient medium contained in a suitable vessel, such as a culture flask or petri dish, and incubated at the required temperature; events are then allowed to run their course. A batch culture is a closed system consisting of a series of transient states difficult to define and even more difficult to control. As cell multiplication proceeds nutrients are consumed and metabolites accumulate, thereby changing the environment of the culture. These changes in turn affect cell metabolism and lead ultimately to cessation of cell multiplication. Under such conditions it is difficult to determine whether a drug-induced effect on cellular metabolism is a direct effect of a drug or is secondary to an inhibition of cell division. Such difficulties can, however, be overcome by the use of an *open* system, such as chemostat continuous-flow culture, in which there is input of substrate and output of cells and cell products.[9,10] Such an open system offers the possibility of obtaining a steady state in which constant conditions can be maintained indefinitely. Under such conditions it is very much easier to detect an effect of a drug on cellular metabolism than in the continuously changing environment of a batch culture. Any event observed before a change in the steady-state cell concentration, while cells are still under controlled conditions, is of particular value in understanding the events that precede an inhibition of cell division.

Theory of Continuous-Flow Culture

Chemostat continuous culture is based on the principle established by Monod[11] that at submaximal growth rates the growth rate of an organism is determined by the concentration of a single growth-limiting substrate. The theory of continuous-flow culture was first established by Monod[12] and independently by Novick and Szlilard,[13,14] who introduced the term chemostat. In practice a chemostat consists of a culture of fixed volume, into which nutrient medium is fed at a constant rate. Medium is mixed with cells (sufficiently so as to approximate to the ideal of perfect mixing),

[9] D. Herbert, R. Elsworth, and R. C. Telling, *J. Gen. Microbiol.* **14,** 601 (1956).
[10] S. J. Pirt, *in* "Principles of Microbe and Cell Cultivation." Blackwell, Oxford, 1975.
[11] J. Monod, "Recherches sur la croissance des cultures bactériennes." Hermann, Paris, 1942.
[12] J. Monod, *Ann. Inst. Pasteur Paris* **79,** 390 (1950).
[13] A. Novick and L. Szilard, *Science* **112,** 715 (1950).
[14] A. Novick and L. Szilard, *Proc. Natl. Acad. Sci. U.S.A.* **36,** 708 (1950).

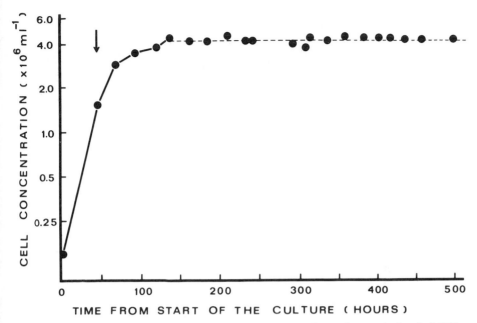

FIG. 1. The establishment of a steady-state chemostat culture of mouse leukemia L1210 cells at a dilution rate of 0.693 day^{-1}. Arrow indicates the start of continuous-flow culture. ----, Steady-state mean cell concentration ± standard deviation, $4.2 \pm 0.2 \times 10^6$ cells/ml.

and cells together with medium leave the culture at the same rate. The culture is first started off batchwise and then, during the period of exponential multiplication, is fed with medium containing a single growth-limiting nutrient, all other nutrients being supplied in excess (Fig. 1). Cell concentration then increases to the value supported by the concentration of the growth-limiting nutrient, that is, provided that the washout rate (i.e., the rate at which the culture is being diluted with fresh medium) is less than the maximum growth rate of the cells. The decrease in the concentration of the growth-limiting nutrient will then slow the growth rate of the cells until the growth rate equals the washout rate. A steady state will thus be established in which both cell density and substrate concentration remain constant. *The steady-state cell concentration is controlled by the concentration of the single growth-limiting component of the medium, and the cell growth rate is controlled by the rate of supply of this component to the culture.*

During the steady state, cell growth rate (μ) is equal to the dilution rate (D), that is, the quotient of the medium flow rate (f) and the culture volume (V)

$$\mu = D = (f/V) \ \mathrm{day}^{-1}$$

Since the culture volume is maintained constant, the cell growth rate can be changed simply by changing the medium flow rate. The doubling time (td) of the culture can be determined from the expression td = ln 2/D.

The chemostat is a self-regulating system; a temporary decrease in the steady-state cell concentration will cause a corresponding increase in cell growth rate that will act to restore steady-state conditions. An increase in the steady-state cell concentration will have the converse effect and will again act to restore steady-state conditions.

Cultivation of Animal Cells in the Chemostat

A number of animal cell lines have been cultivated in the chemostat (Table I). In the majority of these studies, cells were cultivated in serum containing medium either under glucose limitation or under conditions where the growth-limiting nutrient was not identified. Mouse LS cells have, however, been cultivated in the chemostat in a chemically defined protein-free medium under glucose[15] and choline[16] limitation. This system provides a useful means of studying the interaction of various factors with interferon.

The majority of studies on the effect of interferon on cell multiplication in the chemostat have been carried out with mouse leukemia L1210 cells.[17] The mouse L1210 leukemia represents a useful model for human leukemia[18] and is used widely for the study of anticancer agents.[19] Gresser and co-workers have studied extensively the effect of interferon on the multiplication of L1210 cells *in vivo* and *in vitro*.[1,4,20] Mouse L1210 cells are ideally suited for chemostat studies, since they multiply in suspension culture, exhibit a high specific growth rate, and do not form clumps or attach to surfaces. Wall growth can indeed be a major problem in continuous-flow culture, since, apart from the risk of blocking medium lines or overflow pipes, wall growth can cause major deviations from the predicted performance of a chemostat[9,10] owing to the continued reinoculation of cells from the walls of the culture vessel.

[15] M. G. Tovey, G. E. Mathison, and S. J. Pirt, *J. Gen. Virol.* **20**, 29 (1973).

[16] M. G. Tovey, Ph.D. Thesis, University of London, 1971.

[17] M. G. Tovey and D. Brouty-Boyé, *Exp. Cell Res.* **101**, 346 (1976).

[18] H. E. Skipper, F. M. Schabel Jr., and W. S. Wilcox, *Cancer Chemother. Rep.* **35**, 3 (1964).

[19] S. Waxman, *in* "Clinical Cancer Chemotherapy" (E. M. Greenspan, ed.), p. 17. Raven, New York, 1975.

[20] I. Gresser, *in* "Cancer: A Comprehensive Treatise" (F. F. Becker, ed.), p. 521. Plenum, New York, 1977.

TABLE I
ANIMAL CELL LINES CULTIVATED IN THE CHEMOSTAT

Cell type	Species	Growth-limiting nutrient	Reference[a]
L	Mouse	Glucose	1
P 815 Y	Mouse	Glucose	2
LS	Mouse	Glucose	3
		Unknown	4
		Choline	5
L1210	Mouse	Glucose	6
ERK	Rabbit	Unknown	7, 8
		Glucose	1
BHK	Hamster	Phosphate	9
		Unknown	9,10
HeLa	Human	Unknown	11, 12
KB	Human	Unknown	11
Namalva	Human	Glucose	13

[a] Key to references: (1) S. J. Pirt and D. S. Callow, Exp. Cell Res. **33**, 413 (1964); (2) H. Moser, and G. Vecchio, Experientia **23**, 1 (1967); (3) M. G. Tovey, G. E. Mathison, and S. J. Pirt, J. Gen. Virol. **20**, 29 (1973); (4) J. B. Griffiths and S. J. Pirt, Proc. R. Soc. London, Ser. B **168**, 421 (1967); (5) M. G. Tovey, Ph.D. thesis, University of London, 1971; (6) M. G. Tovey, and D. Brouty-Boyé, Exp. Cell Res. **101**, 346 (1976); (7) P. D. Cooper, A. M. Burt, and J. N. Wilson, Nature (London) **182**, 1508 (1958); (8) P. D. Cooper, J. N. Wilson, and A. M. Burt, J. Gen. Microbiol. **21**, 702 (1959); (9) D. G. Kilburn and A. L. Van Wezel, J. Gen. Virol. **9**, 1 (1970); (10) P. Van Hemert, D. G. Kilburn, and A. L. Van Wezel, Biotechnol. Bioeng. **11**, 875 (1969); (11) E. P. Cohen and H. Eagle, J. Exp. Med. **113**, 467 (1961); (12) B. Holmström, Biotechnol. Bioeng. **10**, 373 (1968); (13) M. G. Tovey, Adv. Cancer Res. **34**, 1 (1980).

Establishment and Characterization of Steady-State Chemostat Cultures of Mouse Leukemia L1210 Cells

Mouse leukemia L1210 cells can be readily cultivated under glucose limitation in the chemostat provided that the cells are free from *Mycoplasma* contamination. We have been unable to obtain steady-state cultures from *Mycoplasma*-contaminated clones of L1210 cells.[17] L1210 cells respond well to glucose limitation, which also has the added advantage of stabilizing the pH of the culture. In fact, the pH of the culture can be maintained at pH 7.4 ± 0.1 simply by gassing continuously with the culture vessel and medium reservoir with 5% CO_2 in air.

Animal cells can be cultivated in suitably modified commercially available fermentation equipment such as the New Brunswick Bioflo. Alternatively, mouse L1210 cells can be cultivated in a simple apparatus[17]

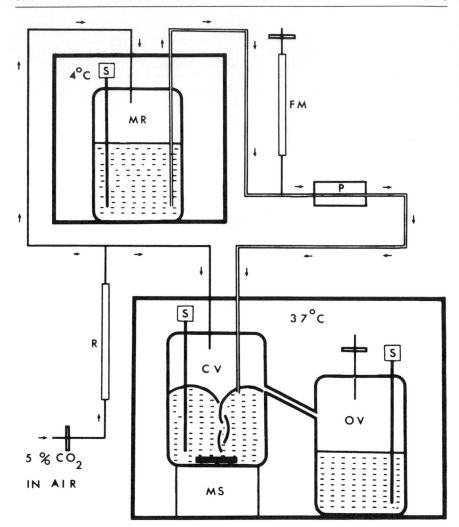

FIG. 2. Apparatus for the chemostat culture of mouse L1210 cells. CV, chemostat culture vessel; MS, magnetic stirrer; OV, overflow vessel; MR, medium reservoir; P, peristaltic pump; FM, medium flowmeter; S, sampling port; R, rotameter. From Tovey and Brouty-Boyé,[17] by permission of Academic Press, New York.

constructed from readily available laboratory glassware, as represented diagrammatically in Fig. 2. A magnetically stirred (150–300 rpm) glass culture vessel is connected via a sidearm overflow to a collection vessel. The sidearm overflow is arranged so that the working volume (300 ml) of the vessel is approximately half the total volume. Nutrient medium stored in a refrigerated (4°) reservoir is supplied to the culture vessel at a constant rate via a variable-speed peristaltic pump (MHRE-2 Watson-

Marlow, Falmouth, UK). The medium reservoir and overflow vessel can be constructed from 1.0-liter flasks with silicone seals and connected to the culture vessel with silicone tubing. All tubing and seals are made either of stainless steel or silicone rubber. The medium reservoir, culture vessel, and overflow vessel are equipped with sampling ports for the aseptic addition of medium and withdrawal of culture. The culture vessel can also be equipped with additional ports for pH and oxygen electrodes. Suitable electrodes and monitoring equipment are commercially available. The apparatus is fitted with autoclavable air filters (Millipore, USA) and can be sterilized as an ensemble.

A batch culture of L1210 cells is initiated in the chemostat culture vessel at 10^5 cells/ml in Eagle's minimal essential medium supplemented with 10% horse serum. During the period of exponential multiplication the culture is fed with nutrient medium containing a growth-limiting concentration of glucose (≤ 1.0 mg/ml) at the required dilution rate. Cell concentration then increases to the value supported by the growth-limiting concentration of glucose in the inflowing medium (Fig. 1).

Steady-state cultures of L1210 cells become established after a period of adjustment ranging from 100 to 450 hr of continuous operation of the chemostat. Adjustment times are proportional to the dilution rate of the chemostat.[21] For example, if one assumes that no cell multiplication occurs, then a flow of medium equivalent to five volume changes is required to reduce cell concentration in the chemostat to 1% of its initial value. This would take a period of 5 days at a dilution rate of 1.0 day^{-1} and 50 days at a dilution rate of 0.1 day.$^{-1}$ The long adjustment periods necessary for the establishment of steady-state cultures of L1210 cells at low dilution rates may explain in part the failure of some investigators to obtain stable steady-state cultures of animals cells in the chemostat.[22-24]

Steady-state cultures of mouse L1210 cells are characterized by constant values of a number of parameters (Table II). The steady states obtained can be considered to be true steady states since the standard deviation of the steady-state means of each of these parameters are of the same magnitude as the standard deviation of their respective assays (Fig. 3). Some parameters, such as the rate of [^{3}H]thymidine incorporation or the intracellular concentrations of adenosine 3',5'-cyclic monophosphate (cyclic AMP) or guanosine 3',5'-cyclic monophosphate (GMP) do not attain steady-state levels until some 100–200 hr after the establishment of a

[21] D. W. Tempest, D. Herbert, and P. J. Phipps, in "Microbial Physiology and Continuous Culture" (E. O. Powell, C. G. Evens, R. E. Strange, and D. W. Tempest, eds.), p. 240. Her Majesty's Stationery Office, London, 1967.

[22] P. D. Cooper, A. M. Burt, and J. N. Wilson, Nature (London) 182, 1508 (1958).

[23] P. D. Cooper, J. N. Wilson, and A. M. Burt, J. Gen. Microbiol. 21, 702 (1959).

[24] E. P. Cohen and H. Eagle, J. Exp. Med. 113, 467 (1961).

TABLE II
CHARACTERIZATION OF STEADY-STATE CHEMOSTAT CULTURES OF MOUSE L1210 CELLS

Constant cell number	Constant rates of incorporation of
Constant cell volume	[³H]Thymidine
Constant intracellular concentrations of	[³H]Uridine
DNA	¹⁴C-labeled amino acids
RNA	[³H]2-Deoxy-D-glucose
Protein	
Adenosine 3′,5′-cyclic monophosphate	Constant percentage of cells labeled by
Guanosine 3′,5′-cyclic monophosphate	autoradiography
Constant concentration of L-lactate in the	
culture supernatant	Constant rate of heat production

constant cell concentration.[25] The incorporation of [³H]thymidine into cellular acid-precipitable material can be used routinely as a sensitive and convenient means of monitoring the steady state. Cultures of mouse L1210 cells have been regularly maintained in the chemostat continuously for periods in excess of 1000 hr, and individual steady states have been maintained for up to 600 hr.[17]

Effect of Interferon on the Multiplication of Chemostat Cultures of L1210 Tumor Cells

The chemostat can be used to study different methods of exposure of tumor cells to a particular drug. Thus interferon may be added directly to the chemostat culture vessel and simultaneously to the supply of medium, thereby causing a stepwise exposure of tumor cells to a constant concentration of interferon. Alternatively, interferon can be added only to the culture vessel, causing a temporary exposure of the cells to the drug, followed by a recovery as interferon is diluted by the incoming medium. If interferon is added only to the medium supply, then the effect of exposing actively multiplying tumor cells to an exponentially increasing concentration of interferon can be studied. The chemostat can be used also as a source of tumor cells for studies performed on cells removed from the chemostat. Although during exposure to interferon the cells are no longer under steady-state conditions, this method does have the advantage of allowing a large number of experiments to be performed with tumor cells produced under controlled and defined conditions.

When L1210 tumor cells are exposed to a constant concentration of interferon, the first detectable inhibition of cell multiplication in the chemostat is usually observed 24 hr after the addition of interferon.[26] The

[25] M. G. Tovey, and C. Rochette-Egly, *Ann. N. Y. Acad. Sci.* **350**, 266 (1980).
[26] M. G. Tovey, D. Brouty-Boyé, and I. Gresser, *Proc. Natl. Acad. Sci. U.S.A.* **72**, 2265 (1975).

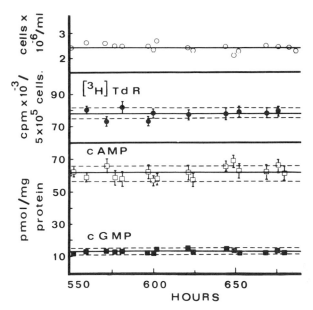

FIG. 3. A chemostat culture of L1210 cells under steady-state conditions at a dilution rate of 0.5 day^{-1} (td 33.2 hr). ○, Cell concentration (steady-state mean ± SD: $2.4 \pm 0.16 \times 10^6$ cells/ml). ●, Incorporation of [³H]thymidine into acid-precipitable material (steady-state mean ± SD: $77,860 \pm 3113$ cpm/5×10^5 cells). □, Intracellular concentration of cyclic AMP (steady-state mean ± SD: $61.8 + 5.0$ pmol/mg protein). ■, Intracellular concentration of cyclic GMP (steady-state mean ± SD: 12.7 ± 1.2 pmol/mg protein). Solid lines with dashed lines above and below represent steady-state mean value with SD. The error bars represent the SDs of the replicates for a particular point. From Tovey et al.,[36] by permission of the editors of *Proceedings of the National Academy of Sciences.*

concentration of cells in the chemostat then decreases progressively in the ensuing days, and the doubling time of the interferon-treated culture can be calculated from the regression curve of cell density.[27]

The environment in the chemostat is strongly selective for cells that can multiply faster than the parental cells, and chemostat culture can be used to isolate L1210 tumor cells resistant to the action of interferon.[26,28] Thus, if the flow of medium into the chemostat is continued after the addition of interferon, then cell concentration increases toward the tenth day despite the continued presence of interferon in the medium. These cells are resistant to interferon and eventually become established at a new steady state (Fig. 4). Chemostat culture can also be used to study factors affecting the selection of interferon-resistant mutants, since mutation rates can be determined very precisely in the chemostat.[14,29]

[27] M. G. Tovey and D. Brouty-Boyé, *Exp. Cell Res.* **118**, 383 (1979).
[28] M. G. Tovey, *Adv. Cancer Res.* **34**, 1 (1980).
[29] A. Novick and L. Szilard, *Cold Spring Harbor Symp. Quant. Biol.* **16**, 337 (1951).

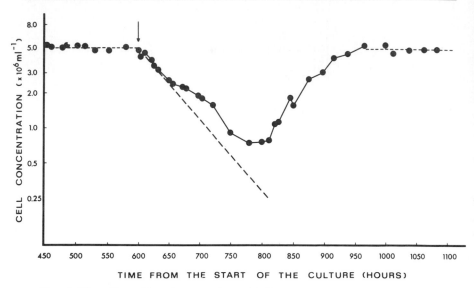

FIG. 4. The effect of interferon on the multiplication of L1210 cells cultivated under steady-state conditions in the chemostat at a dilution rate of 0.30 day^{-1}. Interferon was introduced into the chemostat by injection into the culture vessel and by simultaneous addition to the inflowing medium to maintain a constant concentration of 6400 units/ml for the duration of the experiment. Arrow indicates the time of addition of interferon. ----, Steady-state mean cell concentration ± standard deviation. Steady-state mean cell concentration prior to the addition of interferon was 5.0 ± 0.3 × 10^6 cells/ml. Steady-state mean cell concentration after the reestablishment of a steady state following interferon treatment was 5.0 × 0.4 × 10^6 cells/ml. The inclined dashed line represents the theoretical washout rate. From Tovey,[28] by permission of Academic Press, New York.

Effect of Interferon on Cellular Macromolecular Synthesis and Nucleotide Transport

Macromolecular synthesis in chemostat cultures can be determined either by continuous labeling of the cells with a low level of a radioactive precursor or, alternatively, samples (total volume <0.5% of culture volume) can be removed from the chemostat and pulse labeled under constant conditions, thereby allowing several parameters to be followed simultaneously. Once steady-state conditions have been obtained, constant rates of incorporation of labeled precursors are monitored for the equivalent of at least three doubling times prior to the addition of interferon.

In batch cultures of L1210 cells an inhibition of thymidine incorporation accompanies an inhibition of cell multiplication 18 hr after the addition of interferon.[30] Under such conditions it is difficult to determine

[30] A. Macieira-Coelho, D. Brouty-Boyé, M.-T. Thomas and I. Gresser, *J. Cell Biol.* **48**, 415 (1971).

whether the inhibition of thymidine incorporation is a direct effect of interferon or is secondary to inhibition of cell multiplication. However, in the chemostat inhibition of [³H]thymidine incorporation is observed 2 hr after the addition of interferon, that is, 22 hr prior to inhibition of cell multiplication.[26] We have shown that the inhibition of [³H]thymidine incorporation into acid-precipitable material can be accounted for by an effect of interferon on the transport of thymidine[31] and probably does not reflect an inhibition of the rate of DNA synthesis. The inhibition of thymidine uptake in chemostat cultures of L1210 cells is the first example of an effect of interferon on cellular transport and is in accord with other work showing modifications of the cell surface in interferon-treated cells.[1,7,32−34]

Interferon exerts only a transitory effect on the incorporation of [³H]uridine, and has no detectable effect on the incorporation of ¹⁴C-labeled amino acids in steady-state cultures of mouse L1210 cells.[26] The absence of an effect on protein synthesis in L1210 cells cultivated in the chemostat is in contrast to previous findings using conventional cell cultures.[35] However, it may well be that the inhibition of protein synthesis observed in batch culture occurred in parallel with the decreased rate of cell division and thus would not be observed under the steady-state conditions of the chemostat.

Effect of Interferon on Cyclic Nucleotides

The intracellular concentration of cyclic AMP and cyclic GMP can be determined, using a commercially available protein binding assay (Boehringer Mannheim), on samples removed from the chemostat and extracted with ice-cold 6% trichloroacetic acid without prior separation of the two cyclic nucleotides.[25,36,37]

When mouse leukemia L1210 cells are cultivated under steady-state conditions in the chemostat, the intracellular concentrations of both cyclic AMP and cyclic GMP remain constant (Fig. 3). Chemostat cultures therefore provide an ideal system for the study of the effect of interferon, or other antitumor agents, on the intracellular concentrations of cyclic nu-

[31] D. Brouty-Boyé and M. G. Tovey, *Intervirology* 9, 243 (1978).

[32] E. H. Chang, F. T. Jay, and R. M. Friedman, *Proc. Natl. Acad. Sci. U.S.A.* 75, 1859 (1978).

[33] E. F. Grollman, G. Lee, S. Ramos, P. S. Lazo, R. Kaback, and R. M. Friedman, *Cancer Res.* 38, 4172 (1978).

[34] M. Matsuyama, *Exp. Cell Res.* 124, 253 (1979).

[35] D. Brouty-Boyé, A. Macieira-Coelho, M. Fiszman, and I. Gresser, *Int. J. Cancer* 12, 250 (1973).

[36] M. G. Tovey, C. Rochette-Egly, and M. Castagna, *Proc. Natl. Acad. Sci. U.S.A.* 76, 3890 (1979).

[37] M. G. Tovey, C. Rochette-Egly, and M. Castagna, *J. Cell. Physiol.* 105, 363 (1980).

cleotides. The use of the chemostat has enabled us to show that interferon induces a very rapid and marked increase in the intracellular concentration of cyclic GMP several hours prior to an effect on the concentration of cyclic AMP.[25,36] This increase in the concentration of cyclic GMP, occurring 1–5 min after the addition of interferon, is the earliest effect of interferon on cells described to date and may play a role in the development of some of the diverse effects of interferon on cells.[25,36]

Relationship between Interferon Sensitivity and the Rate of Cell Multiplication

A major reason for the failure to cure some forms of human cancer with current chemotherapy is thought to be the presence in the host of "dormant" tumor cells, which are refractory to cytotoxic agents.[38,39] It is unclear whether "dormant" tumor cells are in a distinct G_0 state or merely in an extended G_1 phase,[40] and therefore still multiplying, albeit slowly. However, whether or not a distinct G_0 state does in fact exist, it is clear that the intermitotic times of populations of tumor cells can vary considerably.[40] It would therefore seem to be of considerable importance to determine the efficacy of an antitumor agent in relation to the rate of tumor cell multiplication. It has been suggested that stationary phase cells offer an *in vitro* model for the G_0 state,[41] and the interferon sensitivity of stationary phase cells has been compared to that of exponential phase cells.[42] But it is highly questionable whether the stationary phase is analogous to the G_0 state.[43] In fact the stationary phase is biochemically anything but stationary, and stationary phase cultures have been shown to contain cells distributed throughout the cell cycle.[44]

The chemostat, however, provides the first *in vitro* cell culture system in which the drug sensitivity of a given cell type can be determined at a wide range of growth rates under otherwise identical conditions. Thus mouse L1210 cells have been cultivated in the chemostat at growth rates ranging from 0.1 day^{-1} (td 166.3 hr) to 2.0 day^{-1} (td 8.3 hr).[17] The doubling time of the interferon-treated culture can be calculated from the regression curve of cell density and can then be compared to the doubling time of the culture prior to interferon treatment. The results of such experi-

[38] M. L. Mendelsohn, *J. Natl. Cancer Inst.* **28**, 1015 (1962).

[39] M. L. Mendelsohn, *Science* **135**, 213 (1962).

[40] B. D. Clarkson, *in* "Control of Proliferation in Animal Cells" (G. Clarkson, and R. Baserga, eds.), p. 945. Cold Spring Harbor Laboratory, Cold Spring Harbor, New York, 1975.

[41] S. C. Barranco, J. K. Novak, and R. M. Humphrey, *Cancer Res.* **33**, 691 (1973).

[42] J. S. Horoszewicz, S. S. Leong, and W. A. Carter, *Science* **206**, 1091 (1979).

[43] C. J. Thatcher and I. G. Walker, *J. Natl. Cancer Inst.* **42**, 363 (1969).

[44] B. K. Bhuyan, T. J. Fraser, and K. J. Day, *Cancer Res.* **37**, 1057 (1977).

ments show that interferon treatment causes the same percentage decrease in the rate of cell multiplication for either fast or slowly multiplying L1210 tumor cells.[27] But at slow growth rates (td > 55 hr) interferon treatment is accompanied by considerable cell death, whereas at fast growth rates (td < 11 hr) cell viability remains high (>99%). These results suggest that, although interferon is not directly cytotoxic, interferon treatment can result in tumor cell death by slowing the rate of cell multiplication below a minimum rate compatible with viability.[27] The results of these chemostat experiments may have some bearing on the mechanism of the antitumor action of interferon. It may well be that interferon treatment reduces the rate of tumor cell multiplication below a critical rate compatible with cell viability, thereby shifting the balance in favor of cell loss over cell proliferation.

Effect of Interferon on Cellular Metabolism

It is difficult to study a biological process, such as tumor cell multiplication, without interfering with the process itself. However, with the advent of microcalorimetry[45,46] it has become possible to measure the heat production of a biological process with a high degree of precision and with a minimum of manipulative interference.

We have shown that under steady-state conditions in the chemostat the observed rate of heat production of L1210 cells remains constant with respect to time,[47] in contrast to batch culture, where heat production changes continuously with cell multiplication.[47-49] Thus microcalorimetry used in conjunction with chemostat culture provides a valuable system for the study of both tumor cell multiplication and the mode of action of antitumor agents, such as interferon. Culture suspension is removed continuously from the chemostat, at a rate less than the medium flow rate, and is pumped at a constant rate through a microcalorimeter fitted with a flow cell (LKB instruments, Stockholm). The culture is then discarded. Once steady-state heat production has been attained, interferon can be added to the chemostat and any resultant change in heat production can then be followed. Using this technique we have shown that an increase in heat production accompanies an inhibition of cell multiplication in interferon-treated chemostat cultures of L1210 cells.[47]

[45] I. Wadsö, Acta Chem. Scand. 22, 927 (1968).
[46] I. Wadsö, in "New Techniques in Biophysics and Cell Biology" (R. H. Pain and B. J. Smith, eds.), Vol. 2, p. 85. Wiley, New York, 1975.
[47] M. Liersch, M. Küenzi, A. R. Schürch, and M. G. Tovey, manuscript in preparation.
[48] R. B. Kemp, Pestic. Sci. 6, 311 (1975).
[49] K. Ljungholm, I. Wadsö, and L. Kjellén, Acta Pathol. Microbiol. Scand. Sect. B: Microbiol. 86B, 121 (1978).

The possible applications of the chemostat culture of animal cells are myriad, and the experiments described are merely examples of the type of study that can be carried out using the chemostat. Such studies are not confined to mouse L1210 cells but are applicable to a wide range of animal cell lines (Table I). Although chemostat culture is based on the cultivation of cells in homogeneous suspension culture, anchorage-dependent cells could possibly be cultivated in the chemostat if attached to a suitable microcarrier.[28,50] If such a system were to prove feasible, it would provide a valuable means of comparing the effect of interferon on the multiplication of normal and tumor cells under identical conditions.

Acknowledgments

I should like to acknowledge the participation of Dr. Danièle Brouty-Boyé in much of the work cited from our laboratory. I am indebted to Dr. Ion Gresser for the continued support and encouragement that made this work possible. The work from our laboratory cited here was aided in part by grants from the C.N.R.S., D.R.E.T. (78-34-210), and I.N.S.E.R.M. (ATP 82-79-114).

[50] D. W. Levine, J. S. Wong, D. I. C. Wang, and W. G. Thilly, *Somatic Cell Genet.* **3,** 149 (1977).

[48] Assay of the Inhibitory Activities of Human Interferons on the Proliferation and Locomotion of Fibroblasts

By IGOR TAMM, LAWRENCE M. PFEFFER, and JAMES S. MURPHY

Study of cell proliferation in monolayer cultures has the advantages that (*a*) diploid untransformed cells can be examined; (*b*) the behavior of individual cells and of their progeny can be followed; and (*c*) the associated morphological and functional changes can be readily investigated.

We will describe two approaches to the investigation of the antiproliferative action of interferon on human fibroblasts: determination of proliferation curves by cell counts or enumeration of mitoses[1,2] and measurement of intermitotic intervals.[2] We will also describe a procedure for the measurement of the inhibitory action of interferon on cell motility, a procedure that can be applied to time-lapse cinemicrographs used for the study of the antiproliferative action of interferon.

[1] I. Tamm and P. B. Sehgal, *J. Exp. Med.* **145,** 344 (1977).
[2] L. M. Pfeffer, J. S. Murphy, and I. Tamm, *Exp. Cell Res.* **121,** 111 (1979).

Cells of human fibroblast strains, such as FS-4[3,4] or ME,[5] are grown in flasks with a 75-cm^2 growth area (3024, Falcon Plastics, Oxnard, California) in Eagle's minimal essential medium (MEM)[6] containing 10% fetal calf serum (FCS). Fibroblasts are passaged using a 4:1 split ratio, and cultures between the 10th and 20th passage are used. One passage at a 4:1 split ratio is equivalent to 2 population doublings, but it is likely to be equivalent to more than 2 generations because of some loss of cells through death and inability to divide.[7] After reaching confluence, fibroblast cultures may be maintained at 37° without change of medium for several weeks without loss of proliferative potential. This can be a considerable experimental convenience in view of the fact that, when passaged serially, diploid fibroblasts undergo a decline in proliferative activity and potential, as they have a limited life-span equivalent to 40–60 population doublings, or approximately 120–160 generations.[7,8]

Serial Cell Counts within Marked Areas of Culture Dishes

Experimental Procedure. Cell proliferation studies by serial cell counts are done in multiwell tissue culture plates (Falcon 3008). Each well has a 2.1 cm^2 growth area, and each plate has four rows of six wells. For convenience in photomicrography, only the central four wells in each row are used. Cells in a flask are dispersed from the monolayer with 0.25% trypsin–0.05% ethylenediaminetetraacetic acid (EDTA) in phosphate-buffered saline (PBS: NaCl, 8 g; KCl, 0.2 g; Na$_2$HPO$_4$, 1.15 g; KH$_2$PO$_4$, 0.2 g in 1 liter of solution)[9] lacking Ca^{2+} and Mg^{2+}. They are diluted to a concentration of 1 to 1.25 × 10^4 cells/ml in Eagle's MEM containing 5–10% of FCS. One milliliter is introduced per well, which gives an initial density of 4.8 to 6.0 × 10^3 cells/cm^2. The cultures are incubated at 37° in a humidified atmosphere of 5% CO$_2$ in air. It can be expected that 1 day after planting the count of adherent spread cells will be approximately one-half of the above value, or 2.4 to 3.0 × 10^3 cells/cm^2.[1,2] A variable time interval (up to 2 days from planting) is required before the rate of increase in cell number becomes maximal. Interferon treatment is begun 24 hr after planting by replacement of medium. Cultures are photographed daily, cells are counted on projected negatives, and the determination of the proliferation rate of cells in control and treated cultures is carried out

[3] E. A. Havell and J. Vilček, *Antimicrob. Agents Chemother.* **2**, 476 (1972).
[4] J. Vilček and E. A. Havell, *Proc. Natl. Acad. Sci. U.S.A.* **70**, 3909 (1973).
[5] L. Pfeffer, M. Lipkin, O. Stutman, and L. Kopelovich, *J. Cell. Physiol.* **89**, 29 (1976).
[6] H. Eagle, *Science* **130**, 432 (1959).
[7] P. J. Good, *Cell Tissue Kinet.* **5**, 319 (1972).
[8] L. Hayflick, *Exp. Cell Res.* **37**, 614 (1965).
[9] R. Dulbecco and M. Vogt, *J. Exp. Med.* **99**, 167 (1954).

between 24 and 72 hr after the beginning of interferon treatment (48–96 hr after planting). Interferon is used at several different concentrations, and in each experiment 8–12 control wells and 4–8 wells per interferon concentration are examined. Each experiment is carried out 3–5 times, and the mean doubling time, based on the slope of increase in cell number, is calculated for each variable.

The entire procedure can also be carried out in small tissue culture flasks, such as Falcon 3012. This avoids pH fluctuations during photography.

Photomicrography. The equipment for photography consists of the following: an inverted (Plankton) microscope (Carl Zeiss, Inc., New York) with phase-contrast optics; a photochanger (Zeiss 473051) to split the image between a binocular eyepiece and a monocular tube that transmits the image to a Nikon 35-mm camera back (M-35 S) through a 10× periplan ocular, a Leitz microibso camera attachment with deflecting prism, focusing telescope, shutter, and a $\frac{1}{3}×$ cone. The cone contains a photocell that can be placed in the light path. The photocell is connected via a cable to a microammeter (Pace Electrical Instruments Co., Inc., Glendale, New York).

Before setting up cultures, the well bottoms of the multiwell plates are marked with $\frac{1}{8}$-inch circles on the outside of the wells, using a Castell technical drawing pen or a Pilot ultrafine-point permanent (SC-UF) marking pen and a RapiDesign template. A circle is brought into concentricity with the microscopic field using a Plan 2.5× objective and a Kpl 8× ocular containing a net micrometer disk (10 × 10 mm; Zeiss 474062). Centering is rapidly achieved by superimposing the sides of the micrometer square on the circle. The drawn circle is outside the field of view of the 10× phase objective used for photography. Phase optics are aligned and the cells brought into focus for photography using the focusing telescope of the Leitz microibso camera attachment. The final magnification on the 35-mm film is 433× (determined by photographing the image of a stage micrometer). The light intensity and exposure times are adjusted to give desirable negatives and photographs. Commonly the transformer (Carl Zeiss RI W30) is set at 5 V, and the exposure time is 0.5 sec. Contact prints are made for survey. For cell counting, the negatives are projected on 8.5 × 13-inch sheets of paper and cells are viewed and counted, or, if desirable, outlined on the paper first and then counted.

Sample Results. The area of culture vessel photographed in this system is 0.46 mm². On the average there are 10–15 human fibroblasts per photograph on the day after planting, and 80–120 cells on day 4 after planting in the generally still subconfluent cultures after incubation of cells in control medium containing 10% FCS.[1] Thus, on the average, the number of cells in control cultures increases eightfold over a 3-day period.

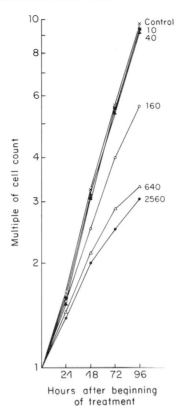

FIG. 1. Rate of proliferation of human fibroblasts (ME strain) in the presence or the absence of human fibroblast interferon at different concentrations based on consecutive cell counts. The numbers in graphs refer to international reference units per milliliter. Cultures were photographed under phase contrast optics (each time return to the same area), and cells were counted on projected negatives. From Pfeffer *et al.*[2]

Figure 1 illustrates proliferation curves of cells of the ME strain of human fibroblasts in control cultures and in the presence of human fibroblast interferon ranging from 10 to 2560 units/ml. The overall doubling times between 24 and 72 hr after the beginning of treatment are as follows: control: 28 hr; interferon: 10 units/ml, 26 hr; 40 units/ml, 28 hr; 160 units/ml 38 hr; 640 units/ml, 58 hr; 2560 units/ml, 59 hr.

Figure 2 records the relationship between the concentration of human fibroblast interferon and the proliferation rates (reciprocals of doubling times) of treated cells expressed as percentage of the rate in controls. The dose-response relationship for inhibition of cell proliferation, as for inhibition of virus multiplication, is approximately exponential rather than linear.

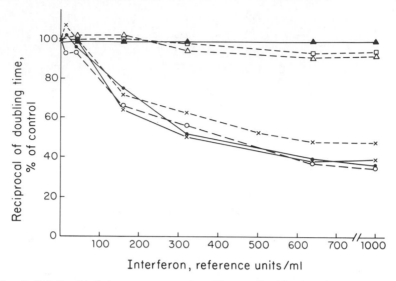

FIG. 2. Relationship between concentration of human fibroblast interferon and doubling time of human, monkey, and mouse cells. Human: ×---×, ME; ●——●, FS-4; ○——○, GM-258; ×——×, HeLa S3; □---□, HeLa monolayer. Monkey: △---△, CV-1. Mouse: ▲——▲, L-929. From Pfeffer *et al.*[2]

The technique described for serial cell counts permits convenient study of multiple cell types and of multiple interferon preparations. Relatively few cells are required for the determination of proliferation rates and of dose-response relationships. The counting technique is nondestructive; this means that the same cultures can be examined repeatedly at intervals of time, thereby avoiding the variation that arises when different sets of culture vessels need to be used for consecutive determinations.

Enumeration of Mitoses by Time-Lapse Cinemicrography

A detailed kinetic analysis of cell proliferation can be obtained by the technique of serial cell counts by doing the counts at frequent intervals. However time-lapse cinemicrography provides a preferable approach to the fine delineation of the proliferation curves of control and treated cells.[2]

Cells are planted in culture flasks with a growth area of 25 cm² (Falcon 3012) at a concentration of 1×10^4 cells/ml in Eagle's MEM containing 15% FCS. Five milliliters of cell suspension gives an initial density of 2×10^3 cells/cm². The flasks are equilibrated with 5% CO_2 and incubated at 37°. One day after planting the cells are refed with MEM containing 15% FCS with or without interferon. The flasks are equilibrated with 5% CO_2 and an 18-gauge 1½-inch long hypodermic needle is thrust through the bottle cap of the closed culture vessel to permit pressure equilibration. The control and treated cultures are observed by time-lapse cinemicrography, using two inverted microscopes (Zeiss), EMDECO control units, and Cine-Kodak Special II 16-mm cameras. The microscopes are encased in Plexiglas boxes maintained at 37° by proportional temperature controllers (Harrell, Inc., East Norwalk, Connecticut). Photomicrographs are taken on Kodak Plus X reversal movie film at 2-min intervals with an exposure of 4 sec. Appropriate neutral density filters (Zeiss) are employed to adjust the illumination (transformer setting of 5.5 V) to yield the proper exposure. For scoring of mitotic cells for proliferation rate analysis, photomicrographs are taken under a 2.5× objective to encompass a larger number of cells in a field (20–40 cells at the start of cinemicrography). Alternatively, a 6.3× phase contrast objective with a long working distance Ph 2 diaphragm-condenser (Zeiss 47 12 91) can be used, which gives superior resolution. Developed films are scored for mitotic cells after examination under a binocular dissecting microscope (magnification 10–30×). One cell is added to the cumulative cell count whenever a mitosis is found. The first several frames of each film are examined and the number of cells enumerated to provide the starting cell count.

Proliferation curves are constructed by plotting the cumulative cell count after each mitotic event versus time after beginning of treatment. When cell proliferation is rapid, only every fifth mitosis may be plotted for clarity. The time after beginning of treatment can be determined by the cumulative frame count from the film start multiplied by the time interval between photomicrographs (in hours, preferably). The generated proliferation curves may be analyzed by least squares linear regression. The slopes of the linear regression curves for any time interval can be used to calculate the population doubling time.

Figure 3 illustrates proliferation curves of human fibroblasts (ME) in the presence or the absence of human fibroblast interferon (640 units/ml) as determined by enumeration of mitoses. It can be clearly seen that there is a *gradual* slowing of the overall proliferative rate in control cultures over the course of 5 days. In interferon-treated cultures the slowing of the overall proliferative rate is more pronounced. There is good agreement between the results in Fig. 3 and those in Fig. 2, as indicated by a comparison of the increments in cumulative indices. For example, the increments between 48 and 72 hr were as follows: mitoses in control culture: 1.74-fold; mitoses in treated culture: 1.42-fold; cells in control culture: 1.73-fold; cells in treated culture: 1.33-fold. Based on the determination at 72 hr after the beginning of treatment, the total numbers of both mitoses and cells were reduced by 50% in interferon-treated cultures. It may be noted that the doubling time of control cells around 36 hr from refeeding

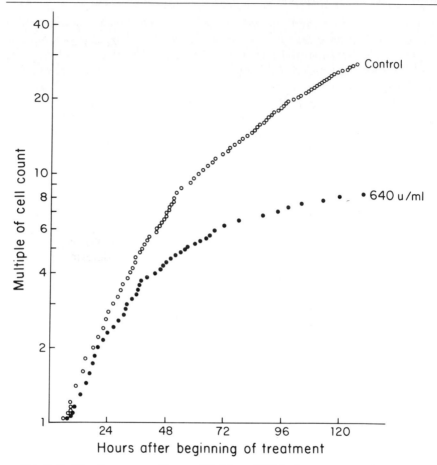

FIG. 3. Proliferation curves of human fibroblasts (ME) in the presence or the absence of interferon (640 units/ml) (based on scoring mitoses). Only every fifth mitosis is plotted for clarity. From Pfeffer et al.[2]

was approximately 18 hr in Fig. 3 (enumeration of mitoses), but approximately 28 hr in Fig. 2 (cell count). This difference is probably due to the fact that 15% FCS was used in the experiment in Fig. 3, but 5% FCS in the experiment in Fig. 2. Thus, inhibition of the rate of cell proliferation by interferon is not dependent on serum concentration.[2]

Determination of Intermitotic Interval

Measurements of doubling time do not distinguish between interferon effects on the duration of the cell cycle and effects on the number of cycling cells in a population. Since it has been demonstrated that interferon treatment causes changes in both parameters in mouse EMT6 tumor cells[10] and in human fibroblasts,[2] it is important to examine the proliferation of interferon-treated cells also by a technique that permits such a distinction to be made, i.e., by following the behavior of individual cells and their progeny.

The cultures for determination of the intermitotic interval in individual cells are set up in 25 cm^2 flasks and incubated as described above for enumeration of mitoses, except that the mitotic behavior of individual cells and their daughter cells is followed using a 10× phase-contrast objective and a long working-distance Ph2 diaphragm-condenser. Under these conditions 6–15 cells are observed per field at the start of cinemicrography. Developed films are examined for the appearance of the first mitosis, and the resulting daughter cells are followed for mitoses. The procedure is repeated for subsequent cell generations during the duration of the film. Cell pedigrees are then constructed and the intermitotic intervals for consecutive descendants are recorded. The time interval between the first and second mitosis for each descendant is considered to be the first intermitotic interval; between the second and third mitosis, the second intermitotic interval; and so on.

Figure 4 illustrates the results of an experiment on the effect of interferon treatment (640 units/ml) on the frequency distribution of intermitotic intervals of human fibroblasts.[2] The progressive prolongation of the intermitotic interval during interferon treatment is clearly demonstrated for successive cell generations for cells that continued to divide. It should be noted in addition that about one-half of the offspring of the cells that had shown an increased first intermitotic interval failed to divide again throughout the course of the experiment. In sharp contrast, the great majority of the one-third of the treated fibroblasts that initially divided at a rate similar to that of control cells continued to cycle and underwent further mitoses. The dividing cells represented a progressively smaller fraction of the population with each succeeding cell generation. However, there was no evidence of degenerative changes such as abnormal rounding or contraction or of cell lysis in interferon-treated cultures.[2]

Determination of the intermitotic intervals for individual cells by time-lapse cinemicrography reveals and makes possible a quantitative definition of the heterogeneous response of cells in a population to the antiproliferative effect of interferon. Results from time-lapse photomicrography can also be used to derive median intermitotic times for succeeding cell generations.[10]

[10] M. C. d'Hooghe, D. Brouty-Boyé, E. P. Malaise, and I. Gresser, *Exp. Cell Res.* 105, 73 (1977).

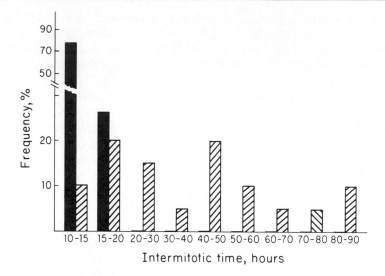

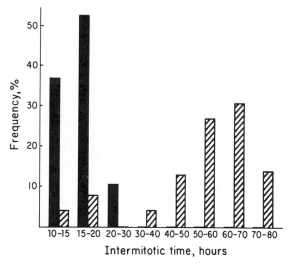

FIG. 4. Frequency distributions of intermitotic intervals of control (filled bars) and interferon-treated (hatched bars) ME cells. *Top:* Intervals between the first and second mitoses; *bottom:* between the second and third mitoses after the beginning of treatment. From Pfeffer *et al.*[2]

Measurement of Cell Locomotion

The films obtained for the determination of intermitotic interval can be used for the measurement of cell motility. Cinemicrographs obtained with the 10× phase objective are projected, and the paths of movement of

LOCOMOTION OF CONTROL AND INTERFERON-TREATED HUMAN FIBROBLASTS

| | Cell locomotion | | | | |
| | Control | | Interferon-treated (640 units/ml) | | |
Time (hr)	μm/min $\pm$ SEM	Cell No.	μm/min $\pm$ SEM	Cell No.	Percent of control[a]
0–12	0.1215 ± 0.0069	44	0.1332 ± 0.0144	35	105
12–24	0.1310 ± 0.0095	48	0.1149 ± 0.0098	35	90.6
24–36	0.1259 ± 0.0073	55	0.0556 ± 0.0054	31	43.8
36–48	0.1290 ± 0.0069	56	0.0549 ± 0.0076	34	43.3
48–60	0.1290 ± 0.0077	62	0.0375 ± 0.0042	39	29.6
60–72	0.1417 ± 0.0110	43	0.0273 ± 0.0019	35	19.3
72–84	0.1257 ± 0.0083	42	0.0247 ± 0.0027	33	19.4
84–96	0.1070 ± 0.0073	40	0.0238 ± 0.0022	37	18.8

[a] A mean control value of 0.1268 μm/min, based on all measurements during the entire period of observation, was used in computing the percentage of control values. The results recorded are based on four separate experiments. From Pfeffer et al.[11]

individual cells are traced onto paper with the aid of a stop-motion projector (photo optical data analyzer Model 224A (MKV), L-W International, Woodland Hills, California). The paths are measured with a distance tracking device (Keuffel and Esser Co., Switzerland), and the measurements are divided by the total magnification, to reduce the path movement to micrometers. The rate of cell movement is determined for a 12-hr time interval and expressed in micrometers per minute.[11]

As shown in the table, the mean rate of cell locomotion in control human fibroblast cultures is relatively constant (range 0.1070–0.1417 μm/min) during the 96-hr period of experiments. The rate of cell locomotion in interferon-treated cultures was similar to control values during the first 24 hr of treatment. However, for the period from 24 to 36 hr after the beginning of interferon treatment, the rate decreased to 44% of the overall mean control value, and it leveled off at ~20% of the control value 60–72 hr after the beginning of treatment.[11]

Acknowledgments

This work was supported by Research Grant CA-18608 and by Program Project Grant CA-18213 awarded by the NCI. L. M. P. was a Postdoctoral Trainee under the Institutional National Research Service Award CA-09256 from the NCI.

[11] L. M. Pfeffer, E. Wang, and I. Tamm, J. Cell Biol. 85, 9 (1980).

[49] Measurement of Antitumor and Adjuvant Effects of Interferon with Transplantable and Other Tumors in Mice

By Michael A. Chirigos

Interferon has a wide variety of inhibitory activities. Of particular interest is its ability to retard tumor cell growth. It has been demonstrated to inhibit the proliferation of malignant cells *in vitro*[1] and *in vivo*.[2,3] Its mechanism of inhibiting neoplastic cell growth does not appear to be simple but may involve several mechanisms. Its activity, however, is definitely associated with the classical interferon molecules.[4] This chapter will deal with measurement of the effect of interferon on experimental mammary carcinoma and leukemia with emphasis on interferon's cytoreductive effect.

Tumor Cell Models

The mammary adenocarcinoma (line 16/C) was isolated by Southern Research Institute, Birmingham, Alabama[5] and maintained in passage by transplantation of metastatic lung foci. When inoculated subcutaneously it will grow progressively with metastasis to lung, lymph nodes, and bone marrow. The MBL-2 tumor line[6] was established from a Moloney leukemia virus-inoculated mouse and maintained in continuous passage as an ascites tumor line. Tumor cells are inoculated intraperitoneally with the development of systemic luekemia within 4–5 days after inoculation. Adult (C57BL/6xC3H)-$B_6C_3F_1$ and C57BL/6 male mice were used for experiments with the mammary adenocarcinoma (line 16/C) and MBL-2 tumor lines, respectively.

Macrophage Assays for Tumoricidal Activity

For the MBL-2 studies, in which macrophage tumoricidal activity was examined, peritoneal macrophages were harvested from treated or normal control mice and washed in RPMI-1640 medium. Approximately 8×10^5 macrophages were seeded into 16-mm wells on tissue culture cluster plates (Costar, Cambridge, Massachusetts) in 1.0 ml of RPMI-1640 me-

[1] S. E. Grossberg, *N. Engl. J. Med.* **287**, 13 (1972).
[2] M. A. Chirigos, *Tex. Rep. Biol. Med.* **35**, 399 (1977).
[3] I. Gresser, *Tex. Rep. Biol. Med.* **35**, 394 (1977).
[4] W. E. Stewart, I. Gresser, and M. G. Tovey, *Nature (London)* **262**, 300 (1976).
[5] T. H. Corbett, D. P. Griswold, B. J. Roberts, J. C. Peckham, and F. M. Schabell, *Cancer Treat. Rep.* **62**, 1471 (1978).
[6] M. A. Chirigos and R. M. Schultz, *Cancer Res.* **39**, 2894 (1979).

METHODS IN ENZYMOLOGY, VOL. 79

dium containing 10% fetal calf serum. The cultures were incubated for 90 min, and macrophage monolayers were washed thoroughly with jets of medium. The macrophage-containing wells were overlaid with 8×10^4 MBL-2 cells. The technique for measuring growth inhibitory macrophages has been described in detail.[7]

Test Agents

Partially purified mouse fibroblast interferon (specific activity 2×10^7 units/mg) was purchased from Dr. Kurt Paucker, Department of Microbiology, Medical College of Pennsylvania, Philadelphia. MVE-6 [Pyran (Pyran-2-succinic anhydride-4,5-dicarboxytetrahydroxy-6-methyl anydride polymer)] was kindly supplied by Dr. David Breslow of Hercules Research Center, Wilmington, Delaware. Ether-extracted *Brucella abortus* (BruPel) was kindly supplied by Dr. Julius S. Youngner, Department of Microbiology, School of Medicine, University of Pittsburgh, Pittsburgh, Pennsylvania.

Measurement of Response of Mammary Adenocarcinoma to Interferon and Interferon Inducers

Mice inoculated subcutaneously with 1×10^7 mammary adenocarcinoma cells were distributed to four groups (Fig. 1). Treatment with placebo, phosphate-buffered saline (PBS), interferon, MVE-6 or BruPel, was by the intralesional route (intradermally around the tumor site), administered daily for a total of 18 treatments. This regimen of treatment resulted in a marked retardation of tumor growth. Control tumor growth was progressive, resulting in a median survival time (MST) to death of 24 days. The range of death occurred from 24 days after tumor inoculation to 46 days later. The ability of interferon and the interferon inducers to retard tumor growth is reflected in the delay in death. Interferon and BruPel treatment resulted in a 66% increase in survival time while MVE-6 treatment resulted in a 129% increase. A significant number of long-term survivors (more than 90 days) occurred in the MVE-6 and BruPel treatment groups. These survivors, however, did have tumors and eventually died. The initially larger tumor mass observed in the BruPel treatment group was ascribed to the retention of the injected material resulting in edema. We previously reported that the subcutaneous mass at the site of tumor inoculation does not necessarily reflect solely tumor cells. Invading macrophages and lymphocytes permeating through the localized tumor along with edema result in an enlarged mass. The results of this model system served to show the efficacy of local administration of interferon

[7] R. M. Schultz and M. A. Chirigos, *Cancer Res.* **38**, 1003 (1978).

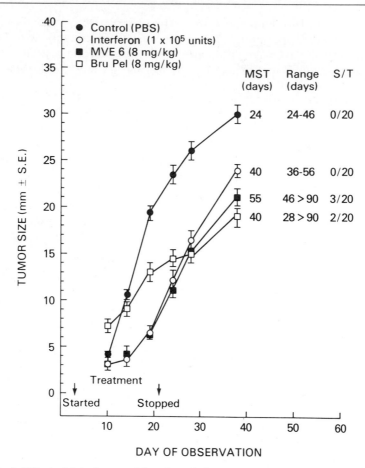

Fig. 1. Effect of interferon and interferon inducers on growth of mammary adenocarcinoma 16/C. $B_6C_3F_1$ female mice were inoculated subcutaneously with 1×10^7 16/C cells. Treatment was started on day 3 and administered daily for a total of 18 treatments to day 21. MST, median survival time; S, survivors; T, total in group.

and other test agents. Whether inhibition of tumor cell replication was due to direct local action of the test agent on tumor cells and/or to activation of host macrophage tumoricidal activity cannot be determined by this method. The B16 murine malignant melanoma was also reported to respond to interferon.[8] Tumor volumes of mice inoculated with the B16 melanoma, and treated with interferon intraperitoneally, were significantly smaller than placebo-treated mice. This tumor was also shown to be inhibited *in vitro* by exposure to interferon. These results indicate a direct

[8] R. S. Bart, N. R. Parzio, A. W. Kopf, J. T. Vilček, E. H. Cheng, and Y. Farcet, *Cancer Res.* **40**, 614 (1980).

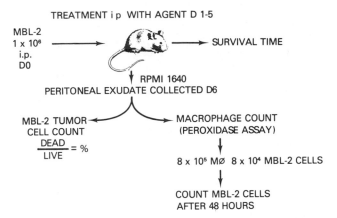

TREATMENT i p WITH AGENT D 1-5

MBL-2
1 x 10⁶ ──────→ ──────→ SURVIVAL TIME
i.p.
D0

↓ RPMI 1640
PERITONEAL EXUDATE COLLECTED D6

MBL-2 TUMOR ←────── ──────→ MACROPHAGE COUNT
CELL COUNT   (PEROXIDASE ASSAY)
$\frac{DEAD}{LIVE}$ = %

8 x 10⁵ MØ 8 x 10⁴ MBL-2 CELLS

COUNT MBL-2 CELLS
AFTER 48 HOURS

FIG. 2. Scheme for evaluating *in vivo* antitumor activity of macrophage activators.

antiproliferative effect of *in vivo* administered interferon on the B16 melanoma tumor.

Assay of in Vivo Antitumor Activity

The MBL-2 leukemic ascites cells are inoculated intraperitoneally, and treatment with test agent was started 1 day after tumor inoculation and continued daily for 5 days. Groups of animals were held for survival time, and similarly treated animals were sacrificed 1 day after the last treatment (Fig. 2). The peritoneal exudate collected was the source for macrophages for *in vitro* testing of their tumoricidal activity on freshly prepared MBL-2 cells and also for counting of viable and dead MBL-2 tumor cells. The experimental model was designed to assess whether the proliferation of MBL-2 tumor cells was being inhibited by direct intraperitoneal treatment with interferon or Pyran (MVE-6) as determined by tumor cell count and whether macrophages were participating in death of tumor cells. Results, typical of the response attained in several experiments, are shown in the table.

The percentage of dead tumor cells, determined by trypan blue exclusion, was significantly higher in those groups that were treated with either interferon of Pyran. In addition, macrophages harvested from these similarly treated groups, and incubated *in vitro* with freshly prepared MBL-2 tumor cells expressed a tumoricidal effect that was significantly higher than the placebo (bovine serum albumin)-treated group. The inhibition of tumor cell proliferation was reflected by a substantial increase in survival time (MST) and number of long-term survivors (>90 days).

In an attempt to define the mechanism(s) by which tumor cell killing occurred in this system, one would have to consider that macrophages,

In Vivo RESPONSE OF MBL-2 TUMOR CELLS TO INTERFERON AND MVE-6 TREATMENT[a]

MBL-2 1×10^6	BSA 0.5 mg	IF 1×10^5 units	MVE-6 0.2 mg	In vitro evaluation[b]		In vivo evaluation	
				Dead tumor cells (%)	Mφ Tumor cell death (%)	MST (days)	Survival total (%)
D0				12		18	0
D0	D 1–5			15	25	23	0
D0		D 1–5		56	52	38	30
D0			D 1–5	48	80	48	40

[a] On day zero (D0) 1×10^6 MBL-2 cells were inoculated intraperitoneally. Mice were treated with bovine serum albumin (BSA), interferon (IF), or MVE-6 on days 1, 2, 3, 4, and 5 (D 1–5).

[b] C57BL/6 mice sacrificed on day 6 for in vitro evaluation of tumor cell viability and testing macrophage (Mφ) tumoricidal activity in vitro.

activated by interferon, or the Pyran interferon inducer, were exerting a tumoricidal effect. The in vivo tumoricidal effect was retained by the macrophages when they were again exposed to fresh tumor cells in vitro. Whether the macrophage tumoricidal activity was the sole mechanism of in vivo tumor cell killing remains to be elucidated, particularly in view of the reports that interferon alone was capable of inhibiting tumor cell replication in vitro.[4,9,10] We also have observed that interferon at 10^3 units/ml directly inhibited MBL-2 leukemia cell replication by 20% in vitro.

Several tumor systems have been reported to respond to interferon treatment.[2,3] Of particular interest is the response of a murine osteogenic sarcoma to both type I and type II interferon[11] and the response of human osteogenic sarcoma to interferon treatment.[12]

In addition to the direct antiproliferative effect of interferon on tumor cells, and the role of interferon on activating macrophages to exert direct tumoricidal activity,[13] natural killer cell activity is augmented in reponse to interferon. Interferon or interferon inducers were found to be capable of

[9] I. Gresser, M. T. Thomas, and D. Brouty-Boyé, Nature (London 231, 20 (1971).

[10] G. B. Rossi, G. Marchegiani, G. P. Matarese, and I. Gresser, J. Natl. Cancer Inst. 54, 993 (1975).

[11] L. A. Glasgow, J. L. Crane, Jr., E. R. Kern, and J. S. Youngner, Cancer Treat. Rep. 62, 1881 (1978).

[12] H. Strander, K. Cantell, S. Ingimarsson, P. A. Jakobsson, V. Nilsonne, and G. Soderberg. Fogarty Int. Cent. Proc. 28, 527 (1976).

[13] R. M. Schultz, J. D. Papamatheakis, and M. A. Chirigos, Science 197, 674 (1977).

enhancing natural killer cell tumoricidal activity on YAC, M109, and MBL-2 tumor cells.[14,15]

Interferon, when used in conjunction with tumor cytoreductive chemotherapy was also found to exert a more beneficial effect against a murine leukemia.[16] The enhanced therapeutic response attained by interferon adjuvant treatment may have been due to its antiproliferative effect on residual leukemia cells that escaped cytoreductive chemotherapy and/or to the interferon-enhanced macrophage and natural killer cell tumoricidal activity.

Evidence indicates that interferon exerts its antiproliferative effect on tumor cells through more than one mechanism: (a) a direct antiproliferative effect; and (b) enhancement of the cellular immune response by direct activation of natural macrophage and natural killer cell tumoricidal activity.

[14] A. Bartocci, V. Papademetriou, and M. A. Chirigos, J. Immunopharmacol. 2, 149 (1980).
[15] J. Y. Djeu, J. A. Heinbaugh, H. T. Holden, and R. B. Herberman, J. Immunol. 122, 175 (1979).
[16] M. A. Chirigos and J. W. Pearson, J. Natl. Cancer Inst. 51, 1367 (1973).

[50] Antiproliferative Assay for Human Interferons

By SHERYL SLIMMER, HIDEO MASUI, and NATHAN O. KAPLAN

Cells

We have used the T-24 clone #9 human astrocytoma for assay of the antigrowth effects of various human interferons. The original tumor tissue was obtained from surgery of an 89-year-old female. The pathological diagnosis of the tumor was glioblastoma multiforme or astrocytoma type IV. The tumor tissue was minced and injected into athymic mice. Human astrocytoma cell lines were established in culture from the tumors at 16 to 22 passages in the mice according to the following procedure. Tumor tissues were minced and plated into culture dishes containing a 1 : 1 mixture of DME and F12 media supplemented with 2.5% fetal calf serum and 5% horse serum. Fibroblasts in primary cultures were killed by addition of antimouse serum. Many clones were obtained from these established human astrocytoma cell lines. Each of these clones was injected back into athymic mice at one million cells per animal. Although many of the clones failed to produce a tumor some were successful. One of these tumorigenic

METHODS IN ENZYMOLOGY, VOL. 79

clones, C#9 was used in the present study. Details of the procedure and characteristics of the cells are elsewhere in this volume.[1]

Reagents

The stock cultures were grown in Falcon T-75 cm^2 tissue culture flasks for 5–7 days using a 1 : 1 mixture of Dulbecco's minimal essential media and Ham's F12 media containing 1.2 g/liter $NaHCO_3$, 0.15 M Hepes, 8 mg/liter ampicillin, 40 mg/liter penicillin, 90 mg/liter streptomycin, 5% horse serum, and 2.5% fetal calf serum at 37° in a 5% CO_2/95% air humidified incubator.

The cells were dispersed using a trypsin solution containing 0.1% trypsin in phosphate-buffered saline, 7.8×10^{-4} M EDTA without Ca^{2+} or Mg^{2+} at pH 7.4. The trypsin solution was inactivated by the addition of culture medium containing serum to the suspension, centrifuged, trypsin solution removed, and the cells resuspended in the culture media, counted on a hemacytometer, and replated at 1×10^5 cells per 60×15 mm Falcon tissue culture dish with 5 ml of media to be used in the assay.

Assay

The growth inhibitory activity of the interferons was measured using a 7–9 day assay comparing the growth of the cells which received (a) nothing additional, (b) the diluents of the interferons, and (c) the different interferon samples at various concentrations.

The cells were plated on day zero and began receiving interferon that same day. The interferon was added daily at the same concentration, with as little change in overall volume as necessary.

Two dishes were counted every other day from each treatment group as follows. The media was aspirated off, the plates were gently washed one time with PBS, and 1 ml of trypsin was added. The plates were returned to 37° and incubated until the cells came off the dish, as determined by microscopic observation. The trypsin was inactivated by the addition of 1 ml of media containing serum. These 2 ml were added to 8 ml of PBS. Each 10-ml sample was evenly dispersed using a bulb and pasteur pipet and then counted on a coulter counter. The coulter counter uses 0.5 ml giving a final dilution of 1 to 20. Each sample was counted twice, these two numbers were averaged with the two obtained from the duplicate sample and then multiplied by 20 to give the total number of cells present on that particular day.

[1] D. Barnes, J. van der Bosch, K. Miyazaki, and G. Sato, this volume [46].

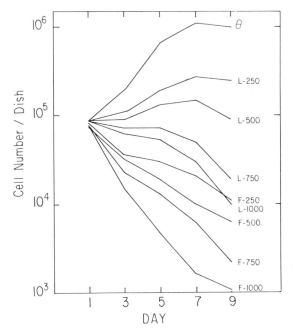

FIG. 1. Comparison of crude leukocyte "CML" interferon with crude fibroblast interferon. θ, Control; L, crude leukocyte "CML" interferon (IFL); F, crude fibroblast interferon (IFF); numbers indicate units per milliliter added daily.

Interferons

The interferons used were obtained from Dr. S. Pestka of the Roche Institute of Molecular Biology. The antiviral activity was measured using the assay from Roche based on reduction of cytopathic effect.[2]

Each sample to be added to the cells was diluted in the culture media at the beginning of the experiment and stored at 4° throughout the experiment.

We believe it is of importance that several concentrations of the interferon be used over an extended period of time in the assay. As can be seen in Fig. 1, the greatest changes, particularly between different concentrations, occur at a period of 5 to 9 days after the interferon has been introduced. This method has allowed us to distinguish between the antigrowth and cytotoxic effects of the interferons. The figure illustrates that crude fibroblastic interferon is more active than the lymphocyte preparation as an antigrowth agent. This assay has been found to be useful in

[2] S. Rubinstein, P. C. Familletti, and S. Pestka, J. Virol., in press.

measuring the antigrowth potential of purified interferon,[3] as well as recombined interferons for their relative antiviral and antigrowth activities; these studies show the antiviral and antigrowth activities are not identical.[4]

We have tested various human cell lines in our assay system and, to date, have found the T-24-9C line preferable to the others due to its ease of manipulation and high responsiveness.

[3] N. O. Kaplan and S. Slimmer, "Cellular Responses to Molecular Modulators." Academic Press, New York, in press.
[4] S. Slimmer, H. Kung, E. Yakobson, T. Staehelin, S. Pestka, and N. O. Kaplan, *Proc. Natl. Acad. Sci. USA.*, in press.

[51] Measurement of Interferon Effects on Murine Leukemia Virus (MuLV)-Infected Cells

By Paula M. Pitha

Highly purified mouse interferon inhibits both replication of MuLV in mouse cells[1-3] and MSV-induced transformation in mouse and rat cells. The interferon-induced inhibition of MuLV replication occurs at two levels. The first block is relatively inefficient and can be demonstrated in exogenous MuLV and MSV infection.[4-6] This effect is irreversible, suggesting that the block occurs before integration of the viral genome. The major interferon effect occurs after synthesis of viral proteins at the level of virus maturation and assembly. This effect is reversible; virus production resumes after interferon is removed.[2,3]

Three types of MuLV infection (exogenous, IdUrd-induced endogenous, and chronic) have been studied, and all were comparably sensitive to interferon. With all three types of infection the amount of infectious virus released into the medium was markedly reduced by interferon and the effect was temporary. In some cell lines, however, a disproportionate reduction in the release of the infectious virus particles and the number of virus particles measured by reverse transcriptase activity was observed, suggesting that defective particles had been formed. In many cells these

[1] A. Billiau, H. Sobis, and P. De Somer, *Int. J. Cancer* **12**, 646 (1973).
[2] R. M. Friedman and J. M. Raumseur, *Proc. Natl. Acad. Sci. U.S.A.* **71**, 3542 (1974).
[3] P. M. Pitha, W. P. Rowe, and M. N. Oxman, *Virology* **70**, 324 (1976).
[4] A. G. Morris and C. Clegg, *Virology* **88**, 400 (1978).
[5] P. M. Pitha and S. P. Staal, *Virology* **90**, 151 (1978).
[6] P. M. Pitha, S. P. Staal, D. P. Bolognesi, T. P. Denny, and W. P. Rowe, *Virology* **79**, (1977).

noninfectious virions accumulate on the cell surface, indicating that the erroneous assembly leads to decreases in both virus release and infectivity. Thus, although in all types of infections the interferon block occurs at the level of maturation and assembly, the exact location and the degree of sensitivity to interferon seems to be determined by the host cell.[3,7-9]

The virions assembled in the presence of interferon do not (with the exception of Friend leukemia virus[11]) differ in morphology and in the size of virion RNA (70 S) from the control virions,[7] but they show altered polypeptide patterns from controls when analyzed on sodium dodecyl sulfate (SDS) gel electrophoresis; in addition to the regular core proteins and envelope glycoprotein, these virions contain proteins and glycoprotein not identified in virions assembled in the absence of interferon.

The molecular basis for the lack of fidelity of virus maturation and assembly in interferon-treated cells remains unknown. Intracellular synthesis and processing of precursors of viral proteins do not seem to be altered greatly by treatment with interferon at concentrations that substantially affect virus maturation. Interferon might inhibit virus assembly by altering cellular membranes, but no interferon-induced changes in the cellular membranes accounting for the lack of fidelity in virus assembly have yet been demonstrated.[13]

Assays of Infectious MuLV

Ecotropic Virus

Cell-Free Virus

Titrations of cell-free virus (N and B tropic) by using standard UV-XC assay can be done in the SC-1 mouse cell line,[14] which does not show Fv-1 restriction; alternatively, N-tropic virus can be assayed on the NIH-3T3 cell line and B-tropic virus on the BALB 3T3 cell line. The cells are plated

[7] P. M. Pitha, N. A. Wivel, B. F. Fernie, and H. P. Harper, *J. Gen. Virol.* **42**, 467 (1978).

[8] P. M. Pitha, B. F. Fernie, F. Maldarelli, T. Hattman, and N. A. Wivel, *J. Gen. Virol.* **46**, 97 (1980).

[9] P. K. Wong, P. H. Yuen, R. Macleod, E. H. Chang, M. W. Myers, and R. J. Friedman, *Cell* **10**, 245 (1977).

[10] P. T. Allen, H. Schlellkaus, L. J. L. D. Van Griensven, and A. Billiau, *J. Gen. Virol.* **31**, 429 (1976).

[11] R. B. Luftig, J. F. Concience, A. Skoreltchi, P. McMiltiau, M. Revel, and F. Ruddle, *J. Virol.* **23**, 799 (1977).

[12] E. H. Chang and R. M. Friedman, *Biochem. Biophys. Res. Commun.* **77**, 392 (1977).

[13] P. M. Pitha, *Ann. N. Y. Acad. Sci.* **350**, 301 (1980).

[14] J. Hartley and W. P. Rowe, *Virology* **65**, 128 (1975).

(3 to 5×10^5 cells per 60-mm plate) in Eagle's minimal essential medium (MEM) with 10% fetal calf serum, which has been preheated to 56° for 20 min to inactivate complement (ΔFCS); 24 hr later DEAE-dextran is added to the medium (final concentration 20 μg/ml). After 2 hr of DEAE-dextran treatment, the medium is removed, cells are washed with MEM and infected with MuLV in 3 ml of MEM and 10% FCS. The cells are incubated for 48 hr, and then the medium is replaced by MEM with 5% ΔFCS. The medium should be replaced every second or third day, especially when SC-1 cells are used. On day 6 after infection the medium is removed, the cells are UV irradiated 3000 erg/cm² per second, and 2×10^6 freshly trypsinized XC cells are added. The medium is replaced on the second day after UV irradiation, and on the third day the culture is fixed with methanol and stained with Giemsa or hematoxylin. An area of infection that has three or more giant cells is counted as a plaque.[15]

Infectious Centers Assay

Interferon seems to have little effect on the number of virus-positive cells as measured by the infectious centers assay.[3] In this technique, the infected cells are dispersed by treatment with a 0.75% trypsin (this concentration of trypsin is also sufficient to remove the cell-associated virus). The cells are counted, and the desired dilution is plated on the appropriate cell culture (SC-1, 3T3 lines) which was seeded at a concentration of 3 to 5×10^5 cells per 60-mm dish 24 hr earlier. The cultures are incubated in MEM and 10% ΔFCS for 48 hr and then the medium is replaced with MEM and 5% ΔFCS. On the fourth day after the addition of the infected cells, the medium is removed and cells are lethally UV irradiated (as in the UV-XC test) and overlayed with 2×10^6 XC cells. The foci of infection are detected by the UV-XC test as described above.

In Situ Infectious Assay

For the determination of the amount of infectious virus present inside the cells at given time points after interferon treatment, the UV-ME test is most useful. This procedure is based on the finding that virus-producing cells remain infectious for some time after lethal UV irradiation. A virus-producing cell will produce a plaque, whereas an infected cell in which virus production has not been initiated will not. Thus, to determine the number of cells producing virus at different times after interferon treatment or removal of interferon, the cells are lethally UV irradiated at the indicated times and overlayed with SC-1 cells (5×10^5 per 60-mm plate). Four days later the cultures are again UV irradiated and overlayed with

[15] W. P. Rowe, W. E. Pugh, and J. W. Hartley, *Virology* **42**, 1136 (1978).

XC cells, and the number of virus-positive cells is determined 3 days later as in UV-XC test.

Syncytia Induction in XC Cells

The cocultivation of cells infected with ecotropic MuLV with the XC cell line (rat tumor cell line) leads to a formation of a larger number of syncytia.[16] It has been shown that treatment of mouse cells with interferon before infection (which prevented virus particle formation in these cells) did not affect the ability of these cells to induce syncytia in XC cells.[6] For the syncytia induction assay, infected cells (5×10^3 cells per 60-mm plate) are grown in 24 hr in MEM and 10% ΔFCS. The medium is then removed, and cells are washed, UV irradiated, and overlayed with 2×10^6 XC cells without irradiation. The number of syncytia is estimated 24 hr later after methanol fixation and Giemsa staining. It is advisable to count only giant mouse cells containing four or more nuclei. Uninfected mouse cells overlayed with XC cells under the identical conditions should always be used as controls.

Xenotropic Viruses

A large class of MuLV viruses present as endogenous viruses in different mouse strains do not replicate in mouse cells. These viruses are called xenotropic viruses and can be assayed on heterologous cells.[17] The mink (S+L−) cells are optimal for the assay. The cells are plated at a density of 2 to 5×10^5 cells per 60-mm dish and the following day are treated with DEAE-dextran (20 μg/ml) for 2 hr, washed, and infected with the virus. The MSV foci are counted 7–10 days later.

Assay of MSV

To divorce the effect of interferon on viral replication from a possible direct effect on the initiation and persistance of the transformed state, it is necessary to use the systems in which the MSV need not spread to induce foci. Transformation in such a system displays single-hit kinetics; i.e., each cell infected with sarcoma virus is capable of giving rise to a focus.[18] Transformation by the MSV in both NRK (rat cells) and in NIH/3T3 cells fulfills this criterion.[4,5] For the transformation assay, the cells are plated at a density of 2 to 3×10^5 cells per 60-mm dish in MEM with 5% FCS and

[16] V. Klement, W. P. Rowe, J. W. Hartley, and W. E. Pugh, *Proc. Natl. Acad. Sci. U.S.A.* **63**, 753 (1969).
[17] J. K. Levy and T. Pincus, *Science* **170**, 326 (1970).
[18] R. Parkman, J. Nevy, and R. C. Ting, *Science* **196**, 387 (1970).

infected 24 hr later with a given concentration of MSV in the presence of 32 μg/ml of Polybrene; 24 hr later the medium is removed and replaced by MEM with 5% FCS and 1% dimethyl sulfoxide (DMSO). Foci are counted on the sixth day after infection; on the ninth day after infection cells are fixed with methanol and foci are counted after staining with Giemsa stain.

Assays of Virus Particles

Virion-Associated Reverse Transcriptase. Virion assay for sedimentable RNA-dependent DNA polymerase can be used for the estimation of both ecotropic and xenotropic MuLV.[19] The collected fluids (10–30 ml) are clarified by centrifugation at 3000 g for 10 min and then at 10,000 g for 30 min at 4°. The supernatants at this point can be frozen at −70° or immediately assayed. Virus is pelleted by centrifugation at 100,000 g for 60 min at 4°. Pellets are resuspended in 100 μl of buffer (100 mM NaCl, 10 mM Tris · HCl, pH 7.4). Then 75 μl of virus suspension are mixed with an equal volume of 100 mM Tris · HCl (pH 8.3) containing 0.4% Triton X-100. A 25-μl sample of the disrupted virus is assayed in a reaction mixture containing 0.1% Triton X-100, 50 mM Tris · HCl, pH 8.3, 100 mM NaCl, 0.5 mM magnesium acetate, 5 mM dithiothreitol, 1 μM [³H]deoxythymidine-5′-triphosphate (approximately 18,000 μCi/mmol), 2 μM oligodeoxythymidylic acid [oligo(dT)], and 8 μM polyriboadenylic acid [poly(A)] in a total volume of 100 μl. The reaction mixture is incubated at 37° for 60 min. Acid-precipitable activity can be measured by spotting 50 μl of sample on a disk of Whatman 3 paper and washing the disks in the solution of cold 10% trichloroacetic acid and 20 mM sodium pyrophosphate.[20] The disks are subsequently gently washed once more with the same solution, twice with cold 5% trichloroacetic acid and 20 mM sodium pyrophosphate, once with ethanol and acetone, then dried and the radioactivity counted. Virus should be also tested for the DNA polymerase activity by adding oligodeoxyadenylic acid [poly(dA)] to the reaction mixture instead of poly(A). The activity of samples containing poly(A) should be 50- to 100-fold higher than those containing poly(dA).[21]

Radioisotope Labeling of Virus Particles. Virus radioactively labeled with [³H]uridine (100 μCi/mmol) can be prepared by incubating infected cells at 90–100% confluency with 100 μCi of [³H]uridine per milliliter in MEM with dialyzed 5% ΔFCS for 2 hr. The medium is then replaced with fresh medium containing 5% ΔFCS and incubated for another hour, after which the medium is harvested and virus particles are purified as

[19] D. Baltimore, *Nature* (*London*) **226,** 1209 (1970).
[20] F. J. Bollum, *Proc. Nucleic Acid Res.* **1,** 296 (1966).
[21] S. Spiegelman, A. Burny, M. R. Das, J. Kendar, J. Schlom, M. Travnicek, and K. Watson, *Nature* (*London*) **228,** 430 (1970).

described below. Alternatively, the cells can be incubated with 20 μCi of [^{3}H]uridine per milliliter for 12–24 hr. The long labeling time is not recommended, however, especially with AKR-L virus, when viral RNA is going to be isolated and analyzed.

To label virion proteins, the infected cells are incubated for 3 hr before labeling in leucine- or methionine-free MEM medium plus 5% dialyzed ΔFCS. The medium is then removed and replaced with leucine-free medium containing 25 μCi of [^{3}H]leucine per milliliter (1 mCi/mmol) or methionine-free medium containing 20 μCi of [^{35}S]methionine per milliliter (1 mCi/mmol). After 12–24 hr, the medium is collected and the virus is purified.

For labeling of virion glycoproteins with [^{3}H]glucosamine, infected cells are incubated with medium containing 25% of the original amount of glucose for 3 hr before labeling and then labeled with 100 μCi of [^{3}H]-glucosamine per milliliter (10 mCi/mmol) for 24 hr. Since [^{3}H]glucosamine is shipped in ethanol solution, it is important to evaporate all the ethanol before use.

Virus Purification

Sucrose Gradient. Harvested medium containing labeled virus particles is clarified at 10,000 g for 10 min and then the supernatant is layered on top of a discontinuous gradient consisting of 20% (w/v) sucrose in standard buffer (TEN : 20 mM Tris · HCl, pH 7.5; 1 mM EDTA; 0.1 M NaCl) and 50% (w/v) potassium tartrate in standard buffer (sucrose : tartrate ratio 5 : 1, v/v).[22] The gradient is centrifuged at 286,000 g in a Beckman SW41 rotor for 1 hr or SW27 rotor at 131,000 g for 3 hr. The virus, which bands at the tartrate–sucrose interface, is collected, diluted twice with standard buffer, and layered on top of a continuous 24% (w/v) to 48% (w/v) sucrose gradient and centrifuged to equilibrium in an SW40 rotor at 286,000 g for 3 hr. Fractions of 250 μl are collected; 50-μl portions are removed and assayed for trichloroacetic acid precipitable radioactivity, and the adsorbance of fractions at 260 and 280 nm is monitored. Fractions containing virus (density 1.16 g/cm^3) are pooled, diluted with standard buffer, and pelleted by centrifugation at 337,000 g for 1 hr in a Beckman SW60 rotor. The $A_{260} : A_{280}$ ratio of purified virus should be in the range of 1.22–1.27.

Sepharose C-14B Chromatography. This method of purification[23] is preferable if the virion glycoprotein gp71 is going to be analyzed, since chromotography preserves more gp71 in the virus particle than does centrifugation. This seems to be important for some MuLV (e.g., AKR L-1

[22] E. Hunter, M. J. Hayman, R. W. Rongey, and P. K. Vogt, *Virology* **69**, 35 (1976).
[23] M. McGrath, O. Witte, T. Pincus, and I. L. Weissman, *J. Virol.* **25**, 923 (1978).

virus) assembled in the presence of interferon, since these viruses are fragile, easily losing gp71 during purification. Culture fluid collected during the 12-hr period is clarified at 10,000 g for 10 min, concentrated with immersible molecular separators (Millipore Corp., Bedford, Massachusetts), and chromatographed on a column of Sepharose C-14B (Pharmacia) equilibrated in TEN. Virus appears in the void volume, and adsorbance or radioactivity can be monitored.

Isolation of Virion RNA

Viral RNA can be isolated from gradient-purified virions by a modification of the Palmiter method.[24] The pooled fractions (1–2 ml) are brought to 4 ml with standard buffer (see above) (pH 9) and extracted with 8 ml of phenol (saturated with standard buffer, pH 9.0), 30 μl of stock mercaptoethanol (Eastman), 0.1 ml of 0.5 M EDTA, 15 μl of rat liver RNA (1 μg/ml), and SDS (final concentration 1%). The mixture is vortexed and shaken at room temperature for 10 min. The phenol–RNA mixture is then centrifuged for 10 min at 8000 g to separate the phases. The water phase (including the interphase) is collected and reextracted twice with 8 ml of chloroform–isoamyl alcohol (24:1). After the final centrifugation, the water phase is collected and precipitated with 2.5 volumes of cold 80% ethanol containing 0.2 M sodium acetate.

Alternatively, the 70 S RNA can be isolated from the virions by disrupting the purified virions in SDS buffer (10 mM Tris · HCl, pH 7.4, 0.14 M NaCl, 1 mM EDTA, 1% SDS) and treating with Pronase (200 μg/ml) for 30 min at 37°. The disrupted virus is then layered on the top of a 15 to 35% sucrose gradient (in 10 mM Tris · HCl, pH 7.5, 0.15 M NaCl, 1 mM EDTA) and centrifuged for 120 min at 38,000 rpm on SW41 rotor. The 0.5-ml fractions are collected and absorbance at 260 nm is determined. The 70 S RNA sediments in the lower part of the gradient.

Gel Electrophoresis of RNA

The viral RNA can be analyzed in agarose gels under denaturing conditions.[25] The RNA pellet (after precipitation with ethanol) is resuspended in 20–30 μl of a 15% glyoxal mixture (150 μl of glyoxal, 500 μl of DMSO, 20 μl of 0.5 M potassium phosphate buffer, pH 6.5, and 330 μl of H$_2$O) and incubated at 50° for 45 min. The sample is then loaded onto a 1.5% agarose gel in 10 mM potassium phosphate buffer, pH 6.5 (which is also the running buffer) and run at 250 V for 2–3 hr. The gel can be dried and

[24] R. D. Palmiter, *Biochemistry* **17**, 3606 (1974).
[25] G. K. McMaster and G. C. Carmichael, *Proc. Natl. Acad., Sci. U.S.A.* **74**, 4835 (1977).

radioactive RNA determined by X-ray fluorography at $-70°$, or the RNA can be transferred by blotting to diazobenzyloxylmethyl paper (DBM paper)[26] and viral RNA can be detected by hybridization with ^{32}P-labeled cDNA probe to MuLV RNA.

Separation and Identification of the Virion Proteins

Purified labeled virions are disrupted in the presence of Nonidet P-40 (1%) 50 mM NaCl, and sodium deoxycholate (0.5%) in 25 mM Tris · HCl, pH 8.0.[8] Aliquots containing 10 to $50 × 10^3$ cpm are adjusted to 20 μl with Laemmli buffer[27] and boiled for 5 min; proteins are separated on a poly-acrylamide (10–15%) gel in the presence of 0.1% SDS and stained with Coomassie Blue R 250. Gels containing ^{35}S or ^3H-labeled proteins are impregnated with Scintillator and dried[8]; radioactive bands are detected by scintillation autoradiography at $-70°$ with preflashed RP X-Omat film.[28] Gels containing ^{125}I-labeled proteins are exposed to Kodak X-Omat film in the presence of a fast tungstate intensifying screen.

Because of the fragility of the MuLV virions assembled in interferon-treated cells, the purification of virions by the sucrose gradients is not always the best choice. Alternatively, the virions can be collected by banding on a sucrose cushion (40% sucrose) and the viral proteins assayed after immunoprecipitation. Labeled virions are disrupted in 25 mM Tris · HCl, pH 8.0, and 50 mM NaCl in the presence of Nonidet P-40 (1%) and sodium deoxycholate (0.5%). Volumes are adjusted to 200 μl with the same buffer, to which 1 μl of the immunized serum is added. The mixture is incubated for 30 min at 37° and then for 2 hr at 4°. The immune complex is precipitated by adding 50 μl of a 10% suspension of *Staphylococcus aureus* (Cowan strain) and incubating for 30 min at room temperature. The precipitate is sedimented in an Eppendorf microcentrifuge (5 min) and washed several times each with 50 mM Tris · HCl, pH 7.0, 0.1 M NaCl, 0.5% Nonidet P-40, and 2.4 M KCl and with 50 mM Tris · HCl, pH 7.0, 0.1 M NaCl, 0.1% Triton X-100, and 5 mM EDTA. After the final wash, 50 μl of a mixture of 6.25 mM Tris · HCl, pH 6.8, 2% SDS, and 10% glycerol is added to the pellet. Samples assayed under reducing conditions are treated with 2-mercaptoethanol (1.25% final concentration), and those assayed under nonreducing conditions are treated with iodoacetamide (0.05 M final concentration); all samples are boiled for 10 min before electrophoresis.

[26] J. C. Alwine, D. J. Kemp, and G. R. Stark, *Proc. Natl. Acad. Sci.* **74**, 5350 (1977).
[27] U. K. Laemmli, *Nature (London)* **227**, 680 (1977).
[28] W. M. Bonner and R. A. Laskey, *Eur. J. Biochem.* **13**, 2633 (1974).

Assays of Viral RNA and Protein in Infected Cells

Preparation of RNA

Interferon-treated infected cells or controls are washed with phosphate-buffered saline (PBS), scraped into a small volume of PBS, and centrifuged at 3000 g for 5 min. RNA can be isolated from the cell pellet by the CsCl centrifugation method.[29] The cell pack is suspended in 5 volumes of TNE buffer (10 mM Tris · HCl, pH 7.6, 100 mM NaCl, and 1 mM EDTA) containing 4% (w/v) Sarkosyl (sodium lauryl sarcosinate) and sonicated briefly. Then 1 g of solid CsCl is added per milliliter, and the suspension is mixed until all CsCl dissolves. The homogenate is then layered on a cushion of 5.7 M CsCl in 10 mM Tris · HCl, pH 7.8 (final density 1.707), and the sample is centrifuged in a Beckman SW41 rotor at 35,000 rpm for 36–48 hr. At the end of the centrifugation, DNA bands at the interface of the CsCl solutions, whereas RNA pellets at the bottom of the tube. The RNA pellet is washed and dissolved in Tris · HCl (10 mM, pH 7.4), and RNA is precipitated with 2.5 volumes of 80% ethanol containing 0.2 M sodium acetate at −20° and collected by centrifugation. RNA prepared by this method is suitable both for nucleic acid hybridization or for analysis on the gels under denaturing conditions. For *in vitro* translation, poly(A)-containing RNA can be isolated by oligo(dT) cellulose chromatography.

Quantitation of Viral RNA by Nucleic Acid Hybridization

The amount of viral RNA in the pool of total cellular RNA can be determined by hybridization with the cDNA prepared from the RNA of the given MuLV virus. The conditions of the reaction for the preparation of cDNA differ, depending on the MuLV used, and it is advisable to follow the original paper.

The hybridization reactions, however, generally are the same for different MuLV probes. The reaction mixture consists of 0.6 M NaCl, 0.02 M Tris · HCl, pH 7.4, 0.001 M EDTA, 0.02% sodium lauryl sulfate, and approximately 500–1000 cpm of the MuLV probe. The cellular RNA is added in different concentrations (10–1000 µg/ml), and the final volume is 50 µl. The reaction mixture is sealed in 50 µl capillary pipettes, boiled for several minutes, and incubated at 65°. To analyze the formation of the RNA–DNA hybrids, the hybridization mixture is diluted into 10 volumes of sodium acetate buffer (30 mM, pH 6.5) containing 200 mM NaCl and 1.8 mM ZnCl₂,[30] and then denatured DNA is added to a final concentration

[29] V. Grisin, R. Crkvenjakov, and C. Byus, *Biochemistry* **13**, 2633 (1974).
[30] J. A. Leong, A. Garapin, N. Jackson, L. Fauscher, W. Levinson, and J. M. Bishop, *J. Virol.* **9**, 891 (1972).

of 50 μg/ml. The sample is divided into two portions, one of which is incubated with S-1 nuclease[31] at 37° for 1 hr. Both samples (treated and untreated) are then spotted on separate disks of Whatman 3 paper, and the disks are washed with cold 10% and 5% trichloroacetic acid, methanol, and acetone as described for the reverse transcriptase assay. The filters are dried and counted in PPO–POPOP toluene-based liquid scintillation fluor. The amount of hybrid formation is expressed as the percentage of counts remaining after S-1 nuclease treatment compared to the untreated control.

This method enables estimation of both the extent of the viral gene transcription and the proportion of total cellular message as viral RNA (comparing the respective $CrT_{1/2}$ values).

ANALYSIS OF VIRAL RNA BY NORTHERN BLOTTING

This method is not quantitative but enables detection of different sizes of viral RNAs present in the cells. The cellular RNA is separated by gel electrophoresis on 1.5% agarose under denaturing conditions. The denaturation is achieved by treatment with glyoxal or methyl Hg as described previously. After electrophoresis (50 V for 2.5 hr) the RNA is transferred by blotting on DBM-paper and hybridized with ^{32}P-labeled cDNA to MuLV or denatured ^{32}P-labeled cloned DNA of the respective virus.[32]

ANALYSIS OF VIRAL PROTEINS BY COMPETITIVE RADIOIMMUNOASSAY

Competitive radioimmunoassay can be used to quantitate the amounts of viral proteins in the infected cells. The optimal reaction condition is determined by a direct radioimmunoassay. To 1–5 μg of ^{125}I-labeled viral protein (e.g., p30 or gp71) in a 10-μl volume a given dilution (in radioimmunoassay buffer: 20 mM Tris · HCl, pH 7.6, 100 mM NaCl, 1 mM EDTA containing 2 mg/ml of bovine serum albumin) of normal rabbit serum (30 μl) and 20 μl of immune serum (e.g., anti p30 or gp71) is added. The mixture is incubated for 3 hr at 37° and at 4° overnight. The samples are then processed as described above.

Radioimmunoprecipitation Assays

The radioimmunoprecipitation assay is used to analyze cell extracts for the presence of material specifically precipitated by the antiserum to purified viral proteins. Cells are labeled with amino acid ([^{3}H]- or [^{14}C]leucine or [^{35}S]methionine, 100–250 μCi/ml) in leucine- or methionine-free medium for 10–20 min. The cells are either lysed immedi-

[31] W. D. Sutton, *Biochim. Biophys. Acta* **240**, 522 (1971).
[32] J. C. Alwine, D. J. Kemp, B. A. Parker, J. Reiser, J. Renart, G. R. Stark, and G. M. Wahl, this series, Vol. 68, p. 220.

ately after labeling or the labeling medium is replaced with nonradioactive growth medium and incubation is continued for 1–6 hr before lysis. Cells are scraped in PBS and centrifuged for 1500 rpm for 5 min; the cell pellet is then lysed at 4° with lysis buffer (PBS, pH 7.4, containing 1% Triton X-100, 0.1% SDS, and 0.5% DOC). For 5×10^6 cells, about 2 ml of lysis buffer are used, and the cell lysate is centrifuged at $2000\,g$ for 15 min. The supernatant (1 ml) is then incubated with 20 μl of nonimmunized serum at 4° overnight, and the 100 μl of protein A Sepharose (Pharmacia) (1 mg/ml in phosphate buffer saline, pH 7.4, freshly washed) is added and incubated at 4° for an additional 3 hr. The mixture is then centrifuged at $100,000\,g$ for 90 min, and the supernatant is collected. To the supernatant 5 μl of immunized serum and 100 μl of cell lysate (10^7 uninfected cells lysed with 1 ml of lysate buffer and centrifuged at $100,000\,g$ for 1 hr) are added, and the mixture is incubated at 4° overnight. The following day, protein A Sepharose is added (20 μl, 4° 3 hr). The protein A is sedimented at 10,000 g for 10 min and washed 3 times with lysis buffer at 4°. After the final wash, 50 μl of the electrophoresis sample buffer (62.5 mM Tris · HCl, pH 6,8, 2% SDS, 10% glycerol and 1.25% mercaptoethanol) is added to the pellet, and the sample is boiled for 10 min before electrophoresis. Radioactive proteins are detected by autoradiography using Kodak RP XO-mat film.[28]

[52] Demonstration of Potentiation of the Antiviral and Antitumor Actions of Interferon[1]

By W. R. Fleischmann, Jr. and L. A. Schwarz

Several reports of potentiation of the antiviral action of interferon have been described in the literature. Additions of actinomycin D at 4–6 hr after interferon treatment have been shown to potentiate the development of the antiviral state of interferon.[2] The antiviral effect of interferon can also be enhanced by the coincubation of cells with interferon and cyclic AMP.[3–6] Further, pretreatment of cells with low levels of interferon (1–4

[1] This work was supported by Department of Health, Education, and Welfare Grants 5-S 07-RR-05427 GRS and 1-R01-CA26475. L. A. S. was supported by a J. W. McLaughlin Predoctoral Fellowship.
[2] C. Chany, F. Fournier, and S. Rousset, *Nature (London)* **230**, 113 (1971).
[3] R. M. Friedman and I. Pastan, *Biochem. Biophys. Res. Commun.* **36**, 735 (1969).
[4] L. B. Allen, N. C. Eagle, J. H. Huffman, D. A. Shuman, R. B. Meyer, Jr., and R. W. Sidwell, *Proc. Soc. Exp. Biol. Med.* **146**, 580 (1974).
[5] J. M. Weber and R. B. Stewart, *J. Gen. Virol.* **28**, 363 (1975).
[6] H. Koblet, R. Wyler, and U. Kohler, Experientia **35**, 575 (1979).

units/ml) primes the cells to give an enhanced response to a subsequent exposure to interferon.[7,8]

This chapter describes another type of potentiation, which occurs when cells are incubated with combined preparations of mouse virus-type and immune interferons. The combined interferon preparations cause a greater than additive, synergistic amplification of the interferon activity. This potentiation phenomenon was originally demonstrated as a greatly enhanced antiviral effect, but recent studies have extended the observation to include the potentiation of the antitumor activity of interferon.[9-11]

Large-Scale Production of Mouse Virus Type Interferon

Reagents and Supplies

Eagle's MEM (Hanks' base, Grand Island Biological)
Fetal bovine serum (Flow Laboratories)
Antibiotics (penicillin, streptomycin)
Newcastle disease virus (Beaudette strain)
Mouse L cells (strain 929)
roller bottles, 2000 ml

Procedure

Any of the current methods of production of mouse virus-type interferon are suitable. In this laboratory, L cell interferon was produced in a large-scale procedure employing roller bottles. L cells (clone 929) were diluted to 5×10^5 cells/ml in growth medium composed of Eagles MEM, 8% fetal calf serum, 0.075% sodium bicarbonate, penicillin (100 units/ml), and streptomycin (100 μg/ml). The cell suspension was added to 2000-ml roller bottles at 100 ml of cells/bottle. The roller bottles were tightly sealed and incubated until they reached confluency. Newcastle disease virus (NDV) was employed for interferon induction. Five milliliters of a virus preparation containing 1×10^8 virus/ml was added to each roller bottle and the virus was allowed to absorb for 60 min. The cells were then washed with growth medium to remove unbound virus and overlaid with

---

[7] E. T. Sheaff and R. B. Stewart, *Can. J. Microbiol.* **15**, 941 (1969).

[8] W. R. Fleischmann, Jr., *Tex. Rep. Biol. Med.* **35**, 316 (1977).

[9] W. R. Fleischmann, Jr., J. A. Georgiades, L. C. Osborne, and H. M. Johnson, *Infect. Immun.* **26**, 248 (1979).

[10] W. R. Fleischmann, Jr., K. M. Kleyn, and S. Baron, *JNCI, J. Natl. Cancer Inst.* **65**, 963 (1980).

[11] M. M. Brysk, E. H. Tschen, R. D. Hudson, E. B. Smith, W. R. Fleischmann, Jr., and H. S. Black, *J. Am. Acad. Dermatol.* **5**, 61 (1980).

100 ml of growth medium. After a 24-hr incubation period, the supernatant fluid was harvested, spun at 2500 g to remove cell debris, adjusted to pH 2 with HCl, and stored at 4° for 5 days to inactivate residual NDV. The interferon was then adjusted to pH 7.2 with NaOH, aliquoted, and stored at −20°. This unpurified virus-type interferon preparation had a specific activity of about 10^3 units/mg protein. It was composed of two components, which have been suggested to be the mouse equivalents of fibroblast and leukocyte interferon.[12]

Large-Scale Production of Immune Interferon

Reagents and Supplies

RPMI-1640 medium containing L-glutamine (Microbiological Associates)
Eagle's MEM (Hanks' base, Grand Island Biological)
Fetal bovine serum (Flow Laboratories)
Tricine [N-tris(hydroxymethyl)methylglycine] (Sigma)
Antibiotics: penicillin, streptomycin, mycostatin, gentamycin
2-Mercaptoethanol
T-cell mitogen: staphylococcal enterotoxin A (SEA) obtained from Microbial Biochemistry Branch, Division of Microbiology, Food and Drug Administration, Cincinnati, Ohio)
C57BL/6 female mice, 8–12 weeks old (Jackson Laboratories)
Plastic petri dishes, 60 mm (Corning)
Centrifuge tubes, 50 ml
Dissecting forceps with teeth
Teflon policeman
Scissors
Ethanol, 70%
Roller apparatus
Roller bottles (Bellco, stock No. 7730-38260)

Procedure

The immune interferon was produced in a large-scale procedure employing roller bottles.[13] C57BL/6 female mice were disinfected with 70% ethanol, and their spleens were excised under aseptic conditions. Generally, six spleens were used for each roller bottle culture. The spleens were placed in 60-mm plastic petri dishes (3–4 spleens/plate) containing 4–5 ml

[12] W. E. Stewart, II, *Nature (London)* **286**, 110 (1980).
[13] L. C. Osborne, J. A. Georgiades, and H. M. Johnson, *Infect. Immun.* **23**, 80 (1979).

of Eagle's MEM (Hanks' base). The spleens were dissociated with dissecting forceps with teeth, leaving a white fibrous matrix. After the spleens had been teased apart, adherent cells were scraped from the plates with a Teflon policeman, and the spleen cell suspension was transferred to a 50-ml centrifuge tube that had been cooled to 4°. After allowing the large tissue clumps to settle, the cell suspension was decanted into another centrifuge tube and spun for 10 min at 1000 rpm in a Sorvall RC-3 centrifuge (HL-8 rotor). The cell pellet was resuspended in 20 ml of RPMI supplemented with 10% fetal calf serum, 0.015 M Tricine (pH 7.2), penicillin (100 units/ml), streptomycin (100 μg/ml), gentamycin (40 μg/ml), and mycostatin (10 units/ml). The cells were then incubated at room temperature for 0.5 hr with 0.8 μg of SEA per milliliter before further dilution. The cells were diluted to 3×10^6 cells/ml in the growth medium, and 200 ml of the cell suspension was added to each 2000-ml roller bottle. 2-Mercaptoethanol was added to the cell suspension at a final concentration of 10 μM. The roller bottles were tightly sealed and rotated on a Bellco cell production roller apparatus at 8 rpm at 37°. After 2 days of incubation, the roller bottle spleen cell cultures were harvested. Cells and cell debris were removed by centrifugation at 2500 g for 20 min. The supernatant fluid was decanted, aliquoted, and stored at $-70°$.

Potentiation of the Antiviral Activity of Interferon

Potentiation of the antiviral activity of interferon can be demonstrated by contrasting the separate protective abilities of immune interferon and of virus-type interferon with the combined protective ability of a mixed preparation of the immune and virus-type interferons in a one-step growth experiment.[9] L cell monolayers (clone 929) were set at 1×10^6 cells per 35-mm petri dish. After a 6–10-hr incubation period for cell attachment, the monolayers were overlaid for 12 hr with 1 ml of medium, immune interferon, virus-type interferon, or a combination of immune and virus-type interferons. The monolayers were challenged with 0.1 ml of mengovirus at a multiplicity of 10 and allowed to incubate for 45 min. After washing to remove unattached virus, the monolayers were overlaid with 2 ml of growth medium and incubated. Virus yields were harvested at 24 hr after infection and quantitated by plaque assay.[14] The table presents the results of a representative experiment. The immune interferon alone provided 3 units of interferon protection. The virus-type interferon alone provided 26 units of interferon protection. The combined immune and virus-type interferon preparation provided 320 units of protection. This represented an 11-fold greater level of protection than expected based on

[14] W. R. Fleischmann, Jr. and E. H. Simon, *J. Gen. Virol.* **20**, 127 (1973).

POTENTIATION OF INTERFERON ACTIVITY BY MIXED PREPARATIONS OF IMMUNE INTERFERON (IF) AND FIBROBLAST INTERFERON[a,b]

IF sample	Virus yield (PFU/ml)	Fold inhibition	IF titer (units/ml)		
			Actual[c]	Expected[a]	Fold potentiation[e]
No IF	$1.0 \times 10^9 \pm 0.1 \times 10^9$[f]	100	—	—	—
Immune IF	$3.8 \times 10^8 \pm 0.4 \times 10^8$	2.6	3	—	—
Fibroblast IF	$2.3 \times 10^7 \pm 0.1 \times 10^7$	43	26	—	—
Immune IF + fibroblast IF	$1.4 \times 10^6 \pm 0.03 \times 10^6$	714	320	29	11

[a] From Fleischmann et al.,[9] with permission.

[b] Mouse L cell monolayers were treated for 12 hr with growth medium (no interferon), immune interferon, fibroblast interferon, and immune interferon and fibroblast interferon in combination. The monolayers were challenged with mengovirus at a multiplicity of infection of 10 PFU/cell, and virus was harvested 24 hr later.

[c] Actual interferon titers were determined by comparison of virus yield with a standard yield reduction curve of interferon activity determined concurrently.

[d] Expected interferon titer was determined by adding the actual titers of the interferons present in the mixed interferon preparation.

[e] The potentiation factor was determined by dividing the actual interferon titer by the expected interferon titer.

[f] Mean ± standard deviation.

the additivity of the two interferons (3 units plus 26 units). Thus, there was a synergistic enhancement of the antiviral activity when the two types of interferons were combined.

An Inhibitor of Interferon Action Affects Antiviral Potentiation

One factor that did affect the level of potentiation was the presence of an inhibitor of interferon action in some immune interferon preparations (see this volume [53]). The inhibitor not only blocked the antiviral activity of the immune interferon in the preparations, but also prevented the expression of the maximum level of potentiation. Immune interferon preparations that contained substantial levels of the inhibitor were unable to demonstrate either antiviral activity or potentiating activity in one-step virus growth experiments. For this reason, only immune interferon preparations that did not contain detectable levels of inhibitor were employed for potentiation studies. Mouse immune interferon induction by SEA peaked 3 days after SEA addition; however, the inhibitor was present in detectable amounts by 3 days. Therefore, 2-day preparations that did not maximize immune interferon production but were free of detectable inhibitor were used. Since inhibitor was present as early as day 1 in phytohemagglutinin P and concanavalin A-induced mouse spleen cell preparations, these mitogens were not good inducers of immune interferon for potentiation studies.

Potentiation of the *in Vivo* Antitumor Effect of Interferon

Potentiation of the antitumor effect of interferon has been demonstrated *in vivo* with two different tumor systems. The first system examined the effect of separate and combined therapies of immune and virus-type interferons on the *development* of P388 tumors in DBA/2 mice.[10] To establish solid tumors, approximately 10^5 tumor cells were injected subcutaneously in a 0.2-ml volume into the right sides of the abdomens of 4-week-old DBA/2 mice. Mice were injected subcutaneously at the approximate tumor site with 0.5 ml of mock interferon, immune interferon, virus-type interferon, or a combination of immune and virus-type interferon. Treatment was begun 3 hr before tumor cell injection and continued for 15 days thereafter. As shown in Fig. 1, tumor development was unaffected by treatment with 25 units of immune interferon alone. Treatment with 25,000 units of virus-type interferon significantly delayed tumor development and increased the survival time of these mice as compared with control mice. Therapy with combined immune and virus-type interferon approximately doubles the delay in tumor development seen for virus-

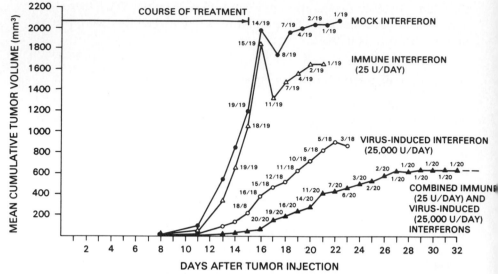

FIG. 1. Effect of combined immune and virus-induced interferons on the rate of P388 tumor development in DBA/2 mice. Each mouse was inoculated subcutaneously with 10^5 P388 tumor cells; the mice were then distributed into four groups of 18–20 each. Mice were inoculated 3 hr before tumor injection and daily for 15 days thereafter at the approximate site of tumor injection with mock interferon (●), immune interferon (△), virus-induced interferon (○), or a combination of immune and virus-induced interferons (▲). The volume of the primary tumor was determined for each mouse on the indicated days after tumor cell injection. Data were plotted as the linear increase in mean cumulative tumor volume. Mean cumulative tumor volumes were determined within each group by averaging the tumor size of all surviving mice with the final tumor size of all mice that had died. Fractions indicate number of survivors per total number of mice at the indicated times after tumor injection for each treatment group. From Fleischmann et al.[10]

type interferon alone. The increased survival time of the mice treated with virus-type interferon was similarly doubled by the combined therapy (Fig. 2). Thus, the addition of 25 units of immune interferon to the 25,000 units of virus-type interferon resulted in a significant potentiation of the *in vivo* antitumor effect.

The second *in vivo* system examined the potentiation of the antitumor effect of interferon on *established* UV light-induced squamous cell carcinomas in hairless mice. The tumors were injected daily for 19 days with 0.2 ml of mock interferon, immune interferon, virus-type interferon, or a combination of immune and virus-type interferons. The results paralleled those described above for the P388 tumors. Treatment with 10 units of immune interferon alone did not affect tumor size. Treatment with 3000

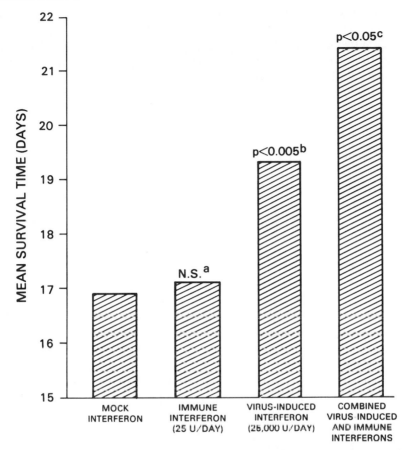

FIG. 2. Effect of combined immune and virus-induced interferons on life-spans of mice inoculated with P388 tumor cells. Each mouse was inoculated subcutaneously with 10^5 P388 tumor cells; the mice were then distributed into four groups of 18–20 each. Mice were inoculated daily for 15 days at the approximate site of tumor injection with mock interferon, 25 units of immune interferon per day, 25,000 units of virus-induced interferon per day, or a combination of 25 units of immune interferon and 25,000 units of virus-induced interferon per day. The day of death of each mouse was noted. Results were plotted as the mean survival time for each of the four groups of mice. Statistical probabilities were determined by the t test. Comparable statistical probabilities were observed when the data were analyzed by the Mann-Whitney U test. Immune interferon treatment was compared to mock interferon treatment (a). NS, not significant. Virus-induced interferon treatment was compared to mock interferon treatment (b). Combined-interferon treatment was compared to virus-induced interferon treatment (c). Fleischmann et al.[10]

units of virus-type interferon gave an apparent low level of tumor regression, but the results were not statistically significant. Combined therapy with immune and virus-type interferon caused a greater than 50% reduction in tumor size. These results also demonstrated that the addition of immune interferon to virus-type interferon potentiated the antitumor activity of the preparation and caused a significant tumor regression.

Preliminary Characterization of the Potentiation Phenomenon

The level of potentiation observed was directly dependent on the concentrations of the two interferons employed. Potentiation was best shown when the interferon concentrations in the mixed preparations were 5 units or more. Therefore, the one-step yield reduction assay was the method of choice for the demonstration of potentiation of antiviral activities. The antiviral potentiation effect was independent of the virus employed for challenge, as vesicular stomatitis virus, vaccinia, and mengovirus all gave comparable results. Similarly, comparable levels of antitumor potentiation were seen for P388 tumors, B-16 melanoma tumors, and UV-light-induced squamous cell carcinomas. Further, the level of potentiation was independent of the purity of the interferons since 10,000-fold purified virus-type interferon and 1000-fold purified immune interferon had the same potentiating effect as the crude virus-type and immune interferons. Thus, the potentiation phenomenon appeared to be a mutually synergistic interaction of the two types of interferon that resulted in a greatly enhanced antiviral and antitumor effect.

[53] Assay and Characterization of an Inhibitor of Interferon Action[1]

By W. R. FLEISCHMANN, JR. and E. J. LEFKOWITZ

Several inhibitors of interferon action have been described in the literature. Some are produced spontaneously,[2–9] whereas others are produced in response to viral infections.[10–12] This chapter describes the identification, large-scale production, partial purification, and preliminary characterization of an inducible inhibitor of interferon action that is produced by mouse spleen cells treated with T cell mitogens.[13]

[1] This work was supported by Department of Health, Education, and Welfare Grant 5 S 07 RR-05427 GRS. E. J. L. was supported by a J. W. McLaughlin Predoctoral Fellowship.
[2] M. Cembrzynska-Nowak, *Arch. Immunol. Ther. Exp.* **25**, 663 (1977).

Large-Scale Production of Inhibitor

Reagents and Supplies

RPMI-1640 Medium containing L-glutamine (Microbiological Associates)

Eagle's MEM (Hanks' base, Grand Island Biological)

Fetal bovine serum (Flow Laboratories)

Tricine: [*N*-tris(hydroxymethyl)methylglycine] (Sigma)

Antibiotics: penicillin, streptomycin, mycostatin, gentamycin

2-Mercaptoethanol

T cell mitogens: staphylococcal enterotoxin A (SEA) (obtained from Microbial Biochemistry Branch, Division of Microbiology, Food and Drug Administration, Cincinnati, Ohio); phytohemagglutinin P (PHA) (Burroughs Welcome); concanavalin A (Con A), twice crystallized (ICN Pharmaceuticals)

C57BL/6 female mice, 8–12 weeks old (Jackson Laboratories)

Plastic petri dishes 60 mm (Corning)

Centrifuge tubes, 50 ml

Dissecting forceps with teeth

Teflon policeman

Scissors

Ethanol, 70%

Roller apparatus

Roller bottles (Belco, stock No. 7730-38260)

Procedure

The interferon inhibitor was produced in a large-scale procedure employing roller bottles.[13,14] C57BL/6 female mice were cleansed with 70%

[3] C. Chany, J. Lemaitre, and A. Grégoire, *C. R. Hebd. Seances Acad. Sci. Ser. D* **269**, 2628 (1969).

[4] L. B. Epstein, *in* "Immunobiology of the Macrophage" (D. S. Nelson, ed.), p. 201. Academic Press, New York, 1976.

[5] F. Fournier, S. Rousset, and C. Chany, *Proc. Soc. Exp. Biol. Med.* **132**, 943 (1969).

[6] S. E. Grossberg and P. S. Morahan, *Science* **171**, 77 (1971).

[7] T. G. Rossman and J. Vilček, *Arch. Gesamte Virusforsch.* **31**, 18 (1970).

[8] M. Rytell, *Arch. Virol.* **48**, 103 (1975).

[9] J. L. Truden, M. M. Sigel, and L. S. Dietrich, *Virology* **33**, 95 (1967).

[10] J. B. Campbell, *Can. J. Microbiol.* **22**, 712 (1976).

[11] C. Chany and C. Brailovsky, *C. R. Hebd. Seances Acad. Sci.* **261**, 4282 (165).

[12] Y. Z. Ghendon, I. G. Balandin, and L. M. Babushkina, *Acta Virol. (Engl. Ed.)* **10**, 268 (1966).

[13] W. R. Fleischmann, Jr., J. A. Georgiades, L. C. Osborne, F. Dianzani, and H. M. Johnson, *Infect. Immun.* **26**, 949 (1979).

[14] L. C. Osborne, J. A. Georgiades, and H. M. Johnson, *Infect. Immun.* **23**, 80 (1979).

ethanol, and their spleens were excised under aseptic conditions. Generally, six spleens were used for each roller bottle culture. The spleens were placed in 60-mm plastic petri dishes (3–4 spleens/plate) containing 4–5 ml of Eagle's MEM (Hanks' base). The spleens were dissociated with dissecting forceps with teeth, leaving a white fibrous matrix. After the spleens had been teased apart, adherent cells were scraped from the plates with a Teflon policeman, and the spleen cell suspension was transferred to a 50-ml centrifuge tube that had been cooled to 4°. After allowing the large tissue clumps to settle, the cell suspension was decanted into another centrifuge tube and centrifuged for 10 min at 1000 rpm in a Sorvall RC-3 centrifuge (HL-8 rotor). The cell pellet was resuspended in 20 ml of RPMI supplemented with 10% fetal calf serum, 0.015 M Tricine (pH 7.2), penicillin (100 units/ml), streptomycin (100 μg/ml), gentamycin (40 μg/ml), and mycostatin (10 units/ml). If SEA was to be employed for induction of inhibitor, the cells were incubated at room temperature for 0.5 hr with 0.8 μg of SEA per milliliter before further dilution. The cells were diluted to 3 × 10^6 cells/ml in the growth medium, and 200 ml of the cell suspension were added to each 2000-ml roller bottle. 2-Mercaptoethanol was added to the cell suspension at a final concentration of 10 μM. If PHA or Con A were to be employed in inhibitor induction, they were added at this time to final concentrations of 25 μg/ml or 2 μg/ml, respectively. The roller bottles were tightly sealed and rotated on a Bellco cell production roller apparatus at 8 rpm at 37°. After 3 days of incubation, the roller bottle spleen cell cultures were fed with 3 ml of 5 times concentrated Eagle's MEM (Hanks' base) supplemented with 10% fetal calf serum. Four days after PHA and Con A treatment or 5 days after SEA treatment, the culture fluid was harvested. Cells and cell debris were pelleted by centrifugation for 15 min at 2000 rpm in a Sorvall RC-3 centrifuge (HL-8 rotor). The supernatant fluid was decanted, divided into aliquots, and stored at −70°.

Inhibitor Detection

The mouse spleen cell supernatant fluid preparations that contain the inhibitor of interferon action also contain immune interferon. The inhibitor can be detected and roughly quantitated by taking advantage of the fact that the inhibitor can be measured in undiluted preparations and, except for highly concentrated inhibitor preparations, the inhibitor activity can be diluted out more rapidly than the interferon activity. Thus, when mixtures of inhibitor and interferon are serially diluted, the more concentrated samples can be assayed by a one-step yield reduction assay for inhibitor activity and the more diluted samples can be assayed by a plaque reduction assay for interferon activity.

For plaque reduction assays (interferon assays), the serially diluted samples were overlaid on confluent mouse L cell monolayers for 12 hr. The fluid was removed and the monolayers were challenged with either mengovirus or vesicular stomatitis virus at a multiplicity of infection of 80–100 plaque-forming units (PFU) per 35-mm petri plate. After a 45-min absorption period, the monolayers were overlaid with a starch or agar overlay.[15] Plaques were counted 20–24 hr after challenge by staining the plates with about 10 drops of a 0.2% (w/v) solution of neutral red. The interferon titer was expressed as the reciprocal of the sample dilution that reduced plaques to 50% of the control number and was then corrected to NIH international standard units.

For yield reduction assays (inhibitor assays), the serially diluted samples were overlaid on confluent mouse L cell monolayers as described above. The fluid was removed, and the monolayers were challenged with virus at a multiplicity of infection of 10 PFU per cell. After a 45-min absorption period, the monolayers were washed twice to remove unabsorbed virus. Progeny virus were harvested 24 hr after challenge and stored at −20° until assayed by plaque assay. Inhibitor activity was detected as a reduced protective capacity of the inhibitor/interferon preparations, relative to that expected on the basis of the amount of interferon present. Figure 1 illustrates this procedure for detecting inhibitor in 3-day SEA-stimulated mouse spleen cell cultures. The inhibitor/immune interferon preparation was 10- to 100-fold less protective against both mengovirus and vesciular stomatitis virus than the fibroblast interferon preparation. Immune interferon preparations that do not contain inhibitor activity are as protective as fibroblast interferon. Thus, the decreased protective capacity of the inhibitor/immune interferon preparation indicates the presence of inhibitor activity. This activity can be roughly quantitated by comparing the relative decrease in the protective capacity of the inhibitor/immune interferon preparation.

Kinetics of Inhibitor Production

As suggested above, not all immune interferon preparations contain inhibitor activity. When the inhibitor was induced by the T cell mitogens SEA, Con A, and PHA, the kinetics of inhibitor production appeared to be related to the kinetics of immune interferon production.[16] For example, interferon production reached a maximum level 1 day after stimulation

[15] W. R. Fleischmann, Jr. and E. H. Simon, *J. Gen. Virol.* **20,** 127 (1973).
[16] W. R. Fleischmann, Jr., E. J. Lefkowitz, J. A. Georgiades, and H. M. Johnson, *in* "Interferon: Properties and Clinical Uses" (A. Khan, N. O. Hill, and G. Dorn, eds), p. 195. Leland Fikes Found. Press, Dallas, Texas, 1980.

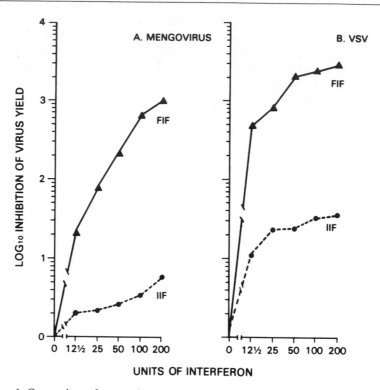

Fig. 1. Comparison of mengovirus and vesicular stomatitis virus (VSV) yield reduction capabilities of immune and fibroblast interferons. L-929 monolayers were treated with various concentrations of fibroblast interferon (FIF) or immune interferon (IIF) for 12 hr before challenge with mengovirus or VSV. One set of monolayers was infected with 100 PFU per plate for a plaque reduction assay. Another set was infected with 10 PFU per cell for a yield reduction assay. The results are plotted as $\log_{10}$ inhibition of virus yield per plaque reduction unit of interferon. (A) Mengovirus-infected monolayers. (B) VSV-infected monolayers. Reproduced from Fleischmann et al.,[13] with permission.

with Con A and 3 days after stimulation with SEA. Figure 2 plots the production of inhibitor following Con A and SEA stimulation of mouse spleen cells. A significant level of inhibitor activity was detectable as a reduced protective capacity of the immune interferon preparation as early as 1 day after stimulation with Con A. Equivalent levels of inhibitor activity were not observed until 3 days after stimulation with SEA. Thus, inhibitor production appears to correlate with interferon production such that immune interferon production may actually trigger the production of the inhibitor, which may in turn regulate the interferon system.

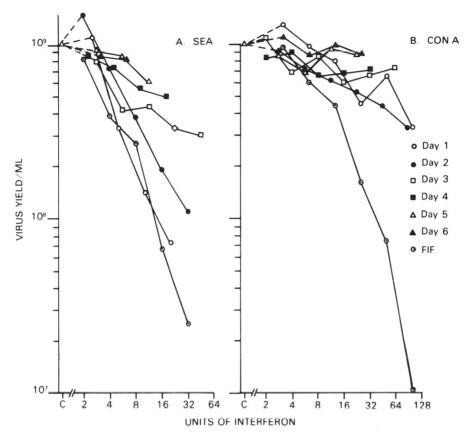

FIG. 2. Kinetics of production of inhibitor. C57BL/6 mouse spleen cells in culture were stimulated with a final concentration of 0.02 μg/ml of staphylococcal enterotoxin A (SEA) or 2 μg/ml of concanavalin A (ConA). Supernatant fluids were harvested at the indicated times after mitogen stimulation and assayed for interferon activity by plaque-reduction assay. The yield-reduction protective capability of serial dilutions of each sample were determined and plotted against the interferon titer of the sample. Inhibitor was detected by the decreased yield-reduction capability of the immune interferon samples. The more shallow the slope of the yield-reduction curve for a given sample, the greater the concentration of inhibitor in that sample. Reproduced from Fleischmann et al.,[16] with permission.

Partial Purification of Inhibitor

Batches of about 1 liter of pooled inhibitor/immune interferon preparation prepared from 4–5-day SEA-stimulated mouse spleen cells were concentrated by a two-step precipitation with ammonium sulfate. Solid ammonium sulfate was added with constant stirring to 55% saturation. The

55% precipitate was removed by centrifugation at 6000 rpm for 15 min in a Sorvall RC5 centrifuge using the GSA head. The carefully decanted supernatant fluid is then raised to 80% saturation by the addition of more solid ammonium sulfate. The 80% precipitates were collected by centrifugation (as above), then resuspended in 50 ml of distilled water and exhaustively dialyzed against distilled water. This represented a 20-fold concentration of the initial preparation. The entire procedure was performed at 4°.

BSA–Affi-Gel was prepared according to the instructions of the manufacturer. Bovine serum albumin (BSA, Nutritional Biochemicals Corporation) was prepared as a 15 mg/ml solution in 0.1 M sodium phosphate, pH 7.0. Of this BSA solution, 25 ml were added to a vial containing 15 g of lyophilized Affi-Gel 10 (Bio-Rad), and the mixture was stirred overnight at 4°. The BSA-Affi-Gel preparation was packed in a 0.9 cm by 5 cm column and washed with 1 M NaCl. The column was then equilibrated with 0.5 M sodium acetate, pH 5.0. The concentrated inhibitor/interferon preparation was dialyzed against 0.5 M sodium acetate, pH 5.0, for 12 hr and loaded onto the BSA–Affi-Gel column. The column was washed with the equilibrating buffer and eluted with a discontinuous salt and pH gradient. The column fractions were assayed for their ability to block the protective effect of 25 units of mouse fibroblast interferon. Fractions eluted with 0.5 M sodium acetate, pH 5.2, and 0.5 M sodium acetate, pH 5.4, contained little or no inhibitor and were discarded. Fractions eluted with 0.08 M sodium chloride, 0.5 M sodium acetate, pH 6.0, and with 1 M sodium chloride, 0.5 M sodium acetate, pH 7.0, contained inhibitor activity and were pooled.

The pooled inhibitor fractions were concentrated 10-fold by dialysis against 25% polyethylene glycol and loaded on an AcA-54 Ultrogel column (2.5 cm by 82.5 cm). The Ultrogel column was washed with phosphate-buffered saline, and 5-ml fractions were collected. The column fractions were assayed for inhibitor activity as described above. As shown in Fig. 3, the inhibitor migrated with proteins of 8000–10,000 daltons. The peak inhibitor fractions from the Ultrogel column were pooled. The two-step chromatographic procedure results in an approximately 1000-fold purification of the inhibitor.

Characterization of the Inhibitor of Interferon Action

The purified inhibitor preparations have been shown completely to block the antiviral activity of both mouse fibroblast interferon and mouse immune interferon. The inhibitor does not reverse an established antiviral

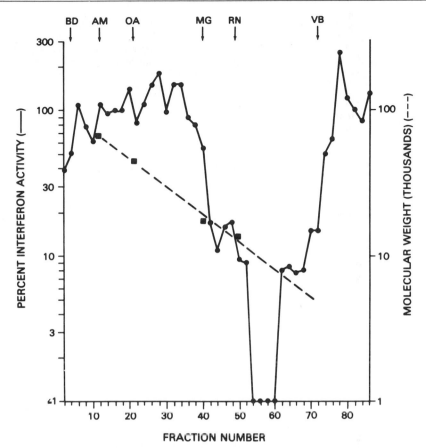

FIG. 3. Ultrogel AcA-54 profile of bovine serum albumin (BSA)–Affi-Gel purified inhibitor activity. Crude immune interferon was absorbed to a BSA-Affi-Gel column (0.9 by 5 cm). The column was eluted with a discontinuous salt and pH gradient. Fractions (2 ml) were collected and stored at $-70°$. Inhibitor-containing fractions eluting with 0.08 M NaCl, pH 6.0, and 1 M NaCl, pH 7.0, were dialyzed against 25% polyethylene glycol to a volume of 5 ml and passed through an Ultrogel AcA-54 gel filtration column (2.5 by 82.5 cm). Fractions (5 ml) were collected. Equal volumes of each fraction were mixed either with growth medium or with fibroblast interferon (final concentration, 25 units of fibroblast interferon per milliliter). This mixture was incubated over cells for 12 hr before mengovirus challenge (10 PFU per cell). Progeny virus yields were harvested at 24 hr after challenge. Fractions containing inhibitor can be detected by their capability to reduce the activity of added interferon. Arrows indicate the elution position of the protein standards used for calibration of the column (BD, Blue Dextran, >70,000 daltons; AM, albumin monomer, 67,000 daltons; OA, ovalbumin, 45,000 daltons; MG, myoglobin, 17,500 daltons; RN, ribonuclease A, 13,700 daltons; VB, vitamin B_{12}, <5000 daltons). Reproduced from Fleischmann et al.,[13] with permission.

INHIBITOR STABILITY: PERCENTAGE OF SURVIVING ACTIVITY[a,b]

Treatment			Fibroblast interferon	Immune interferon	Interferon inhibitor
pH2:	0.25	hr	100	<12	98
	2	hr	100	<12	128
	4	hr	98	<12	110
	12	hr	85	<12	ND[c]
	24	hr	115	<12	62
50°C:	0.5	hr	20	94	100
	1	hr	6	100	122
	2	hr	4	53	132
60°C:	0.5	hr	<4	18	94
	1	hr	<4	<12	122
	2	hr	<4	<12	136
Trypsin:	5	min	7	<12	ND
	30	min	<4	<12	144
	60	min	<4	<12	164

[a] Reproduced, with permission, from Fleischmann et al.[16]

[b] Samples of fibroblast interferon, immune interferon, and inhibitor were treated as indicated and assayed for activity. Values are expressed as percentage of activity of original sample.

[c] ND, not done.

state but can still completely block the development of antiviral activity when added at up to 3 hr after interferon. The inhibitor has also been shown to block the immunosuppressive activity of exogenously added fibroblast interferon in an *in vitro* plaque-forming cell response.

The inhibitor isolated by sequential BSA–Affi-Gel and Ultrogel column chromatography migrates as a small molecule with an apparent molecular weight of 8000–10,000. Some of its physicochemical and biochemical characteristics are presented in the table. It appears to be resistant to trypsin digestion, stable at pH 2 for 24 hr, and stable at 60° for 2 hr. Further characterization of the molecule is underway.

Section V

Effects of Interferon on the Cell Membrane, Cell Surface, and Cytoskeleton

[54] Assay of Effect of Interferon on Viruses That Bud from Plasma Membrane

By RADHA K. MAHESHWARI and ROBERT M. FRIEDMAN

Interferon (IF) has been shown to inhibit the replication of a wide variety of viruses. In most systems studied, virus-directed translation or transcription was inhibited.[1] In IF-treated cells newly infected with murine leukemia viruses (MLV), virus yields are significantly inhibited. Surprisingly, there was also inhibition of virus yields in IF-treated rat and mouse cells chronically infected with a murine RNA tumor virus. This was unexpected because IF was thought to be effective only in situations where cells were treated before virus infection. IF treatment also caused a 10- to 20-fold increase in intracisternal A particles in erythroleukemia cells infected with Friend virus; the A particles may have been immature C particles in these systems. Mouse mammary tumor virus (MMTV) (B particles) production was also inhibited by IF treatment. Inhibition of virus production did not correlate with inhibition of the accumulation of viral RNA or of viral proteins. Studies of Moloney-MLV infection of mouse bone marrow/thymus (TB) cells showed that in IF-treated cells there was a 2000-fold decrease in the production of MMLV, a 10- to 20-fold decrease in the level of virus-specific extracellular reverse trans-criptase activity, and only a 2-fold difference in the number of budding viral particles observed on the plasma membrane as determined by scanning electron microscopy studies. In JLS-V5 cells treated with IF, the number of cell-associated viral particles was increased, while the infectivity : particle ratio was reduced. These results suggested that in these systems the release of virus from the plasma membrane or the production of infectious virus particles was inhibited by IF treatment.

One important question about this work is whether such findings are limited to murine RNA tumor viruses or whether they are operative in the case of viruses from other groups. We have, therefore, studied the effect of low doses of IF (3–30 units/ml) on vesicular stomatitis virus (VSV) infection of L cells. In contrast to previous studies,[2] which employed much higher IF concentration and reported inhibition of VSV-directed transcription or translation, our results showed that in L cells treated with 30 reference units of IF per milliliter there was an approximately 200-fold reduction in the titer of infectious VSV particle production; however, in

[1] R. M. Friedman, *Bacteriol. Rev.* **41**, 543 (1977).
[2] L. A. Ball and C. N. White, *Virology* **84**, 496 (1978).

ACTIONS OF INTERFERON ON MEMBRANE-ASSOCIATED VIRUSES

Class of inhibition	Level of inhibition	Examples of this type of inhibition[a]
I	Increase in intracisternal A-type particles (which may be precursors of C particles)	Friend virus-infected cells[3]
II	Virus particles stick to plasma membrane of cell	AKR-MLV in AKR cells[4] MMTV[5]
III	Particles produced with low infectivity	Moloney MLV in TB cells[6] AKR-MLV in AKR cells[7] VSV in Ly cells[8]

[a] MLV, Murine leukemia virus; VSV, vesicular stomatitis virus; MMTV, mouse mammary tumor virus.

these cultures virus particle production, as measured by VSV particle-associated viral RNA, viral nucleocapsid protein, and viral transcriptase, was inhibited less than 10-fold. We, however, observed a marked inhibition in glycoprotein (G) and membrane protein (M) of VSV released from IF-treated cells. Evidence supporting the selective deficiency of glycoprotein in VSV released from IF-treated cells was also derived from electron microscopy studies. These results suggested that VSV-particles released from IF-treated cells are low in infectivity. The results in many respects resemble those previously reported in IF-treated cells infected with MLV. We concluded, therefore, that such findings on the action of interferon are not limited to murine RNA tumor virus systems; and perhaps the inhibition of some membrane associated viruses (such as VSV) by treatment with relatively moderate concentrations of IF may be widespread phenomenon that is closely related to functional abnormalities in the proteins incorporated into noninfectious virions produced by such cells.

The table summarizes the effect of IF on membrane-associated virus.[3-8] The classes listed may involve successively less severe degrees of alteration in viral proteins; these alterations could result in inhibition of virus assembly or infectivity at later stages in the replication cycle.

[3] C. J. Krieg, W. Ostertag, V. Clauss, I. B. Pragnall, P. Swetly, G. Koesler, and B. J. Weinmann, *Exp. Cell Res.* **116**, 21 (1978).
[4] E. H. Chang, S. J. Mims, T. J. Triche, and R. M. Friedman, *J. Gen. Virol.* **34**, 363 (1977).
[5] V. A. Strauchen, N. A. Young, and R. M. Friedman, *Virology,* **82**, 232 (1977).
[6] E. H. Chang, M. W. Myers, P. K. Y. Wong, and R. M. Friedman, *Virology* **77**, 99 (1977).
[7] R. M. Friedman, E. H. Chang, J. M. Ramseur, and M. W. Myers, *J. Virol.* **16**, 569 (1975).
[8] R. K. Maheshwari, F. T. Jay, and R. M. Friedman, *Science* **207**, 540 (1980).

Materials

Virus Stock and Virus Assays. VSV was the Indiana strain originally obtained from Mr. C. Buckler (National Institutes of Allergy and Infectious Diseases, Bethesda, Maryland). It had been plaque-purified and passaged by us at a multiplicity of 0.01 in Vero cells. The pool of virus used was assayed by plaque titration in Vero cells, where it titered 2×10^9 PFU/ml, by cytopathic effect in Vero, or in Ly cells in microtiter assay plates. Moloney murine leukemia virus was obtained from Dr. P. K. Y. Wong (University of Illinois).

Cells. Ly cells were originally obtained from J. Youngner (University of Pittsburgh). They were grown in monolayers in Eagle's minimum essential medium (MEM) supplemented with glutamine (0.03 g/500 ml), 10% heat-inactivated fetal calf serum (Grand Island Biological Co., Grand Island, New York), penicillin (100 units/ml), streptomycin (100 μg/ml), fungizone (0.25 μg/ml), and gentamicin (5 μg/ml). These cells are very sensitive to the antiviral activity of mouse IF. Vero cells and baby hamster kidney cells (BHK-21) were obtained from the National Institutes of Health Media Section and were also grown in Eagle's MEM supplemented with 10% fetal calf serum.

TB cells, a CFW/D mouse cell line derived from mixed cultures of bone marrow and thymus were obtained from Dr. J. K. Ball. The 15F cell line was obtained from Dr. J. A. McCarter. Both cell lines were maintained in McCoy's 5A medium (Gibco) supplemented with 10% heat-inactivated fetal bovine serum (Gibco). The medium contained streptomycin (100 μg/ml), penicillin (100 units/ml), Aureomycin (50 μg/ml), mycostatin (100 units/ml), and gentamicin (50 μg/ml).

Interferon and Interferon Assays. The specific activity of the preparations employed should be at least 2×10^7 mouse reference units per milligram of protein.[9] We assay IF by inhibition of VSV-induced cytopathic effect or by plaque reduction in Vero cells.

Procedures

Preparation and Purification of Virus Samples

Ly cells were grown in 150-cm² flasks to confluence (2×10^7 cells) and treated with 0–30 units of IF per milliliter for 14 hr. The cell monolayers were washed 3–4 times to remove the residual IF. All the monolayers were then infected with VSV at a multiplicity of 5 PFU per cell; virus was

[9] C. A. Ogburn, K. Berg, and K. Paucker, *J. Immunol.* **111**, 1206 (1973).

adsorbed for 45 min. The unadsorbed virus was removed, and cells were refed with 15 ml of MEM and incubated for 16 hr. The supernatant fluids were then collected and sedimented at 10,000 g to remove cell debris. Aliquots of each sample were saved for assay of virus infectivity. All the supernatant fluids were concentrated 100-fold by sedimenting at 48,000 g for 2 hr; the pelleted virus was further purified in a 20 to 50% (w/w) equilibrium sucrose gradient in SET buffer (50 mM Tris · HCl, pH 7.2, 0.1 M NaCl, and 1 mM EDTA) by centrifuging at 72,000 g for 18 hr in an SW27 rotor. The virus, which banded at 1.16–1.18 g/ml, was sedimented in a Beckman SW27 rotor for 3 hr at 110,000 g. The pellet containing the purified virus was resuspended in 0.1 M NaCl and 10 mM sodium acetate buffer (pH 7.2) and used for analysis of virion transcriptase, protein, or RNA.

Infectivity

The inhibition of the infectious virus by IF has been assayed by the following methods.

Cytopathic Effect Inhibition (CPE) Assay. Microtiter plates were seeded with Vero or L cells. Serial dilutions (10-fold) of virus samples are inoculated into several replicates of L or Vero cells. The cultures were incubated for 24 hr, and cells were observed for cytopathic effect. The "end point" of these is taken to be that dilution of virus which infects 50% of the inoculated cells. The infectivity titer of the original virus suspensions is then expressed in terms of 50% tissue culture infectious dose ($TCID_{50}$) per milliliter.

Plaque Reduction Assay. L or Vero cells were plated in 6 well trays to approximately 90% confluency. The cells were infected with three different dilutions (in duplicate) for each virus sample in a 0.2-ml volume. After 1 hr of adsorption, the virus was removed from the cells and the cells were covered with the overlay of Eagle's No. 2 plaque-agar medium. The medium contained 0.9% Noble agar, 5% fetal bovine serum, 5% tryptose-phosphate broth, streptomycin, penicillin, tetracycline, mycostatin, and 40% Eagle's plaque medium No. 2. After 24–30 hr, the agar overlay was peeled off. The cells were next treated with 10% formalin for 30 min, washed, and stained with hematoxylin–Harris stain (Fisher) for 30 min and washed with water. The unstained round areas were scored as plaques. The titer was calculated, and results were interpreted as number of plaque-forming units (PFU) per milliliter.

Focus Assay. Murine leukemia virus may be assayed by its ability to promote foci in 15F cells. Aliquots of culture fluids to be assayed for infectivity were filtered through a 0.45-mm Millipore filter immediately upon collection and stored at $-20°$ until they were assayed. Virus samples were diluted in McCoy's 5A medium containing 25 μg of DEAE-dextran per milliliter, but without serum. Diluted virus (1 ml) was mixed with 1 ml of 15F cells at a concentration of about 1×10^6 cells/ml in McCoy's 5A medium containing 10% FBS. The 15F cells had been trypsinized and maintained in a spinner culture at a concentration of about 2×10^5 cells/ml for 3–4 hr. The virus–cell mixtures were shaken at $37°$ for 40 min, and then 0.5 aliquots were distributed to 25-cm^2 T flasks containing 4 ml of McCoy's 5A medium. Cultures were incubated at $37°$ in a humid CO_2 incubator. Fluids were changed after 3 days. Usually at this time, cells were rounded up at the future focus-forming site, and eventually some of these slough off the site, leaving small plaques in the cell sheet. After a further period of 2 days, the medium was removed and the cells were fixed with 70% methanol and stained with Giemsa stain. The proportion of colonies infected with MLV (surrounded by syncytia) was measured. The viral titer was obtained by multiplying the MOI by the dilution factor.

Transcriptase and Reverse Transcriptase Assays

Banded virus samples were assayed for transcriptase activity.[10] Standard *in vitro* RNA polymerase mixtures (1 ml) contained the following: 50 mM Tris · HCl (pH 8.0), 0.1 M NaCl, 5 mM MgCl$_2$, 4 mM dithiothreitol, 0.05% Triton X-100, 1 mM ATP, 1 mM CTP, 0.1 mM GTP, 0.1 mM UTP, and 100 μCi of [^{3}H]GTP; unless otherwise noted, a 50-μl virus sample was incubated at $32°$ with the reaction mixture for 2–3 hr, and the reaction was stopped by adding 5% trichloroacetic acid at $0°$. Precipitates were collected on Millipore filters (0.45 μm), washed 5 times with 5% trichloroacetic acid containing 0.02 M sodium pyrophosphate, and dried under an infrared lamp; the radioactivity was quantitated in a liquid scintillation counter.

Tissue culture fluid was prepared for assay of reverse transcriptase activity[11] by sedimentation at 10,000 g for 20 min and then at 40,000 g for 90 min. The viral pellet was resuspended in 0.5 ml of 10 mM Tris, pH 7.4, 1 mM EDTA, 100 mM NaCl (TES). The 100-μl incubation mixtures contained 50 mM Tris · HCl, pH 8.0, 60 mM KCl, 1 mM MnCl$_2$, 2.5 mM dithiothreitol, 0.05% Triton X-100, 0.04 absorbance units of polyribocytidylic acid–oligodeoxyguanylic acid (12–18), and 50 μl of virus sample; 10 μl of 0.01 mM dGTP containing 10 μCi of [^{3}H]dGTP was added to start the reaction. The mixtures were incubated at $37°$ for 30 min. The reaction was stopped by addition of 5% trichloroacetic acid with 0.125 M

[10] R. J. Calonno and A. K. Banerjee, *Virology* **77**, 260 (1977).
[11] R. M. Friedman and J. M. Ramseur, *Proc. Natl. Acad. Sci. U.S.A.* **71**, 3542 (1974).

sodium pyrophosphate and was kept at 40° for 10 min. The acid-insoluble extracts were collected on 0.45-μm Millipore filters, and the tubes and filters were washed several times with 0.02 M sodium pyrophosphate in 5% TCA. The filters were then dried and counted in a liquid scintillation counter.

Analysis of Virion Proteins

Ly cells were treated with IF as described above. After virus adsorption, the monolayers were washed 2–3 times with leucine-free medium, and then [³H]leucine (20 μCi/ml) was added in special MEM devoid of cold leucine and serum. After incubation in serum-free medium for 8 hr at 37°, 5 ml of MEM with 2% dialyzed calf serum were added to each culture. The supernatant fluids were collected at 17 hr, and virus was purified by banding on a sucrose gradient as described above. Analysis of viral proteins in these samples was carried out by polyacrylamide gel electrophoresis[12] in gels composed of 7.5% acrylamide, 0.5 M urea, and 0.1% SDS in 0.1 M phosphate buffer, pH 7.2. Gels were polymerized with ammonium persulfate at a final concentration of 0.07% and $N,N,N,'N'$-tetramethylenediamine at 0.043% (v/v). The electrophoresis buffer contained 0.1 M phosphate (pH 7.2) and 0.1% SDS. Samples of 50 μl were layered on gels 10 cm in length and run at 5 mA/gel for 7–8 hr; after this time the tracking dye had migrated to the bottom of the gels. After electrophoresis, the gels were fixed in 20% (w/v) sulfosalicylic acid for 18 hr at room temperature and stained with 0.25% Coomassie Blue for 4 hr to locate marker proteins (molecular weight): bovine serum albumin (68,000), ovalbumin (45,000), chymotrypsinogen A (25,000), and ribonuclease (13,700). The gels were then sliced into segments of 1.25 mm, and each slice was placed in a scintillation vial with 30% H_2O_2 (100 μl); the vials were kept at 90° for 3 hr, and the radioactivity incoporated into the gel slices was estimated in a liquid scintillation spectrometer.

Virion proteins were also analyzed on SDS–polyacrylamide slab gels, and the incorporation of radioactive precursor was quantitated by fluorography.[13] After electrophoresis the gels were impregnated with 2,5-diphenyloxazole (New England Nuclear, Boston, Massachusetts), dried, and exposed to Kodak X-Omat X-ray film. Fluorograms were scanned with a quick scan Jr. densitometer (Helena Laboratories) for integrating the values.

Extraction and Analysis of Viral RNA

For virion RNA analysis the radioactive virus was prepared as follows: Ly cells were treated overnight with 0–30 units of IF per milliliter,

[12] R. R. Wagner, T. C. Schnaitman, and R. M. Snyder, *J. Virol.* **3**, 395 (1969).
[13] W. M. Bonner and R. A. Laskey, *Eur. J. Biochem.* **46**, 83 (1974).

washed, and then infected with VSV at a multiplicity of 5 PFU per cell. After 4–5 hr of incubation, the medium was decanted and fresh medium containing actinomycin D (2 μg/ml) (Merck, Sharp & Dohme Research Laboratory, Rahway, New Jersey) was added; [³H]uridine (15 μCi/ml) was added after 1 hr and the cultures were incubated overnight. The supernatant fluids were pelleted, and VSV was purified from the pellet by banding on a sucrose gradient as previously described. The RNA was extracted with 0.3% SDS and buffered phenol. The phenol extraction was repeated twice at room temperature, and the residual buffer layer was adjusted to 0.4 M NaCl and mixed with 2 volumes of 99% ethanol. After 16 hr at −20°, the resulting precipitate was sedimented at 10,000 g for 20 min and resuspended in 0.2 ml of SET buffer solution.

Viral RNA was analyzed by electrophoresis on polyacrylamide gels.[14] Gels were prepared by mixing 0.08 g of agarose, 1 ml of 0.5% N,N,N,N¹-tetramethylethylenediamine, 11.3 ml of water, 0.05 ml of 1.6% ammonium persulfate, 0.32 ml of 10% SDS, 1.6 ml of 10× electrophoresis buffer, and 1.6 ml of 19% acrylamide (w/v). The 10× electrophoresis buffer contained 0.4 M Tris · HCl (pH 7.8), 0.05 M sodium acetate, and 0.01 M disodium EDTA. A 50-μl sample of ³H-labeled VSV RNA was mixed with sucrose granules and 5 μl of 0.2% bromophenol blue, loaded on a 10-cm gel, and gently overlayed with electrophoresis buffer. The gels were run at 70 V (3–4 mA/gel) for about 3 hr, depending on migration of the added dye. Ribosomal RNAs were used as markers. After electrophoresis the gels were added to test tubes containing 5% trichloroacetic acid for 1 hr, washed twice, sliced, and incubated for 3 hr at 90° in 100 μl of 30% hydrogen peroxide. Radioactivity was estimated in a liquid scintillation spectrometer.

Viral RNA was also analyzed on 6 to 30% sucrose density gradients made up in SET buffer; a 200-μl containing viral RNA was layered on each gradient and sedimented at 50,000 rpm for 2.5 hr in a Beckman SW65 rotor. Fractions of 0.15 ml were collected, and 100-μl aliquots were spotted on Whatman filter paper disks that were washed three times with cold 5% TCA, followed by alcohol, alcohol-ether (1:1), and ether. Acid-precipitable radioactivity was measured by liquid scintillation spectroscopy.

Electron Microscopy

The viral suspensions were prepared and processed as follows for electron microscopy studies[15]. The medium from cultures infected 18–24 hr previously with VSV or MLV were clarified by centrifugation (1000 g,

[14] A. C. Peacock and C. W. Dingman, *Biochemistry* 7, 668 (1968).
[15] R. K. Maheshwari, A. D. Demsey, S. B. Mohanty, and R. M. Friedman, *Proc. Natl. Acad. Sci. U.S.A.* 44, 2284 (1980).

20 min) to remove large cellular debris. Supernatant fluids containing virus were saved, and a portion was fixed by the addition of one-third volume of 5% glutaraldehyde in Dulbeco's PBS, pH 6.0. In one experiment, the virus from the supernatant was further purified and concentrated by isopycnic banding; 25 ml of clarified supernatant was layered onto a 10-ml potassium tartrate (15 to 45%) or sucrose (15 to 50% w/w) gradient and centrifuged at 25,000 rpm for 3 hr in Beckman SW27 rotor. The virus band was removed and dialyzed against PBS overnight at 4°. In another experiment, virus from the supernatant was concentrated by pelleting at 36,000 g for 2 hr, and the pellet was resuspended in 0.2 ml of TES (pH 6.8) and centrifuged in an Eppendorf centrifuge for 10 min. All these samples were observed with an electron microscope.

Droplets of viral suspensions in Formvar carbon-coated grids were negatively stained with 2–4% sodium phosphotungstic acid at pH 6.5. Unfixed suspensions were stained directly, whereas preparations fixed with 5% glutaraldehyde were washed once with distilled water before staining. All preparations were air-dried and examined with an electron microscope operating at 75 kV.

[55] Cell Surface Alterations Induced by Interferon

By ROBERT M. FRIEDMAN

Interferon treatment induces a number of changes in the surface of cells (see the table). It is not yet established what, if any, relationship these changes have to the known biological activities of interferons, such as the induction of an antiviral state, modulation of the immune system, and inhibition of cell growth. There are, however, examples of polypeptide hormones and other biologically active molecules that induce intracellular changes by first binding to the cell surface and then activating changes on the plasma membrane; these changes in turn result in profound alterations in cell function. What is unknown right now as far as interferon studies are concerned is how changes in the cell surface might result in the types of alterations listed in the table.[1-7] It is also unclear how the alterations of the cell surface are accomplished.

[1] E. H. Chang, F. T. Jay, and R. M. Friedman, *Proc. Natl. Acad. Sci. U.S.A.* **75**, 1859 (1978).

[2] E. Knight, Jr. and B. D. Korant, *Biochem. Biophys. Res. Commun.* **74**, 707 (1977).

[3] L. D. Kohn, R. M. Friedman, J. M. Holmes, and G. Lee, *Proc. Natl. Acad. Sci. U.S.A.* **73**, 3695 (1976).

CHANGES IN THE CELL SURFACE
INDUCED BY INTERFERON TREATMENT

Change	Reference[a]
Alteration in plasma membrane density	1
Increase in intramembranous particles	1
Alteration in cell surface charge	2
Altered capacity to bind thyroid stimulating hormone or cholera toxin	3
Increase in binding of concanavalin A	4
Increase in cytotoxicity of lymphocytes for target cells and in expression of cell surface antigens	5
Altered exposure of surface ganliosides	6
Increase in intracellular levels of cyclic AMP	7

[a] Numbers refer to text footnotes.

Several alterations in the plasma membrane may occur as a consequence of interferon's interaction with cells. These changes are of a chemical, physical, morphological, and immunological nature. Because most of the changes listed in the table involve highly specialized techniques, these will not be described in detail in this section. The reader is referred to the appropriate reference indicated in the table.

Plasma membrane preparations from mouse interferon-treated L cells were analyzed on discontinuous sucrose gradients.[1] In cells not treated with interferon, 77% of the plasma membrane banded in sucrose at a density of 1.22 to 1.23, and 23% at 1.23; in interferon-treated cells, 40% banded at 1.22 to 1.23, and 60% at 1.23. This probably indicated that the protein-to-lipid ratio was increased in the membrane of interferon-treated cells. Since the main components of intramembranous particles are reported to be glycoproteins, these results correlate with those on intramembranous particles.

Interferon treatment resulted in an increase in the concentration of intramembranous particles of mouse cells, a change detected morphologically by freeze-fracture electron microscopy.[1] The number of intramem-

[4] C. Huet, I. Gresser, M. T. Bandu, and P. Lindahl, *Proc. Soc. Exp. Biol. Med.* **147,** 52 (1974).

[5] P. Lindahl, P. Leary, and I. Gresser, *Proc. Natl. Acad. Sci. U.S.A.* **69,** 721 (1972); and **70,** 2785 (1973).

[6] E. F. Grollman, G. Lee, S. Ramos, P. S. Lazo, R. Kaback, R. M. Friedman, and L. D. Kohn, *Cancer Res.* **38,** 4172 (1978).

[7] J. M. Weber and R. B. Stewart, *J. Gen. Virol.* **28,** 363 (1975).

branous particles on both A and B fracture faces increased from two- to sixfold after interferon treatment for 48 hr. The kinetics of the particle increase followed that of the establishment of antiviral activity, and both intramembranous particle density and antiviral activity decreased to control levels by 48 hr after the removal of interferon.

Other methods of analysis have suggested that interferon treatment alters the cell surface. When placed in an electric field, interferon-treated L cells have a higher electrophoretic mobility toward the anode than do untreated L cells.[2] This result indicated that interferon treatment increased the negative net charge on the cell surface and is consistent with observations that interferon treatment alters the electrochemical gradient across cell membranes by increasing net intramembranous negative charge.[6]

Plasma membranes of interferon-treated L cells had an altered capacity to bind TSH or cholera toxin.[3] ^{125}I-labeled TSH and ^{125}I-labeled cholera toxin were specifically bound to preparations of mouse L cell plasma membranes, since binding was prevented by unlabeled thyrotropin or cholera toxin, but not by insulin, glucagon, prolactin, growth hormone, human chorionic gonadotropin, or luteinizing hormone. Mouse interferon also inhibited ^{125}I-labeled TSH binding to L cell plasma membranes. The effect of mouse interferon on ^{125}I-labeled cholera toxin binding was more complex, inhibition occurring only after an initial enhancement at low interferon concentrations. A concentration of interferon 10 times as high was required to inhibit ^{125}I-labeled cholera toxin binding as compared with ^{125}I-labeled TSH binding. Mouse interferon was also able to displace bound ^{125}I-labeled TSH, but not bound ^{125}I-labeled cholera toxin; human interferon could induce changes in binding of [^{125}I]TSH and ^{125}I-labeled cholera toxin to L cell plasma membranes similar to those induced by mouse interferon. In addition interferon treatment has been reported to increase the binding of concanavalin A to the surface of murine leukemia L1210 cells.[4]

Immunological studies have suggested an alteration in cell surfaces following interferon treatment.[5] Interferon induces an enhancement of the specific cytotoxicity of sensitized mouse lymphocytes. When suspensions of splenic lymphocytes from C57BL/6 mice that had been immunized with L1210 cells were incubated with mouse-brain interferon or with medium that did not contain interferon, the cells incubated with interferon had enhanced cytotoxic effects against L1210 cells. Interferon had no effect on the cytotoxic properties of nonsensitized lymphocytes. The factor responsible for the enhanced cytotoxicity of lymphocytes could not be separated from the antiviral factor in interferon preparations. These results suggested that interferon in this system induced a surface alteration in the splenic lymphocytes, and this alteration increased their cytotoxic properties. Mouse interferon has also been shown to enhance the expression of cell surface histocompatibility antigens in L1210 cells. Like some of the other systems reviewed in this section, these seem to be examples of an action of interferon that may not be directly related to antiviral activity.

Interferon treatment has also been shown to alter the surface exposure of gangliosides and other components of the plasma membrane.[6] G_{M2} and G_{M3}, the predominant gangliosides in the plasma membranes of interferon-sensitive L cells, have been implicated as interferon-specific receptor components. No difference in ganglioside pattern could be detected in membranes isolated from L cells before or after interferon treatment or when labeled with [^{3}H]sodium borohydride after mild sodium metaperiodate treatment; however, differences were found in the pattern of incorporation of tritium after galactose oxidase treatment. The decrease in tritium incorporation in interferon-treated membranes indicated that less of the oligosaccharide moieties was exposed on the surface of the cell membrane. This suggested a change in orientation of these gangliosides in the membrane, and since G_{M1} and G_{D1b} have been implicated as components of the receptors for cholera toxin and TSH, respectively, this result was consistent with previous observations that pretreatment of L cells with interferon altered the binding of both TSH and cholera toxin.

Studies on cyclic AMP and interferon action[7] are discussed in detail in Section III,B of this volume.

[56] Assays to Measure Plasma Membrane and Cytoskeletal Changes in Interferon-Treated Cells

By LAWRENCE M. PFEFFER, EUGENIA WANG, FRANK R. LANDSBERGER and IGOR TAMM

Although interferons were originally characterized as antiviral proteins, many studies have established that interferons exert multiple effects on cells *in vivo* and *in vitro*.[1,2] Interferons inhibit the proliferation of normal and transformed cells; however, the underlying mechanism of interferon action is unclear. Interferon-treated cells continue macromolecular syn-

---

[1] P. S. Sehgal, L. M. Pfeffer, and I. Tamm, *in* "Chemotherapy of Viral Infections" (P. E. Came and L. A. Caliguiri, eds.), Springer-Verlag, Berlin, in press (1982).

[2] The New York Academy of Science Conference on Regulatory Function of Interferons (J. Vilček, ed.) (1980), in press.

METHODS IN ENZYMOLOGY, VOL. 79

thesis at a slightly reduced overall rate. Interferon treatment enhances the production of a number of cellular proteins and increases the levels of several enzymes.

The phenotype of interferon-treated human cells is characterized by several outstanding structural and functional features. Treatment of human fibroblasts with purified human fibroblast interferon (640 units/ml) for 3 days results in increased mean cell surface area, volume, and mass.[3,4] In the cytoplasm of these enlarged and well spread fibroblasts are distinct actin-containing microfilament bundles, which are markedly increased in size and number.[4] Fibronectin is arranged on the surface of interferon-treated cells in long filamentous arrays.[4] The motility of interferon-treated cells is markedly reduced.[4] In related studies it has been shown that interferon also inhibits the ability of human tumor cells to redistribute cell surface receptors for concanavalin A[5] and increases the rigidity of the plasma membrane lipid bilayer.[5a] The action of interferon on human cells appears to entail a coordinated cellular response involving the cell membrane with its associated proteins and the cytoplasmic actin-containing microfilaments.[4]

A part of the altered phenotype of interferon-treated cells may be secondary to the slowing and, ultimately, the arrest of cell cycle traverse. However, changes in the plasma membrane–cytoskeletal complex, be they secondary to inhibited cell cycling or directly caused by interferon action, may in and of themselves reduce the proliferative capacity of the cells.

In this chapter we describe techniques for the measurement of interferon-induced changes in cell size, cytoskeleton, cell surface fibronectin distribution, lateral mobility of cell surface receptors, and fluidity of the plasma membrane.

Cell Cultures

Human diploid fibroblast strains, such as FS-4[6] and ME,[7] are used between the 10th and 20th passage levels in culture. The rationale behind the use of these cultures is discussed elsewhere in this volume.[8] Fibroblasts are grown as monolayer cultures in Eagle's minimal essential medium (MEM)[9] supplemented with 10% fetal calf serum (FCS) (Grand Is-

[3] L. M. Pfeffer, J. S. Murphy, and I. Tamm, *Exp. Cell Res.* **121**, 111 (1979).

[4] L. M. Pfeffer, E. Wang, and I. Tamm, *J. Cell Biol.* **85**, 9 (1980).

[5] L. M. Pfeffer, E. Wang, and I. Tamm, *J. Exp. Med.* **152**, 469 (1980).

[5a] E. Wang, L. M. Pfeffer, and I. Tamm, *Proc. Natl. Acad. Sci. U.S.A.*, in press (1981).

[6] J. Vilček and E. A. Havell, *Proc. Natl. Acad. Sci. U.S.A.* **70**, 3909 (1973).

[7] L. M. Pfeffer, M. Lipkin, O. Stutman, and L. Kopelovich, *J. Cell. Physiol.* **89**, 29 (1976).

[8] I. Tamm, L. M. Pfeffer, J. S. Murphy, this volume [48].

[9] H. Eagle, *Science* **130**, 432 (1959).

land Biological Co., Grand Island, New York) and incubated at 37° in a humidified atmosphere of 5% CO_2 in air. Monolayers are routinely passaged weekly at a subculture ratio of 4:1 by removal from the substrate with 0.25% trypsin–0.05% ethylenediaminetetraacetic acid (EDTA) in phosphate-buffered saline (PBS) lacking Ca^{2+} and Mg^{2+} (PBS-def.: NaCl, 8 g; KCl, 0.2 g; Na_2HPO_4, 1.15 g; KH_2PO_4, 0.2 g in 1 liter of solution).

HeLa cell subclone S_3 (human epithelioid carcinoma line) is grown in suspension at cell concentrations between 2×10^4 and 1×10^6 cells/ml in the spinner modification of Eagle's medium[9] supplemented with 4% FCS.

Measurement of Cell Surface Area

Human fibroblasts are planted into 35-mm petri dishes (3001 Falcon Plastics, Oxnard, California) at a density of 2×10^3 cells/cm² in 2 ml of MEM supplemented with 10% FCS (cell concentration = 2×10^4 cells/ml). The tissue culture dishes contain three to five sterile No. 1 thickness round cover glasses, 12 mm in diameter (SGA Scientific, Inc., Bloomfield, New Jersey). Cover glasses are readily sterilized by rinsing in 70% ethanol and flaming over a Bunsen burner. At 1 day after planting the medium is decanted and monolayers are refed with 2 ml of MEM supplemented with 10% FCS without interferon or containing interferon at varying concentrations. After incubation for varying lengths of time, cover glasses are removed from the 35-mm dishes with sterile technique and placed in a new set of 35-mm dishes. The cover glasses are rinsed twice with PBS and fixed for 10–15 min at room temperature in 1 ml of 3.7% formaldehyde (reagent grade, Fisher Scientific Co., Fair Lawn, New Jersey). Each cover glass is inverted and mounted over a drop of glycerol-2× PBS (1:1, v/v) on clean glass slides. Residual liquid is removed with a Kimwipe, and each cover glass is sealed in place by coating the edges with nail polish.

Slides are examined in a Zeiss Model III RS photomicroscope (Carl Zeiss, Inc., New York) with a 16× phase-contrast objective (total magnification: 200×) and photomicrographs taken with a 35 mm Nikon camera equipped with an automatic exposure meter. At least 100 photomicrographs are taken per variable. The negatives are developed, enlarged 5.7 times, and printed on high-contrast photographic paper. The periphery of each cell is outlined on the prints with a fine-point marking pen. The areas encompassed by individual cells are cut out with a sharp razor blade or scalpel and weighed on an analytical Sartorius balance (Model 3716). As a standard of reference, photomicrographs are taken of a calibrated electron microscope grid (300-mesh copper grid, Electron Microscopy Sciences, Fort Washington, Pennsylvania). To obtain the conversion factor from

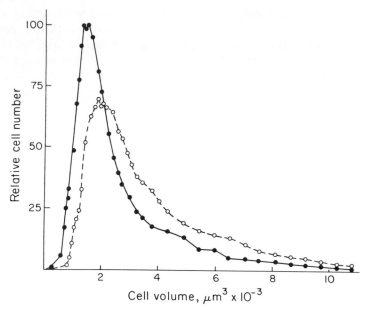

FIG. 1. Cell volume distributions of control and interferon-treated human fibroblasts (ME strain). ●———●, Control cells after 3 days of mock treatment; ○---○, interferon-treated cells after 3 days of treatment at 640 units/ml. The distribution of interferon-treated cells is broadened and shifted to larger cell volumes as compared to the control cells. Cell volume analysis was kindly performed on several of our samples by Dr. R. Zucker (Papanicolaou Cancer Research Institute, Miami, Florida).

milligrams to square micrometers, photographs of a known area of the reference grid are taken at a total magnification of 200× and the known area is cut and weighed.

Cell Volume Analysis

Human fibroblasts are planted into 75-cm² flasks (Falcon 3024) at a density of 2×10^3 cells/cm² in 15 ml of MEM supplemented with 10% FCS (1×10^4 cells/ml). Cultures are refed at 1 day after planting with MEM containing 10% FCS with or without human interferon, 640 units/ml. This interferon concentration is sufficient to cause essentially maximal reduction in the rate of proliferation of several strains of human fibroblasts.[3] After incubation at 37° for varying periods, monolayers are rinsed twice with PBS and then incubated for 5–10 min at 37° in 1 ml of 0.25% trypsin–0.05% EDTA in PBS. Cells removed from the substrate are gently resuspended in the trypsin solution with a Pasteur pipette and transferred to tubes containing 4 ml of 4% glutaraldehyde (EM Laboratories, Inc.,

Elmsford, New York) in PBS chilled in ice. Electronic volume analysis is performed as described previously using a Coulter H4 Channelyzer equipped with a 70 μm in diameter orifice, 84 μm in length tube.[10] The apparatus is programmed to give the mean, standard deviation, coefficient of variation, and skewness for the cell population. The instrument is calibrated with polystyrene beads with fixed diameters of 9.96 μm and 18.23 μm (Coulter Electronics, Inc., Hialeah, Florida).

Figure 1 shows the results of a representative experiment on the effect of interferon treatment on cell volume. After 3 days of treatment, the cell population is considerably more heterogeneous, as indicated by a coefficient of variation (21.5%) approximately twice that of the value for the control population (11.5%). The coefficient of variation can be calculated as the ratio of the peak width at one-half maximum amplitude to the maximum amplitude $\times$ 100. It is also evident that there is considerable overlap in cell volumes distributions of the interferon-treated and control populations of fibroblasts. The mean volume of interferon-treated cells is 4.0×10^3 μm^3, whereas that of control cells is 3.0×10^3 μm^3, representing a 33% increase. However, these estimates do not represent the actual volumes of fibroblasts in monolayer culture because these measurements are made after suspension of cells following removal from the growth surface.

Indirect Immunofluorescence Staining for Microtubules, 10-nm Filaments, Microfilaments, and Fibronectin

Human fibroblasts are seeded onto 12-mm glass cover glasses as described above. After a 3-day incubation in the presence or the absence of interferon, 640 units/ml, the cell monolayers are rinsed in PBS (PBS-def. supplemented with 0.1 g of CaCl$_2$ and 0.1 g of MgCl$_2 \cdot$ 6 H$_2$O per liter), fixed for 10–15 min at room temperature in 3.7% formaldehyde in PBS, and then rinsed again in PBS. The cell membrane is then made permeable to antisera by plunging the cover glasses into acetone in glass petri dishes at $-20°$ for 2 min. The cover glasses are allowed to dry in air at room temperature on Kimwipes, and the cell monolayers are rehydrated in PBS for 10 min at room temperature. The cover glass cultures are then ready for incubation with antisera specific for the different cytoskeletal proteins and for cell surface fibronectin.

Cytoplasmic microtubules (cf. Fig. 2A) are localized by using rabbit antiserum prepared against 6 S tubulin, a monomeric form isolated from calf brain.[11] Ten-nanometer filaments (cf. Fig. 2B) are visualized with

[10] R. M. Zucker, R. M. D'Alisa, and E. L. Gershey, *Exp. Cell Res.* **122**, 1 (1979).
[11] E. Wang, R. Cross, and P. W. Choppin, *J. Cell Biol.* **83**, 320 (1979).

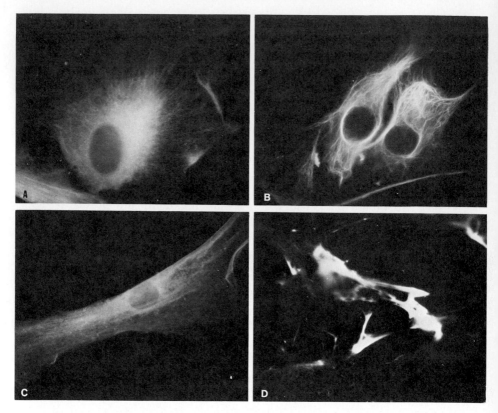

FIG. 2. Distribution of microtubules, 10-nm filaments, intracellular actin, and extracellular fibronectin in human fibroblasts. Cells of strain ME human fibroblasts were grown on cover glasses and stained for tubulin (A), 10-nm filaments (B), intracellular actin (C), and extracellular fibronectin (D). Cells were examined by epifluorescence microscopy using a 63× objective. Magnification, ×900.

rabbit antiserum directed against the 53,000- and 54,000-dalton protein subunits of this cytoplasmic fiber.[12] The actin-containing microfilament bundles (cf. Fig. 2C) are localized by incubation of the cover glasses with deoxyribonuclease I (DNase I, Worthington Biochemical Corp., Freehold, New Jersey), which reacts specifically with actin[13] and antiserum to DNase I. Fibronectin (Fig. 2D) is localized on the fibroblast cell surface by incubation of the cover glass cultures with rabbit antiserum directed against purified cold-insoluble globulin. The procedure for ob-

[12] J. Starger, W. E. Brown, A. E. Goldman, and R. D. Goldman, *J. Cell Biol.* **78**, 93 (1978).
[13] E. Wang and A. R. Goldberg, *J. Histochem. Cytochem.* **26**, 745 (1978).

taining high-titer monospecific antibody preparations has been described in detail.[14] In the prepartion of cover glass cultures for fibronectin staining, the acetone extraction step is omitted.

Indirect immunofluorescence staining of the cover glass cultures is performed at room temperature with the specific antisera in a moist atmosphere to prevent evaporation. In general, 50 μl of antiserum is layered over the samples on cover glasses, and the cover glasses are incubated for 30 min. The monolayers are washed for 60 min with three changes of PBS. The samples are then incubated for an additional 30 min with 50 μl of fluorescein-conjugated goat antiserum directed against rabbit globulin (Antibodies, Inc., Davis, California). The goat antiserum has previously been absorbed with formaldehyde-fixed and acetone-treated cultured rat kidney (NRK) cells in order to prevent nonspecific binding. The cover glasses are washed again for 60 min in three changes in PBS. After washing, the cover glasses are mounted on glass slides and sealed with nail polish.

Immunofluorescence Microscopy

Immunofluorescence microscopy of stained cells is performed with a Zeiss photomicroscope III RS equipped with epifluorescence and phase-contrast optics. A 63× planapochromatic phase-contrast lens is used. This microscope uses an illumination system consisting of an excitation filter (BP450-490) that selects light in the wavelength range of 450–490 nm and a barrier filter (LP520) that blocks out nonspecific light scattering >520 nm. This arrangement of filters is optimal for fluorescein isothiocyanate-conjugated molecules. The light source for epifluorescence is a xenon arc lamp (Zeiss Model 1162) that emits light within the UV range. The patterns of immunofluorescence staining are recorded with Kodak Tri-X film exposed at ASA 1600 and processed with Diafine developer (Acufine, Inc., Chicago, Illinois).

Lateral Mobility of Cell Surface Receptors for Concanavalin A

Human HeLa S_3 cells are seeded in spinner bottles at a cell concentration of 5×10^4 cells/ml in spinner medium supplemented with 4% FCS with or without interferon, 640 units/ml. At varying times after incubation at 37°, aliquots of $2–5 \times 10^5$ cells are removed and pelleted at 800 g. The cell pellets are resuspended in 1 ml of medium containing 2 μg of fluorescein-conjugated concanavalin A (fl-Con A) (Calbiochem-Behring

[14] L. B. Chen, P. H. Gallimore, and J. K. McDougall, *Proc. Natl. Acad. Sci. U.S.A.* **73**, 3570 (1976).

Corp., LaJolla, California) and placed in an ice bath for 45 min. The optimal concentration of fl-Con A is markedly dependent on the type of cell studied. The redistribution of Con A receptors is inhibited at high concentrations of fl-Con A. Thus it is important to investigate a wide range of concentrations of fl-Con A, i.e., 1 pg/ml–500 μg/ml, before selecting a concentration for experiments.

After incubation on ice, the cells are pelleted and resuspended either in 3.7% formaldehyde in PBS or in spinner medium in which they are warmed to 37° for 60–90 min before fixation in formaldehyde. Cells are fixed at room temperature for 10–20 min, pelleted, and resuspended in 50 μl of PBS and 50 μl of glycerol–2× PBS (1 : 1, v/v). The cell suspension is pipetted onto clean glass slides; a 22 × 30 mm cover glass is placed on top and sealed in place with nail polish. The stained cells are examined by epifluorescence microscopy.

When either control (cf. Fig. 3A) or interferon-treated (cf. Fig. 3B) HeLa cells are incubated at 4° in the presence of fl-Con A prior to fixation, a diffuse pattern of fluorescence is observed over the entire cell surface.[5] When control HeLa cells are warmed to 37° and incubated for 60–90 min, the diffuse surface fluorescence redistributes, forms patches, and ultimately concentrates at one pole of the cell, forming a cap (cf. Fig. 3C). In contrast, incubation of interferon-treated HeLa cells at 37° results in a patchy distribution of surface fluorescence (Fig. 3D), which in most cases does not progress to cap formation even after 180 min of incubation at 37°.[5] To quantitate the interferon-mediated inhibition of capping, at least 500 cells are scored per variable, such as interferon concentration or duration of treatment.

Structure of the Plasma Membrane Lipid Bilayer

Electron spin resonance (ESR) techniques in combination with the use of probe molecules (spin labels) have been useful in the elucidation of the structural properties of membranes[15] and have been applied to the study of the plasma membrane of interferon-treated cells. These techniques are sensitive to changes in membrane structure and permit measurements on relatively small quantities of viable cells.

An ESR signal can be detected from molecules that have an unpaired electron. In the study of the interaction of interferon with cell surfaces, the nitroxide derivatives of stearic acid, with the nitroxide moiety at the C_n position have proved to be very useful. Spin label molecules in this class have an unpaired electron on the nitrogen. Generally, we have used

---

[15] L. J. Berliner, (ed.), "Spin Labeling: Theory and Applications." Academic Press, New York, 1975.

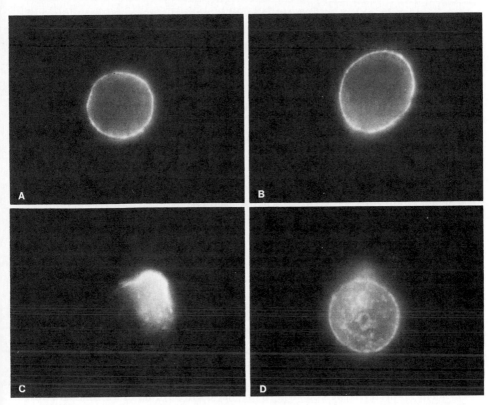

FIG. 3. Effect of interferon treatment on the redistribution of Con A receptors on the cell surface. HeLa-S$_3$ cells were grown in suspension in the presence or the absence of interferon, 640 units/ml. At 72 hr after the beginning of treatment, cells were pelleted, resuspended in 1 ml of spinner medium containing 2 μg of fluorescein-conjugated Con A (fl-Con A), and held at 4° for 45 min. The cells were pelleted and either fixed (A and B) or resuspended in spinner medium and incubated at 37° for 60–90 min (C and D) before fixation. The cells were mounted on slides and examined by epifluorescence microscopy. Control (A) and interferon-treated (B) cells held at 4° with fl-Con A; control (C) and interferon-treated (D) cells held at 4° with fl-Con A and then incubated at 37°. From Pfeffer *et al.*[5]

the stearic acid derivative with the nitroxide moiety coupled to the fifth carbon of the fatty acyl chain (C5). The spin labels C5, C12, and C16 are available from Syva (Palo Alto, California) and Molecular Probes (Plano, Texas), which sells the C7 probe as well.

$$CH_3-(CH_2)_{17-n}-C-(CH_2)_{n-2}-COOH$$

C$_n$

The ESR spectra of C_n spin-labeled intact cells are similar to those of purified plasma membrane fractions and are significantly different from those of various cytoplasmic membrane fractions.[16] The spectral line shape is sensitive to the motion of the spin label fatty acyl chain incorporated into the plasma membrane. As is implied in the structural formula of C_n, the structural properties of the lipid bilayer can be studied as a function of distance from the polar head group region.[17]

Of central importance in the use of nitroxide derivative spin labels in the study of interferon interaction with the cell surface is the method of addition of the label to the cells and the identification of the labeling site(s). Since these labels have only limited water solubility, they are added to cells either complexed to bovine serum albumin (BSA) or in an ethanol solution. The spin-label BSA complex is prepared by incubation overnight at room temperature of 250 mg of defatted BSA (Sigma Chemical Co., St. Louis, Missouri) in 5 ml of PBS over a thin film of 5 mg of C_n. Although ethanol is a known membrane perturbant, the small amounts (5–20 μl) used are probably of little significance.

Human HeLa S_3 cells are seeded in spinner bottles and incubated with or without interferon, 640 units/ml as was described above. At varying times after incubation at 37°, aliquots of 5×10^5–2×10^6 cells are collected and pelleted at 700 g, resuspended in 15 ml of PBS, and pelleted again. The cell pellet is resuspended in 100 μl of PBS and 20 μl of spin-labeled stearic acid complexed with BSA and incubated at 37° for 5 min. After incubation, 15 ml of PBS are added at room temperature and the cells are pelleted. The cells are resuspended in a minimal volume of PBS, taken up in a 100-μl disposable capillary pipette, and pelleted in the pipette. The pipette is then inserted into the cavity of an ESR spectrometer at room temperature. We have used a Varian E-12 spectrometer. The time elapsed from the start of harvesting the aliquot of cells to the end of recording the ESR spectrum is 20 min. Representative ESR spectra of C5-labeled control and interferon-treated HeLa cells are shown in Fig. 4.

In experiments with fibroblasts, cells are seeded onto 100-mm petri dishes at a density of 2×10^3 cells/cm^2 in 15 ml of MEM supplemented with 10% FCS and refed at 1 day after planting with medium with or without interferon. At varying times after incubation at 37°, cultures are washed twice with 10 ml of PBS and incubated for 5 min at 37° with 2 ml of PBS and 0.5 ml of C5–BSA complex. Monolayers are washed with PBS and cells collected into 10 ml of PBS by scraping with silicone strips. Cells are pelleted and processed prior to insertion into the ESR spectrometer as described above.

[16] F. R. Landsberger and R. W. Compans, *Biochemistry* **15**, 2356 (1976).
[17] P. E. Godici and F. R. Landsberger, *Biochemistry* **14**, 3927 (1975).

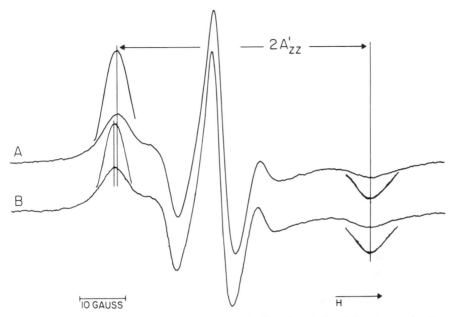

FIG. 4. Electron spin resonance spectra of HeLa-S_3 cells spin-labeled with C5 using the C5-BSA labeling procedure. Spectrum A, Control cells; B, cells treated with interferon, 640 units/ml, for 2.5 hr. The parameter $2A'_{zz}$ (the distance between the outermost peaks) is indicated for spectrum A. The superimposed peaks were obtained by recording the corresponding peaks at an increased modulation level. The observed $2A'_{zz}$ value for spectrum B is larger than that for spectrum A indicating that the plasma membrane bilayer fluidity is decreased upon interferon treatment.

An optimal ESR spectrum is obtained with the greatest incorporation of spin label into the membrane consistent with minimal structural perturbation. A spin label-to-lipid ratio of about 1 : 100 avoids the spin–spin interactions that distort the shape of the ESR spectrum. The amplitude of the ESR spectrum decreases with time as the nitroxide group of the spin label is reduced, i.e., no ESR signal, when the label is internalized into the cell. The ESR spectrometer settings (modulation amplitude, scan rate, and time constant) are selected to optimize the signal-to-noise ratio. Microwave power levels exceeding 5–10 mW result in detectable heating of the sample, altering the shape of the ESR spectrum.[18]

Of utmost importance is the washing of the cells prior to incubation with the spin label–BSA complex. All serum must be carefully removed

[18] B. J. Gaffney, *Proc. Natl. Acad. Sci. U.S.A.* **72**, 664 (1975).

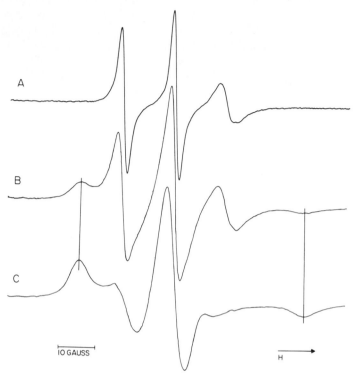

10 GAUSS

H

FIG. 5. Electron spin resonance spectra of HeLa-S_3 cells spin labeled with C16 using the C16-BSA labeling procedure. Spectrum A, Control HeLa-S_3 cells; B, control HeLa-S_3 cells without adequate washing after labeling; C, C16–BSA complex alone. The vertical lines indicate that spectrum B contains spectral features characteristic of spectrum C. Comparison of spectrum A of Fig. 4 with that of Fig. 5 indicates that the observed spectral splitting for C16-labeled cells is considerably less than that for C5-labeled cells. This indicates that the plasma membrane lipid bilayer is more fluid in the region probed by the C16 spin label than in that probed by the C5 spin label. The greater motion of the C16 spin label in comparison with the C5 spin label is referred to as the fluidity gradient of a lipid bilayer.

before addition of the spin label because serum components effectively bind the spin label. Inadequately washed cells will have unincorporated spin label–BSA complex associated with the cells resulting in a spectrum as shown in Fig. 5. In this case, signal from the unincorporated spin label–BSA complex is superimposed on the sharp spectrum characteristic of C16 spin label incorporated in a lipid bilayer.

The interpretation of ESR spectra of lipid membrane structures labeled with C_n probes have been reviewed elsewhere.[15] Briefly, in a lipid bilayer composed of a single lipid species the distance between the outermost peaks of a spectrum, $2A'_{zz}$, decreases with increasing motion of the nitroxide moiety. However, a plasma membrane has a heterogeneous lipid composition and can have domains that differ in the fluidity of the fatty acyl chain.[19,20] The ESR spectrum of C_n spin labels incorporated into a plasma membrane reflects the sum of these domains weighted by the amount of spin label in each. Therefore, a change in the ESR spectrum of a C_n-labeled plasma membrane reflects a change in the weighted or "effective" fluidity of the lipid bilayer.

Figure 4 shows that interferon treatment of HeLa-S_3 cells results in an increase in the value of $2A'_{zz}$ as compared to control cells. The increased distance between the outermost peaks reflects a decreased freedom of motion of the spin label within the plasma membrane, or an increased membrane rigidity. If it is assumed that the spin label interacts with specific membrane components, the increased "effective" rigidity of the plasma membrane lipid bilayer of interferon-treated cells is due to a change in cell surface lipid–protein interaction.

Acknowledgments

This work was supported by Research Grants CA-18608, AI-14040, and NSF PCM 7922956, and by Program Project Grant CA-18213 awarded by the NCI, NIAID, and National Science Foundation. L. M. P. is a Postdoctoral Trainee under the Institutional National Research Service Award CA-09256 from the NCI, and F. R. L. is an Andrew W. Mellon Foundation Fellow.

[19] D. S. Lyles and F. R. Landsberger, *Proc. Natl. Acad. Sci. U.S.A.* **73**, 3497 (1976).
[20] K. Tanaka and S. Ohnishi, *Biochim. Biophys. Acta* **426**, 218 (1976).

Section VI

Relationships of Interferon to the Immune System

[57] Assay of Augmentation of Natural Killer Cell Activity and Antibody-Dependent Cell-Mediated Cytotoxicity by Interferon

By RONALD B. HERBERMAN, JOHN R. ORTALDO, and TUOMO TIMONEN

Natural killer (NK) cells are a subpopulation of lymphocytes that are found in a variety of species, including mice, rats, and humans, and have appreciable spontaneous cytotoxic activity against a wide range of tumor cells and some normal cells.[1-3] There is increasing evidence that NK cells play a role in resistance against tumor growth.[1,3]

Extensive studies have been performed on mouse NK cells. These effector cells are distinct from mature T cells, B cells, or macrophages, but bear the T cell-associated antigen, Thy 1.[4] In addition, they have been found to possess cell surface receptors for the Fc portion of IgG ($Fc_\gamma R$).[5] The NK activity becomes readily detectable in normal mice about 3 weeks old. After peak levels of reactivity are reached at 5–8 weeks, there is a decline to low spontaneous levels by about 12 weeks of age. However, mouse NK activity is readily augmented by inoculation with tumor cells, allogeneic normal cells, bacterial products, or viruses.[1,3] With each type of inoculation, the peak level of NK activity is usually found to coincide with the time of appearance of serum interferon (IF). With almost all types of stimuli, the augmentation of NK activity appears to be mediated through the induction of IF.[3]

Human NK cells can be readily detectable in the peripheral blood of most normal individuals. These cells are characteristically $Fc_\gamma R$ positive, complement-receptor negative, negative for surface immunoglobulin, nonphagocytic, and nonadherent.[1,3] Human NK cells express T cell-related antigens, and an appreciable portion rosette weakly with sheep erythrocytes, and, therefore, they appear to be in the T cell lineage. It has been found that human NK cells have physical characteristics that distinguish them from most peripheral blood lymphocytes, being large cells with prominent azurophilic granules (large granular lymphocytes,

[1] R. B. Herberman, J. Y. Djeu, H. D. Kay, J. R. Ortaldo, C. Riccardi, G. D. Bonnard, H. T. Holden, R. Fagnani, A. Santoni, and P. Puccetti, *Immunol. Rev.* **44**, 43 (1979).

[2] R. B. Herberman and H. T. Holden, *Adv. Cancer Res.* **17**, 305 (1978).

[3] R. B. Herberman, ed., "Natural Cell-Mediated Immunity against Tumors." Academic Press, New York, 1980.

[4] J. Mattes, S. O. Sharrow, R. B. Herberman, and H. T. Holden, *J. Immunol.* **23**, 2851 (1979).

[5] R. B. Herberman, S. Bartram, J. S. Haskill, M. Nunn, H. T. Holden, and W. H. West, *J. Immunol.* **119**, 322 (1977).

LGL).[3,6] Based on these characteristics, it has been possible to enrich for LGL on discontinuous Percoll density gradients.[7] As with mouse NK activity, human NK activity can be augmented significantly above the high levels that spontaneously exist in peripheral blood lymphocytes (PBL), by treatment with IF *in vitro*.[3] Furthermore, the effector cells after treatment with IF had the same cell surface characteristics as spontaneous NK cells. Thus, in addition to its influence on other immune functions, IF clearly is involved in the positive regulation of NK activity.

Antibody-dependent cell-mediated cytotoxicity (ADCC) is another cytotoxic phenomenon, that was described several years earlier than NK activity.[8,9] ADCC against tumor cell targets has been shown to be mediated mainly by a subpopulation of lymphocytes, called "killer" or K cells, which have $Fc_\gamma R$ and thereby can bind and interact with target cells coated with IgG antibodies. It has been recently shown that NK and K cells have similar surface markers, and it seems likely that NK and ADCC activities are different functional properties of the same cells.[1,3] It therefore has been of considerable interest to determine whether IF could augment ADCC as well as NK activity. Although this issue is still somewhat controversial, the weight of evidence indicates that treatment of human peripheral blood lymphocytes with IF can induce significant augmentation of ADCC activity.[3]

The methods described here are for augmentation of human NK and ADCC activities by IF. Similar procedures have been used to augment NK activity of mouse spleen cells.[10]

Materials

Medium: A variety of media can be utilized for suspending the target cells and effector cells. The tissue culture medium in which the target cells are normally passaged is suitable for the assay. Our assays have been done with RPMI-1640 medium supplemented with 5% fetal bovine serum (horse serum has been found to be a satisfactory alternative), 25 mM HEPES buffer, penicillin (100 μg/ml), and streptomycin (100 μg/ml).

For enrichment of NK and K cells, density gradients of Percoll (Pharmacia Fine Chemicals, Piscataway, New Jersey) have been used. The osmolality of the Percoll must be adjusted to 285 mOsm of H_2O per kilogram with 10× concentrated phosphate-buffered saline.

[6] T. Timonen, E. Saksela, A. Ranki, and P. Hayry, *Cell. Immunol.* **48**, 113 (1979).

[7] E. Saksela and T. Timonen, *in* "Natural Cell-Mediated Immunity against Tumors" (R. Herberman, ed.), p. 173. Academic Press, New York, 1980.

[8] I. C. M. MacLennan, *Transplant. Rev.* **13**, 67 (1972).

[9] P. Perlmann, H. Perlmann, and H. Wigzell, *Transplant. Rev.* **13**, 19 (1972).

[10] J. Y. Djeu, J. A. Heinbaugh, H. T. Holden, and R. B. Herberman, *J. Immunol.* **122**, 175 (1979).

Interferon: Crude or partially purified type I human IF preparations (either fibroblast, leukocyte, or lymphoblastoid) have been found to be satisfactory. In addition, comparable results have been obtained with purified, homogeneous human leukocyte IF.[11]

Methods for Cytotoxicity by Peripheral Blood Lymphocytes

Effector Cells. Human peripheral blood mononuclear leukocytes of normal, adult donors are separated from heparinized blood by centrifugation on a Ficoll-Hypaque density gradient,[12] in a 50-ml graduated conical plastic centrifuge tube (Falcon Plastics, Los Angeles, California). This procedure provides mononuclear cells with a yield of greater than 95% and viability greater than 90%. The cells are washed twice with Hanks' balanced salt solution (BSS, without serum) and resuspended to the desired concentration in assay medium. Monocytes can be effectively removed from the mononuclear cell suspension by adherence to Sephadex G-10 (Pharmacia). Up to 1×10^8 leukocytes can be added to a column of 5 ml of packed Sephadex G-10, prewashed with BSS. After incubation at room temperature for 15 min, the column is flushed with 50 ml of BSS. The nonadherent cells (containing <1% monocytes) are collected and washed.

Treatment with IF. A total of 2 to 10×10^6 lymphocytes in 1 ml of assay medium are incubated at 37° with IF. Incubation periods as short as 5 min have given significant boosting of cytotoxic activity, but higher and more consistent levels of augmentation have been achieved with incubation for 60–180 min. Routinely, 1000 units of IF are used, but lower concentrations can be effective. Dose responsive effects are usually between 100 and 1000 units. With some donors, as little as 1 unit of pure leukocyte IF has significantly augmented activity. After incubation with IF, the cells are washed with medium and resuspended at the concentrations needed for the cytotoxicity assay. The presence of IF during the assay is not desirable, since it may decrease the sensitivity of the target cells to lysis.[13]

Assay for NK Activity. The untreated or IF-treated effector cells are tested for cytotoxic activity in a 4-hr ^{51}Cr release assay, as previously described in detail.[14] Various target cells are suitable, but K-562, a myeloid tumor cell line derived from a patient with chronic myelogenous leukemia in blast crisis, has usually been found to be most sensitive for measurement of NK activity. The target cells (2 to 4×10^6 in 0.6 ml of

[11] M. Rubinstein, S. Rubinstein, P. C. Familletti, R. S. Miller, A. A. Waldman, and S. Pestka, *Proc. Natl. Acad. Sci. U.S.A.* **76**, 640 (1979).

[12] A. Böyum, *Scand. J. Clin. Lab. Invest.* (Suppl. 97) **21**, 1 (1968).

[13] G. Trinchieri, B. Perussia, and D. Santoli, *in* "Natural Cell-Mediated Cytotoxicity against Tumors" (R. Herberman, ed.), p. 655. Academic Press, New York, 1980.

[14] J. R. Ortaldo, G. D. Bonnard, and R. B. Herberman, *J. Immunol.* **119**, 1351 (1977).

assay medium) are labeled with 150 μCi ^{51}Cr (as sodium chromate, New England Nuclear, Boston, Massachusetts) for 30–45 min. Cells are washed, resuspended in assay medium, and adjusted to the desired concentrations. Target cells (5×10^3 per reaction) are mixed with effector PBL (ranging from 5×10^5 to 5×10^3 cells/reaction), with resultant 100:1 to 1:1 attacker to target cell ratios (A/T), in Microtest plates (No. 3040, Falcon Plastics). Reaction mixtures of 0.1 ml of targets and 0.1 ml of effectors are incubated in a 5% CO_2 in air atmosphere for 4 hr at 37°. All reactions are performed at least in triplicate. Autologous controls (unlabeled target cells being added, in place of effector cells, to the labeled targets) are employed for baseline release values in all experiments. The supernatants can be conveniently harvested with the Titertek Harvesting System (Flow Laboratories, Rockville, Maryland). The percentage of isotope released is calculated by the formula

$$\text{Percent release} = \frac{\text{cpm released from cells during incubation}}{\text{total cpm incorporated into cells}} \times 100$$

The percentage of specific cytotoxicity (a) is calculated as $a = b - c$, when b is the percentage of release in the experimental group and c is the percentage of release in the autologous control.

Assay for ADCC. Untreated or IF-treated effector cells are tested for ADCC in the same 4-hr ^{51}Cr release assay described above for NK activity.[15] The differences between the tests involve the use of other target cells, coated with IgG antibodies. The percentage of ADCC is calculated by subtracting the percentage of cytotoxicity against the non-antibody-coated target cell from that obtained with the antibody-coated target.

A variety of target cells can be used for ADCC. However, for optimal interpretation of results, a target cell line that is highly resistant to NK activity should be used. When target cells with some degree of susceptibility to NK activity are used, higher NK levels will be seen in the interaction of IF-treated PBL with the non-antibody-coated target cells. Thus, it would be difficult to discriminate the contribution of IF-boosted NK and ADCC activities against antibody-coated target cells. For our studies of ADCC, some mouse T cell lymphomas (YAC-1, RBL-5, RL♂1, EL-4) have been very useful, since they are completely resistant to NK activity of IF-treated human PBL and are highly sensitive to ADCC when coated with a rabbit anti-mouse T cell serum.[15]

The source of antibody used to sensitize the target cells may also be an important factor in the demonstration of IF augmentation of ADCC. Consistent boosting of activity has been seen in our experiments with rabbit

[15] J. R. Ortaldo, S. Pestka, R. B. Slease, M. Rubinstein, and R. B. Herberman, *Scand. J. Immunol.* **12**, 365 (1980).

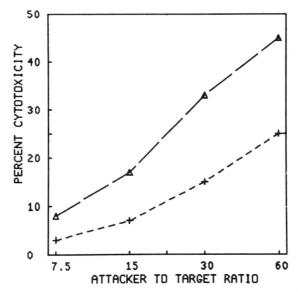

FIG. 1. Augmentation of cytotoxicity with interferon (IF). Shown are the results of a typical experiment of untreated (+) and IF-treated (Δ) human peripheral blood lymphocytes in a 4-hr ^{51}Cr release assay, with percentage of lysis of the target cells at various attacker-to-target cell ratios (5000 K-562 target cells in each group).

antibodies. Lower degrees of augmentation have been seen with human or mouse antisera.

Even with suitable target cells and antibodies, the degree of augmentation of ADCC is often not as great as for NK activity. After treatment of PBL with IF for 1 hr, ADCC activity is usually increased to 150–200% of the levels seen with untreated PBL. In contrast, NK activity of the same PBL may be increased to 200–500% of the control levels. A larger increase in ADCC activity can usually be obtained by pretreatment of PBL for 18 hr. The quality and degree of purity of the IF preparation may also be more critical for augmentation of ADCC activity than for NK activity. For example, if the preparations with IF also contain immune complexes or aggregated immunoglobulin, these contaminants could preferentially inhibit ADCC.

Analysis of Data. Pretreatment of PBL with IF usually results in appreciable augmentation of NK activity, at all ratios of effector cells to target cells (Fig. 1). The degree of boosting of activity varies substantially among normal donors and also is influenced by the target cell used in the assay. Various methods can be used to analyze the degree of augmentation by IF.

1. The percentage cytotoxicity of treated and untreated cells, at one effector : target cell ratio, can be compared. For example, in Fig. 1, at a 30 : 1 ratio, the IF-treated cells produced 33% lysis, whereas the untreated PBL gave 15%. Such a comparison lends itself well to determination of the significance of the difference between the two points, by a Student's t test.
2. Lytic units (LU) for each cell population can be calculated. A lytic unit is defined as the number of effector cells needed to give 30% cytotoxicity.

In Fig. 1, a 100 : 1 ratio of the untreated PBL was required to give 30% lysis, indicating that these cells had 19 LU/10^7 cells. With the IF-treated cells, a 26 : 1 ratio gave 30% lysis and thus there were 75 LU/10^7 cells. These calculations provide an overall quantitative expression of the data for the entire dose-response curve of each group. Then the differences between the IF-treated and untreated cells can be expressed as (a) the percentage of activity in the treated population relative to the untreated (in the example in Fig. 1, the lytic activity of the IF-treated cells was 390% of that of the untreated PBL); or (b) the difference in LU between the IF-treated and untreated cells (in Fig. 1, the increase is 56 LU). It is not yet possible to state which of these expressions more closely reflects the biology of the effects of IF on this effector cell.

Comparison of LU between groups provides the best quantitative assessment of the degree of augmentation by IF. Comparison of percentage cytotoxicity values at one effector : target cell ratio can give misleading results, particularly if one or both of the points are not on the linear portion of the dose-response curve. However, there are also some limitations to the use of the LU method. Valid calculation of LU can be determined only from the linear portion of the cytotoxic dose-response curve. Also, LU from two experimental groups can be compared only if the slopes of the dose-response curves are parallel. Usually, treatment by IF does result in a curve of augmented cytotoxicity that is parallel with that for the untreated PBL.

Method for Cytotoxicity by Enriched Effector Cell Population

For detailed studies of the effects of IF on NK and ADCC activities, it is very helpful to work with enriched populations of effector cells. The LGL, separated on discontinuous Percoll density gradients, have been found to account for virtually all of the NK and ADCC activities.[3] Separation of the LGL, with high cytotoxic activity and almost devoid of other lymphoid cell types, is particularly advantageous for examination of the mechanism of augmentation of function, and the associated biochemical changes, induced in these cells by IF.

Separation of LGL. A discontinuous gradient of Percoll is prepared by carefully layering a series of fractions with varying concentrations of Percoll (from 37% to 60%) into a 15-ml conical test tube (No. 2095, Falcon). These fractions are made by mixing Percoll (osmolality adjusted to 285 mOsm of H_2O per kilogram with RPMI-1640 medium, supplemented with 10% fetal bovine serum and 5 mM HEPES (adjusted to 280–285 mOsm). The following are the proportions of Percoll and medium in each fraction. It is essential that the density of each fraction, as measured with a refractometer, be correct. If the density of the osmolality-corrected Percoll is 1.125 g/ml and that of the medium is 1.0284 g/ml, then the fractions can be made by mixing the indicated volumes of Percoll with medium to give a total volume of 6.0 ml (see tabulation below).

Fraction[a]	Density (refractive index at 25°)	Usual volume of Percoll (ml) in 6 ml of mixture
1	1.3432	2.50
2	1.3436	2.70
3	1.3440	2.85
4	1.3443	3.00
5	1.3446	3.15
6	1.3450	3.30
7	1.3470	4.00

[a] The gradient is constructed with 2.5 ml of the first three fractions and 1.5 ml of the last four, added in order of decreasing density.

For separation of peripheral blood lymphocytes, 5×10^7 cells in 1.5 ml of medium are placed on the top of the gradient, which is then centrifuged at 550 g for 30 min at room temperature. Acceleration and deceleration should be slow. Fractions are collected from the top with a Pasteur pipette. Most LGL and the associated NK and ADCC activities should be in fractions 2 and 3 (peak in fraction 3), whereas most of the cells recovered should be in fraction 6 or 7 (5–15% of input cells in fractions 2 and 3 and 40–70% in fractions 6 and 7, with usual overall recovery of 90%).

Cytotoxicity Assays with LGL. Fractions 2 and 3 are pooled in a 50-ml conical centrifuge tube and washed with RPMI-1640 medium plus 2% fetal bovine serum (first 10 ml medium with 10% serum and then 40 ml of serum-free medium added). The tube is centrifuged at 300 g for 10 min,

AUGMENTATION OF NATURAL KILLER CELL (NK) ACTIVITY
AFTER TREATMENT OF PERIPHERAL BLOOD LYMPHOCYTES
(PBL) AND PERCOLL-SEPARATED LARGE GRANULAR
LYMPHOCYTES (LGL) WITH INTERFERON

Cells	Treatment with interferon	NK activity $(LU/10^7 \text{ cells})^a$
PBL	−	41.2
	+	84.5
LGL fractions	−	359
	+	621

^{a} LU, lytic units

and the cells are resuspended in assay medium to the desired concentration, and the above-described procedures are used for IF treatment and measurement of cytotoxic activity.

The table shows the data from a typical experiment with LGL. Although the cells in the LGL fractions represented only 8% of the input PBL, they contained about 70% of the total activity and showed a ninefold increase in relative activity over that of PBL. After treatment with IF, both populations showed about a twofold increase in NK activity. However, the absolute increase in lytic units in the LGL fractions was clearly much higher than that seen with the unseparated PBL.

[58] Procedures for the Mixed Lymphocyte Reaction and Cell-Mediated Lympholysis

By IVER HERON and KURT BERG

A number of nonantiviral effects of human type I interferons have been established. Since T cells play a major role in viral immunity as well as in the defense against neoplasms, we examined the effects of interferon on human T cell responses. Mixed lymphocyte cultures lead to generation of cytotoxic T cells specific for the immunizing alloantigens, and Heron et al.[1] showed that the addition of interferon during the mixed lymphocyte culture period suppressed the proliferative responses moderately, yet the killer cell product, measured by cell-mediated lympholysis against chromium-labeled lymphoblast targets, was enhanced.

[1] I. Heron, K. Berg, and K. Cantell, J. Immunol. **117**, 1371 (1976).

This observation was later confirmed by Zarling et al.[2] and was further analyzed for the mechanisms involved by Heron and Berg.[3] In the following the methodology employed in these systems will be outlined.

Mixed Lymphocyte Culture Period (Sensitization Phase)

Isolation of Lymphoid Cells. Essentially, the method of Böyum[4] was used. Normal, healthy, nonrelated volunteers served as donors. Heparinized blood (20 units/ml) or glass bead defibrinated blood is mixed with an equal volume of PBS, pH 7.4. Six milliliters of this solution are gently layered upon 3 ml of Ficoll-Isopaque ($d = 1.077$) or lymphoprep ($d = 1.076$, Nygård, Oslo, Norway) in a 10-ml test tube. The tubes are centrifuged for 25 min at room temperature at 650 g. The cells at the interface are pipetted off and washed once (350 g, 10 min) in medium 199 supplemented with heparin (10 units/ml), 10 mM HEPES, penicillin (100 units/ml), streptomycin (100 μg/ml) and 5% human serum. The following wash is performed at 200 g for 10 min with pooled serum from 4–10 normal male A Rh-positive donors. After pooling and heat inactivation (56°, 30 min) the serum is filtered through a 0.22 μm Millipore filter and stored in aliquots at $-20°$. A given pool will normally be used within 3 months.

The cells are resuspended in culture medium and counted. Counting is performed on unstained cell preparations under a phase contrast microscope. The cell suspensions are adjusted to the desired concentration, normally 2×10^6 lymphocytes per milliliter of culture medium. Culture media used are (A) Eagle's MEM or RPMI-1640 with bicarbonate or (B) Iscove's modified Dulbecco medium (Gibco) supplemented with (A) 15% human serum, 2 mM glutamine, penicillin (100 units/ml), and streptomycin (100 μg/ml), and (B) 15% serum and antibiotics as above.

Production of Stimulator Cells. Approximately 4000 rad are delivered at a rate of 450 rad/min by a 137 Cs source (Risoe, Denmark).

Production of the Cultures. Two types of systems can be used depending on the exact aim of the experiment: (1) microplate cultures in which subsequent thymidine incorporation and cell-mediated lympholysis assays can be directly applied on individual cultures; and (2) bulk mixed lymphocyte cultures, which are sampled for thymidine incorporation, and counted and redistributed in smaller cultures for the subsequent cell-mediated lympholysis.

Normally responder cells at a concentration of 2×10^6 per ml are mixed with an equal volume of irradiated stimulator cells at a concentra-

[2] J. M. Zarling, J. Sosman, L. Eskra, E. C. Borden, J. S. Horoszewicz, and W. A. Carter, *J. Immunol.* **121**, 2002 (1978).

[3] I. Heron and K. Berg, *Scand. J. Immunol.* **9**, 517 (1979).

[4] A. Böyum, *Scand. J. Clin. Lab. Invest.* **21**, 31 (1968).

tion of 0.5 to 1 × 10⁶ per milliliter. If other concentrations of responder cells are employed, the responder to stimulator cell ratio is changed (see below).

For system (1), 0.2 ml of the cell mixtures is distributed per well of the round-bottom microtiter plate (Nunclon microtiter plate, Roskilde, Denmark); and for system (2), 2 ml of the mixed cell suspension are seeded into the wells of Costar plates (Nunclon 24 wells) or 10–15 ml are cultured in a 50-ml Falcon flask (No. 3013) that is left in the upright position during incubation.

Interferon preparations are added in microliter quantities to cells after the mixing of responder and stimulator lymphocytes. Microtiter plates are covered with lid and wrapped in adhesive kitchen film. Incubation takes place at 37° (or in selected instances at 38° and 39°[5]) in a humidified atmosphere of air containing 5% CO_2. After 4 days, aliquots of resuspended suspension from the bulk cultures (system 2) are distributed in 0.2-ml volumes in microtiter plates. To each well of the microtiter plates are added 0.04 μCi of [2-¹⁴C]thymidine (CFA 219, specific activity 60 mCi/mmol; Amersham, England) in 10 μl of saline, and incubation is continued for another 16–18 hr. Thymidine incorporation is stopped by placing the plates at 4°.

Harvesting and Scintillation Counting. The cultures are harvested with a semiautomatic multisample precipitator (Scatron). The cells are collected on glass fiber paper, Whatman GF/A, which is dried and transferred to polypropylene counting tubes. Scintillation fluid is added, and counting is performed in a liquid scintillation counter.

Microtiter plates for cell-mediated lympholysis and bulk cultures are left in the incubator for 6 days for standard cell-mediated lympholysis.

Cell-Mediated Lympholysis (Effector Phase)

Production of Target Cells. Cell suspensions containing 1 × 10⁶ nonirradiated cells per milliliter of culture medium, taken from the cell preparations discussed above at the stage before mixing cells for mixed cultures, are incubated at 37° for 3 days and 5 μg of concanavalin A (Pharmacia, Sweden) are added per milliliter. After a further 3 days of incubation, cells are washed once in medium 199 containing 5% human serum and resuspended in 0.1 ml of this medium. Then 100 μCi of Na_2 ⁵¹CrO_4 (CJS-1P Amersham 100–300 mCi per milligram of Cr) is added, and cells are incubated at 37° for 60–90 min. The labeled lymphoblasts are washed 3 or 4 times in cold medium. Resuspension of cells between the centrifugations is done by careful pipetting to disintegrate cell clumps. The concentration

[5] I. Heron and K. Berg, *Nature* (*London*) **274**, 508 (1978).

of viable cells is determined by counting under phase contrast micros-copy, and the suspension is adjusted to 1×10^5 cells per milliliter of medium 199 with 15% human serum. The target cells are kept at 4° until added to the effector cells.

Effector Cell Preparation. System 1: Microtiter plates containing the mixed lymphocyte cultures at day 6 of incubation are used. Most of the supernatant of each well is removed. Wells are refilled with medium 199 containing 5% human serum, and the plates are centrifuged. After one more washing cycle, the cellular content of the wells is resuspended in 0.1 ml of medium 199 with 15% serum. The contents of single, representative wells, are removed and counted for determination of viable cell recov-eries.

System 2: The bulk cultures are transferred to centrifuge tubes and cells are washed twice in medium 199 with 5% serum. The pellets are resuspended in medium 199 with 15% serum, and the number of viable cells is determined using phase contrast microscopy. The cell concentra-tion is adjusted.

The Lympholysis Reaction. System 1: One-tenth milliliter of the appro-priate target cell suspension containing 10^4 ^{51}Cr-labeled cells is added per well of the microtiter plate, which already contains effector cells in a 0.1-ml volume. Triplicates or quadruplicates are preferred. Target cells are also added to wells containing culture medium but no effector cells. These serve to determine spontaneous release and maximum release, respectively. Each plate is covered with a lid and incubated for 4–6 hr at 37°. Cells are resuspended by tapping cautiously the edges of the plate. After centrifugation, 0.150 ml of supernatant is removed from each well and transferred to a counting vial. Maximum ^{51}Cr content is determined by resuspending the target cells and withdrawing 0.150 ml for counting. Close to 80% of the incorporated ^{51}Cr can be released by freeze-thawing target cells. Counting is carried out in a gamma counter.

System 2: Effector cells in the appropriate numbers, depending on which ratios between effector and targets are chosen, are suspended in 1-ml volumes in small test tubes (Nunc 160-41661); 10^4 target cells are added in 0.1 ml. Spontaneous release is determined in tubes containing 1 ml of culture medium and target cells. Maximum ^{51}Cr release is deter-mined in tubes containing 1 ml of water with added 10^4 target cells after three times freezing and thawing.

After centrifugation for 5 min at 100 g, incubation is performed at 37° for 4–6 hr. Cultures are resuspended and centrifuged after the incubation period, and the supernatants are harvested into counting vials simply by pouring all of the supernatants from the pelleted cells.

The standard formula for calculation of cytotoxicity was used. Killing is expressed (in percent) as

$$\frac{{}^{51}\text{Cr release with effector cells} - \text{spontaneous } {}^{51}\text{Cr release}}{\text{maximal } {}^{51}\text{Cr release} - \text{spontaneous } {}^{51}\text{Cr release}} \times 100$$

Specific killing is cytotoxicity expressed on target cells from the person used as stimulator in mixed lymphocyte culture, and autokilling is cytotoxicity on target cells autologous to the responder.

General Comments

Choice of Medium. The choice of medium to assess the proliferative responses in the mixed lymphocyte cultures has not been found to be very critical. Medium 199, Eagle's MEM, and RPMI-1640 have thus been useful. From controlled comparative studies it has, however, become clear that, as far as the ability of media to support the generation of cytotoxic T lymphocytes (CTL) is concerned, the choice of medium is of major importance. The table gives the results of experiments in which effector cells were generated in mixed lymphocyte cultures performed in medium 199 and Eagle's MEM in parallel. These media were also compared for usefulness in the effector phase. Medium 199 was found to be clearly suboptimal during the CTL generation phase, but fully supportive for the effector phase. RPMI-1640 has been found to be comparable to Eagles MEM.

Effect of Cell Densities in the Mixed Cultures for Total Cell Recovery and for Cytotoxic T Lymphocyte (CTL) Generation. Two types of cultures were set up. In Fig. 1, the ratio between responder and stimulator cell numbers

TWO MEDIA COMPARED FOR SUPPORTIVE CAPACITY DURING
KILLER CELL GENERATION AND DURING THE EFFECTOR PHASE

| Expt. No. | Cr released (%) | | | | Phase |
| | Medium 199 | | Eagle's MEM | | |
	Spec.[a]	Auto.[b]	Spec.	Auto.	
1	10	−1.5	28	0.4	During mixed lymphocyte culture
2	0	0.6	49	−0.2	
3	2.3	−4.5	22	−3.8	
4	46	−1.1	91	−1.3	
5	−2.5	−1.7	19	0.5	
1	35	−4.8	33	−3	During effector phase
2	28	1.2	28	1.0	
3	39	0.8	41	1.7	

[a] Specific target cell.
[b] Autologous target cell.

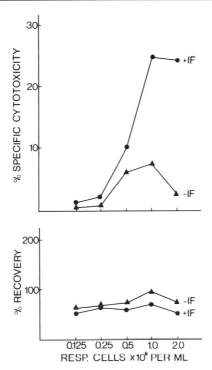

FIG. 1. Interferon was added (+IF) at a concentration of 500 units/ml to mixed lympho
cyte cultures, and the cell-mediated cytotoxicity generated was assessed at a fixed effector-
to-target cell ratio (10:1) and compared with cytotoxic T lymphocytes (CTL) generated in
the absence of exogenous IF. Different responder cell densities during the mixed cultures
were assessed with a fixed responder-to-stimulator cell ratio, namely 1:0.25. In the lower part
of the figure, the viable cells recovered after the mixed cultures as percentage of the responder cell input 6 days earlier is depicted. The curves represent the means from three experiments on different allogeneic combinations. Spontaneous ⁵¹Cr release from target cells was
8.7% (mean).

was kept constant, namely 1:0.25, and cell density in cultures was varied.
In Fig. 2, the number of irradiated stimulator cells was kept constant,
namely 0.5×10^6/ml and the responder cell density was varied.

Viable cell recoveries were found to be very dependent on the total cell
density (responder plus stimulator cells), which is most easily seen when
the low responder cell concentration is assessed with stimulator cells in a
1:0.25 ratio (Fig. 1) versus the relatively high number of stimulators in
Fig. 2. Recoveries also dropped if total density was raised above $1.5 \times$
10^6/ml, even though such cultures had their medium changed once due to
acidity developing around days 4–5. Optimal cell recoveries were obtained with densities from 0.5 to 1×10^6/ml and were even higher in cases

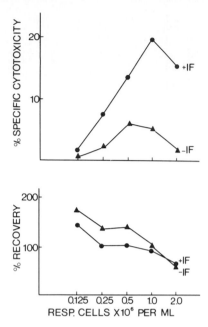

FIG. 2. See legend to Fig. 1. The only difference has been that a fixed number of stimulator cells, namely 0.5×10^6 per milliliter, has been employed in these experiments during the mixed lymphocyte reaction.

of low responder cell concentrations. Interferon-induced inhibition of cell proliferation and total cell recovery was found to be greatest at the lowest responder cell concentrations, whereas in the more dense cultures cell yields without and with interferon present were the same. This was like-wise observed when thymidine incorporation was measured, as shown in Fig. 3.

The CTL activity and recoveries were found to depend mostly on the *responder* cell concentration (Figs. 1 and 2); $0.5-1 \times 10^6$ responder cells per milliliter was optimal. The interferon-induced augmented CTL genera-tion at different densities was found to vary in parallel with the control CTL generation; the highest IF-induced CTL increase as percentage of control CTL was found at onefold higher responder cell density than the optimum for control CTL generation.

The Target Cells and Specificity. Lymphocytes stimulated by phytohemagglutinin (PHA) were initially used as described by Lightbody *et al.*[6] Later we preferred Con A blasts for two reasons. Allogeneic PHA

[6] J. Lightbody, D. Bernoco, V. C. Miggiano, and R. Ceppelini, *J. Bacteriol. Virol. Immunol.* **65**, 243 (1971).

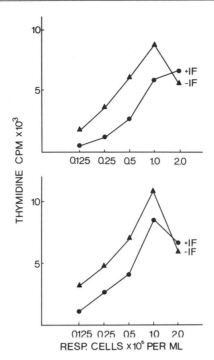

FIG. 3. Thymidine uptake in mixed lymphocyte cultures of varying cell densities. In the upper part of the figure responder-to-stimulator cell ratio was constant (1 : 0.25), and in the lower part a fixed number of stimulator cells (0.5 × 10⁶ cells/ml) had been employed. Thymidine incorporation in the presence of 500 units of interferon (+IF) and in the absence of interferon (−IF) were compared. Inhibition of thymidine uptake by interferon is relatively most pronounced at the lower cell densities.

blasts are usually sensitive targets, but when high levels of killing are found, there would often be a parallel rise, although at a lower level, in lysis of autologous PHA blasts (autokill), which would be looked upon as non-specificity. This problem does not occur when Con A blasts are used. Second, a portion of Con A blasts are different from HLA known to express also the DR antigens, which have been shown to be recognized as target structures in CTL.[7] Nonstimulated lymphocytes can also be used as target cells, but the sensitivity is often lower when freshly prepared lymphocytes are used.

Lymphoblast targets from donors unrelated to the actual responder and stimulator (third party) would often be killed by CTL, depending on the number of target determinants shared with the specific stimulator.

[7] C. F. Feighery and P. Stastny, *Immunogenetics* **10**, 31 (1980).

Interferon-induced enhanced specific CTL reactivity was also found to influence third-party cytotoxicity in parallel.[3] The fine specificity of the CTL generated during interferon-containing mixed lymphocyte reaction appeared to be the same as without interferon.

Effector-to-Target $(E:T)$ *Ratio in CML and Length of Incubation*. If system 2 is employed, different $E:T$ ratios can be applied. This is mostly done by keeping the number of target cells constant and varying the number of effector cells present in the reaction mixture in a constant volume. When numbers of effectors are plotted on a log scale, Cr release percentage will mostly fall on a straight line unless very high ratios are included. Curves like that from CTL generation with and without interferon have been published previously.[1,5] If CTL responses are generated in the absence of interferon, most authors have used $E:T$ ratios of 50:1 or higher, and many groups use 6-hr assays to improve the kill percentages in CTL in man. These conditions are used to maximize ^{51}Cr release in tubes containing effector cells but spontaneous ^{51}Cr release will also increase. The calculated kill percentage will in this way be "artificially" high when the standard formula for calculation is used, particularly if targets with high levels of spontaneous release are employed. When augmented CTL generated with interferon is assessed, we chose 4-hr incubations and often quite low $E:T$ ratios (Figs. 1 and 2). This ensures a system with spontaneous ^{51}Cr release of about 10% and that there will be no "plateau" effect of release with the killer cells. Under these conditions the control CTL will thus often be low.

Kinetics of the CTL Response and Time of Interferon Addition. The cytotoxic response in most experiments[3] peak at day 6, after an initial rise at day 5, and will decrease after day 8 of culture. The kinetics in the absence and in the presence of interferon was found to be similar.

Time of Addition of Interferon. Treatment, for 1–2 hr, of responder cells, stimulator cells, or both, with interferon, which is subsequently washed away, before initiation of the mixed lymphocyte cultures has been found not to enhance CTL generation. Addition of interferon at the beginning of the culture period gives rise to the augmentation of CTL as well as inhibition of thymidine uptake; the addition of interferon with a delay of 1, 2, 3, or 4 days[1] resulted in a gradually declining augmentative effect on CTL generation, and interferon added during only the last days had no effect. Interferon added only during the cell-mediated lympholysis culture did not have any effect[1,8]; this is in marked contrast to similar additions to natural killer cell cultures.[8,9]

[8] I. Heron, M. Hokland, A. Møller-Larsen, and K. Berg. *Cell Immunol.* **42,** 183 (1979).
[9] G. Trinchieri and D. Santoli, *J. Exp. Med.* **147,** 1314 (1978).

Interferon Dose-CTL Response. System 1 is most useful for this type of investigation. Addition of varying concentrations of leukocyte interferon to mixed lymphocyte cultures, under conditions described as optimal in this chapter, have led to different degrees of CTL enhancement.[1,3] The range of concentrations found most stimulating was 200–1000 units/ml. Concentrations as low as 50 units/ml significantly enhanced CTL, but at lower levels. However, concentrations of interferon above 5000 units/ml have been found not only to suppress the mixed lymphocyte reaction quite extensively, but also to inhibit generation of CTL compared to control cultures not containing interferon.

Acknowledgments

This work has been supported by the Danish Medical Research Council and the Danish Cancer Association. The competent technical assistance by Mrs. Inger Sørensen is sincerely appreciated.

[59] Assay of Immunomodulation by Type I Interferon *in Vivo*

By Edward De Maeyer

Interferon affects many parameters of immunity, and since the experimental conditions vary according to the type of immune reaction studied, it is not possible to prescribe a general recipe applicable to every particular situation. Therefore, this chapter presents various points found to be of practical importance while studying modulation by interferon of the delayed-type hypersensitivity reaction in the mouse.

Selection of Animals

Pooling. For a given experiment, mice should be of the same age and sex, unless the latter represent the very factors studied. Females can be pooled and then randomly distributed into different cages according to treatment groups; this is usually not possible with males because of fighting. One can, however, circumvent this problem by having males belonging to different experimental groups in the same cage (they can easily be distinguished by ear markings or toe clipping). This procedure ensures that mice that have been raised in a given cage are distributed among the different treatment groups, and eliminates cage effects.

Origin. The origin of the animals is of the utmost importance, and breeding one's own mice remains the best solution to this problem provided one has the proper breeding facilities and animal quarters. Mice commercially available do not always live up to standards, and significant variations have been observed in the extent and duration of delayed hypersensitivity reactions in mice belonging to the same inbred strain but of different commercial origins. It is, therefore, recommended that only animals of the same origin be used for the duration of a given research project; switching dealers in the middle of an investigation would be ill-advised.

Age. Very young mice will not mount as strong a delayed hypersensitivity reaction as older animals. Generally mice $2\frac{1}{2}$–3 months old are used.

Genotype. The effects of interferon on delayed hypersensitivity are influenced by the genotype of the mouse; for example, more interferon is required to inhibit expression of delayed hypersensitivity to sheep red blood cells (SRBC) in C57BL/6 mice than in BALB/c mice. It is therefore preferable to use inbred mice of defined genotype, since such genetic effects can be masked when using random-bred strains.

Antigen Dosage

In most studies sheep erythrocytes (SRBC) commercially available from the Pasteur Institute (Institut Pasteur, 25 rue du Dr. Roux, 75015 Paris, France) are used as antigen. The method of sensitization and challenge is derived from Lagrande and Mackanness[1] with minor modifications. The optimal antigen dose cannot be extrapolated from the literature but has to be determined for the particular animals used. With SRBC, for example, the optimal sensitizing dose is 10^5 SRBC according to Lagrange and Mackanness, whereas in my hands, at Orsay, it is 10^6 SRBC both for BALB/c and C57BL/6 mice.

Suboptimal doses of antigen are usually required in order to bring out immune-enhancing effects of interferon. Again, these can be determined only by trying various quantities of SRBC.

Administration of Interferon

Route of Inoculation. Interferon can be administered both intravenously and intraperitoneally. The intravenous route is preferred because it gives less individual variation. The retroorbital sinus offers a convenient route of intravenous injection of SRBC or of interferon. With some experience, up to 50 mice can be inoculated in about 1 hr, using a 0.45×13 mm gauge needle, provided one does not administer more than 0.4 ml per

[1] P. H. Lagrange, G. B. Mackanness, and T. E. Miller, *J. Exp. Med.* **139,** 528 (1974).

animal, preferably only 0.2 ml; for larger volumes, which should be avoided if possible, a slower inoculation time is needed.

Timing of Interferon Administration. This is a highly critical variable, as opposite results (i.e., stimulation of sensitization instead of suppression) can be observed when interferon is given after, rather than prior to, sensitization.

Stabilizing Interferon. Many experiments are now carried out with pure mouse interferon. Since the last purification step often implies elution from a column with high salt or acid buffer, care should be taken to dialyze the preparation against distilled water or PBS if the dilution factor is not big enough to ensure isotonicity. To avoid loss of activity during dialysis one can add 1% of serum derived from mice of the same genotype as those in which the experiment is to be done. Under these conditions, pure interferon can be dialyzed without any appreciable loss of activity. It should be distributed into small aliquots, stored at $-70°$, and thawed only once.

[60] General Considerations in Measurement of Interferon-Induced Antibody Suppression

By ROGER J. BOOTH and JOHN MARBROOK

In addition to being released by virus-infected cells, interferon (IF) can be produced by lymphoid cells that have been stimulated by antigens or mitogens. These conditions are precisely those that lead to the production of immune regulatory substances or "lymphokines."

Lymphokines, by definition, have a positive and/or negative influence on immune reactions, and two main points need to be considered in studying the effect of IF on antibody production. First, the purity of the IF preparation must be such that it is only the biological activity of the IF molecule that is being assayed. Second, a large number of lymphokines have been described and these are mostly defined by their biological activity in *different* assay systems. There is a real possibility that one factor will function in more than one assay, thus creating the illusion of the presence of multiple factors. Similarly, the exposure of cells to one factor may cause the production of a second factor, and it may be the second factor that is assayed under particular conditions.

The measurement of IF-induced immunosuppression is complicated by the fact that most antibody responses involve multiple cell types and even the precursors of antibody-forming cells (AFC) consist of subpopula-

Copyright © 1981 by Academic Press, Inc.
ISBN 0-12-181979-5

tions that represent discrete stages in the differentiation pathway of B cell lineage (see below). Interferon-induced antibody suppression is also considered in this volume [61] and [62], and the assay systems are summarized below. In general, insufficient attention has been paid to the detailed analysis of the kinetics of immunosuppression, particularly in relation to the heterogeneity of responding cells. As this directly affects the level of suppression observed, the following discussion will concentrate on the means whereby an assay system can be devised to minimize the problems that result from having a complex cell system to assay possible complex effects of an immunosuppressive agent such as IF.

The Assay System

The main approach has been to stimulate the appropriate precursor cells with antigens or mitogens in the presence of IF and to measure the number of specific AFC which are produced (see Table I). This has been carried out *in vivo* where the AFC are assayed in the spleen (or lymph nodes). As relatively large amounts of IF are required to maintain suppressive circulating concentrations, this approach does not lend itself readily to being a standard assay system. In addition, circulating lymphoid cells may redistribute during the assay period, and the inability to assay the immunosuppressive activity of IF in a "closed" system renders *in vivo* assay systems generally unacceptable.

The assay method for measuring immunosuppression can depend on the question under investigation. A routine assay method should allow the measurement of replicate assays of many samples of IF, particularly if a range of samples of fractionated material is being examined. A variety of culture techniques have been used to generate immune responses, and these are summarized in Table I. Most of the reported studies involve the response of mouse spleen cells, and, in general, the culture systems require 10–20 million spleen cells in 1-ml cultures, although microculture systems have been devised for measuring the response of smaller numbers of lymphoid cells.[1]

The antiviral activity of IF is measured in units, where an IF unit is that concentration which reduces some measurable parameter of viral growth by 50%. Ideally the immunosuppressive activity could be expressed in the same way, where one suppressive unit would be the amount of IF reducing an immune response by 50%. Experimental results that illustrate the relationship between IF units and immunosuppression have been summarized in Table II. When compared on this basis it is immediately apparent that large differences in immunosuppressive activity of IF

[1] G. J. Finlay, R. J. Booth, and J. Marbrook, *Eur. J. Immunol.* **7**, 123 (1977).

TABLE I

ASSAY SYSTEMS FOR MEASURING INTERFERON (IF)-INDUCED ANTIBODY SUPPRESSION

Assay system	IF introduced	Mouse strain	Antigens used	Culture conditions	Parameters of antibody response measured
In vivo[a]	Type I intraperitoneally (ip) at time of immunization	6–8 wk CFW	10^8 sheep red blood cells (SRBC) intravenously (iv)	—	Direct plaque-forming cells (PFC) in spleen 48 hr after antigen
In vivo[b]	Type I iv 2–4 days before antigen	6 wk Swiss-Webster	, 2×10^7 SRBC iv	—	Direct PFC in spleen 6 days after antigen
In vivo[c,d]	Type I iv 2–0 days before antigen	6–8 wk Swiss-Webster ♀	1×10^7 SRBC iv	—	Serum hemagglutinins and hemolysins 4 and 6 days after antigen
In vivo[e]	Type II induced with BCG/PPD 24 hr before antigen	5–8 wk CBA	5×10^7 SRBC ip	—	Direct PFC in spleen 4 and 5 days after antigen
In vitro[c,d]	Type I iv prior to culture	6–8 wk Swiss-Webster ♀	3×10^6 SRBC/culture	5×10^6 viable, nonadherent spleen cells or 1×10^7 unseparated spleen cells in 1 ml of Eagle's MEM + 10% fetal calf serum (FCS) cultured in 35-mm petri dishes in a humid atmosphere of 10% CO_2, 7% O_2, 83% N_2 at 37° with gentle rocking (7 cycles/min) (method of Mishell and Dutton[f]).	Direct PFC after 3–5 days

(continued)

TABLE I (*continued*)

Assay system	IF introduced	Mouse strain	Antigens used	Culture conditions	Parameters of antibody response measured
In vitro[e,g]	Type II added at the time of antigen or 24 hr earlier	5–8 wk C57BL/6J,[e] 6–8 wk BALB/c ♀[g]	3×10^6 SRBC/ culture	Mishell and Dutton[f] (as described above)	Direct PFC after 4 and 5 days
In vitro[h]	Type I present throughout culture period or cells pretreated and washed before antigenic challenge	C57BL/6♀	3×10^6 SRBC/ culture	1.5×10^7 spleen cells in 1 ml of Eagle's MEM + 10% FCS in 35-mm petri dishes in a humid atmosphere of 10% CO_2, 7% O_2, 83% N_2 at 37° with gentle rocking[f]	Direct PFC after 3–5 days
In vitro[i]	Selected cell populations pretreated with type I for 6 hr and then washed and cultured with antigen	BALB/c, nu/nu, (nu/nu × BALB/c)F₁	3×10^6 SRBC/ culture	10^7 F₁ cortisone-resistant thymocytes (T cells or 10^7 nu/nu mesenteric lymph node cells (B cells), or 4×10^5 adherent peritoneal exudate cells (macrophages) per 1 ml of Eagle's MEM + 10% FCS were pretreated for 6 hr with IF, washed 3 times, and cultured. Mishell and Dutton[f] type cultures contained a mosaic of 3×10^5 macrophages + 3×10^6 T cells + 7×10^6 B cells in 1 ml of Eagle's MEM + 10% FCS	Direct PFC after 3–5 days

In vitro[j]	Type I present throughout culture period or cells pretreated and washed before antigenic challenge	10–14 wk CBA/J ♀	5 × 10^6 SRBC/culture	1.4 × 10^7 spleen cells in 1 ml RPMI-1640 + 10% FCS in Marbrook vessels[k] in a humid atmosphere of 10% CO_2 in air at 37°	Direct PFC after 2–3 days
In vitro[l]	Type I present throughout culture period	11–13 wk CBA/J ♂	8 × 10^5 SRBC/culture	4 × 10^5–1.5 × 10^6 density fractionated spleen cells in 0.2 ml of RPMI-1640 + 10% FCS in 96-well flat-bottom microtiter wells[m] in a humid atmosphere of 10% CO_2 in air at 37°	Direct PFC after 2–3 days
In vitro[n]	Type I present throughout culture period	10–14 wk CBA/J ♀	2 × 10^7 SRBC/culture + 10 μg/ml Escherichia coli lipopolysaccharide	2.5 × 10^6–6 × 10^6 spleen cells + 1.4 × 10^7 thymocytes in 3.5 ml of RPMI-1640 + 10% FCS in 64-well polyacrylamide cultures[o] in a humid atmosphere of 10% CO_2 in air at 37°	Direct PFC after 2–4 days

[a] W. Braun and H. B. Levy, Proc. Soc. Exp. Biol. Med. 141, 769 (1972).
[b] T. J. Chester, K. Paucker, and T. C. Merigan, Nature (London) 246, 92 (1973).
[c] B. R. Brodeur and T. C. Merigan, J. Immunol. 113, 1319 (1974).
[d] B. R. Brodeur and T. C. Merigan, J. Immunol. 114, 1323 (1975).
[e] J. R. Virelizier, E. L. Chan, and A. C. Allison, Clin. Exp. Immunol. 30, 299 (1977).
[f] R. I. Mishell and R. W. Dutton, J. Exp. Med. 126, 423 (1967).
[g] G. Sonnenfeld, A. D. Mandel, and T. C. Merigan, Cell Immunol. 34, 193 (1977).
[h] H. M. Johnson, B. G. Smith, and S. Baron, J. Immunol. 114, 403 (1975).
[i] R. H. Gisler, P. Lindahl, and I. Gresser, J. Immunol. 113, 438 (1974).
[j] R. J. Booth, J. M. Rastrick, A. R. Bellamy, and J. Marbrook, Aust. J. Exp. Biol. Med. Sci. 54, 11 (1976).
[k] J. Marbrook, Lancet 2, 1279 (1967).
[l] G. J. Finlay, R. J. Booth, and J. Marbrook, Eur. J. Immunol. 7, 123, (1977).
[m] B. L. Pike, J. Immunol. Methods 9, 85 (1975).
[n] R. J. Booth, J. M. Booth, and J. Marbrook, Eur. J. Immunol. 6, 769 (1976).
[o] J. Marbrook and J. S. Haskill, Cell. Immunol. 13, 12 (1974).

TABLE II
A Comparison of Antiviral and Immunosuppressive Activities of Interferon (IF)

System used	Parameter measured	International IF units to give 50% inhibition of immunity
Mouse spleen cells cultured with sheep red blood cells (SRBC)	Direct PFC at 5 days	Type I: 10,000[a]
	Direct PFC at 5 days	Type I: 10[b]
	Direct PFC at 4 days	Type I: 100[c]
Mouse spleen cells cultured with bacterial lipopolysaccharide	Direct PFC at 5 days	Type I: 100[d]
	Direct PFC at 3 days	Type I: 800[c]
Mouse spleen cells cultured with SRBC; IF added 24 hr before antigen	Direct PFC at 4 days	{Type I: 2000[e] Type II: 2.5[e]

[a] R. H. Gisler, P. Lindahl, and I. Gresser, J. Immunol. 113, 438 (1974).

[b] H. M. Johnson, B. G. Smith, and S. Baron, J. Immunol. 114, 403 (1975).

[c] R. J. Booth, J. M. Rastrick, A. R. Bellamy, and J. Marbrook, Aust. J. Exp. Biol. Med. Sci. 54, 11 (1976).

[d] H. M. Johnson, J. A. Bukovic, and S. Baron, Cell. Immunol. 20, 104 (1975).

[e] G. Sonnenfeld, A. D. Mandel, and T. C. Merigan, Cell. Immunol. 34, 193 (1977).

occur in similar work by different investigators. Possible reasons for these discrepancies are considered below.

The Target for IF-Induced Antibody Suppression

Whereas assays of the antiviral activity of IF rely on its effect on a homogeneous, often clonally derived, population of cells in tissue culture, no such simple system is available for studying the immunosuppressive activity. The generation of an antibody response is a complex process involving interaction of antigen with a network of cells of different lineages. Antigen-sensitive B lymphocytes respond to appropriate stimulation by proliferating and differentiating into specific AFC. (AFC can be measured by the hemolytic plaque assay, where erythrocyte antigens or antigens that can be chemically coupled to erythrocytes have been used to induce the response.) In addition, the activation process requires the participation of cells of other lineages, such as macrophages and subpopulations of T lymphocytes. Furthermore, there is heterogeneity within the population of B cells such that the antigen-sensitive cells may be present as a number of discrete maturation stages of the B cell lineage. Finally, the activation of a population of B cells is not a synchronized process but involves continual recruitment from the pool of antigen-sensitive precursors over a period of time during the early stages after antigenic stimulation.

In studying the effect of IF (or any other soluble mediator) on such a complex, multicomponent system, two important points emerge: (a) that there are a number of possible targets on which IF might act; and (b) that its effect will depend not only on the time at which it is added, but also on the time at which the response is measured.

Cell Lineage Affected by IF. Because most antibody responses require a cooperative interaction of B and T lymphocytes and macrophages, IF could suppress the generation of such responses by affecting any of these populations. By assessing the response to a "thymus-independent antigen" that activates B lymphocytes directly without cooperation with other lymphocytes[2] [e.g., bacterial lipopolysaccharide (LPS)], IF-induced suppression can still be observed. This indicates that IF either acts directly on B cells or induces within the cell population the production of a second factor, which in turn suppresses B cell responsiveness. An alternative approach to this question is to separate populations of B cells, T cells, and macrophages, treat each with IF, and then mix them in culture with antigen. By using this technique Gisler *et al.*[3] were able to show that only when B cells were pretreated with IF was there any suppression of the subsequent antibody response. Although this result strongly suggests that IF acts directly on B lymphocytes, it does not exclude the additional possibility that IF induces the production of a suppressor cell within the population.[4]

Kinetics of Antibody Suppression. As mentioned above, it is possible to pretreat lymphocytes for a few hours with interferon before antigenic challenge and still observe suppression. In fact, both *in vivo*[5-7] and *in vitro*[8,9] studies have demonstrated that IF is more effective in suppressing antibody responses if added at or before the time of antigenic stimulation rather than later, suggesting that it affects some early event(s) in the generation of the response. Paradoxically, however, time course studies[10,11] reveal that, in the presence of IF, the PFC response to SRBC appears to develop normally for the first 2–3 days and then to decline, suggesting either that IF causes premature abortion of the response in

[2] J. Andersson, O. Sjöberg, and G. Möller, *Eur. J. Immunol.* **2**, 349 (1972).

[3] R. H. Gisler, P. Lindahl, and I. Gresser, *J. Immunol.* **113**, 438 (1974).

[4] R. K. Gershon, *Transplant. Rev.* **26**, 170 (1975).

[5] T. J. Chester, K. Paucker, and T. C. Merigan, *Nature (London)* **246**, 92 (1975).

[6] B. R. Brodeur and T. C. Merigan, *J. Immunol.* **113**, 1319 (1974).

[7] B. R. Brodeur and T. C. Merigan, *J. Immunol.* **114**, 1323 (1975).

[8] J. R. Virelizier, E. L. Chan, and A. C. Allison, *Clin. Exp. Immunol.* **30**, 299 (1977).

[9] G. Sonnenfeld, A. D. Mandel, and T. C. Merigan, *Cell Immunol.* **34**, 193 (1977).

[10] H. M. Johnson, B. G. Smith, and S. Baron, *J. Immunol.* **114**, 403 (1975).

[11] R. J. Booth, J. M. Rastrick, A. R. Bellamy, and J. Marbrook, *Aust. J. Exp. Biol. Med. Sci.* **54**, 11 (1976).

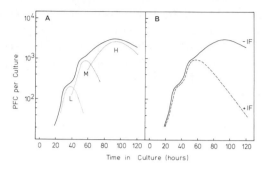

FIG. 1. Time course of the antibody response of mouse spleen cells to sheep red blood cells. (A) The normal response (——) is shown to be a composite of the individual responses of discrete subpopulations of B cell precursors (· · · · ·), which segregate with low (L), medium (M), and high (H) buoyant densities.   (B) A comparison of the response in the presence (-----) and in the absence (——) of interferon. The interferon-induced suppression observed late in the course of the response is due to a selective effect on high density precursors only.

some way, or that it affects only the antigen-sensitive cells that give rise to PFC late in the course of the response.

The population of B cells that respond to SRBC is known to be heterogeneous, so that the precursors that contribute to the PFC observed early in the response are distinguishable from those that generate PFC observed later. These discrete subpopulations can be separated by buoyant density fractionation techniques described elsewhere.[12,13] Using these subpopulations, IF can be shown to suppress only the high-density precursors that give rise to PFC late in the response, whereas the low-density precursors, which are predominantly responsible for the early part of the response, are unaffected by IF.[1] These results are shown schematically in Fig. 1 and explain the apparently paradoxical kinetics of antibody suppression by IF. They also highlight the need for caution when interpreting results from such a complex system, and indicate that the time of assay may be a critical factor that possibly contributes toward the variability observed in assays of immunosuppression in different laboratories (Table II).

Clonal Analysis of IF-Induced Antibody Suppression

The production of AFC involves a combination of two processes: the activation of antigen-specific precursor B cells, and the clonal expansion of each activated precursor. The AFC measured at any time during the response therefore comprise the progeny of a restricted number of dis-

[12] K. Shortman, *Aust. J. Exp. Biol. Med. Sci.* **46**, 375 (1968).
[13] J. S. Haskill and J. Marbrook, *J. Immunol. Methods* **1**, 43 (1971).

crete clones of cells. In its suppressive interaction with B lymphocytes, IF could affect either or both of the above processes, resulting in a reduction in the number of proliferating clones and/or the size of the clones. To distinguish between these two possible sites of IF action requires a culture system in which individual precursors are able to become activated, and develop into clones of homogeneous AFC that are segregated from each other.

Such culture systems have been developed and fully described elsewhere.[14,15] We have employed the polyacrylamide system of Marbrook and Haskill[14] (Table III) to study the clonal aspects of IF-induced suppression of the antibody response against SRBC.[16] In this system, splenic lymphocytes are seeded into the culture vessel at a concentration such that, when the cells settle into the bottom, not every well contains a SRBC-specific B cell. When the contents of each well is assayed for PFC after 3–4 days, there will be a very high probability that the PFC in any one well are clonally derived (i.e., they are the progeny of a single SRBC-specific precursor B cell).

By studying the nature of the clones that develop in the presence of IF, it is possible to determine whether IF suppresses B cells by interfering with the specific activation of precursor B cells or by inhibiting the subsequent clonal expansion of activated precursors. However, in conducting such investigations it is necessary to ensure that the number of clones that develop is proportional to the number of spleen cells (B cells) in the culture and that the average number of AFC per clone remains constant as the number of clones is decreased. This avoids the possible artifactual situation where any treatment, such as IF, that reduces an immune response will also cause a diminished cloning efficiency at lower levels of *in vitro* responsiveness.

In general, T cell-dependent responses are markedly nonlinear with respect to cell density, and consequently the number of clones is not proportional to the number of spleen cells cultured. This deficiency in the assay system can be overcome however by using a thymus-independent antigen or by incorporating the B cell mitogen LPS into the cultures.[16] In addition, it is often necessary to include a filler cell population (e.g., syngeneic thymocytes or irradiated spleen cells) to ensure that the cloning efficiency remains constant irrespective of the number of clones per culture. When these conditions are satisfied it can be demonstrated that IF-induced antibody suppression, at least at the moderate IF concentrations used, is apparently an all-or-none phenomenon resulting in a reduction in the number, but not in the average size, of the clones.[16]

[14] J. Marbrook and J. S. Haskill, *Cell. Immunol.* **13**, 12 (1974).
[15] I. Lefkovits, *Eur. J. Immunol.* **2**, 360 (1972).
[16] R. J. Booth, J. M. Booth, and J. Marbrook, *Eur. J. Immunol.* **6**, 769 (1976).

TABLE III
CLONAL NATURE OF THE ANTIBODY RESPONSE[a]

Polyacrylamide culture containing spleen cells, filler cells (thymocytes), antigen (SRBC), and LPS

Contents of each well assayed for SRBC-specific PFC

Results of a Typical Assay

24 Positive wells	0	3	0	0	0	0	8	0
40 Negative wells	2	0	0	2	0	2	0	0
Number of clones (c) corrected for coincidence is given by $n = t(1 - 1/t)^a$, where n = number of negative wells, t = total number of wells	0	0	6	0	0	0	1	3
	8	0	0	9	0	4	0	0
	0	0	2	0	10	0	0	4
	3	4	0	0	0	9	0	0
	0	0	3	0	2	7	0	0
	5	0	12	1	0	0	2	0

	Control culture without interferon (example above)	Culture containing interferon
Number of clones (c)[b]	30	10
Total number of PFC	112	36
Mean clone size (PFC/clone)	3.7	3.6

[a] Response as revealed by the polyacrylamide culture system of Marbrook and Haskill.[14] Because of the possibility of coincidence of two or more clones within a single positive well, the number of clones per culture is not directly equal to the number of positive wells but may be calculated using the given formula. The mean clone size can then be found by dividing the total number of plaque-forming cells (PFC) by the number of clones in that culture. The examples shown illustrate the differences that may be observed in cultures with or without interferon.

[b] In the presence of interferon the number of clones is reduced, but the mean clone size does not change significantly.

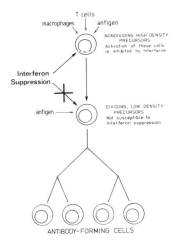

FIG. 2. The mechanism of interferon-induced antibody suppression. The population of antigen-sensitive B cells is composed of various subpopulations that represent discrete differentiation states within the B cell lineage. Available evidence suggests that only the activation of nondividing, high-density precursors can be inhibited by interferon.

Thus, IF appears to prevent the activation of antigen-sensitive B cells, perhaps by blocking them in a resting phase as has been recently suggested.[17] Those cells that already are in cell cycle at the time of IF treatment (i.e., the low-density precursors that contribute mainly to the early part of the response) are unaffected and respond normally (Fig. 2).

Concluding Remarks

By employing various lymphocyte fractionation techniques and clonal analyses, it is possible to gain a clearer understanding of the mechanics of IF-induced antibody suppression (Fig. 2). Although it is possible to assay immunosuppression by simply adding a putative IF sample to a population of responding spleen cells, the results outlined above stress the importance of timing in adding the IF and assaying AFC. Spleen cells from aging mice with, for example, a cryptic infection may have AFC precursors that are predominantly in a cycling state. Such cells would be mostly unresponsive to IF suppression and yield a poor assay system.

The results obtained to date are consistent with the view of IF as a molecule intimately involved in immunoregulation and justify its inclusion in the group of immunological mediators known as lymphokines. Moreover, the mechanism by which IF inhibits the antibody response has

[17] D. Monard, *Nature* (*London*) **282**, 365 (1979).

suggested to us how this immunosuppressive phenomenon might be an advantage to an animal undergoing a viral infection by "conserving" the humoral immune system at a time when an active antibody response would be harmful to the host.[16]

Much of the investigation of IF-induced immunosuppression has been derived from work with mouse interferon and murine *in vitro* responses. There is no reason why similar approaches to those described above could not be adopted to examine the parameters relating to IF-induced antibody suppression in other species (particularly man) provided that adequate, well-defined culture systems were available.

[61] Measurement of Interferon Modulation of the *in Vitro* Antibody Response

By Howard M. Johnson

Preliminary Considerations

The *in vitro* system for antibody production in the mouse consists of the use of dissociated spleen cells, which are made up of B and T lymphocytes and macrophages, all of which are normally required for an antibody response.[1] Addition of a suitable antigen, such as sheep red blood cells (SRBC), to the spleen cells in culture media under defined conditions stimulates the production of antibody-forming cells. The antibody-forming cells are enumerated by mixing them with a high concentration of SRBC (10%) in a soft agar medium at 37° for 1 or 2 hr. During this incubation, the antibody that is produced by the cells binds to surrounding SRBC. Addition of guinea pig serum, which contains complement, to this complex results in lysis of the SRBC. Thus, a plaque or zone of hemolysis surrounds a cell that is producing antibody to SRBC. The antibody-forming cell is frequently referred to as a plaque-forming cell (PFC), and the antibody response to SRBC is spoken of as the anti-SRBC PFC response. This system has been of considerable value in looking at a variety of events related to regulation of the antibody response by interferon. Advantages over *in vivo* experiments are economy of reagents and better control of the microenvironment of the antibody-forming cell.

[1] E. S. Golub, "The Cellular Basis of the Immune Response," 2nd ed. Sinauer Assoc., Sunderland, Massachusetts, 1981.

The *in Vitro* System

Several factors are of considerable importance in establishing a successful *in vitro* antibody system. These include the strain of mice, the lot of fetal calf serum, and the source of the SRBC antigen. C57BL/6 mice have been found to be the preferred strain for consistently high or moderate PFC responses. Suitable fetal calf serum for high PFC responses has to be determined empirically. It may require the testing of 5 to 10 lots before a suitable lot is obtained. High-responding SRBC have to be identified in a similar fashion. One way of circumventing the strict requirements for reagents is to use 2-mercaptoethanol in the PFC system. Concentrations of 1×10^{-6} to 5×10^{-5} M 2-mercaptoethanol will enhance the PFC response. 2-Mercaptoethanol, however, at concentrations above $1 \times 10^{-6} M$ will block or reverse the suppressive effects of interferon,[2] so it should not be used at concentrations above $1 \times 10^{-6} M$.

Procedures

Dissociated normal mouse spleen cells are cultured for *in vitro* anti-SRBC PFC responses by modification of a previously described procedure.[3] Mice are killed by cervical dislocation, and the spleens (4–6 spleens) are removed and placed in a 60-mm culture dish containing 5–10 ml of modified Eagle's minimal essential medium (MEM).[3] Small forceps are used to tease the cells gently from the spleen. The suspended cells are transferred to a 50-ml conical tube in an ice bath and left for 3–5 min to allow particles to settle. The suspension of dissociated cells is transferred to another conical tube and centrifuged at 200 g in an RC-3 refrigerated centrifuge (4°). Normally, the packed cells are suspended to a concentration of 1.5×10^7/ml in a modified culture media[3] or in RPMI-1640; 1-ml volumes are added to 35-mm plastic petri dishes. The antigen, a 1% suspension of SRBC, is added at 2 drops/plate with a Pasteur pipette. 2-Mercaptoethanol may be added at a concentration of $1 \times 10^{-6} M$. The cultures are incubated at 37° on a rocker platform (Bellco) at 8–10 cycles/minute in an airtight humidified box in an atmosphere of 7% O_2, 10% CO_2, and 83% N_2. Cells may also be incubated in 0.25-ml volumes in multiwell plates (Falcon 3008) with concentrations remaining the same as for 1-ml volumes. Incubations are normally for 5 days with daily feeding with two drops of a nutritional mixture[3] and one drop of fetal calf serum/culture dish.

[2] H. M. Johnson, *Cell. Immunol.* **36**, 220 (1978).
[3] R. I. Mishell and R. W. Dutton, *J. Exp. Med.* **126**, 423 (1967).

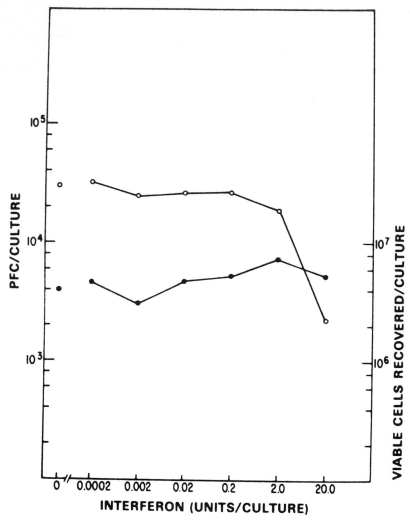

FIG. 1. The effect of highly purified mouse ascites tumor virus-type interferon on the primary *in vitro* plaque-forming cell (PFC) response. Direct sheep red blood cell PFC per culture (O——O) and viable cells recovered per culture (●——●) were determined on day 5.

Enumeration of PFC

To enumerate the PFC in the *in vitro* response, scrape the culture dishes with a plastic Teflon policeman. Transfer the cells to a 17 by 100 mm plastic tube and centrifuge for 10 min at 200 g (7°). Suspend the cells to the original volume in modified MEM. Add 20–100 μl of the cell suspension to plastic tubes in a 45° waterbath containing 0.5 ml of melted agarose (0.5% in modified MEM) and 0.1 ml of a 10% suspension of washed SRBC. Mix well and pour the contents onto microscope slides that had previously been coated with 0.5% agarose. Place slides in a humidified chamber and incubate at 37° for 1 hr. Reconstitute lyophilized guinea pig serum (complement) and dilute to a 1 : 30 dilution in modified MEM. Place the slides with the agar sides face down in the complement on Jerne racks. Incubate in a humidified culture box for an additional 2–3 hr at 37°, then remove the slides; the holes or plaques are counted to indicate the direct (IgM antibody) PFC response.

Interferon Suppressions

Mouse virus-induced interferon is normally added to the plates at the initiation of culture.[4] Results of a typical experiment are presented in Fig. 1. Suppression of the PFC response was approximately 90%. Interferon is responsible for the suppression, since it is observed with interferon purified to homogeneity and is specifically blocked by pretreatment of the interferon with specific antibody. In general, 20–60 units of interferon per milliliter will suppress the *in vitro* PFC response by 90% or more. This represents less than 0.05 ng of interferon per milliliter. Interferon similarly suppresses the antibody response in the mouse, but this requires relatively more interferon (50,000 units or more),[5] and it is impossible to control the concentrations of interferon in the microenvironment of the antibody-producing cell.

Variations of the *in Vitro* System

Antigens other than SRBC can be used in the *in vitro* PFC system. SRBC are thymus-dependent antigens, requiring the presence of T cells for an antibody response.[1] Some antigens, such as lipopolysaccharides from gram-negative bacteria and dinitrophenyl coupled to the carbohydrate Ficoll,[6] do not require T cells for an antibody response. Interferon will also suppress these systems.[7] With variations and manipulations such as these, it is possible to determine the role of interferon in immunoregulation.

[4] H. M. Johnson, B. G. Smith, and S. Baron, *J. Immunol.* **114**, 403 (1975).
[5] T. J. Chester, K. Paucker, and T. C. Merigan, *Nature (London), (New Biol.)* **246**, 92 (1973).
[6] T. M. Chused, S. S. Kassan, and D. E. Mosier, *J. Immunol.* **116**, 1579 (1976).
[7] H. M. Johnson, J. A. Bukovic, and S. Baron, *Cell. Immunol.* **20**, 104 (1975).

[62] Assay of Effect of Interferon on *in Vitro* Primary Plaque-Forming Response to Sheep Red Blood Cells

By WOLF H. FRIDMAN and ERNESTO FALCOFF

Interferon-induced inhibition of *in vitro* antibody synthesis was its first described immunoregulatory effect.[1] This effect has relevance in studying the effect of interferon on the differentiation of specialized cells. In fact, antibody production to a majority of antigens (thymus-dependent antigens) is the consequence of a series of interactions involving T (thymus-derived) lymphocytes, macrophages, and B (bone-marrow derived) lymphocytes, the latter being the antibody-synthesizing cells.[2] T cells, and to a lesser degree macrophages, regulate antibody synthesis by B cells either in a positive way (T helper cells) or in a negative way (T suppressor cells). Inhibition of antibody production by any substance may therefore result in an inactivation of the functions of T helper on B lymphocytes, or in an enhancement of suppressive functions. The establishment of *in vitro* culture systems in which a primary antibody response can be induced in murine splenocytes rendered possible the study of the different cellular compartments involved in the reaction and the effect of a given molecule on each compartment. The original procedure of Mishell and Dutton[3] has been simplified[4] and used to show that early addition of virus-induced[1] and immune[5] interferons inhibit antibody synthesis *in vitro* by acting directly on the precursors of the antibody-producing cells. Under certain conditions, late addition of viral interferon may enhance the response.[1]

Mice

Nearly all conventional strains of mice can be used as donors of spleen cells. The most commonly used strains are C3H, C57BL/6, BALB/c, DBA/2, and even more often a hybrid strain between C57BL/6 and DBA/2 called B6D2F1 or BDF1. To assess the effect of interferon on B cells, congenically athymic nude (*nu/nu*) mice are used. These mice may be outbred or bred on a defined background. To elicit good antibody re-

[1] R. H. Gisler, P. Lindahl, and I. Gresser, *J. Immunol.* **115**, 438 (1974).
[2] D. H. Katz, "Lymphocyte Differentiation, Recognition and Regulation." Academic Press, New York, 1977.
[3] R. W. Mishell and R. W. Dutton, *J. Exp. Med.* **126**, 443 (1967).
[4] R. H. Gisler and W. H. Fridman, *J. Exp. Med.* **142**, 507 (1975).
[5] M. A. Lucero, J. Wietzerbin, S. Stefanos, C. Billardon, E. Falcoff, and W. H. Fridman, *Cell. Immunol.*, in press.

sponse, mice should be male animals, pathogen free, and kept in good animal house conditions.

Antigens

Various antigens can be used to induce specific primary *in vitro* antibody production. They are of three classes: (*a*) particular antigens, such as sheep red blood cells (SRBC) (the most commonly used), burro red blood cells, etc.; (*b*) carrier–hapten complexes, in which a small molecule (hapten) is coupled to a carrier protein. Under these circumstances the antibodies produced are directed against the hapten. In this class of antigens, the most commonly used are dinitrophenyl-keyhole limpet hemocyanin (DNP-KLH) or dinitrophenyl-human γ-globulin (DNP-HGG); and (*c*) antigens, usually haptens, coupled to a B cell mitogen. These antigens directly stimulate B cells and the response does not require T cell help. DNP-aminoethyldextran (DNP-AE-dextran) or TNP-lipopolysaccharide (TNP-LPS) are such antigens.

The technique described herein describes a typical stimulation of murine spleen cells by SRBC. A batch of SRBC should be selected for its good stimulating capacity and the same sheep used throughout the study.

Media, Buffers, Antibiotics, and Sera

The cultures are performed in RPMI-1640 (Gibco). Spleen cell preparation and the plaque-forming cell (PFC) assay are performed in balanced salt solution (BSS) prepared from two stock 10 × solutions (see accompanying tabulation).

Solution I	Solution II
Glucose, 10 g	$CaCl_2$, 2 H_2O, 1.86 g
KH₂PO4, 0.6 g	KCl, 4 g
Na_2HPO_4, 12 H_2O, 5.23 g	NaCl, 80 g
Phenol red, 0.1 g	$MgCl_2$, 6 H_2O, 2 g
Complete to 1 liter with distilled water	Complete to 1 liter with distilled water

The BSS buffer is prepared by adding 10 ml of solution I plus 10 ml of solution II to 80 ml of distilled water. Sodium bicarbonate buffer (BIC 3000, Eurobio) is added to the cultures.

Fetal calf serum (FCS) is an important agent for the induction of *in vitro* primary antibody synthesis. Some FCS exert inhibitory behavior; others induce nonspecific polyclonal B cell activation. Different batches of

FCS should therefore be tested, and a batch should be chosen that sustains a good response without inducing antibody production in unstimulated cultures. Once this batch is selected, it should be bought and used as long as possible to ensure the reproducibility of the cultures.

Horse serum is sometimes used; it ameliorates the reproducibility of the cultures.

Culture Conditions for in Vitro Induction of a Primary Antibody Response to SRBC

Animals are sacrificed in the animal house and brought dead to the laboratory. This is an important point, since stress can profoundly depress the antibody response. For the same reason, if the animals travel by train, car, or plane from the animal house to the laboratory, they should be kept at least 48 hr before sacrifice.

Spleens are taken and teased in BSS with two curved needles to liberate the spleen cells, which are then resuspended with a Pasteur pipette. The cell suspension is left for 3 min in a conical tube to allow debris to sediment. The supernatant containing the spleen cells is then washed once in BSS, and the pellet is resuspended in culture medium containing RPMI-1640, 10% FCS, 1% horse serum, 3% bicarbonate, 200 mM glutamine (Eurobio), and antibiotics (penicillin and streptomycin, 5000 units of each). All these procedures are done in the cold.

The cells are counted in a Mallasez chamber, adjusted to 8×10^6 cells/ml in the culture medium, and incubated in 12×75 mm Falcon plastic tubes (Falcon, Oxnard, California) in a final volume of 1 ml. One drop of SRBC (1% solution in BSS) from a Pasteur pipette is added to each culture, except to the unstimulated controls; the tubes are incubated for 5 days at 37° in an incubator containing a CO_2 atmosphere (5% CO_2 in air). Various products to be tested can be added to the culture, but the final volume should always be about 1 ml. Each test combination should be done at least in duplicate. The duration of the procedure, from sacrifice of the mice to the start of the culture, should not exceed 60 min.

Induction of a Primary Immune Response to SRBC in T-Deprived Cells

The culture procedure is basically similar to the procedure described above except that spleen cells from nude mice are used as responder cells and T cells are replaced by a culture supernatant (10–50 μl) of spleen cells stimulated by concanavalin A (Con A) added 24 hr after antigen.[6] The Con A supernatant is prepared by incubating 10^7 spleen cells from a conventional mouse with 2 μg of Con A for 24 hr at 37° in RPMI-1640 medium.

[6] M. Yagello, C. Rabourdin-Combe, and W. H. Fridman, Transplant. Proc. 11, 895 (1979).

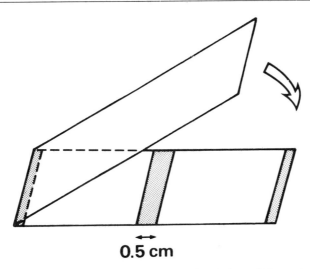

0.5 cm

FIG. 1. Procedure used to prepare Cunningham slides.

Plaque-Forming Cell (PFC) Assay

The most sensitive assay for measurement of antibody production induced in a primary culture is the local hemolysis technique in liquid medium described by Cunningham.[7] The principle of this technique is to enumerate the number of antibody-producing cells by counting the number of plaques produced by these cells. Plaques are due to lysis of SRBC by anti-SRBC antibodies in the presence of complement. The technique is as follows. After 5 days of incubation at 37°, which corresponds to the peak response, the tubes are removed from the incubator. At this time the cells are in the pellet. The supernatants are removed and are replaced by BSS; the cells are washed once in BSS in the cold. They are then resuspended in BSS, in a volume usually of 2–4 ml per tube chosen to have about 50 plaques per counting chamber. The following mixtures are then prepared in U25 microtiter plates (Orcineir): 100 μl of cell suspension, 25 μl of guinea pig serum (as source of complement) previously adsorbed on SRBC to remove heterologous antibodies, and 25 μl of SRBC (20%) in BSS. Of this suspension, 75 μl are then poured into the counting chamber, which has been prepared by sealing two slides with double-faced Scotch tape (Fig. 1).

The chambers are sealed with melted paraffin and horizontally incubated for 1 hr at 37°. The plaques are then counted with a magnifying lens under dark field illumination. Each positive plaque contains a single cell in its center; this is the antibody-producing cell and is called a plaque-

[7] A. J. Cunningham, *Nature (London)* **207**, 1106 (1969).

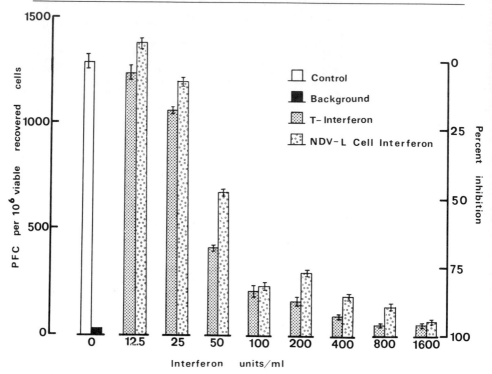

FIG. 2. Dose effect of viral and T interferon on the number of anti-sheep red blood cell (SRBC) plaque-forming cells. Interferons were added simultaneously with SRBC. From Lucero *et al.*[5]

forming cell (PFC). Each test should be made at least in duplicate. The number per PFC per chamber is counted, and the results are expressed as PFC/10^6 viable recovered cells (as counted on the lymphocytes rewashed in the culture tubes) according to the formula:

$$\text{PFC}/10^6 \text{ cells} = \frac{\text{PFC/per chamber} \times 150}{75} \times \frac{10}{\text{number of } 10^6 \text{ cells/ml}}$$

For example, if 50 PFC per chamber are found in a tube containing 10^6 cells/ml,

$$\text{PFC}/10^6 \text{ cells} = \frac{50 \times 150}{75} \times \frac{10}{1} = 1000 \text{ PFC}/10^6 \text{ cells}$$

The factor 10 comes from the fact that only 0.1 ml of cells is placed in the reaction and that the calculation determines the number of cells per milliliter.

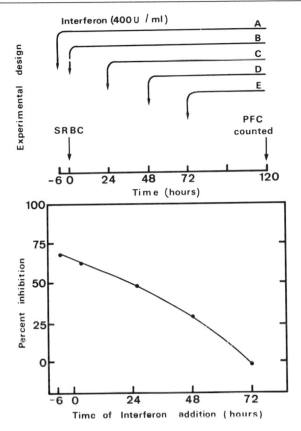

FIG. 3. Kinetics of the interferon effect on anti sheep red blood cell (SRBC) plaque-forming cells (PFC). Control: 720 ± 41 PFC/10⁶ viable cells; background: 20 ± 3 PFC/10⁶ viable cells.

Interferon-Induced Suppression of *in Vitro* Primary Antibody Production

Both viral (NDV-L cell) and T interferons (see other chapters) inhibit equally well PFC formation in spleen cell cultures stimulated with SRBC. In the experiment shown in Fig. 2,[5] 8×10^6 spleen cells from B6D2F1 mice were stimulated with SRBC, and graded doses of viral or T interferon were added at the same time as antigen. Both interferon preparations exerted similar suppressive activity. The effect was significant (over 50% inhibition) when 50 units of interferon were used.

Figure 3 shows a kinetics curve obtained with 400 units of T interferon. Maximum suppression was observed when interferon was added prior to or together with antigen; the effect is still significant if the addition of

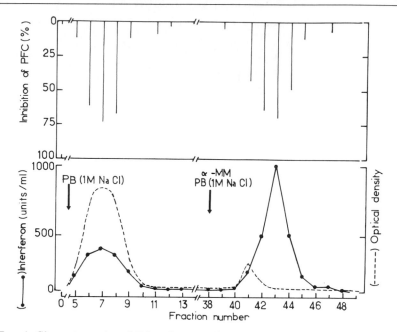

FIG. 4. Chromatography of T interferon on Con A–Sepharose. Effect of fractions on the number of anti-sheep red blood cell (SRBC) plaque-forming cells. Six milliliters of phytohemagglutinin-induced interferon containing 600 units of interferon activity and 5 mg of protein per milliliter was dialyzed against 0.002 M phosphate buffer, pH 7.4, containing 1 M NaCl and applied onto a Con A–Sepharose column (0.9 × 5 cm). The column was washed with the same buffer and eluted with 0.1 M methyl α-mannoside the same buffer. The volume of each fraction was 1 ml. Optical density was followed by UV absorption (LKB Uvicord 4701).

interferon is delayed for 24 hr and becomes insignificant if interferon is added to the cultures 48–72 hr after antigen.[5] The kinetics are identical if viral interferon is used.[1] The best way to obtain a good inhibition of PFC formation with interferon is therefore to add over 100 units to stimulated cultures prior to, or together with, the stimulating antigen.

Finally, Fig. 4 shows that interferon is indeed the immunosuppressive molecule. As an example, in this experiment T interferon has been chromatographed on a Con A–Sepharose column (see Volume 78 [78]) and each fraction has been tested for antiviral and antibody suppressing activity. A perfect correlation is found between both activities. Only fractions containing antiviral activity (interferon) suppress the *in vitro* primary antibody response to SRBC.

In conclusion, the measure of the interferon-induced suppression of the *in vitro* PFC response can easily been done in the highly reproducible culture system described here if interferon is added at the beginning of the culture. It is useful to use partially purified interferon, since, especially when immune interferons are tested, cell culture supernatants may contain factors influencing the *in vitro* antibody response, such as amplifying the helper factors.[8] The reproducibility of the technique is also dependent on the use of selected batches of FCS and SRBC.

[8] J. J. Farrar, P. L. Simon, W. J. Koopman, and J. Fuller-Bonar, *J. Immunol.* **121**, 1353 (1978).

[63] Measurement of Effects of Interferon on Hemopoietic Colony-Forming Cells: Effects of Mouse Interferon

By T. A. McNeill, W. A. Fleming,* and M. Killen

The hemopoietic system, which consists of pluripotent stem cells, progenitor cells restricted to particular lines of differentiation, and various maturing cell populations, has been the subject of intensive study since the development of culture systems in which precursor cells can generate colonies of differentiating progeny. Colonies of all hemopoietic classes can now be grown *in vitro*, and the cells from which these colonies arise are described as colony-forming cells (CFC) and designated with an appropriate prefix, e.g., GM-CFC referring to the progenitor cell of the granulocyte-macrophage series. Work referred to in this paper is concerned solely with murine GM-CFC.

Colony growth from murine GM-CFC is dependent upon the addition of a glycoprotein colony-stimulating factor (GM-CSF). Commonly used sources of GM-CSF are media conditioned by tissue culture cells, e.g., embryo fibroblasts[1] or L cells[2]; media conditioned by tissue fragments, e.g., lung[3]; and mouse serum, particularly after inoculation of bacterial endotoxin.[4]

The GM-CFC culture method provides an experimental system that allows (*a*) enumeration of GM-CFC in different organs; (*b*) assay of GM-CSF levels; (*c*) assay of colony-inhibiting factors; (*d*) studies on multiplication rates and patterns of differentiation of GM cells in culture; and (*e*) studies on the responsiveness of GM-CFC to GM-CSF after exposure to various agents *in vivo* and *in vitro*.

* Deceased.

[1] T. R. Bradley and M. Sumner, *Aust. J. Exp. Biol. Med. Sci.* **46**, 607 (1968).
[2] P. E. Austin, E. A. McCulloch, and J. E. Till, *J. Cell. Physiol.* **77**, 121 (1971).
[3] J. W. Sheridan and D. Metcalf, *J. Cell. Physiol.* **81**, 11 (1973).
[4] D. Metcalf, *Immunology* **21**, 427 (1971).

The Culture System

The agar culture system exists as numerous variations on a basic theme. Essentially, double-strength culture medium is mixed with double-strength agar solution in distilled water, the cells are added, and aliquots are pipetted into small plastic petri dishes to which a small volume of GM-CSF has already been added. The mixture gels by cooling, and the cultures are incubated for 7 days in a fully humidified atmosphere of 10% CO_2 in air.

A detailed account of the system and the technical problems involved has been given by Metcalf.[5] It seems likely that complex interrelationships exist between different constituents of the medium, and it is therefore impossible to be precise about medium requirements for optimum colony growth. In our experience the simple Earle's balanced salt solution gave results as good as the more complex Eagle's medium. As far as supplements are concerned, fetal calf serum and sodium pyruvate were found to be essential whereas other commonly used supplements, such as trypticase soy broth, horse serum, L-asparagine, L-serine, and DEAE-dextran, usually made little difference in the number or size of colonies.[6] Fetal calf serum is the most critical supplement, and new batches must be pretested against a known "good" batch in the agar culture system itself. A batch of calf serum found to be suitable in some other culture system is not necessarily suitable for culture of GM-CFC.

Quantitative Aspects and Assay of Inhibitors

Considerable heterogeneity exists within the GM-CFC population in normal mouse bone marrow. One aspect of this is seen in the responsiveness of the population to stimulation by GM-CSF where colony number is related, in sigmoidal fashion, to the concentration of stimulating factor up to a maximum number where all GM-CFC are stimulated. This, combined with the fact that murine colonies will not grow in the absence of added GM-CSF, has important implications for the assay of an inhibitor such as interferon.

1. Inhibitory effects have to be assayed against a background of stimulation. With interferon, the degree of inhibition by a given concentration is inversely related to the concentration of GM-CSF in the culture; indeed, whenever high GM-CSF concentrations are used, the inhibitory effect of interferon may be masked.[7]

---

[5] D. Metcalf, "Haemopoietic Colonies." Springer-Verlag, Berlin and New York, 1977.
[6] M. Killen, unpublished observations.
[7] T. A. McNeill and I. Gresser, *Nature* (*London*), (*New Biol.*) **244**, 173 (1973).

2. When inhibitory effects are measured against a low level of background stimulation, it is only those GM-CFC that are most sensitive to GM-CSF against which it is being tested.

Effects of Interferon

The inhibitory effects of interferon on mouse GM-CFC were demonstrated by the following series of observations.

1. When mice were given intraperitoneal inoculation of polyribo-inosinic-polyribocytidylic acid, there was a close temporal relationship between the interferon response and a loss of colony-stimulating activity in the serum.[8]
2. Fractionation of such sera on Sephadex gave peaks that contained both antiviral and colony-inhibiting activity. These activities were closely associated in terms of molecular size, sensitivity to pH and heat, and in species specificity for target cells.[9,10]
3. Highly purified interferon showed strong colony-inhibitory activity.[7]

Complicating Factors: Effects of Interferon Inducers and Existence of Other Inhibitors

Interferon Inducers. Interpretation of the effects of interferon on GM-CFC can be complicated by effects of some of the materials used to induce interferon. *In vivo* inoculation of 5 μg of bacterial lipopolysaccharide causes a marked rise in serum GM-CSF, which can reach 100 times control levels within 2–3 hr.[4] Inoculation of poly(I) · poly(C) can cause alterations in GM-CSF levels as can several viruses, such as influenza and Newcastle disease virus.[8] These agents may also stimulate increases in GM-CFC levels in hemopoietic tissues. *In vitro* release of GM-CSF has also been associated with exposure of various cell cultures to interferon inducers. Lipopolysaccharide and poly(I) · poly(C) have been shown to stimulate such release in cultures of blood leukocytes, bone marrow cells, peritoneal and spleen cells,[11,12] and poly(I) · poly(C) and NDV increase the amount of GM-CSF in the medium of L cell cultures.[13] It is well known that lymphoid cells undergoing responses to

[8] T. A. McNeill and M. Killen, *Immunology* **21**, 751 (1971).
[9] T. A. McNeill and W. A. Fleming, *Immunology* **21**, 761 (1971).
[10] T. A. McNeill, W. A. Fleming, and D. J. McCance, *Immunology* **22**, 711 (1972).
[11] A. C. Eaves and W. R. Bruce, *Cell Tissue Kinet.* **7**, 19 (1974).
[12] E. W. Ruscetti and P. A. Chervenick, *J. Lab. Clin. Med.* **83**, 64 (1974).
[13] M. Havredaki and T. A. McNeill, *J. Gen. Virol.* **27**, 107 (1975).

antigens or mitogens may release many factors, including interferon and GM-CSF.

Other Colony Inhibitors. It is obvious that, in experiments undertaken to investigate the effects of a colony inhibitor such as interferon, account must be taken of the possibility that other factors present in the preparation may have colony-inhibiting effects. It is therefore relevant in the present context to refer briefly to information concerning other known inhibitors for the GM-CFC system. Two types of inhibitor have been described. One type is thought to affect the GM-CFC or their immediate progeny and the other type is thought to exert its effect by modulating the production of GM-CSF by macrophages. Interferon falls into the first category. Whether or not it has any effect on GM-CSF production has not been determined.

Inhibitory serum "lipoproteins"[14] have been described in human and mouse serum. Granulocyte chalones have been proposed as tissue-specific species-nonspecific inhibitors of granulopoiesis[15] although despite a considerable amount of work, their existence still remains uncertain.[16] Prostaglandins of the E series have been shown to inhibit growth of GM-CFC when added to agar cultures in extremely low concentrations.[17] Like interferon, their inhibitory effects can be counteracted by high concentration of GM-CSF.

[14] S. H. Chan, D. Metcalf, and E. R. Stanley, *Br. J. Haematol.* **20,** 329 (1971).
[15] T. Rytömaa, *Br. J. Haematol.* **24,** 141 (1973).
[16] S. P. Herman, D. W. Golde, and M. J. Cline, *Blood* **51,** 207 (1978).
[17] J. I. Kurland and M. A. S. Moore, *Exp. Hematol.* **5,** 375 (1977).

[64] Measurement of Effects of Human Leukocyte Interferon on Human Granulocyte Colony-Forming Cells *in Vitro*

By DHARMVIR S. VERMA, GARY SPITZER, and JORDAN U. GUTTERMAN

In 1966 Bradley and Metcalf[1] successfully adapted the *in vitro* semisolid agar culture system to clone the granulocyte-macrophage progenitor cell (GM-CFC), a cell that gives rise only to granulocytes and monocytes. Since then, this system has been advantageously used to explore various mechanisms involved in the proliferation and differentiation not only of GM-CFC, but also of other progenitor cells, such as BFU-E (burst-forming unit erythroid) and CFU-E (colony-forming unit

[1] T. R. Bradley and D. Metcalf, *Aust. J. Exp. Biol. Med. Sci.* **44,** 287 (1966).

erythroid) that give rise to mature erythrocytes and megakaryocyte progenitor cells (MEG-CFC), etc.[2,3]

Therapeutic use of various interferon preparations in viral infections and malignant disorders has generally been associated with neutropenia.[4-6] Although McNeill and Fleming[7] had found an association between a poly(I) · poly(C)-induced increase in serum interferon levels in mice and inhibition of marrow GM-CFC almost a decade ago, it is only recently that *in vitro* experiments in our laboratory have suggested a regulatory role for human leukocyte interferon (HLIF) in granulopoietic proliferation and differentiation. In this chapter we describe the use of *in vitro* semisolid agar culture system in delineating the HLIF effects on human granulopoiesis.

Preparation of Culture Medium. To prepare culture medium, 10.24 g of alpha-modified minimum essential medium (K. C. Biological, Inc., Lenexa, Kansas 66215) was dissolved in 375 ml of sterile distilled water (Abbott Laboratories, North Chicago, Illinois 60064). Subsequently, 5 ml of glutamine solution (3%; Difco Laboratories, Detroit, Michigan), 45 ml of sodium bicarbonate (5%; Abbott Laboratories, North Chicago, Illinois 60064), and 1 ml of penicillin–streptomycin solution containing 10,000 units/ml of each (Grand Island Biological Co., Grand Island, New York 14072) were mixed. Then, the pH of this mixture was adjusted to 7.2–7.4 and osmolality to 704 mM. Prior to use, the culture medium was enriched with 30% fetal calf serum (FCS; Gibco Laboratories, Grand Island, New York 14072) by adding 15 ml of FCS to 35 ml of α-MEM. To obtain single-strength culture medium, an equal volume of distilled water was added to the above mixture. Heretofore this was referred to as α-MEM plus 15% FCS.

Preparation of Ficoll–Hypaque Solution. To prepare the solution of density 1.077 g/ml, 100 g of Ficoll 400 (Pharmacia Fine Chemicals, Uppsala, Sweden) was dissolved in 1250 ml of distilled water. Subsequently, 310 ml of 50% Hypaque sodium (Winthrop Laboratories, Division of Sterling Drug Inc., New York 10016) was added to the Ficoll solution. This solution was then used for density gradient separation of marrow cells.

[2] J. R. Stephenson, A. A. Axelrad, D. L. McLeod, and M. M. Shreeve, *Proc. Natl. Acad. Sci. U.S.A.* **68**, 1542 (1971).
[3] D. Metcalf, H. R. McDonald, N. Odartchenko, and B. Sordat, *Proc. Natl. Acad. Sci. U.S.A.* **72**, 1744 (1975).
[4] H. B. Greenberg, R. B. Pollard, L. I. Lutwick, P. B. Gregory, W. S. Robinson, and T. C. Merigan, *N. Engl. J. Med.* **295**, 517 (1976).
[5] T. C. Merigan, K. H. Rand, R. B. Pollard, P. S. Abdallah, G. W. Jordan, and R. P. Fried, *N. Engl. J. Med.* **298**, 981 (1978).
[6] T. C. Merigan, *Tex. Rep. Biol. Med.* **34**, 541 (1977).
[7] T. A. McNeill and W. A. Fleming, *Immunology* **21**, 761 (1971).

Collection of Light-Density Human Marrow Cells. Bone marrow speci-mens obtained from normal human volunteers were collected in poly-styrene centrifuge tubes (Corning Glass Works, Corning, New York 14830) containing a solution of preservative-free heparin (1000 units/ml; Gibco Laboratories, Grand Island, New York 14072) in phosphate-buffered saline (PBS). To obtain light-density marrow cells, the specimen was cen-trifuged over a column of Ficoll–Hypaque solution (density 1.077 g/ml) at 200 g for 35 min. The interface cells were aspirated with a Pasteur pipette and washed twice with PBS and once again with single-strength α-MEM plus 15% FCS. Subsequently, the cells were resuspended in α-MEM plus 15% FCS at an approximate concentration of 1-1.5 × 10⁶ cells/ml. These cells were then used as target cells for GM-CFC assay.

Preparation of Human Placental Conditioned Medium (HPCM). To pre-pare HPCM, a source of colony-stimulating activity (CSA), a freshly de-livered human placenta was washed with sterile saline and then placed on a sterile stainless steel tray inside a laminar vertical-flow hood. The outer membranes were teased off using scissors and forceps. Pieces of 1 cm³ were cut free, placed in sterile saline in 100 × 15 mm petri dishes (Fisher Scientific Company, Pittsburgh, Pennsylvania 15219), and washed vigor-ously by agitation. To completely eliminate the blood, the procedure was repeated three times by transferring the tissue pieces sequentially to fresh saline-containing petri dishes. Subsequently six to eight pieces of the placental cubes were placed in 75 cm² plastic tissue culture flasks (Corning Glass Works, Corning, New York 14830), each containing 20 ml of RPMI-1640 culture medium (K. C. Biological Inc., Lenexa, Kansas 66215) with 5% FCS. The flasks were incubated in a horizontal position with their caps loose in a fully humidified atmosphere of 7% CO_2 in air for 7 days. The culture media were then filtered through a double layer of cotton gauze to remove tissue fragments and centrifuged at 2200 g for 20 min; the supernatant was filtered through 0.22 μm Millipore filters (Milli-pore Corporation, Bedford, Massachusetts 01730). These conditioned media referred to as HPCM were stored in aliquots of 50 ml at −80° until their use as a source of CSA.

Granulocyte-Macrophage Progenitor Cell Assay in Semisolid Agar. One-milliliter underlayers of a mixture of 0.5% Bactoagar (Difco Labora-tories, Detroit, Michigan) and α-MEM plus 15% FCS were prepared in 35-mm plastic petri dishes (Corning Glass Works, Corning, New York 14830) and allowed to dry. To stimulate GM-CFC proliferation and differ-entiation, HPCM was incorporated in the underlayers at an optimum con-centration of 20% (v/v). Subsequently, these underlayers were overlaid with a 1-ml mixture of 0.3% Bactoagar, α-MEM, and 15% FCS containing 1 or 2 × 10⁵ human marrow target cells. These cultures were then incubated at 37° in a fully humidified atmosphere of 5% CO_2 in air. After 8 days of incubation, the dishes were scored under an Olympus dissecting microscope ($\times 25$ magnification) for the number of colonies (aggregates of ≥ 40 cells) and clusters (aggregates of 3–39 cells) per dish. The final results were noted as the mean from triplicate dishes $\pm$ one standard deviation.

Morphological Examination of Cell Aggregates. Individual colonies or clusters were picked up with a Unopette pipette (Becton and Dickinson Co., Rutherford, New Jersey 07070) and placed on egg albumin-smeared microscope slides that were etched in small squares with a diamond pencil (one cell aggregate per square) as described by Verma *et al.*[8] After they were dried, the slides were stained with Luxol Fast Blue MBS (Hartman-Leddon Co., Philadelphia, Pennsylvania) for 2 hr, then rinsed in distilled water for 2 hr more and counterstained with Harris' hematoxylin for 2 min. The percentage differential counts were calculated from the cumulative data on differentials of 50 clusters from each pertinent observation.

Use of HLIF. The preparation of HLIF was obtained from Doctor Cantell. It was provided in 0.5-ml vials containing a total of 3×10^6 units of interferon and 4 mg of protein with no preservative chemicals. For use, a whole vial was thawed and stored at 4° as a stock solution in α-MEM at a concentration of 10^5 International Reference Units per milliliter for up to 2 weeks.

Effects of HLIF on GM-CFC Growth in Vitro. The various concentrations of HLIF were incorporated in the upper layers of the agar culture dishes, and the target marrow cells were stimulated by HPCM present in the underlayers. Table I presents the number of colonies and clusters noted in the presence of 0, 10, 100, 1000, and 10,000 units/ml. A progressive decrease in the size of cell aggregates was noted with increasing concentrations of HLIF in the culture dishes, leading to a decline in the colony incidence and corresponding increase in the cluster incidence. This resulted in nearly a constant plating efficiency, regardless of the HLIF concentrations used. The HLIF effect, though variable on different marrow target cells, remained consistent. This variability could be due to a variable sensitivity of marrow cells from different donors to HLIF.

Since the HLIF preparations used in these experiments did not stimulate any colony or cluster formation in the absence of HPCM, the increase in clusters noted with increasing HLIF concentrations suggested either an inhibition of proliferation at a latter stage of the granulopoietic maturational process or a prolongation of the generation time of each successive

[8] D. S. Verma, G. Spitzer, A. R. Zander, R. Fisher, K. B. McCredie, and K. A. Dicke, *Exp. Hematol.* **8,** 32 (1980).

TABLE I
EFFECT OF HUMAN LEUKOCYTE INTERFERON ON COLONY AND CLUSTER FORMATION
WHEN INCORPORATED IN AGAR OVERLAYERS
WITH HUMAN PLACENTAL CONDITIONAL MEDIUM[a,b]

Expt. No.	Cell aggregates scored	Plating efficiency without interferon (control)	Number of colonies and clusters as percentage of control at various concentrations of interferon per dish (units/ml)			
			10	100	1000	10,000
1	Colonies[c]	171.3 ± 3.5	24.7	9.5	0.6	0.0
	Clusters[d]	370.0 ± 10.0	124.3	138.0	138.7	155.4
	Total	541.3	93.0	97.0	95.0	106.0
2	Colonies	124.3 ± 5.1	16.6	1.8	0.0	0.0
	Clusters	326.7 ± 15.0	135.7	135.0	122.0	119.4
	Total	451.0	103.0	98.0	88.7	86.5
3	Colonies	35.0 ± 4.6	14.2	2.8	0.0	0.0
	Clusters	68.0 ± 6.5	139.7	147.0	152.5	147.0
	Total	103.0	97.0	98.0	100.0	97.0
4	Colonies	26.7 ± 4.0	87.2	93.6	36.3	37.4
	Clusters	253.0 ± 5.8	102.0	104.0	110.7	109.4
	Total	279.7	101.3	103.0	107.0	102.5

[a] Reproduced, with permission, from *Blood,* Grune & Stratton.
[b] Number of light density (<1.077 g/ml) cells plated per dish was 2×10^5 for the first two experiments and 10^5 for the last two. The colony and cluster incidences are means ± SD from triplicate culture dishes.
[c] Aggregates of ≥ 40 cells.
[d] Aggregates of 3–39 cells.

mitosis. When the cultures were incubated for 4 more days, no significant difference in colony or cluster numbers was noted. This suggested that either prolonged generation time was not the cause of our observations or, if it was, the prolongation was of a rather significant degree or was increasing progressively with each successive mitosis. To explore these points further, we examined the clusters morphologically. Table II presents the differential counts of clusters from one of these experiments. It is clear from these data that increasing HLIF concentrations induced a progressive differentiation block.

Taking these and other data[9] into consideration, we suggest a model of granulopoiesis. The colony-stimulating activity stimulates the proliferation and differentiation at each successive stage of the granulopoietic maturational process. The prostaglandins of E series may inhibit this process

[9] D. S. Verma, G. Spitzer, A. R. Zander, J. U. Gutterman, K. B. McCredie, K. A. Dicke, and D. A. Johnston, *Exp. Hematol.* **9,** 63 (1981).

TABLE II

EFFECT OF HUMAN LEUKOCYTE INTERFERON ON CELL COMPOSITION OF AGGREGATES GROWN IN AGAR CULTURE[a]

Interferon concentration (units/ml)	Differential count of cells contained in clusters (%)[b]						
	Myeloblasts	Promyelocytes	Myelocytes	Metamyelocytes	Bands	Polymorphs
0	0.0	0.0	5.7	22.8	23.0	48.5
10	3.0	10.0	27.0	27.2	22.4	10.4
100	9.6	19.2	27.0	27.8	10.6	4.8
1,000	6.0	12.7	37.0	30.0	9.2	4.1
10,000	8.4	16.2	40.8	28.0	6.6	0.0

[a] Reproduced, with permission, from *Blood*, Grune & Stratton.

[b] Average percentage from 50 clusters examined for morphological type. Morphological examination of the colonies from dishes with and without interferon revealed normally differentiating cells.

by blocking GM-CFC proliferation,[10] and HLIF acts by counteracting the stimulatory effect of CSA with progressively increasing responsiveness of the successive generations. It is of interest that virtually all microorganisms may induce interferon elaboration.[11] It is possible, therefore, that the dose and/or immunochemical determinants of a microorganism may govern the elaboration of CSA, prostaglandins E, and interferon to different extents, and thus align the granulopoietic response according to the needs.

Acknowledgments

The authors express their appreciation for the technical assistance of Sherrie Smith and Anne McCrady and the secretarial assistance of Kathi Orsak. This work was supported in part by Grants CA-11520, CA-14528, RR-05511-16, and RR-5511-17 from the National Institutes of Health, Bethesda, Maryland.

[10] J. I. Kurland, R. S. Bockman, H. E. Broxmeyer, and M. A. S. Moore, *Science* **199**, 552 (1978).

[11] M. Ho and J. A. Armstrong, *Annu. Rev. Microbiol.* **29**, 131 (1975).

Section VII

Somatic Cell Genetics of Interferons

[65] Methods for Mapping Human Interferon Structural Genes

By DORIS L. SLATE and FRANK H. RUDDLE

Studies on the human interferon system have revealed a complex set of genetic and epigenetic factors that influence the synthesis and action of interferon. Interferon induction requires *de novo* RNA and protein synthesis as determined through the use of transcription and translation inhibitors.[1] Cells that are actively synthesizing interferon can be enucleated, and the resulting cytoplasts can still produce interferon.[2] The yield of interferon can vary over a wide range for human cells, depending on such factors as the "age" of a culture, the inducing stimulus, and whether the cells have a limited life-span or are "transformed" to indefinite growth in culture.

Not only do the cell type and induction regimen affect the amount of interferon made, they can also affect the type of interferon made. Three major types of human interferon have been described: fibroblast (F), leukocyte (Le), and immune (sometimes called T or type II) interferons. F type is usually produced in fibroblast cultures treated with poly(I) · poly(C); Le type is usually made in buffy coat cells or in virus-transformed established lymphoblastoid lines after viral induction. Immune interferon is produced by various cells of the immune system in response to several types of stimulation (for example, treatment with mitogens or exposure to antigens that trigger immune recognition). These human interferon species differ in a number of respects, including stability at pH 2 and high temperature, antigenically, in molecular weight, and in cross-species antiviral activity. Havell, Hayes, and Vilček[3] have reported that human fibroblasts can synthesize both F and Le type interferons after viral induction.

The human interferon system is also amenable to various treatments that can modulate the amount or time course of interferon synthesis. Two major examples are priming and superinduction. In priming, low doses of interferon are used to pretreat cells, which are then normally induced; interferon is produced earlier than in unprimed control cells treated with just the inducer.[4] Superinduction involves the use of suitably timed doses of translation and transcription inhibitors and induction with poly(I) ·

[1] M. Ho and J. A. Armstrong, *Annu. Rev. Microbiol.* **29**, 131 (1975).

[2] D. C. Burke and G. Veomett, *Proc. Natl. Acad. Sci. U.S.A.* **74**, 3391 (1977).

[3] E. A. Havell, T. G. Hayes, and J. Vilček, *Virology* **89**, 330 (1978).

[4] W. E. Stewart, II, L. B. Gosser, and R. Z. Lockart, *J. Virol.* **7**, 792 (1971).

poly(C). This technique lengthens the period of interferon synthesis and increases final yields. On a molecular level, superinduction increases the amount of interferon messenger RNA synthesized and also stabilizes this messenger.[5,6]

The genetic control of human interferon synthesis and mechanism of action has been examined in tissue culture systems via somatic cell hybridization and the study of aneuploid or aneusomic human cell lines. We will attempt to summarize the major techniques and findings in the mapping of genes governing human interferon production.

Somatic Cell Hybrids in Human Interferon Gene Mapping

Most of the somatic cell genetic studies that have resulted in human gene assignments have been performed with human/rodent hybrid cells. Human/mouse and human/Chinese hamster hybrid cell lines tend to lose human chromosomes preferentially, making it possible to correlate the presence or the absence of a specific human phenotype with the presence or the absence of a particular human chromosome. We will not describe the generation of hybrids, or the isozyme and karyotype analysis, required for gene mapping. Detailed procedures for these techniques can be found in the literature (for example, see Kennett[7] for a discussion of hybridization methods, Nichols and Ruddle[8] and Kozak et al.[9] for starch gel electrophoretic conditions to distinguish mouse and human isozymes, and Kozak, Lawrence, and Ruddle[10] and Worton and Duff[11] for karyotype procedures for analyzing hybrid cells).

The first assignment for human genes involved in the synthesis of interferon was made by Tan, Creagan, and Ruddle.[12] They characterized a large number of human/mouse hybrid lines and found that the ability to produce human interferon could not be associated with the presence of any one human chromosome. When they examined the correlation between human interferon production and the presence of two human chromosomes together, however, they found that when a hybrid con-

[5] P. B. Sehgal and I. Tamm, *Virology* **92**, 240 (1979).
[6] R. L. Cavalieri, E. A. Havell, J. Vilček, and S. Pestka, *Proc. Natl. Acad. Sci. U.S.A.* **74**, 4415 (1977).
[7] R. H. Kennett, this series, Vol. 58, p. 345.
[8] E. A. Nichols and F. H. Ruddle, *J. Histochem. Cytochem.* **21**, 1066 (1973).
[9] C. A. Kozak, R. E. K. Fournier, L. A. Leinwand, and F. H. Ruddle, *Biochem. Genet.* **17**, 23 (1979).
[10] C. A. Kozak, J. B. Lawrence, and F. H. Ruddle, *Exp. Cell Res.* **105**, 109 (1977).
[11] R. G. Worton and C. Duff, this series, Vol. 58, p. 322.
[12] Y. H. Tan, R. P. Creagan, and F. H. Ruddle, *Proc. Natl. Acad. Sci. U.S.A.* **71**, 2251 (1974).

tained both human chromosomes 2 and 5, human interferon synthesis could be induced with proper stimulation. They also noted that the Giemsa banding pattern of human chromosome 5 resembled that published by Stock and Hsu[13] for an African green monkey small subtelocentric chromosome, which had been implicated in monkey interferon synthesis several years earlier by Cassingena et al.[14]

More recently, however, the two-chromosome requirement for human interferon production has been questioned. In human/Chinese hamster hybrids, apparently only the presence of human chromosome 5 is necessary for human interferon synthesis.[15] We have concluded that both chromosomes 2 and 5 contain genes for human interferon[16]; below we will briefly describe the interferon induction and assay procedures that we routinely use for studying interferon synthesis in somatic cell hybrids.

For viral induction, we expose confluent monolayers of cells to Newcastle disease virus at 10 PFU/cell for 1 hr in 1 ml of serum-free medium. Cells are then washed with medium, and refed with fresh medium, containing 2% fetal bovine serum (FBS), for 24 hr. Medium is harvested and dialyzed against 0.15 M KCl pH 2 for several days at 4° to inactivate residual virus, and then against serum-free medium for 24 hr before assay.

For simple poly(I) · poly(C) inductions, 100 μg of poly(I) · poly(C) (P-L Biochemicals) per milliliter in phosphate-buffered saline (PBS) is added to confluent monolayers for 2 hr, after which the cells are washed and refed with medium containing 2% FBS. Medium is collected after 18–20 hr, centrifuged at low speed to remove any cell debris, and assayed.

For poly(I) · poly(C) superinductions, 100 μg of poly(I) · poly(C) per milliliter are added in PBS to cells for 2 hr along with 20 μg of cycloheximide (Sigma) and 50 μg of DEAE-dextran (Pharmacia) per milliliter. After removal of the PBS and washing with serum-free medium, the cells are treated with cycloheximide (20 μg/ml) for 3 hr in serum-free medium. The cycloheximide is then removed; the cell sheet is washed, and the cells are treated with 2 μg of actinomycin D (Calbiochem) per milliliter for 1 hr. After washing, the cells are refed with medium containing 2% FBS; medium is collected after 18 hr for interferon assay.

Medium incubated with hybrid cells after induction is assayed for interferon activity on both mouse (A9) and human (FS7 or GM 2504 Trisomy 21) fibroblasts. The assays are performed in 96-well microtiter trays (Linbro) by treating ~2 × 10⁴ cells/well with various dilutions of

[13] A. D. Stock and T. C. Hsu, Chromosoma 43, 211 (1973).
[14] R. Cassingena, C. Chany, M. Vignal, H. Suarez, S. Estrade, and P. Lazar, Proc. Natl. Acad. Sci. U.S.A. 68, 580 (1971).
[15] M. J. Morgan and P. Faik, Br. J. Cancer 35, 254 (1977).
[16] D. L. Slate and F. H. Ruddle, Cell 16, 171 (1979).

interferon preparations for 18–24 hr. The tray is vigorously inverted over paper toweling to remove residual interferon, and then 0.1 PFU of vesicular stomatitis virus per cell is added for 24 hr, at the end of which virus-induced cytopathic effects (CPE) are scored by examining the cells under an inverted phase contrast microscope. Cell controls (no interferon, no virus), virus controls (no interferon, + virus), and interferon controls (treated with a known amount of human leukocyte or fibroblast interferon) are included in each assay.

To determine the antigenic type of human interferon produced in a hybrid, antisera against human leukocyte and fibroblast interferons were obtained from Dr. J. K. Dunnick (Antiviral Substances Program of the National Institute for Allergy and Infectious Diseases). They are titered against human interferon preparations of known specific activity in a modification of the CPE assay described above. Various antiserum dilutions (25 μl) are mixed with 25-μl samples containing 10 units of human interferon and incubated at 37° for 30 min. Human FS7 or GM 2504 cells are then added, and the assay is completed as usual. In neutralization tests for interferon produced by hybrid cells, approximately 10 units of human interferon are added to the dilution series of antibody preparations as above.

In human/mouse hybrid cells containing human chromosomes 2 and 5, chromosome 5 can be eliminated by treatment with diphtheria toxin (0.25 L_f units/ml, Connaught) for at least 48 hr. This selection procedure cannot be used with human/Chinese hamster hybrids, since the two species have approximately the same sensitivity to the toxin. Human/Chinese hamster hybrids tend not to retain human chromosome 2, and lines with 2^-5^+ karyotypes can be easily obtained.

In our studies, we found that the presence of either human chromosome 2 or 5 could direct the synthesis of human interferon. With antibody neutralization tests performed on the interferon produced by hybrid cells, we determined that the human interferon was of the fibroblast (F) antigenic type. This was true when the human parental input was a fibroblast or when it was a peripheral blood leukocyte. The mouse or Chinese hamster parents of our hybrid cells were always fibroblast lines. The antibody neutralization data may therefore indicate that the epigenetic state of the parental rodent cell may influence the production of interferon of the segregating parental type.

Meager et al.[17] have reported that human chromosome 9 bears the gene for human fibroblast interferon. Human/mouse hybrid cells that retain an X/9 translocation chromosome produce human interferon, but lose this

[17] A. Meager, H. Graves, D. C. Burke, and D. M. Swallow, *Nature* (*London*) **280**, 493 (1978).

ability when selected in 8-azaguanine. Meager *et al.*'s interpretation of their data suggests that chromosomes 2 and 5 do not carry additional interferon loci, but some of their hybrids with 2 or 5 present (and 9 absent) produce low levels of human interferon.

Since the hybrids we used for the mapping studies described above did not contain human chromosome 9, we attempted to confirm the results of Meager *et al.*[17] by using a chromosome-mediated gene transfer line retaining only an X/9 translocation on a mouse A9 background (C. Miller and F. H. Ruddle, unpublished data). This line was generated by fusing isolated metaphase HeLa chromosomes to A9 cells. The procedure is outlined below (see also, Miller and Ruddle[18]).

Approximately 10^8 mitotic cells are collected, pelleted gently (800 rpm, 5 min), and resuspended at 5×10^6 cells/ml in $0.075\ M$ KCl, $0.1\ \mu g$ of vinblastine per milliliter at room temperature for about 5 min. Cells are pelleted at $4°$, resuspended in 15 mM Tris · HCl, pH 7, 3 mM CaCl$_2$ at 5×10^6 cells/ml, and put on ice for 15 min. Triton X-100 is added to 1%, and the cells are incubated at $37°$ for 10 min. Cells are then homogenized in a Dounce homogenizer (10–12 strokes) on ice and diluted with an equal volume of cold 15 mM Tris · HCl pH 7, 3 mM CaCl$_2$. The suspension is then centrifuged at 800 rpm at $4°$ for 10 min, and the supernatant containing chromosomes is collected. (The pellet may be redissolved and centrifuged to recover any additional chromosomes.) Chromosomes are pelleted at 2000 rpm for approximately 30 min and then resuspended in HEPES buffer (NaCl, 8 g/liter; KCl, 0.37 g/liter; Na$_2$HPO$_4$ · 2 H$_2$O, 0.125 g/liter; dextrose, 1 g/liter; HEPES, pH 7.1, 5 g/liter) in about 30 ml per 10^8 donor cells. An aliquot for chromosome counts is removed, and the rest is pelleted. The chromosomes should be resuspended at 2×10^8/ml, with the suspension done in 7 : 1 HEPES buffer : 2 M CaCl$_2$. Half the required amount of HEPES buffer is added; and the chromosomes are vigorously resuspended by vortex mixing, then the rest of the HEPES is added. Then the 2 M CaCl$_2$ is added, and the chromosomes are left at room temperature for about 30 min. Recipient cells should have been plated the previous day to be approximately 25% confluent when chromosomes are added. The medium is removed from the recipient cells, and 2 ml of chromosome suspension are added per 75-cm^2 culture flask. Cells are incubated 30 min at room temperature. Next 20 ml of medium with fetal bovine serum and antibiotics are added to the flask and incubated for 4 hr at $37°$. Dimethyl sulfoxide (DMSO) is added to a final concentration of 10%, and the cells are incubated for 30 min at room temperature. The DMSO medium is removed and replaced with fresh medium; the cells are incubated at $37°$, and selective medium is added after 24–48 hr.

[18] C. L. Miller and F. H. Ruddle, *Proc. Natl. Acad. Sci. U.S.A.* **75**, 3346 (1978).

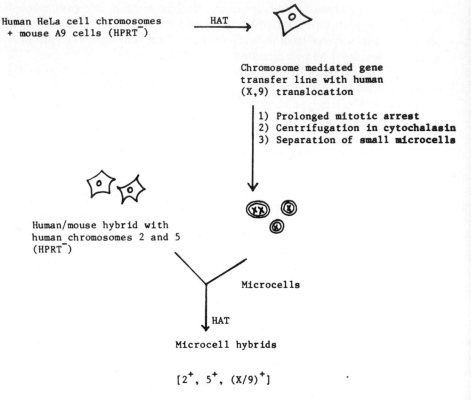

FIG. 1. Generation of microcell hybrids containing human chromosomes 2 and 5 and an (X/9) translocation.

The chromosome-mediated gene transfer line retaining only the human X/9 translocation produced human interferon under virus induction and poly(I) · poly(C) superinduction.[19] In order to study the effect of the X/9 translocation chromosome on the other interferon loci, we transferred this chromosome to a human/mouse $2^{+}5^{+}$ hybrid via the microcell procedure published by Fournier and Ruddle.[20] Interferon production in the resultant microcell hybrids is still being evaluated (Fig. 1).

It has not been possible to determine whether the three human interferon loci described thus far are identical. Antigenically the interferons made by cells with various combinations of the three implicated chromosomes is fibroblast (F) type, but there may be differences in the molecules coded by each chromosome. It is highly unlikely that genes with exactly

[19] D. L. Slate and F. H. Ruddle, *Ann. N.Y. Acad. Sci.* **350**, 174 (1980).
[20] R. E. K. Fournier and F. H. Ruddle, *Proc. Natl. Acad. Sci. U.S.A.* **74**, 319 (1977).

the same nucleotide sequence exist on the three chromosomes implicated in human interferon synthesis. It is possible that regulatory genes and/or the epigenetic state of a cell may determine which gene(s) to transcribe in response to a given stimulus.

Use of Aneuploid or Aneusomic Human Cells in Gene Mapping

Another method for studying the genetic control of human interferon synthesis employs aneuploid or aneusomic human cell lines and depends on observing differences in interferon production depending on the dosage of specific chromosomes or portions of chromosomes. Tan and his colleagues[21] have reported studies with human fibroblasts with different numbers of copies of chromosome 5, some with more short arms than long arms of this chromosome, and others with this situation reversed. Their data suggest that the amount of human interferon produced was inversely related to the number of chromosome 5 short arms, and proportional to the number of chromosome 5 long arms. Slate and Ruddle[16] found that cri-du-chat fibroblasts, with deletions of part of the short arm of chromosome 5, produce less interferon than other human fibroblasts. It is unclear whether these results with aneuploid cell lines reflect true gene dosage relationships or merely represent the results of aberrant regulation in cells lacking a balanced set of genes required for the battery of steps involved in what is termed interferon induction. Interpretation of data obtained from established cell lines with rearranged karyotypes is very difficult when the parameter to be measured (interferon yield) varies over a wide range owing to a combination of genetic, epigenetic, and cell culture conditions.

Mapping Other Human Interferon Genes

As mentioned earlier, most of the somatic cell genetic studies on human interferon synthesis have been performed with human (aneuploid) fibroblasts or with rodent fibroblasts as the nonsegregating parent in human/rodent hybrids. These factors may be responsible for the fact that the only gene assignments thus far have been for human fibroblast interferon. Construction of leukocyte × leukocyte hybrids or hybrids with the nonsegregating parent being a tumor cell derived from the immune system may allow the mapping of leukocyte and immune interferons. It will be extremely interesting to see whether these interferons are coded by the same chromosomes involved in fibroblast interferon synthesis. It may also

[21] Y. H. Tan, *in* "Interferons and Their Actions" (W. E. Stewart, II, ed.), pp. 73–90. CRC Press, Cleveland, Ohio, 1977.

be possible to exploit the finding that small amounts of leukocyte interferon can be made by human fibroblasts induced with viruses; using suitable purification schemes or antigenic methods, this small amount of leukocyte interferon could be detected after induction of fibroblast-derived cell hybrids and associated with the presence of a particular human chromosome.

It is clear that we have only just begun to decipher the genetics behind human interferon synthesis. With cloned interferon gene probes, it will be possible to confirm or refute the current gene assignments and to learn in molecular detail about the organization of the human interferon genes.

Acknowledgments

We would like to thank Carol Miller for her gene transfer cell line. This work was supported by NIH Grant GM 09966 and a grant from the American Cancer Society to F.H.R.

[66] Somatic Cell Genetic Methods for Study of Sensitivity to Human Interferon

By Doris L. Slate and Frank H. Ruddle

The genes governing production of interferon and sensitivity to its action have proved to be asyntenic in the species that have been studied.[1] Somatic cell genetic analyses of interspecific heterokaryons and hybrids have indicated that interferon production and response genes from two different species can function in a single cell. If the two parental species produce interferons that do not show significant cross-species activity, then it is possible to map genes in interspecific hybrids selectively segregating the chromosomes of one input species.

Chromosome assignments for genes governing sensitivity to interferon have been made in man[2,3] and mouse (this volume [67]). In this chapter we will concentrate on the mapping of genes involved in the antiviral response to human interferon and discuss the recent somatic cell genetic studies that have attempted to determine the nature of the gene product(s) responsible for conferring interferon sensitivity to a cell.

[1] D. L. Slate and F. H. Ruddle, *Pharmacol. Ther.* **4,** 221 (1979).

[2] Y. H. Tan, J. A. Tischfield, and F. H. Ruddle, *J. Exp. Med.* **137,** 317 (1973).

[3] C. Chany, M. Vignal, P. Couillin, N. V. Cong, J. Boue, and A. Boue, *Proc. Natl. Acad. Sci. U.S.A.* **72,** 3129 (1975).

Assignment of Human Gene(s) Governing Interferon Sensitivity to Chromosome 21

Human/mouse somatic cell hybrids segregating human chromosomes were used by Tan, Tischfield, and Ruddle[2] and by Chany et al.[3] to establish the linkage of human interferon sensitivity and cytoplasmic superoxide dismutase genes. These genes were assigned to human chromosome 21.

Figure 1 shows how this mapping was accomplished. Since mouse and human interferons show very little cross-species activity, human/mouse hybrid cells that segregate human chromosomes can be used for mapping the gene governing sensitivity to human interferon. By using a combination of isozyme[4,5] and karyotype[6] analysis with interferon assays, the chromosomal localization of the gene can be determined.

Most workers use a microtiter inhibition of cytopathic effects assay for interferon sensitivity. This assay requires small numbers of cells and can be used to screen large numbers of hybrid clones for sensitivity. Briefly, hybrid cells (2×10^4/well) are exposed to a dilution series of human interferon concentrations in wells of a 96-well microtiter tray, and then challenged with vesicular stomatitis virus (0.1 PFU/cell) for approximately 24 hr. Sensitivity to human interferon is manifested by resistance of cells to the cytolytic effect of the virus. In general, human/mouse hybrid cells are less sensitive to human interferon than are the human fibroblasts used as parents of the hybrids.

Effect of Chromosome 21 on Sensitivity to Human Interferon

Tan et al.[7] examined the effect of chromosome 21 dosage on the attainment of the antiviral state after treatment of cells with human interferon. Monosomy 21 human fibroblasts are less sensitive to human interferon (i.e., require more interferon to attain the antiviral state) than are disomy 21 or trisomy 21 (Down's syndrome) cells. This gene dosage relationship has been exploited by employing trisomy 21 cells in human interferon assays to provide greater sensitivity in detecting small amounts of interferon (for example, in in vitro translation experiments involving interferon messenger RNA).

[4] E. A. Nichols and F. H. Ruddle, J. Histochem. Cytochem. 21, 1066 (1973).
[5] C. A. Kozak, R. E. K. Fournier, L. A. Leinwand, and F. H. Ruddle, Biochem. Genet. 17, 23 (1979).
[6] C. A. Kozak, J. B. Lawrence, and F. H. Ruddle, Exp. Cell Res. 105, 109 (1977).
[7] Y. H. Tan, E. L. Schneider, J. Tischfield, C. J. Epstein, and F. H. Ruddle, Science 186, 61 (1974).

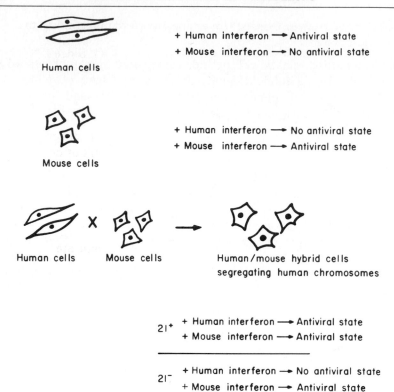

FIG. 1. Mapping the human interferon sensitivity gene to chromosome 21 depends on the lack of cross-species activity between mouse and human interferons.

Regional Localization of Interferon Sensitivity Gene(s)

Using human cell lines with deletions or extra copies of regions of chromosome 21, regional assignments can be made for the gene(s) responsible for interferon sensitivity. Tan and Greene[8] and Epstein and Epstein[9] have assigned the sensitivity gene to the distal portion of the long arm of chromosome 21. Sensitivity to both type I (leukocyte and fibroblast) and type II (immune) interferons is controlled by chromosome 21.

Human Chromosome 21 Alone Is Sufficient for Sensitivity to Human Interferon

Chany[10] reported that human/mouse hybrid cells that retained only human chromosome 21 would not respond to human interferon; the pres-

---

[8] Y. H. Tan and A. E. Greene, *J. Gen. Virol.* **32,** 153 (1976).
[9] L. B. Epstein and C. J. Epstein, *J. Infect. Dis.* **133,** A56 (1976).
[10] C. Chany, *Biomed. Express* **24,** 148 (1976).

ence of any other human chromosomes in a hybrid could (nonspecifically) render the hybrid sensitive if chromosome 21 was present. However, a more recent study by Slate et al.[11] has demonstrated that a hybrid cell line retaining human chromosome 21 as the only detectable human genetic material could respond to both human leukocyte and fibroblast interferons; the presence of other human chromosomes increased interferon sensitivity. The different conclusions of these two studies probably reflect differences in the concentrations of interferon used and the frequency of chromosome 21 in hybrids when the interferon assays were performed.

One interesting finding reported by Slate et al.[11] was that human interferon could trigger a change in a mouse component of the interferon response system in the human/mouse hybrid retaining only human chromosome 21. Several reports have shown that incubation of extracts from interferon-treated cells with double-stranded RNA and $[\gamma\text{-}^{32}P]ATP$ results in the phosphorylation of two major proteins[12-14]: a subunit of eukaryotic protein initiation factor 2 with a molecular weight of about 35,000, and another larger protein of unknown function. The mouse form of this larger protein has a molecular weight of approximately 67,000, and the human form is slightly larger at 69,000. In extracts from a human/mouse hybrid cell retaining only human chromosome 21 treated with human (or mouse) interferon, the MW 67,000 mouse protein is phosphorylated, but the MW 69,000 human is not (see the table).

Sensitivity to Nonantiviral Activities of Interferon Is Also Controlled by Chromosome 21

Interferon exhibits not only antiviral effects, but also triggers a number of other changes in treated cells (see review by Stewart[15]). Chromosome 21 appears to confer sensitivity to these other effects as well. For example, monosomy, disomy, and trisomy 21 cells differ in their susceptibility to the growth-inhibitory effect of human interferon,[16] and sensitivity to the "priming" of interferon is also governed by chromosome 21.[17]

[11] D. L. Slate, L. Shulman, J. B. Lawrence, M. Revel, and F. H. Ruddle, J. Virol. 25, 319 (1978).

[12] A. Zilberstein, P. Federman, L. Shulman, and M. Revel, FEBS Lett. 68, 119 (1976).

[13] B. Lebleu, G. C. Sen, S. Shaila, B. Carbrer, and P. Lengyel, Proc. Natl. Acad. Sci. U.S.A. 73, 3107 (1976).

[14] W. K. Roberts, A. Hovanessian, R. E. Brown, M. J. Clemens, and I. M. Kerr, Nature (London) 264, 477 (1976).

[15] W. E. Stewart, II, "The Interferon System." Springer-Verlag, Vienna and New York, 1979.

[16] Y. H. Tan, Nature (London) 260, 141 (1976).

[17] H. M. Frankfort, E. A. Havell, C. M. Croce, and J. Vilček, Virology 89, 45 (1978).

APPEARANCE OF HUMAN OR MOUSE PHOSPHOPROTEINS IN CELL EXTRACTS OF
INTERFERON-TREATED CELLS INCUBATED WITH DOUBLE-STRANDED RNA AND [γ-^{32}P]ATP

		Phosphoprotein	
Cell extract	Type of interferon	MW 67,000 (mouse)	MW 69,000 (human)
Mouse L	Mouse	+	−
	Human	−	−
Human FS7	Mouse	−	−
	Human	−	+
Mixture of mouse L and human FS7	Mouse for L, human for FS7	+	+
Human/mouse hybrid with chromosome 21 as the only human chromosome	Mouse	+	−
	Human	+	−

Nature of Gene Product Responsible for Interferon Sensitivity

All the findings discussed above have led to the formulation of a model wherein the gene product coded by chromosome 21 responsible for conferring human interferon sensitivity to cells is a cell surface receptor. Chany[10] has suggested that interferon binding to cells is accomplished through a two-component receptor system: a "binding" site would be responsible for relatively nonspecific interactions and may involve gangliosides, while an "activator site" would be involved in determining the species specificity of interferon action after binding. Wiranowska-Stewart and Stewart[18] have found that human fibroblasts bind different amounts of human interferon depending on the dosage of human chromosome 21.

Kohn et al.[19] have reported that several hormones, cholera toxin, and heterologous interferons can block the action of human interferon in human cells. This effect may be due to the occupation or blocking of sites in the interferon receptor complex or to the inaccesssibility of receptors after membrane rearrangements following binding of these agents.

Antibodies against Chromosome 21-Coded Cell Surface Components Can Block Human Interferon Action

Revel, Bash, and Ruddle[20] have studied the effect of allowing human cells to react with antibodies raised against human cell surface compo-

[18] M. Wiranowska-Stewart and W. E. Stewart, II, *J. Gen. Virol.* **37,** 629 (1977).

[19] L. D. Kohn, R. M. Friedman, J. M. Holmes, and G. Lee, *Proc. Natl. Acad. Sci. U.S.A.* **73,** 3695 (1976).

[20] M. Revel, D. Bash, and F. H. Ruddle, *Nature (London)* **260,** 139 (1976).

nents coded by human chromosome 21. They injected human/mouse hybrid cells containing human chromosomes 4, 21, and 22 into mice of the same strain (C3H/HeJ) as the mouse parental A9 cell line. The antiserum collected from the injected mice could block the action of human interferon in human cells. Treatment of cells with antiserum before exposure to interferon resulted in higher virus yields than were obtained with cells treated only with interferon. This blocking activity was more pronounced on monosomy 21 than on disomy or trisomy 21 cells. Preadsorption of antiserum onto hybrid cells retaining human chromosomes 4 and 22 (but lacking 21) did not remove interferon blocking activity, but preadsorption onto hybrids retaining human chromosome 21 could remove this activity.

Our current protocols for production and assay of blocking antibodies are outlined below. Approximately 10^7 human/mouse hybrid cells are injected intraperitoneally into C3H/HeJ mice at weekly intervals. Cells are harvested in calcium-, magnesium-free phosphate-buffered saline with 1 mM EDTA and washed twice in physiological saline before injection. After 4 weeks, mice are tail bled, and a 1:50 dilution of serum in medium is added to human FS7 fibroblasts in microtiter tray wells for 30 min at 37°. Cells are then treated with a dilution series of human interferon for 18 hr and then challenged with 0.1 PFU/cell of vesicular stomatitis virus for approximately 24 hr. Cytopathic effects are scored, and/or the virus yields from each well are quantitated on mouse A9 cells. After blocking activity has been found in a serum sample, the mouse is usually given a booster injection of 2×10^7 hybrid cells and then sacrificed a week later. Serum is collected by heart puncture, and recently spleens have been removed for the generation of "hybridomas" (spleen cell × myeloma hybrid cells) producing monoclonal antibodies with interferon blocking activity (M. Kamarck, P. D'Eustachio, D. L. Slate, and F. H. Ruddle, unpublished results). A human/mouse hybrid cell line retaining only human chromosome 21 can elicit the production of blocking antibodies and can be used to adsorb such blocking activity in antiserum preparations.[21]

It should be possible to use antibodies directed against the putative human cell surface receptor for interferon to quantitate such receptors. Thus far there have been technical problems with this approach combined with indirect immunofluorescence procedures. The results of Kohn et al.[19] concerning hormone and heterologous interferon blocking of interferon action may have a parallel in the studies using antibodies against chromosome 21-coded cell surface proteins. Chromosome 21 may code for other cell surface determinants besides an interferon receptor, with the blocking effect of antiserum on interferon action due to either specific or non-

[21] D. L. Slate and F. H. Ruddle, *Cytogenet. Cell Genet.* **22**, 265 (1978).

specific binding effects. Monospecific antibodies to chromosome 21-coded determinants should allow us to identify the interferon receptor and quantitate it, and help to resolve the problems mentioned above.

Acknowledgments

We wish to thank Drs. P. D'Eustachio and M. Kamarck for their help in studying hybridomas. This work was supported by NIH Grant GM09966 and an American Cancer Society grant to F.H.R.

[67] Methods for Mapping a Murine Gene Governing Sensitivity to Interferon

By PIN-FANG LIN, DORIS L. SLATE, and FRANK H. RUDDLE

The techniques of somatic cell genetics have permitted the genetic analysis of interferon production and sensitivity in heterokaryons and hybrid cells. Fusions between cells of two species for which cross-species interferon activity is not observed have indicated that interferon production and response to interferon (as measured by the attainment of the antiviral state) are separate genetic functions, controlled by genes on different chromosomes.[1,2] Human/mouse and human/Chinese hamster hybrids have been employed for mapping human genes involved in the interferon system. These hybrids tend to segregate human chromosomes preferentially, allowing human genes to be assigned to specific chromosomes.[3]

Mouse gene mapping has been accomplished through a combination of breeding studies and somatic cell genetic techniques. Mouse/Chinese hamster cell hybrids usually lose mouse chromosomes during prolonged cultivation and can therefore be used for mapping murine genes. In many cases, however, mouse and Chinese hamster isozymes cannot be distinguished upon electrophoresis, and in these cases, mouse/human hybrids would be useful for mouse mapping studies. "Reverse segregant" mouse/human hybrid cells, which lose mouse chromosomes, can be generated by fusing primary mouse cells with established human cell lines.[4]

[1] R. P. Creagan, Y. H. Tan, S. Chen, and F. H. Ruddle, *Fed. Proc., Fed. Am. Soc. Exp. Biol.* 34, 2222 (1975).

[2] D. L. Slate and F. H. Ruddle, *Pharmacol. Ther.* 4, 221 (1979).

[3] F. H. Ruddle and R. P. Creagan, *Annu. Rev. Genet.* 9, 407 (1975).

[4] C. M. Croce, *Proc. Natl. Acad. Sci. U.S.A.* 73, 3248 (1976).

Using such hybrids for mapping has two major problems: (*a*) the loss of mouse chromosomes is rapid; and (*b*) occasionally this loss stops when a haploid set of mouse chromosomes is left. Microcell hybrids can also be used for mouse gene mapping.[5] Such hybrids are constructed by fusing "microcells" containing only a few mouse chromosomes to human cells, thereby controlling the direction of chromosome segregation.

We will briefly describe the generation and analysis of mouse/human whole cell and microcell hybrids that were used to map the murine gene conferring sensitivity to mouse interferon to chromosome 16. Epstein and her colleagues[6] have also made this gene assignment using mouse/Chinese hamster hybrid cells.

Generating Mouse/Human Hybrid Cells for Mapping Murine Genes

The human parental cell line employed was a xeroderma pigmentosum (complementation group A) fibroblast designated CRL1223, which was obtained from the American Type Culture Collection (Rockville, Maryland). It has been transformed with SV40 and was maintained in Dulbecco's modified Eagle's medium (DME) supplemented with 10% fetal bovine serum (FBS). Several different fusion procedures involving mouse primary cells were utilized, with spleen cells or embryo fibroblasts used as murine parental material.

Mouse spleen cells were obtained by dissecting spleens from C57BL/6J mice (Jackson Laboratories) and gently pressing cells from the tissue into DME. Cells were washed twice in serum-free DME; for some fusions red blood cells were removed by lysing them in 0.85% NH_4Cl. The hybridizations were performed by a modification of the procedure of Davidson and Gerald.[7]

Briefly, the mouse spleen cells and human 1223 cells were mixed at a 10 : 1 ratio and pelleted together. Polyethylene glycol 1540 (PEG, Baker Chemical), 40% (w/w) in 1 ml of DME was then added dropwise to the cell pellet. After gently stirring the cells in PEG for 2.5 min, the concentration of PEG was gradually reduced by adding DME to a final volume of 50 ml. After incubation of the cell suspension at 37° for 30 min, the cells were pelleted, resuspended, and incubated overnight in DME with 10% FBS at 37°. Hybrid cells were selected by adding $10^{-6} M$ ouabain[8] 48 hr after PEG fusion. The average yield of hybrids was about 2.5×10^{-5} per recipient human cell. An alternative hybridization procedure developed by Dr. A.

[5] R. E. K. Fournier and F. H. Ruddle, *Proc. Natl. Acad. Sci. U.S.A.* **74**, 319 (1977).
[6] D. R. Cox, L. B. Epstein, and C. J. Epstein, *Proc. Natl. Acad. Sci. U.S.A.* **77**, 2168 (1980).
[7] R. L. Davidson and P. S. Gerald, *Somatic Cell Genet.* **2**, 165 (1976).
[8] L. H. Thomson and R. M. Baker, *Methods Cell Biol.* **6**, 209 (1973).

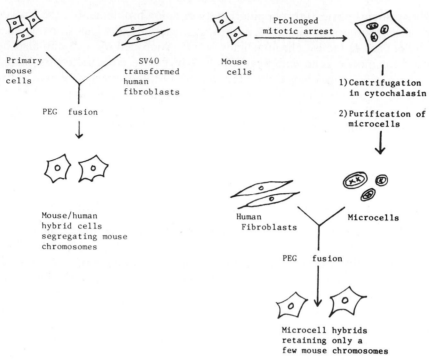

FIG. 1. Generation of whole cell and microcell hybrids for mouse gene mapping. PEG, polyethylene glycol.

Lalazar in this laboratory (unpublished data) was also used for some fusions. A monolayer culture of human 1223 recipient cells was treated with concanavalin A (Con A, 100 μg/ml) prior to the addition of donor mouse spleen cells. The monolayer-suspension fusion was facilitated by 40% PEG; this modified fusion procedure increased the yield of hybrids twofold.

Another set of hybrids was generated by fusing karyoplasts from C57BL/6J mouse embryo fibroblasts and 1223 human cells. Karyoplasts were prepared by growing mouse embryo fibroblasts on small plastic disks. The disks were spun cell side down in DME containing 10 μg of cytochalasin B per milliliter at 39,000 g for 30 min at 35°. The pellets containing karyoplasts were pooled and fused to the 1223 cells at a ratio of 1 : 1 with the aid of 40% PEG. Hybrid cells were selected by exposure to 10 J/m² of ultraviolet light.

Microcell hybrids were generated by the method of Fournier and Ruddle[5] using mouse embryo fibroblasts as the microcell donor. A summary of the procedures used in generating our hybrids is presented in Fig. 1.

Isozyme and Karyotype Analysis of Hybrids

Twenty-three mouse/human hybrid cell lines segregating mouse chromosomes were analyzed for the expression of mouse and human isozymes by starch gel electrophoresis.[9,10]

Karyotypes were performed on approximately 30 cells for each clone by a combination of Giemsa banding and Hoeschst 33258 centromeric staining.[11] Since our hybrid cells segregated mouse chromosomes rapidly, chromosome spreads and harvesting of cell pellets for isozyme analysis were generally performed within 1 week of each other.

Interferon Assay

Mouse A9 control cells and 20 mouse/human hybrid cell lines were tested for their sensitivity to partially purified mouse interferon (provided by Dr. P. Lengyel). Cells were exposed to serial dilutions of mouse interferon and challenged with vesicular stomatitis virus (VSV, 0.1 PFU/cell) in a cytopathic effects assay.[12] Cells were examined for cytolysis and protection from cell killing by interferon. Hybrid cells were generally less sensitive to mouse interferon than were mouse A9 cells. Interferon sensitivity assays were performed within 1 week of isozyme and karyotype analysis.

Linkage of the Genes for Mouse Cytoplasmic Superoxide Dismutase and Interferon Sensitivity and Assignment to Chromosome 16

Examination of the isozyme and interferon sensitivity data for the mouse/human hybrid cell lines that we studied revealed that the genes for mouse cytoplasmic superoxide dismutase (SOD-1) and interferon sensitivity were linked. Extensive karyotype analysis resulted in the assignment of these two genes to murine chromosome 16.[13]

These gene assignments are especially interesting since the same linkage relationship (SOD-1 and interferon sensitivity) has been described for the analogous human genes.[14,15] The two genes have been mapped to human chromosome 21, and a gene dosage relationship has been observed

[9] E. A. Nichols and F. H. Ruddle, *J. Histochem. Cytochem.* **21,** 1066 (1973).

[10] C. A. Kozak, R. E. K. Fournier, L. A. Leinwand, and F. H. Ruddle, *Biochem. Genet.* **17,** 23 (1979).

[11] C. A. Kozak, J. B. Lawrence, and F. H. Ruddle, *Exp. Cell Res.* **105,** 109 (1977).

[12] D. L. Slate and F. H. Ruddle, *Cell* **16,** 171 (1979).

[13] P.-F. Lin, D. L. Slate, and F. H. Ruddle, *Science* **209,** 285 (1980).

[14] Y. H. Tan, J. A. Tischfield, and F. H. Ruddle, *J. Exp. Med.* **137,** 317 (1973).

[15] C. Chany, M. Vignal, P. Couillin, N. V. Cong, J. Boue, and A. Boue, *Proc. Natl. Acad. Sci. U.S.A.* **72,** 3129 (1975).

for human interferon sensitivity in human fibroblasts with 1, 2, or 3 copies of chromosome 21.[16]

Many other examples of evolutionarily conserved linkages have been described. The linkage of SOD-1 and interferon genes in species as divergent as man and mouse suggests that these genes may be linked in other species as well.

The nature of the gene product responsible for conferring interferon sensitivity is not known. A large body of experimental evidence suggests that a cell surface receptor is a likely possibility.[17] Since *in vivo* genetic analysis in the mouse is more advanced than in humans, it may be possible to study mutations or variants at the murine interferon sensitivity locus and see whether a receptor is involved.

Acknowledgments

We wish to thank D. Bonito, F. Lawyer, and Dr. D. Pravtcheva for assistance in cell culture and karyotype analysis. This work was supported by Grant GM09966 from the NIH and a grant from the American Cancer Society.

[16] Y. H. Tan, E. L. Schneider, J. Tischfield, C. J. Epstein, and F. H. Ruddle, *Science* **186**, 61 (1974).
[17] C. Chany, *Biomed. Express* **24**, 148 (1976).

[68] Procedures for the Isolation and Characterization of Cell Lines Resistant to Interferon

By T. KUWATA

It has been reported that interferons exhibit pleiotropic actions on various mammalian cells.[1,2] Such actions have been verified to be the expression of interferon molecules themselves, not due to contaminants in the interferon preparation.[3] Therefore, cell lines resistant to interferon are considered to be those which are resistant to various actions of interferon. However, it is at present not clear whether the response of cell lines to such pleiotropic actions of interferon are expressed in parallel or not. There were several reports which indicate that cells resistant to cell

[1] I. Gresser, *Cell Immunol.* **34**, 406 (1977).
[2] W. E. Stewart, II, "The Interferon System." Springer-Verlag, Vienna and New York, 1979.
[3] I. Gresser, J. De Maeyer-Guignard, M. G. Tovey, and E. De Maeyer, *Proc. Natl. Acad. Sci. U.S.A.* **76**, 5308 (1979).

growth inhibiting action of interferon are at the same time resistant to antiviral action of interferon.[4,5] However, another group of workers[6] and we[7] have obtained different results. Human melanoma cell lines were compared as to their sensitivity to antiviral and anticellular actions of interferon, and it was revealed that sensitivities to these two actions of interferon are not always identical.[8] Thus, in this chapter procedures for the isolation and characterization of cell lines resistant to the cell growth inhibiting action of interferon are described. At the same time, sensitivity of such resistant cell lines to the antiviral action of interferon is also discussed.

Human interferons may be used widely as antiviral and antitumor agents in patients. Therefore, it seems important to isolate and study cells that show resistance to interferon. Such study may give useful information on clinical application of interferon and at the same time clarify the mode of action of interferon.

Procedures for the Isolation of Cells Resistant to Interferon

Isolation without Using Mutagen

The method of isolating resistant cell lines from an interferon-sensitive cell population is in principle identical with that of isolating cells resistant to growth-inhibiting drugs. General procedures for isolating mutant cells were described in detail by Thompson.[9] As a model system we used chiefly human transformed cell lines RSa and RSb. These cell lines are characterized by their high sensitivity to the anticellular action of human interferon.[7]

Effects of interferon on the growth of cells are influenced by the density of cells and units of interferon applied. Generally cell populations of high density are phenotypically resistant to the anticellular action of interferon. For example, monolayers of 2×10^5 cells of RSa or RSb in 60-mm plastic petri dishes were sensitive to 500 units of leukocyte interferon (IFN-α) per milliliter, but 10^6 cells/dish were less sensitive to the same concentration of IFN-α. Therefore, one must use a low cell density before the start of the interferon treatment. The amount of interferon should be determined in preliminary tests. In our case, for the isolation of

[4] I. Gresser, M.-T. Bandu, and D. Brouty-Boyé, *J. Natl. Cancer Inst.* **52**, 553 (1974).

[5] J. Hilfenhans, H. Damm, and R. Johannsen, *Arch. Virol.* **54**, 271 (1977).

[6] N. Fuchsberger, L. Borecky, and V. Hajnicka, *Acta Virol (Engl. Ed.)* **18**, 85 (1974).

[7] T. Kuwata, A. Fuse, and N. Morinaga, *J. Gen. Virol.* **33**, 7 (1976).

[8] A. A. Creasey, J. C. Bartholomew, and T. C. Merigan, *Proc. Natl. Acad. Sci. U.S.A.* **77**, 1471 (1980).

[9] L. H. Thompson, this series, Vol. 58, p. 308.

interferon-resistant cells from RSa cells, we started with about 2 to 4 × 10⁵ cells and 200 units of Le-IF per milliliter. If the continuous presence of interferon results in marked decrease in the cell number observed under microscopic examination, then interferon treatment should be stopped intermittently. After a few passages, when cells are found to grow in the presence of interferon, the amount of interferon should be increased gradually, for example, from 200 up to 1000 or 2000 units/ml. By these procedures, interferon-resistant cell populations are obtained. Then, by plating single-cell suspensions onto petri dishes, clonal cell strains are isolated. The IF^r cell line was thus isolated. When 2 × 10⁵ cells were seeded on petri dishes 60-mm in diameter, IF^r cells are fairly resistant to 1000–2000 units of IFN-α per milliliter. The resistance of this strain is not absolute as reported for L1210 R cells which are resistant to 100,000 units of interferon per milliliter.[4]

Frequency of isolating mutant cells is apparently dependent on each cell strain. Gresser and his associates[10,11] used mouse leukemic cell strain L1210 for their experiments. From this strain resistant mutant strains were isolated in high frequency. The strain was suggested to contain spontaneously occurring interferon-resistant cells in 1–5% of the parent cell population. By adopting continuous treatment with a low dose of mouse interferon (40–100 units/ml), resistant cell strains were isolated from L cells[6] after 17–50 passages and also from transformed BALB/c mouse embryonic cells.[12] In the latter case, resistance to the antiviral action of interferon was recognized after 200 passages, and then resistance to anticellular action was reported.[13] If the concentration of interferon can be increased more rapidly, it may not be necessary to make many cell passages with interferon before obtaining resistant cells.

Interferon-resistant cell clones were isolated from cell colonies by inoculating an appropriate number of cells in soft agar or agarose medium, in which interferon was incorporated.[4] It is also possible to isolate interferon-resistant clones from selected colonies after plating cells in plastic petri dishes and then adding certain amounts of interferon.[8] However, the colony-forming ability of cells is more sensitive to interferon action than growth of monolayer cells. Therefore, the choice of interferon concentration should be carefully determined.

[10] I. Gresser, D. Brouty-Boyé, M.-T. Thomas, and A. Macieira-Coelho, *Proc. Natl. Acad. Sci. U.S.A.* **66,** 1052 (1970).

[11] I. Gresser, D. Brouty-Boyé, M.-T. Thomas, and A. Macieira-Coelho, *J. Natl. Cancer Inst.* **45,** 1145 (1970).

[12] C. Chany and M. Vignal, *J. Gen. Virol.* **7,** 203 (1979).

[13] P. Lindahl-Magnusson, P. Leary, and I. Gresser, *Proc. Soc. Exp. Biol. Med.* **138,** 1044 (1971).

On the other hand, some cell strains are more sensitive to cell growth-suppressing action of human fibroblast interferon (IFN-β) than to IFN-α.[14,15] Thus, attempts were made to isolate mutant strains resistant to IFN-β from our RSa cell population by the method described above, but they were not successful.

Isolation Using Mutagens

If it is difficult to isolate interferon-resistant mutants from a sensitive population, then attempts should be made to use a mutagen for such purpose. Commonly used mutagens for mammalian cells are ethyl methane sulfonate (EMS), methyl-N'-nitrosoguanidine (MNNG), and others. Procedures for isolating mutant cells by using mutagens were described in detail in this series.[9] We adopted EMS for isolating IFN-β resistant mutant cells from RSa. To exponentially growing RSa cells in 100-mm petri dishes, application of 100 μg of EMS per milliliter for about 24 hr was the optimal condition for such purpose. Then cells were washed three times with serum-free MEM, and the growth medium was added to the cells. For the period of recovery from the cytotoxicity of the mutagen and at the same time for the fixation of the mutation, surviving cells must be further incubated for about 2–3 weeks. This period of culture is dependent on cell type. Direct selection of interferon-resistant mutants from growing colonies by adding the appropriate concentration of interferon is desirable, but not always successful. Owing to the high sensitivity to interferon and low colony-forming ability of RSa cells, we preferred to isolate resistant cells by intermittent exposure to interferon and subculturing in the presence of interferon, for example, 20 units of fibroblast interferon per milliliter. Thus, we could isolate 3 fibroblast interferon-resistant cell lines from the RSa population.[15] On the other hand, by using a melanoma cell line treated with 200 μg of EMS per milliliter for 24 hr, we could easily isolate clones, which are resistant to 1000 units of fibroblast interferon per milliliter, from colonies grown in petri dishes (unpublished data).

Spontaneously Resistant Strain

Since human tumor cell lines are divergent in their sensitivities to interferon,[5,8,16] some strains are spontaneously resistant to interferon. Therefore cell lines should be examined by estimating the sensitivity of cells to 100, 1000, and 10,000 units of IFN-α or IFN-β per milliliter. In this

[14] S. Einhorn and H. Strander, *J. Gen. Virol.* **35**, 573 (1977).
[15] T. Kuwata, A. Fuse, N. Suzuki, and N. Morinaga, *J. Gen. Virol.* **43**, 435 (1979).
[16] A. Adams, H. Strander, and K. Cantell, *J. Gen. Virol.* **28**, 207 (1975).

way, interferon-resistant cell strains will be incidentally found. We have found a human adenocarcinoma cell line named HEC-1 to be resistant to the antiviral and anticellular action of even 5000 units of IFN-α and IFN-β.[16a,16b]

Characterization of Cell Lines Resistant to Interferon

Two strains of interferon-resistant cells, mouse L1210 R and human IF[r], have been maintained and their characteristics were analyzed. Therefore, the characterization of interferon-resistant cell lines is described by referring to these cells.

At first, sensitivities of isolated cell strains to antiviral and cell growth-inhibiting action of interferon must be estimated quantitatively. Procedures for these assays are described in other sections of this volume. We seed 2×10^5 cells into 60-mm petri dishes, and the next day the cells are treated with various concentrations of interferon for 6 days. Then, survival percentages are calculated by counting viable cells in each dish. For a cell line of high plating efficiency, it is better to evaluate resistance of cells to interferon by estimating reduction in colony formation.

Stability of interferon resistance should be also examined. This can be followed by making cell passages *in vitro* without adding interferon. L1210 R cells are quite stable in interferon resistance, and one such subline was cultured in the absence of interferon for 4 years and found still to be interferon resistant.[4]

IF[r] cells are not as stable in their interferon resistance. Without addition of interferon, IF[r] cells reverted to parental type sensitivity to interferon over a period of 1 year.

L1210 R cells were at first isolated as a mutant strain resistant to the cell growth-suppressing action of interferon. They are at the same time resistant to the antiviral action of interferon. This fact has been considered to be the evidence that L1210 R cells are resistant to interferon itself, not to other contaminants in the interferon preparation.

Resistance of IF[r] cells to the anticellular action of human interferon is not absolute, and the cells are only partially resistant to 2000 units/ml or higher amounts of IFN-α. IF[r] cells are still sensitive to the antiviral action of human interferon despite their resistance to the anticellular action of interferon. One of the reasons for such a discrepancy may be that in IF[r] cells there still remain receptors for interferon in the cell membrane. When IF[r] cells were treated with a total of 10,000 units of IFN-α, 0.1% of

[16a] M. Verhaegen, M. Divizia, P. Vandenbussche, T. Kuwata, and J. Content, *Proc. Natl. Acad. Sci. U.S.A.* **77**, 4479 (1980).

[16b] H-Y. Chen, T. Sato, A. Fuse, T. Kuwata, and J. Content, *J. Gen. Virol.* **52**, 177 (1981).

COMPARISON OF INTERFERON-SENSITIVE AND RESISTANT CELLS

Cell lines	Sensitivity to cell growth inhibitory action	Sensitivity to antiviral action	Induction of (2'-5')-oligo(A) synthetase and protein kinase
L1210 S	+++	+++	++++[a]
L1210 R	−	−	−[a]
RSa	++++	++++	++++[27]
IF[r]	+	++~+++	++++[27]
HEC-1	−	−	(+)[16a]

[a] Data supplied by J. Content.

the applied interferon was recovered.[7] This fact may indicate that the interferon receptor in IF[r] cells was reduced, but not absent. From L1210 R, cells applied interferon was not recovered at all.[4] To recover the cell-bound interferon, cells grown in plastic petri dishes 100–150 mm in diameter are treated with various amounts of interferon for 2–4 hr, then washed three times with MEM. The cells are scraped off into 10 ml of MEM with a rubber policeman. The suspension is centrifuged at 800 rpm for 5 min, and the cell pellet is resuspended in 2–4 ml of MEM, sonicated, and then assayed for antiviral activity.[7,17]

Receptors for interferon are considered at least in part to be composed of gangliosides.[18,19] Glycolipid composition of RSa and IF[r] cells were compared and no appreciable difference in the amount of GM_2 and GM_3 was found.[20] However, the existence of differences in glycolipid or glycoprotein content between interferon sensitive and resistant cells cannot be excluded.

Sensitivity of cells to interferon action has been demonstrated to be coded for by chromosome 21.[21,22] Therefore, karyotypes of parental and mutant cells should be examined with special attention on chromosome 21. Karyotypes of interferon sensitive RSa and resistant IF[r] cells were compared, but specific chromosomal changes were not identified that

[17] W. E. Stewart, II, E. De Clercq, and P. De Somer, *J. Virol.* **10**, 707 (1972).

[18] F. Besancon and H. Ankel, *Nature (London)* **252**, 478 (1974).

[19] V. E. Vengris, F. H. Reynolds, Jr., M. D. Hollenberg, and P. M. Pitha, *Virology* **72**, 486 (1976).

[20] T. Kuwata, S. Handa, A. Fuse, and N. Morinaga, *Biochem. Biophys. Res. Commun.* **85**, 77 (1978).

[21] Y. H. Tan, in "Genetics of the Human Interferon System" (W. E. Stewart, II, ed.), p. 73. CRC Press, Cleveland, Ohio, 1977.

[22] D. L. Slate and F. H. Ruddle, *Cytogenet. Cell Genet.* **22**, 265 (1978).

could explain interferon resistance.[23] The method of karyotyping has been described in this series.[24]

Interferon treatment induces (2'-5')-oligo(A) synthetase and protein kinase in interferon-sensitive cells,[25] but not in interferon-resistant cells, such as L1210 R cells.[26] However, it has been demonstrated that an IF[r] cells interferon enhances these enzyme activities.[27] A detailed method for assay of these enzymic actions is included in other sections of this volume. Comparison of three types of interferon-resistant cells are given in the table.

---

[23] T. Kuwata, A. Fuse, and A. Tonomura, *in* "Proceedings of the Second International Workshop on Interferons" (M. Krim and W. E. Stewart, II, eds.). Rockefeller Univ. Press, New York, 1980.

[24] R. G. Worton and C. Duff, this series, Vol. 58, p. 322.

[25] C. Baglioni, *Cell* **17**, 255 (1979).

[26] P. Vandenbussche, J. Content, B. Lebleu, and J. Werenne, *J. Gen. Virol.* **41**, 161 (1978).

[27] P. Vandenbussche, M. Divizia, M. Verhaegen, A. Fuse, T. Kuwata, E. De Clercq, and J. Content, *Virology* **111**, 11 (1981).

[69] Methods for Enucleation and Reconstruction of Interferon-Producing Cells

By DEREK C. BURKE

A number of different cultured mammalian cells can be enucleated by centrifugation in the presence of cytochalasin B. At low doses (1–5 μg/ml), this fungal metabolite affects cell attachment, endocytosis, and membrane permeability, and also cytokinesis, whereas at higher doses (>10 μg/ml) it causes some cell enucleation (for a review see Tanenbaum[1]). The extent of enucleation can be increased by centrifugation in the presence of the drug, probably because the nucleus is forced into a cytoplasmic stalk, which subsequently severs spontaneously. In this way, anucleate cells consisting of most of the cytoplasm of the intact cell, surrounded by an intact plasma membrane, and extruded nuclei, consisting of the nucleus surrounded by a small amount of cytoplasm and a plasma membrane, can be prepared. The anucleate cells are also termed cytoplasts; and the extruded nuclei, minicells or karyoplasts (for a review see Ringertz and Savage[2]). The terms cytoplast and karyoplast will be used in this chapter.

---

[1] S. W. Tanenbaum, "Cytochalasins." North-Holland Publ., Amsterdam, 1978.

[2] N. R. Ringertz and R. E. Savage, "Cell Hybrids." Academic Press, New York, 1976.

METHODS IN ENZYMOLOGY, VOL. 79

In spite of the drastic conditions used to prepare these cell fragments, the cytoplasts rapidly resume their normal shape, which has been greatly altered during the cytochalasin treatment, and synthesize protein for about 12–18 hr. The karyoplasts are fragile and easily damaged, but will synthesize RNA and DNA for some time after enucleation, and about 10% of the karyoplasts will ultimately form viable cell colonies.[3]

It is also possible to reconstruct viable cells by fusion of cytoplasts and karyoplasts with inactivated Sendai virus.[2] Such cells will form viable cell colonies, and these may be selected for by use of culture conditions that inhibit division of the parental cells.[4–7]

Preparation of Cell Fragments

Principle

Cells are centrifuged in the presence of cytochalasin B to separate the karyoplasts from the cytoplasts. This is usually most conveniently carried out by treating a cell monolayer so that the cytoplasts remain attached to a solid support while the karyoplasts are recovered from the bottom of the centrifuge tube. However, methods for enucleation of cells in suspension, using centrifugation in a discontinuous Ficoll gradient or in a colloidal silica gradient, have also been described.[8,9] My limited experience with such techniques has been unsuccessful, and the techniques will not be discussed here.

The efficiency of enucleation varies from one cell type to another, and depends on cell density, dose of cytochalasin B, speed of centrifugation and temperature.[10–13] Cells with a fibroblastic morphology tend to be enucleated poorly, since the karyoplasts are trapped in the fibroblastic processes, and some cell type adhere so loosely that they leave the solid support without undergoing enucleation. This may sometimes be

[3] G. A. Zorn, J. J. Lucas, and J. R. Kates, *Cell* **18**, 659 (1979).

[4] G. Veomett, D. M. Prescott, J. Shay, and K. R. Porter, *Proc. Natl. Acad. Sci. U.S.A.* **71**, 1999 (1974).

[5] T. Ege, U. Krondahl, and N. R. Ringertz, *Exp. Cell Res.* **88**, 428 (1974).

[6] T. Ege and N. R. Ringertz, *Exp. Cell Res.* **94**, 469 (1975).

[7] U. Krondahl, N. Bols, T. Ege, S. Linder, and N. R. Ringertz, *Proc. Natl. Acad. Sci. U.S.A.* **74**, 606 (1977).

[8] M. H. Wigler and I. B. Weinstein, *Biochem. Biophys. Res. Commun.* **63**, 669 (1975).

[9] D. M. Prescott and J. B. Kirkpatrick, *Methods Cell Biol.* **7**, 189 (1973).

[10] W. E. Wright, *Methods Cell Biol.* **7**, 203 (1973).

[11] G. Poste, *Methods Cell Biol.* **7**, 211 (1973).

[12] R. D. Goldman and R. Pollack, *Methods Cell Biol.* **8**, 123 (1974).

[13] T. Ege, H. Hamburg, U. Krondahl, J. Ericsson, and N. R. Ringertz, *Exp. Cell Res.* **87**, 365 (1974).

prevented by pretreating the surface with collagen,[12] concanavalin A,[14] fibronectin, polylysine, or protamine sulfate (G. Veomett, personal communication). Under optimal conditions, close to 100% of the cells can be enucleated. Mouse L cells have been used extensively and enucleate well; they are recommended for establishing the technique. The percentage of efficiency of enucleation is readily estimated by staining the cytoplasts with crystal violet after recovery; any nucleated cells are then obvious, and both nucleated and enucleated cells can be counted.

Procedure

Cells are grown either on round glass disks (diameter = 25 mm with 2 to 3×10^5 cells/disk) or in 25-cm^2 tissue culture flasks (with 1×10^6 cells/ flask) for 24 hr to allow the cells to attach firmly. The only flasks that I have found to be satisfactory are those manufactured by Falcon, the type 3012 flasks, which have a small black cap that is not at an angle to the flask. All the other types of flasks I have used either crack or break at the base of the neck during the centrifugation procedure.

Enucleation Using Glass Disks[7]

The tissue culture medium is removed, and the disks are placed, cell-side down, in a centrifuge tube for the SS34 rotor of a Sorvall R2C centrifuge. A plastic plug may be inserted immediately above the disk to prevent it from tilting during centrifugation, and the tube is filled with warm medium containing 10 μg of cytochalasin B per milliliter (Aldrich Chemical Co., Milwaukee, Wisconsin). Alternatively several disks can be placed in a single centrifuge tube separated by plastic annular spacers, but if this is done many of the karyoplasts from the upper disks will be lost. The tube is centrifuged at 23,500 g (10,000 rpm) for 20 min at 37°. The cytoplasts remain attached to the glass disks and will recover their normal morphology when placed in normal growth medium for 1 hr at 37°. The karyoplasts are obtained by gentle resuspension of the pellet in growth medium. The pellet also contains some intact cells, the number depending on the cell type. The number of whole cells may be reduced either by prior centrifugation of the disks in medium without cytochalasin B, or by incubation of the resuspended pellet in a tissue culture dish for 4 hr at 37° when the intact cells, but not the karyoplasts, attach.[15]

[14] T. V. Gopalakrishnan and W. F. Anderson, *Proc. Natl. Acad. Sci. U.S.A.* **76**, 3932 (1979).
[15] J. J. Lucas, E. Szekely, and J. R. Kates, *Cell* **7**, 115 (1976).

Enucleation Using Tissue Culture Flasks[16]

The tissue culture medium is removed, and the flasks are completely filled with warm growth medium containing 10 μg of cytochalasin B per milliliter (about 40 ml/flask). The cap is replaced and tightened. The enucleation procedure is carried out in a Sorvall RC-2B centrifuge using the GSA rotor, both of which are warmed to 37°. Approximately 125 ml of water at 37° are added to each rotor well, and then a filled flask is placed in each well, with the cell side nearest to the center. The rotor, with its cover off, is then centrifuged at approximately 500 rpm, and water is squirted into the rotor wells until all the wells contain the maximum volume of water. This is apparent when water is ejected from the rotor. The rotor is stopped, the cover is replaced, the centrifuge bowl is dried, and centrifugation is performed at 8500 rpm for 30 min. The karyoplasts form a pellet in the corner of the flask and are easily recovered. Since six flasks can be enucleated at the same time, 6 to 9 $\times$ 10^6 cytoplasts and karyoplasts can be recovered from a single run. If more karyoplasts are required, then the resuspended karyoplasts may be kept at 37° while a second run is performed. It should be noted that other types of centrifuge rotors may not be suitable; I have had no success with the MSE-18 centrifuge and the 6 $\times$ 250 ml rotor.

Fusion of Cell Fragments—Reconstruction

Principle

Karyoplasts and cytoplasts are incubated in the presence of UV-inactivated Sendai virus when fusion occurs with the production of viable nucleated cells. Since interferon production by the reconstituted cells utilizes the cells produced by the fusion rather than cells that have been grown in selective media to eliminate parental cells, it is essential that enucleation be complete, since otherwise it will not be possible to determine whether the interferon is formed by unenucleated cells or reconstructed cells. Any contribution due to unenucleated cells can be measured by fusing cytoplasts and karyoplasts using totally inactivated Sendai virus; any interferon produced will then be due to unenucleated cells.[17]

Preparation of Inactivated Sendai Virus

Sendai virus is inoculated into the allantoic cavity of 10-day-old eggs with 10^4 egg-infective doses$_{50}$ (EID$_{50}$) of Sendai virus in 0.1 ml (using a

[16] G. Veomett, J. Shay, P. V. C. Hough, and D. M. Prescott, *Methods Cell Biol.* **13**, 1 (1976).
[17] D. C. Burke and G. Veomett, *Proc. Natl. Acad. Sci. U.S.A.* **74**, 3391 (1977).

standard allantoic fluid preparation, this corresponds to about a 10^3 dilution in medium containing 2% calf serum). The eggs are incubated in a humidified incubator at 37° for 48 hr and then chilled at 4° overnight. The allantoic fluid is harvested by removing the air-sac end of the egg, cutting carefully through the membrane (so as to avoid bleeding), and removing the fluid with a Pasteur pipette, holding the embryo to one side with a spoon spatula. The infected allantoic fluid is clarified by centrifugation at 3000 rpm for 15 min, and the amount of virus measured by hemagglutinin titration. To do this, doubling dilutions of virus (volume 0.25 ml) are made in saline in a disposable multiwell tray, and then an equal volume (0.25 ml) of 0.5% fowl red cells is added. After standing for 30 min, the highest dilution showing partial agglutination is determined, and the reciprocal of the dilution giving the number of hemagglutining units (HA units) per 0.25 ml. The infectivity titer may also be determined by injecting 0.05 ml of serial 10 × dilutions into 10-day-old fertile eggs (6 eggs per dilution), incubating at 37° for 3 days, and determining in which eggs virus has multiplied by harvesting a small amount of allantoic fluid and testing for the presence of viral hemagglutinin by adding an equal volume of 0.5% fowl red cells. The infectivity titer is then calculated by a standard method (e.g., as described by Goodheart[18]). Some virus should be stored in small aliquots at −70° as stock seed, and the remainder is inactivated.

Virus may be inactivated either by ultraviolet (UV) irradiation or by treatment with β-propiolactone, but since the latter is carcinogenic, the UV irradiation procedure is recommended. It is essential to dialyze the virus against phosphate-buffered saline before irradiation, since this removes soluble material that adsorbs UV light (mainly salts of uric acid) and ensures that the optimal UV irradiation condition is the same for each batch of allantoic fluid. It is best to determine the optimal time of UV irradiation by direct testing, since the intensity of irradiation and the amount of UV-adsorbing impurities can vary so widely, and, in any case, the relationship between the inactivation of virus infectivity by UV irradiation and the production of virus with maximum fusing ability is not understood. The simplest procedure is to irradiate a thin layer of virus (1 ml in a 5-cm petri dish) at a standard distance from the UV light source for different times, and determine the conditions that give virus with maximum fusing ability, i.e., by determining the fusion index as described by Ringertz and Savage (pp. 40–45).[2] A batch of inactivated virus can then be prepared and stored in aliquots at −70°.

[18] C. R. Goodheart, "An Introduction to Virology." Saunders, Philadelphia, Pennsylvania, 1969.

Cell Reconstruction

Cell fusion requires calcium, which is supplied at near the optimal concentration by the use of either Hanks' or Earlcs' buffered salt solution. The optimal pH for fusion is near pH 8.0.

Cell reconstruction can be separated into two stages: the agglutination of cytoplasts and karyoplasts by the Sendai virus and the actual fusion. The former occurs as well at 4° as at 37°, and it is usual to include a 4° incubation in the process so that metabolic activity slows and the viral enzyme neuraminidase, whose action leads to elution of the virus, is inactive. After this period at 4°, the temperature is raised to 37° so that fusion (which does not occur at 4°) may occur. Warming to 37° will also permit virus elution from the cell surfaces. The usual conditions involve incubating cytoplasts, attached to a solid support, with UV-inactivated Sendai virus (about 200 HA units/10^6 cells) at 4° for 15–20 min. Then excess karyoplasts (~10 karyoplasts per cytoplast) are added, and the mixture is incubated at 37° for 30–45 min.[7] I found it advantageous to remove excess Sendai virus and then to wash these cells with ice-cold buffered salt solution before addition of karyoplasts, since this prevented agglutination of karyoplasts by the excess Sendai virus.[17] In this way yields of reconstructed cells were as high as 3–8%.

Production of Interferon by Enucleated and Reconstructed Cells

Mouse L cells produce high titers of interferon (up to 10^5 international research units) when induced with NDV strain F. However, enucleated cells do not produce any interferon, whereas reconstructed cells produce as much interferon per cell as do normal L cells.[17] Thus the interferon system, requiring a nuclear gene and cytoplasmic protein synthesis, has also been reconstructed.

Section VIII

Preparation and Assay of Antibodies against Interferons

[70] Preparation and Absorption of Antiserum against Human Leukocyte Interferon

By BARBARA DALTON and KURT PAUCKER*

Specific antisera to interferons provide a valuable tool for their purification and characterization. When anti-interferon sera are available in sufficient quantity, antibody affinity chromatography can be used to increase the purity of an interferon preparation up to 10,000-fold. The specificity of antibodies has also made possible studies on the antigenic similarities and differences in interferons.

This chapter describes the production and absorption of an antiserum to Sendai virus-induced human leukocyte interferon. A sheep was used for antiserum production in the work described below; however, similar procedures can be employed in other animal species.

Production of Antiserum to Human Leukocyte Interferon

Principle. Ideally, a fairly pure preparation should be used for immunization, but since large quantities of highly purified interferon were unavailable, partially purified materials were used. A sheep was utilized for the production of the antiserum in order to prepare a large quantity of serum. Although the immunogen was not highly purified, a large volume of high-titered antiserum was obtained. Antibodies to contaminants were removed by the immunoabsorption procedure described below.

Preparation of Inoculum. Although interferons of varying purity have been used for immunization, Mogensen et al.[1] have reported that the best response is obtained when a highly purified interferon is used for inoculation. There are many different purification procedures available for human leukocyte interferon, and interferon purified by any of these methods can be used for immunization. It has been noted that the inducer has an effect on the antigenic type of interferon produced[2]; therefore, interferons produced by the same method should be used in an immunization schema.

The Sendai virus-induced human leukocyte interferon used for the immunization of the sheep represented in Fig. 1 was of two different degrees

---

* Deceased.

[1] K. E. Mogensen, L. Pyhälä, and K. Cantell, *Acta Pathol. Microbiol. Scand. Sect. B* **83**, 443 (1975).

[2] E. A. Havell, T. G. Hayes, and J. Vilček, *Virology* **89**, 330 (1978).

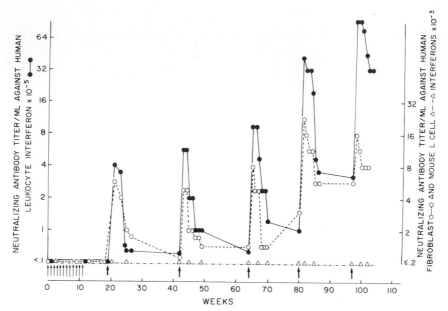

Fig. 1. Development of antibodies to human leukocyte and fibroblast interferons in a sheep immunized with human leukocyte interferon. The small arrows indicate inoculation with 2.5×10^6 units of interferon (specific activity = 1×10^4 units/mg). Large arrows indicate booster inoculations containing 2.5×10^7 units of interferon (specific activity = 1×10^6) mixed with an equal volume of Freund's complete adjuvant. The neutralizing activity against the human interferons was measured with human FS-4 cells, and the activity to mouse L cell interferon was determined with mouse L-929 cells.

of purity. Initial immunization utilized crude interferon[3] with a specific activity (SA) of 1×10^4 units/mg of protein. The interferon utilized for booster inoculations was partially purified by antibody affinity chromatography[4], or by physicochemical means[5] and had specific activities in the range of 1×10^6 units per milligram of protein. The antibody-purified material was dialyzed against 0.01 M sodium phosphate buffer, pH 7.0, and then lyophilized in the presence of 0.5% sheep albumin. The material was reconstituted in a small volume of phosphate-buffered saline (PBS), pH 7.0, for inoculation.

Choice of Animal for Immunization. A variety of animals have been used for the production of antisera to interferons. The sheep used in this

[3] H. Strander and K. Cantell, *Ann. Med. Exp. Biol. Fenn.* **44,** 265 (1966).
[4] K. Berg, C. A. Ogburn, K. Paucker, K. E. Mogensen, and K. Cantell, *J. Immunol.* **114,** 640 (1975).
[5] K. Cantell, S. Hirvonen, K. E. Mogensen, and L. Pyhälä, *In Vitro* **3,** 35 (1974).

study is a Suffolk-Hampshire female, 1 year old at the start of the immunization schedule. The animal chosen for immunization is dependent upon the quantity of antiserum desired and the amount of interferon available for inoculation. Large animals require more interferon to induce an antibody response than do small animals. For example, approximately a total of 1×10^6 units, administered in multiple injections, is necessary to obtain detectable antibody titers in a mouse, whereas 2×10^7 and 5×10^7 units are needed to stimulate rabbits and sheep, respectively. Pyhälä et al.[6] have reported that 10 injections of 1 to 2.5×10^5 units/kg animal weight are desirable.

Immunizations. The immunization schedule illustrated in Fig. 1 was based on the procedure described by Mogensen et al.[1] Figure 1 illustrates the development of antibodies to human leukocyte interferon over a 110-week period, arrows indicate the times of injection. After 12 weekly injections of 2.5×10^6 units of interferon (specific activity $= 1 \times 10^4$ units per milligram of protein), the animal was boosted following a 6-week period with a more purified preparation (specific activity $= 1 \times 10^6$ units per milligram of protein) containing 2.5×10^7 units of interferon mixed with an equal volume of Freund's complete adjuvant (FCA). Other booster inoculations were given periodically, usually allowing a rest period of 6 weeks or longer between the last bleeding and the next inoculation.

All the weekly injections as well as subsequent injections containing FCA were administered by subcutaneous inoculation in four or five sites. Immunogen was injected into the soft tissue areas just above the fore and hind limbs on the ventral surface, proximal to the superficial inguinal and axillary lymph nodes to facilitate the rapid processing of the antigen.

Schedule of Bleedings. The sheep was bled from the jugular vein at the times indicated by the circles in Fig. 1. The blood was allowed to clot at 4° overnight, and the serum was collected. One milliliter samples were removed for assay and the bulk of the serum stored at $-20°$. Initially small quantities of blood (25 ml) were taken to monitor the rise of antibody titers. When sufficiently high antibody levels had developed, larger volumes of blood were collected. The antiserum titers to human leukocyte interferon were maximal 1 week after each booster inoculation, and the level remained within twofold of the maximum for approximately 5 weeks (Fig. 1).

When large volumes were repeatedly drawn from the animal, the hematocrit was monitored. Removal of 300 ml of blood biweekly lowered the hematocrit significantly; therefore, weekly bleedings of 200 ml of

[6] L. Pyhälä, *Acta Pathol. Microbiol. Scand. Sect. C* **36,** 291 (1978).

blood were taken for 6 weeks after each booster. During this period, injections of iron and a B complex vitamin mixture were administered on the day blood was drawn. When such a protocol was followed, the hematocrit levels remained in the normal range.

Interferon Neutralization Assay. Serum that had been heated for 30 min at 56° to inactivate complement was tested for the ability to neutralize human leukocyte and fibroblast interferons. The assay used was described by Paucker *et al.*[7] Briefly, dilutions of the antiserum were titrated against several dilutions of interferon in a checkerboard fashion; after incubation for 1 hr at 37°, 30,000 FS-4 cells contained in 0.1 ml were added to each well, and the plates were returned to the incubator. The next day 0.05 ml of encephalomyocarditis virus was added to each well, to give a multiplicity of infection equal to 0.2. The following day the plates were examined microscopically for the extent of cytopathic effect. The highest dilution of antiserum that neutralized eight reference units of interferon by partially restoring viral cytopathic effect, corrected to 1.0 ml volume, represents the titer of the antiserum.

In the same manner the antibody titers to mouse L cell interferon were measured with 50,000 mouse L-929 cells per well rather than human FS-4 cells as the target cell.

Figure 1 illustrates the development of antibodies to both the leukocyte (Le) and fibroblast (F) antigenic types, contained in the human leukocyte interferon use for immunization. The titers to these antigens increased after each booster inoculation. In an effort to detect cross species antigenic similarities, the neutralization of mouse L cell interferon was attempted. The serum contained no detectable neutralizing antibodies to mouse L cell interferon by the assay described above.

Passive Hemagglutination Test. Antibody titers to known contaminants found in the interferon preparation were measured by passive hemagglutination by the procedure of Boyden[8] as modified by Ogburn *et al.*[9] Washed sheep red blood cells were treated with tannic acid (25 μg/ml) and incubated with the test antigens for 30 min at 37° to facilitate binding of antigens to the surface of the red blood cells. The cells were then washed to remove unbound antigen and used as the indicator in the assay. Serial dilutions of the heat-inactivated antiserum were made in a round-bottom microtiter plate (Linbro Scientific No. 76-311-05), and 0.05 ml of a 0.6% solution of the antigen-coated sheep red blood cells was added to each well. Controls included known positive sera and antigen-coated sheep red

[7] K. Paucker, B. J. Dalton, C. A. Ogburn, and E. Törmä, *Proc. Natl. Acad. Sci. U.S.A.* **72**, 4587 (1975).
[8] S. V. Boyden, *J. Exp. Med.* **93**, 107 (1951).
[9] C. A. Ogburn, K. Berg, and K. Paucker, *J. Immunol.* **111**, 1206 (1973).

blood cells in diluent alone (PBS containing 1% normal rabbit serum). The mixture was incubated at room temperature for 4 hr, and the pattern of hemagglutination was recorded. The presence of antibodies to the test antigen is indicated by agglutination of the antigen-coated red blood cells. The titer of the serum is defined as the reciprocal of the highest dilution showing complete hemagglutination, corrected for a 1 ml volume.

Three known contaminants of Sendai virus-induced human leukocyte interferon were chosen to monitor the success of the absorption procedure presented: (a) egg albumin, since the inducer virus was produced in embryonated eggs; (b) human albumin, present in the serum used to cultivate the cells; and (c) human leukocyte extract, a mixture of antigens from sonicated human leukocytes.

Absorption of Antiserum

Principle. Animals inoculated with impure interferon preparations produce antibodies to contaminants as well as to the interferon. If these contaminants are known and if they can be obtained, the removal of antibodies to these contaminants from the anti-interferon serum by immunoabsorption techniques can be attempted. In this absorption procedure antigen is bound to a solid support and subsequently mixed with the antibody preparation in order to remove antibodies to the insolubilized antigen.

Absorptions. The antigens used for the immunoabsorption of the sheep anti-human leukocyte interferon serum were those antigens contained in the unbound fraction from a previous interferon purification procedure. In some cases it was necessary to supplement with known antigens. For example, the initial absorption of the anti-human leukocyte interferon serum with unbound material indicated that very little, if any, antibodies to egg albumin were removed, so egg albumin was added to the unbound material for use in the absorption procedures.

The unbound fraction and egg albumin were bound to cyanogen bromide-activated Sepharose 4B (Pharmacia Fine Chemicals, Piscataway, New Jersey) by the manufacturer's procedure. Heat-inactivated antiserum was mixed with the antigen-bound Sepharose and stirred at 4° overnight. The absorbed serum was collected by filtration. The Sepharose was then washed three times with PBS. The absorbed antiserum and washes were tested individually to determine the titer of antibodies to the three test contaminants by the passive hemagglutination procedure and for anti-interferon activity in a neutralization test. The washes containing neutralizing activity and with reduced antibody titers to the test antigens were pooled with the absorbed serum. The bound antibodies were

ANTIBODY TITERS IN UNABSORBED AND ABSORBED SHEEP ANTI-HUMAN LEUKOCYTE
INTERFERON SERUM

		Antibody titer per ml[a] against			
Sample	Treatment	Human leukocyte interferon	Egg albumin	Human albumin	Human leukocyte extract
1	Unabsorbed	3.0×10^6	8.0×10^3	1.3×10^5	8.0×10^3
	Absorbed	3.0×10^6	$4.0 \times 10^2 \ (95)^b$	$4.0 \times 10^2 \ (>99)$	$4.0 \times 10^2 \ (95)$
2	Unabsorbed	6.0×10^5	6.4×10^4	8.0×10^6	1.3×10^5
	Absorbed	3.0×10^5	$4.0 \times 10^2 \ (>99)$	$1.6 \times 10^3 \ (>99)$	$6.4 \times 10^3 \ (95)$
3	Unabsorbed	4.0×10^5	3.2×10^5	3.2×10^5	3.2×10^5
	Absorbed	2.0×10^5	$6.4 \times 10^3 \ (98)$	$1.6 \times 10^3 \ (99)$	$1.6 \times 10^3 \ (>99)$
Pool	Absorbed	1.0×10^6	4.0×10^2	4.0×10^2	1.6×10^3

[a] Expressed as reciprocal of end-point dilution. Titers to interferon were determined by neutralization; others by passive hemagglutination.
[b] Values in parentheses indicate percentage removed by absorption.

removed from the Sepharose-antigen complex by high salt-low pH buffer or urea, and the complex was reused for future absorptions.

The antibody titers of three serum samples, before and after absorption, are presented in the table. The data indicate that in all cases the titers against the three test antigens were reduced by 95% or greater while the anti-interferon titer remained within a twofold range of the titer in the unabsorbed serum. The level of antibodies to the test contaminant antigens that remained after absorption were reduced to background titers found in control serum taken prior to immunization.

Further Processing of the Antiserum

The absorbed serum, pooled with the appropriate washes, was precipitated with 50% ammonium sulfate to obtain the globulin fraction, reconstituted to the original serum volume with distilled water, and dialyzed against an appropriate buffer. The absorbed globulin can then be stored frozen or lyophilized. The globulin fraction can be further purified by isolating the IgG portion by any of various available chromatographic procedures.

Serum collected from the sheep illustrated in Fig. 1 was absorbed as indicated in the table; globulin was precipitated and dialyzed against 0.01 M phosphate buffer. After sterilization by filtration the material was lyophilized. The immunoglobulin G fraction was not isolated from the globulins in this case.

Concluding Comment

The wide use of interferon in research made it desirable to have antiserum standards that antigenically identify a variety of interferon preparations used in different laboratories. For this reason NIH supported an effort to produce antiserum standards. This chapter and chapter [74] illustrate our methods for the production and absorption of two such antisera under NIH contract NOI AI 82568.

[71] Preparation of Antibodies to Human Lymphoblastoid Interferon

By Mark E. Smith

Antibodies to interferons have been used to a great extent in immunoadsorbent affinity chromatography, a key step in several interferon purification procedures.[1-4] The high degree of purification attained by this method and the ability of specific antisera to differentiate between the various types of interferons make the preparation of anti-interferons essential. Standard immunization techniques using human leukocyte or lymphoblastoid interferons of varying degrees of purity as immunogen have resulted in activities in the range of 10^3-10^6 interferon units neutralized per milliliter of serum.[5-7] The methods reported here have been successful in producing antibodies to partially purified and homogeneous human lymphoblastoid interferon by immunization of sheep and rabbits.[8,9]

Immunization Procedures

All interferon samples used for immunogen were prepared as water-in-oil emulsions with complete Freund's adjuvant (Difco Laboratories, Detroit, Michigan). The emulsion was made using the double syringe

[1] Y. Iwakura, S. Yonehara, and Y. Kawade, *J. Biol. Chem.* **253**, 5074 (1978).
[2] K. C. Zoon, M. E. Smith, P. J. Bridgen, D. zur Nedden, and C. B. Anfinsen, *Proc. Natl. Acad. Sci. U.S.A.* **76**, 5601 (1979).
[3] Y. H. Tan, H. Okamura, and H. Smith-Johannsen, *Ann. N. Y. Acad. Sci.* in press (1980).
[4] K. Berg, *Scand. J. Immunol.* **6**, 77 (1977).
[5] K. Berg, C. A. Ogburn, K. Paucker, and K. E. Mogensen, *J. Immunol.* **114**, 640 (1975).
[6] K. E. Mogensen, L. Pyhälä, and E. Torma, *Acta Pathol. Microbiol. Scand.* **83**, 443 (1975).
[7] J. Vilček, E. A. Havell, and S. Yamazaki, *Ann. N. Y. Acad. Sci.* **284**, 703 (1977).
[8] C. B. Anfinsen, S. Bose, L. Corley, and A. Gurari-Rotman, *Proc. Natl. Acad. Sci. U.S.A.* **71**, 3139 (1974).
[9] M. E. Smith and K. C. Zoon, unpublished results.

technique described in detail by Herbert.[10] Sheep were maintained and plasmapheresis was performed at the Ungulate Unit, NIH Animal Center, Poolesville, Maryland under the supervision of Mr. L. Stuart. Plasmapheresis of immunized sheep had no apparent harm to the animals and was capable of generating 1–1.5 liters of plasma per animal every 2 weeks. Rabbits were maintained, immunized, and bled under NIH contract No. 263-79-D-0329 with Meloy Laboratories, Springfield, Virginia.

Precipitation Assay for Anti-interferon Serum

Two-hundred microliters of dilutions of anti-interferon serum were incubated with 1000–5000 units (100 μl) of human lymphoblastoid interferon and made to a total volume of 1.0 ml with preimmunized sheep or rabbit serum diluted 1 : 50 in phosphate buffered saline, pH 7.4. After 2–18 hr of incubation at 37°, second antibody, either burro anti-sheep γ-globulin in the case of sheep anti-interferon or goat anti-rabbit γ-globulin in the case of rabbit anti-interferon was added in amounts sufficient precipitate all test antibody. This mixture was incubated at 4° for 2 hr, centrifuged, and the supernatants assayed for antiviral activity. Interferon units precipitated were quantitated and determined as the difference between control normal sheep serum and test anti-interferon. The average of two or three dilutions of anti-interferon was used to determine the precipitating activity as units of interferon bound per milliliter anti-interferon serum.

Immunization with Partially Purified Interferon

Crude human lymphoblastoid interferon was produced by induction of the Burkitts' lymphoma derived cell line Namalva with Newcastle disease virus strain B1. This material was partially purified by immunoadsorbent affinity chromatography with antibodies raised against partially purified human leukocyte interferon.[8] Two sheep were immunized with biweekly intramuscular injections of 10^6 units of interferon (specific activity 10^5 units per milligram of protein) in 1.5 ml of complete Freund's adjuvant for 16 weeks. At the end of this time interferon binding activity was observed in the sera at a concentration of 200,000 units neutralized per milliliter of serum. Preparative amounts of serum were obtained from 400–800 ml of clotted whole blood, and the γ-globulin was prepared from the hyperimmunized serum by precipitation twice with ammonium sulfate at 50% saturation at 25°.

[10] W. J. Herbert in "Handbook of Experimental Immunology" (D. M. Weir, ed.), p. A3.1. Blackwell, Oxford, 1978.

The specificity of the partially purified γ-globulin fraction was enhanced by adsorption to an "impurities" column. This column was prepared by binding all known impurities in the interferon immunogen preparation (fetal calf serum, soluble extracts of Namalva cells or leukocytes, and egg allantoic fluid containing NDV B1) to CNBr-activated Sepharose 4B. The anti-interferon preparation was passed through the column 6–8 times and the column was regenerated each time by washing with 4.0 M guanidine · HCl in phosphate-buffered saline pH 7.4. This adsorption process resulted in an increased volume of the γ-globulin preparation, which required concentration by ultrafiltration over an Amicon PM-10 membrane (Amicon Corp., Lexington, Massachusetts) to 10–20 mg/ml. The γ-globulin was then coupled to CNBr-activated Sepharose 4B-CL at 20–25 mg per milliliter of packed gel and used as the first step in purification of lymphoblastoid interferon.

When highly purified interferon became available for use in immunization, the same sheep previously used for antibody production were again immunized. Eighteen months after the initial immunization, interferon of higher specific activity ($10^{7.5}$–10^8 units per milligram of protein) was prepared in complete Freund's adjuvant, and 10^6 units in 1.5 ml were given intramuscularly biweekly for 6 weeks. The interferon binding activity observed from these animals was 3×10^7 units neutralized per milliliter of serum. The γ-globulin fraction from this serum was purified by the method of Joustra and Lundgren[11] using ion exchange chromatography on QAE Sephadex. The same types of "impurities" adsorbent column was used to increase the specificity of the γ-globulin fraction.

One-year-old female New Zealand white rabbits were immunized with the same high specific activity ($10^{7.5}$–10^8 units/mg) interferon preparation. Biweekly subdermal injections of 0.1 ml of interferon in complete Freund's adjuvant were given at four sites (total 10^5 units per animal per four injections) on each animal's dorsal surface. After 10 weeks, the maximum interferon binding activity of 200 units/ml was observed in serum from two of four animals. IgG was obtained from serum of these two animals by ion exchange chromatography with QAE Sephadex.[11]

Immunization with Homogeneous 18,500 Dalton Component of Human Lymphoblastoid Interferon

The 18,500 dalton component of human lymphoblastoid interferon[12] was used as immunogen with a single sheep not previously immunized.

[11] M. Joustra and H. Lundgren, *Protides Biol. Fluids* **17**, 511 (1969).
[12] K. C. Zoon, M. E. Smith, P. J. Bridgen, M. W. Hunkapiller, and L. E. Hood, *Science* **207**, 527 (1980).

PREPARATION OF ANTIBODIES TO HUMAN LYMPHOBLASTOID INTERFERON (IF)

Specific activity of immunogen (IF units/mg protein)		Total IF immunogen[a] (IF units)	Maximum IF binding activity (units bound/ml serum)
Sheep 1283, 1284	10^5	8×10^6	$>2 \times 10^{5b}$
Sheep 1283, 1284[c]	5×10^7	6×10^6	3×10^7
Sheep 1487	$2-4 \times 10^8$	2×10^7	6×10^5
Rabbits 3, 4	5×10^7	5×10^5	2×10^2

[a] Quantity of immunogen required for each animal to attain maximum binding activity.
[b] Determined by direct neutralization without anti-sheep γ-globulin.
[c] Sheep boosted 18 months after initial immunization.

The animal was given biweekly intramuscular injections of 10^6 units of pure 18,500 dalton component (2×10^8 units per milligram of protein) in 1.5 ml of complete Freund's adjuvant for 22 weeks. By week 10 neutralization activity of 200,000 units/ml was observed. The level increased to 600,000 units/ml at week 12 and remained for 10 weeks with continued immunization. The sheep was subjected to plasmapheresis biweekly, and 1–1.5 liters of plasma were obtained regularly over a 10-week period for a total accumulation of 12 liters of serum with consistent neutralizing titer of 600,000 units/ml.

Summary

Antiserum to partially purified human lymphoblastoid interferon has been produced by typical immunization procedures in both sheep and rabbits (see the table). The material has been shown to suffice for use in immunoadsorbent affinity chromatography for purification of human lymphoblastoid interferon.[2,12] Monospecific antibodies to the 18,500 dalton component of human lymphoblastoid interferon have also been prepared in large quantities by immunization of a single sheep.

[72] Production and Quantitation of Neutralizing Antibodies for Human Fibroblast Interferon

By EDWARD A. HAVELL

Three classes of human interferons are recognized based on antigenic differences.[1,2] Antisera raised by immunization with human interferon synthesized by peripheral blood buffy coat cells (leukocyte interferon) are capable of neutralizing the antiviral activity of leukocyte interferon but are generally incapable of neutralizing the antiviral activities of interferons derived from human cells of nonlymphoid origin (fibroblast interferon) or of interferons produced by antigen-sensitized lymphocytes stimulated by contact with the sensitizing antigens or produced following nonspecific stimulation of T lymphocytes by mitogens, such as phytohemagglutinin (immune, type II or T interferons). Conversely, antisera raised against human fibroblast interferons have been shown to neutralize the antiviral activities of fibroblast interferons, but not those of either human leukocyte or immune interferon preparations. The availability of specific antisera have proved to be useful for both the immunological characterization of human interferons and the purification and isolation of interferons by antibody affinity chromatography.

Human fibroblast interferons of varying purity have been used to elicit neutralizing antibodies in rabbits,[1] chickens,[3] and sheep.[4] Pure fibroblast interferon has a specific activity of about 2 to 5×10^8 interferon units per milligram of protein; however, neutralizing antibodies have been raised using crude fibroblast interferons having specific activities as low as 10^4 units per milligram of protein. It has been our experience that the greater the quantity of interferon administered and the higher the purity of the preparation, the more rapidly rabbits respond with high-titered anti-human fibroblast interferon neutralizing antibodies. Y. H. Tan and his colleagues (personal communication) have successfully immunized sheep with what they believe to be a pure preparation of human fibroblast interferon. The immunized animal has produced very high-titered neutralizing antibody, which should also be monospecific for interferon, and is being used in antibody affinity chromatography procedures for the purification of interferon prior to amino acid sequencing. In this chapter, I have

[1] E. A. Havell, B. Berman, C. A. Ogburn, K. Paucker, and J. Vilček, Proc. Natl. Acad. Sci. U.S.A. **72**, 2185 (1975).
[2] M. J. Valle, G. W. Jordan, S. Haahr, and T. C. Merigan, J. Immunol. **115**, 230 (1975).
[3] V. E. Vengris. B. D. Stollar, and P. M. Pitha, Virology **65**, 410 (1975).
[4] Y. H. Tan, personal communication.

outlined the procedures that we have used to (a) obtain rabbit anti-human fibroblast serum with partially purified human fibroblast interferon; and (b) determine the titer of the neutralizing anti-interferon serum.

Production of Antibody

Antigen. Human fibroblast interferon used for immunization was produced in FS-4 fibroblast cultures stimulated with polyinosinic · polycytidylic acid under the conditions previously described for the "superinduction" of interferon.[5] Crude interferon (specific activity of about 10^4 reference units per milligram of protein) was concentrated by precipitation with perchloric acid, followed by ion-exchange chromatography on CM-Sephadex C-25. The final step in the purification was chromatography on concanavalin A–agarose according to the method of Davey *et al.*[6] The specific activity of the fibroblast interferon used for the immunizations was $>10^7$ reference units per milligram of protein.

Immunization Procedure. Three young adult female New Zealand white rabbits (2.2–2.4 kg) were immunized. The decision to immunize several animals was made because previous immunization studies clearly established that there was considerable variability in the antibody response among individual animals. The rabbits were bled via the lateral ear vein for 50 ml of blood prior to the onset of the immunization, and the sera were collected. These preimmunization sera have proved to be valuable as control sera in various studies.

Each rabbit was initially immunized with a total of 1.1×10^6 reference units of fibroblast interferon administered three times at 14-day intervals. Prior to the injections, the backs of the rabbits were shaved and the interferon was administered subcutaneously at multiple sites along the spine. The first two series of injections were in Freund's complete adjuvant (Perrin's modification, purchased from Calbiochem, La Jolla, California) which was thoroughly emulsified in an equal volume of the interferon preparation. A total of 2 ml of the emulsion was administered subcutaneously at four sites on the back of each rabbit. The third, biweekly series of injections, and all subsequent booster immunizations were carried out with Freund's incomplete adjuvant (Difco Laboratories, Detroit, Michigan). Immediately before the injections, each rabbit was bled via the ear for 40–50 ml of blood; the blood was allowed to clot and the serum was collected and heat-inactivated at 56° for 30 min. The sera were stored at −20° until assayed for interferon neutralizing antibody. Interferon neutralization assays revealed that all three rabbits responded with high-

[5] E. A. Havell and J. Vilček, *Antimicrob. Agents Chemother.* **2**, 476 (1972).
[6] M. W. Davey, E. Sulkowski, and W. A. Carter, *Biochemistry* **15**, 704 (1976).

SERUM LEVELS OF RABBIT ANTI-HUMAN FIBROBLAST
INTERFERON NEUTRALIZING ANTIBODIES

Days from beginning of immunization	Neutralizing titer[a] of serum for human fibroblast interferon
0	<32
14	<32
28	25
42	1600
56	2500
70	5000
113	5000
190	2000
239	6500
315	2500
405	3500
498	3000
567	5000

[a] Reciprocal of the highest dilution of serum which, when mixed with an equal volume of interferon (20 units/ml), neutralizes 50% of the antiviral activity as determined by the development of cytopathic effect.

titered neutralizing antibody specific for human fibroblast interferon. Subsequent booster immunizations at 30–60-day intervals resulted in the maintenance of peak levels of anti-human fibroblast interferon neutralizing antibodies (see the table).

Titration of Anti-Human Fibroblast Interferon Neutralizing Serum

The titration of an anti-interferon serum is based on the ability of the immune serum to neutralize the interferon's protective effect on cells, which is determined by the development of viral cytopathic effect. The potency or titer of an antiserum is defined as the reciprocal of the highest dilution of antiserum which when mixed with an equal volume of interferon (20 units/ml) neutralizes 50% of the antiviral activity of the interferon. Since the final concentration of interferon in the antiserum–interferon mixture is 10 units/ml, this would mean that 1 ml of antiserum having a neutralizing titer of 2000/ml would be capable of neutralizing (to 50%) a total of 20,000 interferon units.

The method of titration used to quantitate the neutralizing activity of anti-interferon serum is based on an adaptation of the microassay procedure for the quantitation of interferon.[5] Micro Test II 96-well plastic tissue culture plates (Falcon Plastics, Oxnard, California) are seeded 1 day prior to the start of the assay with a suspension of human foreskin (FS-4) fibroblast cells (20,000 cells/well) in 0.2 ml of Eagle's minimal essential medium (MEM) containing 5% fetal bovine serum. Serial twofold dilutions of the antiserum to be tested are made in culture medium, to which an equal volume of interferon (20 units/ml) is added. The interferon-antibody mixtures are then mixed and incubated at 37° for 1 hr. After this incubation period, the medium is aspirated from the fibroblast cultures and 0.2 ml of each of the antibody dilutions containing interferon mixtures are added to duplicate wells. The test interferon being used in the neutralization study is simultaneously assayed to determine the exact quantity of interferon being reacted with the dilutions of antiserum. After an 18-hr incubation, the antibody–interferon mixtures are removed from the cultures, and each well receives 0.2 ml of medium containing 4000 plaque-forming units of vesicular stomatitis virus; then the cultures are incubated at 37°. The viral cytopathic effect is scored 48 hr later, and the neutralizing titer is taken as the well containing the highest antiserum dilution in which 50% or more of the cells are destroyed by the virus. The concomitant titration of the interferon used in the neutralization assay enables the determination of the exact quantity of interferon being reacted with the antiserum so that the neutralizing titers can be corrected to an arbitrary standard concentration of interferon (usually 20 interferon units/ml prior to mixing with an equal volume of each dilution of serum). Since a linear relationship exists between the neutralizing titer of an antiserum and the quantity of interferon being reacted in the neutralization assay (author, unpublished observation), a neutralization assay performed with half the standard interferon concentration will result in an observed serum neutralization titer twice the value that would have been obtained with the standard amount of interferon.

Differences in the neutralizing titer of an antiserum for human interferons of the same antigenic class have been reported by Dalton and Paucker.[7] These investigators found that the corrected neutralizing titer of an antiserum for different preparations of human lymphoblastoid interferon varied by 10-fold. A lower neutralizing titer indicates that more antibody is required to neutralize a standard amount of antiviral activity. Several possible explanations can be offered to explain the reason why an antiserum would have a lower neutralizing titer for an interferon, when compared to another interferon, these include: (a) the presence of biologically inactive molecules that can bind antibody and thus compete with active molecules for antibody; (b) the possibility that the interferon molecule has more antibody binding sites; or, (c) alteration of the interferon in

[7] B. J. Dalton and K. Paucker, *Infect. Immun.* **23**, 244 (1979).

such a way that the antibody has less avidity for the antigenic determinants.

Failure of an antiserum to neutralize the antiviral activity of interferon preparations could result from any of three possible causes.

1. The serum does not possess neutralizing antibodies.
2. The interferon reacted with the specific antiserum is antigenically distinct and is not recognized by the neutralizing antibodies.
3. The interferon preparation reacted with the antiserum contains a mixed population of antigenically distinct interferons, and, while the antiserum neutralizes the antiviral activity of the interferon for which it is specific, the interferon molecules not neutralized render the assay cells resistant to viral destruction.

Preparations of both human leukocyte and fibroblast interferons have been shown to possess low levels of the other antigenically distinct interferon species.[8,9] Chapter [73] in this volume describes a highly sensitive anti-interferon serum neutralization assay that measures the neutralization of interferon activity by virus yield and can be used to identify and quantitate antigenically distinct types of interferons within interferon preparations.

[8] E. A. Havell, B. Berman, and J. Vilček, in "Proceedings of Symposium on Clinical Use of Interferon," p. 49. Ygoslav Academy of Sciences and Arts, Zagreb, 1975.
[9] E. A. Havell, T. G. Hayes, and J. Vilček, *Virology* **89**, 330 (1978).

[73] Qualitative and Quantitative Virus-Yield Assay for Antibody Neutralization of Interferons

By EDWARD A. HAVELL

Interferon has become a generic term used to describe a class of glycoproteins capable of rendering cells resistant to virus multiplication. Three distinct human interferons (leukocyte, fibroblast, and immune) are recognized on the basis of differences in physiochemical, antigenic, and biological properties. Both the degree of antiviral activity on cells of other species and the neutralization of interferon activities by specific antisera are useful criteria for differentiating human interferons. Gresser and his colleagues[1] found that human leukocyte (Le) interferon exhibited considerably more antiviral activity on porcine and bovine cells

[1] I. Gresser, M.-T. Bandu, D. Brouty-Boyé, and M. Tovey, *Nature (London)* **251**, 543 (1974).

than on homologous human cell cultures. However, human fibroblast (F) interferon was much less active on the bovine cells than homologous cell cultures. Human interferons have also been characterized on the basis of antigenic differences. Antisera capable of neutralizing the antiviral activity of human F interferon fail to neutralize human Le interferon; and, antisera raised against human Le interferon generally do not neutralize human F interferons.[2] Neither anti-F nor anti-Le antisera neutralize human immune (also called type II or T) interferons.[3]

Interferon preparations often consist of mixtures of antigenically distinct interferon species. Human Le interferon preparations were found to contain low levels of F interferon[4]; and, the induction of human fibroblast cultures with virus resulted in the synthesis of both F interferon and low levels of Le interferon.[5] Interferons produced by human lymphoblastoid cells were found to consist of Le interferon and variable quantities of F interferon that could constitute as much as 20% of the total antiviral activity of the preparation.[6] Murine L cell interferons have been found to contain a minor interferon population that is active on human and bovine cultures.[7] This minor L cell interferon component was shown to possess common antigenic determinants with human Le interferon.

The presence of two interferon species that differ in their activities could complicate studies designed to characterize the biological activities of an interferon preparation. In addition, antisera raised against such an interferon preparation may possess neutralizing antibodies that are specific for each antigenically distinct interferon in the mixture. The use of such antisera for the serological classification of interferons could give conflicting results as to the antigenic nature of the interferon.

An antibody neutralization assay for interferon activity has been developed that enables the resolution and quantitation of antigenically distinct interferons in a preparation. This method of assay measures the neutralization of antiviral activity by virus yield and is both objective and a highly sensitive means for determining the degree to which an interferon is neutralized. This procedure is outlined using anti-Le interferon serum to neutralize the antiviral activities of Le interferon on three different cell types that differ in their sensitivities to human interferons. The antibody neutralization curves reflect the amount of antibody needed to neutralize

[2] E. A. Havell, B. Berman, C. A. Ogburn, K. Paucker, and J. Vilček, *Proc. Natl. Acad. Sci. U.S.A.* **72,** 2185 (1975).

[3] M. J. Valle, G. W. Jordan, S. Haahr, and T. C. Merican, *J. Immunol.* **115,** 230 (1975).

[4] E. A. Havell, B. Berman, and J. Vilček, *in* "Proceedings of Symposium on Clinical Use of Interferon," p. 49. Yugoslav Academy of Sciences and Arts, Zagreb, Yugoslavia, 1975.

[5] E. A. Havell, T. G. Hayes, and J. Vilček, *Virology* **89,** 330 (1978).

[6] B. J. Dalton and K. Paucker, *Infect. Immun.* **23,** 244 (1979).

[7] E. A. Havell, *Virology* **92,** 324 (1979).

the antiviral activity of the interferon on each cell type and can reveal the presence of antigenically distinct interferons in the interferon preparation. This means of assay can also detect neutralizing antibodies with different specificities in the antiserum.

Materials

Cells. Two human cell types were employed: the FS-4 strain of diploid human fibroblasts, derived in Dr. Jan Vilček's laboratory (New York University, New York City) from a neonatal foreskin; and the GM-2504 strain, derived from the skin of a child with Down's syndrome, trisomic for chromosome 21 (obtained from the Human Genetic Mutant Cell Repository, Camden, New Jersey). It is important to note that chromosome 21 carries the locus controlling interferon sensitivity, and human cells trisomic for this chromosome are more sensitive to interferon action than are diploid cells.[8] The bovine kidney cell line (MDBK) was provided by Dr. Pravin Sehgal (Rockefeller University, New York City). Cultures of FS-4 and GM-2504 cells were seeded in 24-well (17 mm) plastic plates (Falcon Plastics, Oxnard, California) with 50,000 cells per well; and bovine MDBK cells were seeded at 100,000 cells per well. Cells were grown to confluence in Eagle's minimal essential medium (MEM) supplemented with 5% fetal bovine serum (FBS) and routinely used 4–5 days after seeding.

Interferon and Anti-Interferon Sera. The human leukocyte interferon was the kind gift of Dr. Kari Cantell (State Serum Institute, Helsinki, Finland). This interferon was acidified (pH 2.0) for 5 days and then centrifuged at 100,000 *g* for 1 hr prior to use. This interferon preparation had a titer of 65,000 reference units/ml when assayed on diploid FS-4 fibroblasts by the microtiter method of Havell and Vilček.[9] The reference fibroblast interferon standard G-023-902-527 was used to standardize the potency of the leukocyte interferon.

Rabbit anti-human fibroblast interferon antibody was prepared and assayed according to the procedure outlined in this volume [72]. Sheep anti-human leukocyte interferon antibody was the kind gift of Dr. L. Borecky (Institute of Virology, Bratislava, Czechoslovakia).

Interferon Assays

In order to ensure that the quantity of Le interferon to be used in the antibody neutralization assays elicited a maximum degree of virus inhibi-

[8] Y. H. Tan and A. E. Greene, *J. Gen. Virol.* **32**, 153 (1976).
[9] E. A. Havell and J. Vilček, *Antimicrob. Agents Chemother.* **2**, 476 (1972).

tion in the human and bovine cell cultures, an interferon dose-response study was carried out. Duplicate cultures of the three cell types (FS-4, GM 2504, and MDBK) were incubated at 37° with 0.5 ml of MEM with 1% FBS containing increasing concentrations (units/ml) of human Le interferon. Control cultures received 0.5 ml of MEM with 1% FBS. After 18 hr, the cultures were washed once with Hanks' balanced salt solution (HBSS) and then infected with vesicular stomatitis virus (VSV) at a multiplicity of infection (MOI) of 1. Virus was allowed to adsorb for 1 hr, after which unadsorbed virus was removed by washing the cultures three times with HBSS. Interferon-treated and nontreated control cultures received 0.4 ml of MEM containing 1% FBS; 9 hr later, after a single cycle of VSV replication, the culture fluids from the duplicate cultures were collected, pooled, and frozen at −70° until they were titrated for virus content by plaquing on monkey Vero cells. It is important to point out that the titration of virus yield after one replication cycle eliminates the possibility that endogenously produced interferon, synthesized by the assay cells in response to the VSV, could contribute to the antiviral activity being quantitated.[10]

The dose-response curves of Le interferon assayed on the three cell types were obtained by plotting the virus yields ($\log_{10}$ VSV/ml) against the interferon concentration. These studies resulted in a family of curves that exhibited similar slopes, were regularly sigmoidal, and have a maximum level of viral inhibition of approximately 5 logs. The Le interferon was found to be 18 times more active on the human trisomic GM-2504 cultures than on either the human diploid (FS-4) or the heterologous bovine (MDBK) cultures. Human F interferon assayed on the same three cell types exhibited similar antiviral activities as the Le interferon on the two human cells; but unlike the Le interferon, the F interferon was 40 times less active on the bovine cultures as compared to its activity on the human diploid FS-4 cells. These antiviral activities of both human interferons are in agreement with previous reports as to their relative activities on these three cell types.[1,11] Antibody neutralization assays of Le interferon activity were performed in a similar manner by measuring virus yield with a constant interferon concentration of 20 international units/ml. At this concentration of interferon it was found that a maximum degree of virus inhibition was induced in all three cell types.

Antiserum Neutralization of Interferon Activities As Measured by Virus Yield

Sheep anti-Le interferon serum was used to neutralize the antiviral activities of human Le interferon on human FS-4, GM-2504, and bovine

[10] J. Vilček, S. Yamazaki, and E. A. Havell, *Infect. Immun.* **18,** 863 (1977).
[11] E. A. Havell, Y. K. Yip, and J. Vilček, *Arch. Virol.* **55,** 121 (1977).

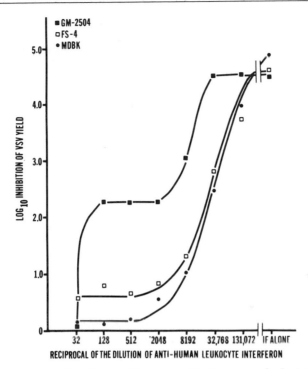

FIG. 1. Anti-human Lc interferon antibody neutralization curves for Le interferon antiviral activities on human GM-2504, human FS-4, and bovine MDBK cultures. A constant quantity of human Le interferon [40 international interferon (IF) reference units as determined on FS-4 cells] was allowed to react with an equal volume of each dilution of antibody and placed on the cultures for 18 hr. Virus yields were determined after a single cycle of vesicular stomatitis virus (VSV) replication.

MDBK cultures. Serial fourfold dilutions of the antiserum were made in MEM containing 1% FBS and mixed with an equal volume of human Le interferon (40 international reference units/ml as determined by assay on diploid FS-4 cells) and the mixture was incubated for 1 hr at 37°. The interferon–antibody mixtures (0.5 ml) were added to duplicate cultures of confluent cells and incubated for 18 hr at 37° in a 5% CO_2 humidified incubator. After this incubation period, the cells were washed once with HBSS, infected with VSV (MOI of 1), and the virus yield following a single-cycle of virus replication was determined as described above for the interferon assay.

The antibody neutralization curves for the Le interferon activity on the different cell types are shown in Fig. 1. The Lc interferon neutralization curves are plotted as the log_{10} inhibition of VSV yield per milliliter, which is determined by subtracting the log_{10} VSV yield obtained with each antiserum dilution containing a final concentration of 20 Le interferon units from the $\log_{10}$ VSV yield obtained from nontreated control cultures. The data presented in Fig. 1 provide an example of multiple interferon species within a single preparation. It is apparent from the data that the neutralization by antibody of Le interferon activity on trisomic human GM-2504 cultures resulted in a biphasic neutralization curve: at the lowest dilution of antibody (1/32) there was neutralization of all antiviral activity; however, at the next three successive antibody dilutions there was only a partial neutralization of interferon activity, which is shown as a plateau in the interferon neutralization curve at 2 logs of VSV inhibition. In addition, a partial neutralization of Le interferon antiviral activity was noted at the lower antibody dilutions in assays conducted with diploid FS-4 fibroblasts, but with these cells the level of antiviral interferon activity was much less pronounced, being maximal at only 0.75 log of VSV inhibition (Fig. 1). The antiviral activity that was not neutralized by the anti-Le interferon antibody and detected on both human cell types was not expressed on the bovine MDBK cultures. The anti-Le interferon antibody neutralization curves for Le interferon activity on the human diploid FS-4 and bovine MDBK cultures were sigmoidal, and the antibody neutralized the antiviral activities on each of these cell cultures to the same degree.

It has previously been demonstrated that human Le interferon preparations can contain small quantities ($\leq 1\%$) of antigenically distinct human F interferon.[4] Thus, the most likely explanation for the plateaus in the antibody neutralization curves at 2.0 and 0.75 logs of VSV inhibition on the human GM-2504 and FS-4 cells, respectively, is that human F interferon is present in the Le interferon preparation. This possibility was tested by allowing diminishing dilutions of anti-Le interferon antibody to react with Le interferon in the presence of a constant amount (1/64 dilution) of high-titered anti-F interferon antibody. The mixtures were then assayed on GM-2504 cultures for neutralization of antiviral activity by inhibition of VSV yield (Fig. 2). The inclusion of anti-F interferon antibody in the interferon–antibody mixtures eliminated the plateaus in the neutralization curves and resulted in the complete neutralization of the antiviral activity at all dilutions of anti-Le antibody, whereas without the anti-F interferon there was only partial neutralization of antiviral activity.

The position of the Le interferon neutralization curve on GM-2504 cultures obtained with the two antibodies (Fig. 2) indicates that more anti-Le interferon antibody was required to neutralize the Le interferon on these cells than on the FS-4 or MDBK cultures (Fig. 1). Human Le interferon exhibited similar antiviral activities on both the human FS-4 and bovine MDBK cells, and the antibody neutralization curves for the antiviral activities of a constant amount of Le interferon on these two cells were almost identical. The antibody neutralizing titer, which is the recip-

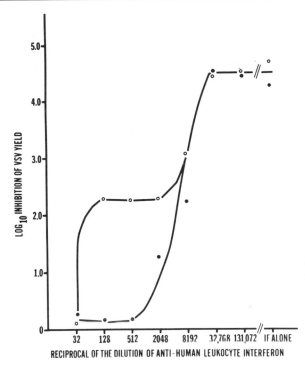

FIG. 2. Antibody neutralization curves of human Le interferon activity on human GM-2504 cultures by anti-human Le interferon antibody alone (○) or in the presence of a constant quantity of high-titered anti-human F interferon serum (●).

rocal of that dilution of antibody corresponding to the midpoint on a $\log_{10}$ scale of maximum inhibition of virus yield induced by the interferon, was 30,000 on the FS-4 and MDBK cells. However, the antibody neutralizing titer for the same quantity of Le interferon on the human GM-2504 cells, which were approximately 18 times more sensitive than the human FS-4 or bovine MDBK, was 8000. These findings indicate that more antibody is required to neutralize the same amount of Le interferon on cells exhibiting a greater degree of sensitivity to the antiviral action of interferon.

The amount of F interferon present in the Le interferon preparation can be determined from a F interferon dose-response curve on GM-2504 cultures. The quantity of F interferon causing 2 logs of VSV inhibition in GM-2504 cultures is 0.1 unit/ml. This concentration of F interferon represents 1% of the total Le interferon activity in the final antibody–interferon mixture. It is important to point out that the complete neutralization of the Le interferon activity on GM-2504 cultures at the 1/32 dilution of anti-Le interferon (Fig. 1) is probably due to the presence of low levels of anti-F interferon antibodies in the anti-Le interferon serum, since it has been previously demonstrated that animals immunized with crude human Le interferon preparations, having low levels of human F interferon, can produce specific antibodies to both Le and F interferons.

The procedure described above allows for the immunological identification of antigenically distinct interferons in a preparation. It is also possible to combine this procedure with the biological and physiochemical properties that differentiate the three antigenically distinct human interferons (Le, F, and immune) to identify the presence of either of the two other interferons in various interferon preparations. It should be noted that if anti-F interferon serum had not been available for the immunological characterization of the minor interferon component in the Le interferon preparation, it would have been possible to reach a tentative conclusion that this minor component was F interferon based on the differences in sensitivities exhibited by the three cell types to this interferon and on the fact that the Le interferon preparation had been previously acidified (immune interferon is destroyed by low pH). We believe that analyses such as these may provide the basis for identifying distinct interferons in preparations of interferons from species other than man and in particular should prove useful for characterizing murine interferons, where at present only one interferon-neutralizing antiserum is available but at least three antigenically distinct interferons have been described.[7,12]

[12] J. S. Youngner and S. B. Salvin, *J. Immunol.* **111**, 1914 (1973).

[74] Preparation and Absorption of Antiserum against Mouse L Cell Interferon

By BARBARA DALTON and KURT PAUCKER*

This chapter describes the production and absorption of sheep antiserum to live Newcastle disease virus (NDV) induced mouse L cell interferon. The production and processing of such antiserum utilizes the same principles and general procedures as outlined in this volume [70]. Readers should refer to that chapter for the principles and procedures involved. Specific information as it pertains to an antiserum to mouse interferon is given below.

Preparation of Inoculum. Mouse L cell interferon, induced by live NDV[1] (multiplicity of infection = 1), precipitated with ammonium sulfate[2]

* Deceased.
[1] C. A. Ogburn, K. Berg, and K. Paucker, *J. Immunol.* **111**, 1206 (1973).
[2] E. Knight, Jr., *J. Biol. Chem.* **250**, 4139 (1975).

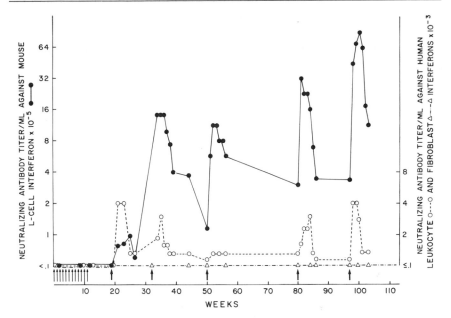

FIG. 1. Development of antibodies to mouse and human interferons in a sheep immunized with mouse L cell interferon. Small arrows indicate inoculation with 1.3×10^7 interferon units each of bovine serum albumin-purified mouse L cell interferon (specific activity = 1.0×10^6 units per milligram of protein). The large arrows indicate booster inoculations containing 2.5×10^7 units each of antibody affinity chromatography-purified mouse L cell interferon (specific activity = 2×10^7 units per milligram of protein) mixed with an equal volume of Freund's complete adjuvant. The neutralizing activity against mouse L cell interferon was measured with mouse L-929 cells, and the activity against human interferons was measured with human FS-4 cells.

and partially purified by bovine plasma albumin Sepharose[3] chromatography to a specific activity (SA) of 1×10^6 units per milligram of protein, was utilized for the first series of injections. Subsequent booster inoculations utilized ammonium sulfate precipitated interferon, purified by antibody affinity chromatography[1] to a specific activity of 2×10^7 units per milligram of protein. All interferon preparations were stabilized with 0.5% sheep albumin prior to lyophilization and reconstituted with a small volume of phosphate-buffered saline (PBS), pH 7.0, for inoculation.

Immunizations. The sheep immunized with mouse L cell interferon was a female of Suffolk-Hampshire stock, 1 year old at the start of the immunization. The immunization schedule illustrated in Fig. 1 was based on

[3] J. S. Erickson and K. Paucker, *J. Gen. Virol.* **43**, 521 (1979).

the procedure described by Mogensen *et al.*[4] The small arrows in Fig. 1 indicate 12 weekly injections; each injection contained 1.3×10^7 units of interferon (SA $= 1 \times 10^6$ units per milligram of protein). Six weeks after the 12th injection, a booster inoculation of 2.5×10^7 units of L cell interferon (SA $= 2 \times 10^7$ units per milligram of protein), mixed with Freund's complete adjuvant (FCA) was given. Other booster inoculations, indicated by the large arrows in Fig. 1, were administered periodically, usually allowing a rest period of 6 weeks or longer between the last bleeding and the next inoculation. All injections were administered as described in this volume [70].

Schedule of Bleedings. The sheep was bled from the jugular vein at the times indicated by the circles in Fig. 1. The blood was handled as described in this volume [70]. The antiserum titers to mouse L cell interferon were maximal 1 week after a booster inoculation.

Interferon Neutralization Assay. The assay utilized to detect neutralizing antibodies is presented in this volume [70]. Serum was tested for the ability to neutralize mouse L cell interferon as well as human leukocyte and fibroblast interferons. Figure 1 illustrates the development of antibodies to these interferons. The serum from the sheep immunized with mouse L cell interferon had high antibody titers to mouse L cell interferon, low titers to human leukocyte interferon, and no detectable titers to human fibroblast interferon.

Passive Hemagglutination Test. Antibody titers to known contaminants found in the interferon preparation were measured by a passive hemagglutination test as detailed in this volume [70]. Three contaminants of NDV-induced mouse L cell interferon were chosen to monitor the success of the absorption procedure: (*a*) egg albumin, since the inducer virus was produced in embryonated eggs; (*b*) bovine serum albumin, present in the serum used to culture the mouse cells; and (*c*) L cell extract, a mixture of antigens from sonicated mouse L cells.

Absorption of Antiserum

Absorption. The anti-mouse interferon serum was absorbed by the same procedure described for the absorption of anti-human leukocyte interferon serum in this volume [70]. As in the case of the absorption of antiserum to human leukocyte interferon, the unbound material also had to be supplemented with egg albumin in order to remove a large portion of the antibodies to this contaminant.

The antibody titers of three serum samples, before and after absorp-

---

[4] K. E. Mogensen, L. Pyhälä, and K. Cantell, *Acta Pathol. Microbiol. Scand. Sect. B* **83**, 443 (1975).

ANTIBODY TITERS IN UNABSORBED AND ABSORBED SHEEP ANTI-MOUSE L CELL INTERFERON SERUM

		Antibody titer per ml[a] against			
Sample	Treatment	L cell interferon	Egg albumin	Bovine plasma albumin	L cell extract
1	Unabsorbed	1.3×10^6	6.5×10^5	2.5×10^4	4.0×10^2
	Absorbed	6.4×10^5	4.0×10^2 (>99)[b]	1.0×10^3 (95)	1.0×10^2 (75)
2	Unabsorbed	4.8×10^5	2.6×10^4	6.4×10^3	1.6×10^3
	Absorbed	2.4×10^5	4.0×10^2 (99)	4.0×10^2 (94)	1.0×10^2 (94)
3	Unabsorbed	9.6×10^5	5.1×10^4	2.6×10^4	1.6×10^3
	Absorbed	6.4×10^5	1.0×10^2 (>99)	4.0×10^2 (98)	4.0×10^2 (75)
Pool	Absorbed	5.0×10^5	2.0×10^2	2.0×10^2	2.0×10^2

[a] Expressed as reciprocal of end-point dilution. Titers to interferon were determined by neutralization; others by passive hemagglutination.

[b] Values in parentheses indicate percentage removed by absorption.

tion, are presented in the table. The data indicate that in most cases the titers against the three test antigens were reduced by 95% or greater while the anti-interferon titer remained within a twofold range of the titer in the unabsorbed serum. The level of antibodies to the test contaminant antigens that remained after absorption were reduced to background titers found in control serum taken prior to immunization.

Further Processing of the Antiserum

The absorbed sheep anti-mouse L-cell interferon serum was handled in the same manner as the sheep anti-human leukocyte interferon, detailed in this volume [70].

[75] Preparation of Antisera to Interferon by Interferon · Blue Dextran Complexes

By ANNA D. INGLOT

It is well established that the antigenicity of a given antigen is dependent on its molecular size and on prolonged and proper contact of it with the immunocompetent cells. Interferons are proteins of relatively small molecular size; they are active in nanogram amounts and are cleared

METHODS IN ENZYMOLOGY, VOL. 79

rapidly from both the blood and tissues of animals (for review see Stewart[1]).

It occurred to us that the immunogenicity of interferon could be enhanced by coupling it with high molecular weight carrier.[2]

Blue Dextrans as Carriers for Interferon

Dextran T fractions (Pharmacia) appear to be almost ideal neutral polymers. They are commercially available in well defined fractions, have consistent physical and chemical properties, can be sterilized by autoclaving, and have been safely used in medicine as blood plasma expanders. However, although interferons do not bind to unsubstituted dextrans, three groups of investigators independently reported that human and mouse interferons bind to Blue Dextran 2000 (BD-2000, i.e., dextran of MW 10^6, bound to Cibacron Blue F3 GA) or to the dye immobilized on agarose or other matrix.[3-5]

Cibacron Blue F3 GA [4 - chloro - 2 phenyloamine - 6 - (1''' - aminoanthraquinone - 4''' - aminophenylamino)-1,3,5 - triazine - 3',2'',2''' - trisulfonic acid] belongs to a large family of common industrial dyes. The noncovalent binding between interferons and this chromophore is strong, so that $1 M$ NaCl or KCl together with ethylene glycol was required for their displacement.

We have confirmed and extended these findings.[2] BD-2000 · interferon complex was found to have very low toxicity for tissue cultures and for animals when injected intraperitoneally (ip) or subcutaneously (sc). However, the molecular weight of the BD-2000 · interferon complex was too high to be recommended for intravenous (iv) administration. For this reason, a novel series of Blue Dextrans covering the molecular range 10,000 to 2,000,000 were synthesized and tested.[6] Mouse interferon was found to bind firmly to the Blue Dextrans of various molecular weights. It has been determined that BD-70 · interferon complex (MW of the carrier = 7×10^4) can be safely injected iv into mice.[6]

The BD · interferon complexes of various molecular weights were found to have antiviral and anticellular activity equal to or even higher

[1] W. E. Stewart, II, "The Interferon System." Springer-Verlag, Vienna and New York, 1979.

[2] A. D. Inglot, B. Kisielow, and O. Inglot, *Arch. Virol.* **60**, 43 (1979).

[3] T. C. Cesario, P. Schryer, A. Mandel, and J. G. Tilles, *Proc. Soc. Exp. Biol. Med.* **153**, 486 (1976).

[4] J. De Maeyer-Guignard and E. De Maeyer, *C. R. Hebd. Seances Acad. Sci.* Paris *Ser. D* **283**, 709 (1976).

[5] W. J. Jankowski, W. Von Muenchhausen, E. Sulkowski, and W. A. Carter, *Biochemistry* **15**, 5182 (1976).

[6] A. D. Inglot, M. Konieczny, D. Charytonowicz, T. Chudzio, B. Kisielow, O. Inglot, and E. Oleszak, *Arch. Immunol. Ther. Exp.* **28**, 313 (1980).

than that of native interferon. They were stable for months when stored at 4° (freezing is not recommended). The rate of clearance of interferon bound to BD 2000 is significantly lower than that of the native interferon.[2,6] Therefore, the BD-2000 · interferon complex was chosen for the immunization of animals to raise antibody to interferon.

Immunization of Animals with BD · Interferon Complex

Preparation of Antigen. Mouse type 1 interferon is derived from suspension culture of C-243 cells induced with Newcastle disease virus.[7] It is concentrated by ultrafiltration and partially purified by precipitation with ammonium sulfate at pH 2.0.[1] Interferon can be efficiently further purified by affinity chromatography on Affi-Gel 202, controlled-pore glass and by other methods.[1]

Purified interferon is first dialyzed against Dulbecco's phosphate-buffered saline (PBS), pH 7.2. One volume of the solution containing interferon (10^4–10^6 units/ml) is mixed with 0.05 volume of 2% BD 2000 (Pharmacia) solution in water. The final concentration of BD in the preparation is 0.1%. Then 0.5 volume of a 30% solution of polyethylene glycol (PEG, MW 6000, Serva) is added, and the same is incubated overnight at 4°. The final concentration of PEG in the reaction mixture is 10%. The sample is centrifuged at 700 g for 30 min at 4°. The blue sediment is collected and dissolved in 0.1 of the initial volume of PBS. Almost all the initial interferon activity in the preparation is bound to the Blue Dextran. This "blue" interferon is used for the immunization.

The interferon · BD-2000 complex was found to be fully active in the antiviral and antiproliferative assays for mouse interferon. It was not necessary to elute the interferon from the complex for assay. However, the interferon · BD complex required prolonged solubilization in PBS (at least 24 hr at 4°), since it became aggregated after precipitation with PEG. The BD · interferon complex after solubilization was titrated in the same way as native uncomplexed interferon.

Immunization. Five rabbits were immunized with mouse interferon bound to BD-2000, titering approximately 10^5 units/ml. Rabbits received a total of 19 sc injections of 1 ml of the "blue" interferon over a period of 11 months. They were bled thereafter. Two sheep received 37 sc injections of 5 ml of the "blue" interferon (titering from 10^5 to 10^6 units/ml) over a period of 2 years. First, 15 injections were given at 2-week intervals; later, the injections were given at intervals of 6–8 weeks. Sheep were repeatedly bled 10 days after the last injection of the "blue" interferon. All five rabbits and two sheep immunized with BD · interferon complex over

---

[7] I. Gresser, M. G. Tovey, M.-T. Bandu, C. Maury, and D. Brouty-Boyé, *J. Exp. Med.* **144**, 1305 (1976).

ACTIVITY OF THE ANTI-INTERFERON SERUM OBTAINED FROM ANIMALS
IMMUNIZED WITH BLUE DEXTRAN · INTERFERON COMPLEX

Animal No.	No. of injections of interferon	Duration of immunization (months)	Interferon neutralizing titer[a]
Rabbits 1–5	19	11	3200–6400
Sheep 1	27	19	48,000
Sheep 1	40	31	48,000
Sheep 2	27	19	96,000
Sheep 2	40	31	2,048,000

[a] Assayed against 10 units of interferon. Sheep 1 and 2 received the same total amount of interferon.

a period of 1–2 years produced anti-interferon antibody. The neutralizing titers of the antisera depended on the time of immunization and on the individual response of animals (see the table).

Comments. BD · interferon complex injected into mice, rabbits, and sheep stained the skin and tissue blue at the site of administration and along draining lymph nodes. However, it was remarkably well tolerated by the animals. No signs of toxicity or allergic reactions have been noted during administration of the BD · interferon complex to mice, rabbits, and sheep over periods of 1 month, 1 year, and 2 years, respectively.

When the BD · interferon complex was used for immunization, it was not necessary to use Fruend's adjuvant to boost the antibody response. It is worth pointing out that the adjuvants are often poorly tolerated by the hyperimmunized animals.

Since the BD · interferon complex is cleared more slowly than the native interferon, it could efficiently react with the immunocompetent cells. Thus the immunization procedure appears to require less antigen (interferon) than the more classic methods applied previously.[7,8] The antigenicity of the carrier (BD-2000) is very weak. We have not been able to detect antibodies precipitating BD-2000 in the sheep anti-interferon sera by immunodiffusion or immunoelectrophoresis.

Applications of Antisera to Interferon

Anti-mouse interferon serum has been used in studying the role of endogenous interferons in acute and chronic virus infections *in vitro*[8–10]

[8] B. Fauconnier, *Arch. Gesamte Virusforsch.* **31**, 266 (1970).
[9] A. D. Inglot, E. Albin, and T. Chudzio, *J. Gen. Virol.* **20**, 105 (1973).
[10] J. Vilček, S. Yamazaki, and E. A. Havell, *Infect. Immun.* **18**, 863 (1977).

and *in vivo*[7,11-13] as well as in antitumor effects of several interferon-inducing substances.[14]

For this purpose anti-interferon serum is applied in the form of purified immunoglobulin that is extensively absorbed prior to use.[7,13] The general finding is that the administration of anti-interferon globulin results in the enhancement of virus and/or neoplastic disease.[12-14] However, under certain circumstances anti-interferon globulin can suppress the symptoms of chronic viral infections.[15] In addition, anti-interferon sera are widely used by investigators purifying interferons and studying their action as well as their molecular and antigenic heterogeneity.[1,16-20]

[11] B. Fauconnier, *Pathol. Biol.* **19**, 575 (1971).

[12] J. L. Virelizier and I. Gresser, *J. Immunol.* **120**, 1616 (1978).

[13] A. D. Inglot, O. Inglot, A. Zółtowska, and E. Oleszak, *Int. J. Cancer* **24**, 445 (1979).

[14] I. Gresser, C. Maury, and M. G. Tovey, *Int. J. Cancer* **17**, 647 (1976).

[15] I. Gresser, L. Morel-Maroger, P. Verroust, Y. Rivière, and J. C. Guillon, *Proc. Natl. Acad. Sci. U.S.A.* **75**, 3413 (1978).

[16] K. Paucker, *Tex. Rep. Biol. Med.* **35**,25 (1977).

[17] N. Fuchsberger, B. Styk, L. Borecky, and V. Hajnická, *Acta Virol. (Engl. Ed.)* **20**, 107 (1976).

[18] L. J. Gardner and J. Vilček, *J. Gen. Virol.* **44**, 161 (1979).

[19] K. Berg, *Scand. J. Immunol.* **6**, 77 (1977).

[20] S. V. Skurkovich, A. J. Olshansky, R. S. Samoilova, E. I. Eremkina, *J. Immunol. Methods* **19**, 119 (1978).

[76] A Rapid Quantitative Assay of High Sensitivity for Human Leukocyte Interferon with Monoclonal Antibodies

By THEOPHIL STAEHELIN, CHRISTIAN STÄHLI, DONNA S. HOBBS, and SIDNEY PESTKA

In an earlier report,[1] we have described 13 hybridomas (LI-1 to LI-13) which were obtained from one fusion and produce monoclonal antibodies against human leukocyte interferon (IFL). Here, we describe the use of two monoclonal antibodies in a simple and rapid solid phase sandwich immunoassay for the detection and quantitative determination of human leukocyte interferon. The sensitivity of the assay approaches that of the much more cumbersome and lengthy biological assays such as the cytopathic effect inhibition test.[2]

[1] T. Staehelin, B. Durrer, J. Schmidt, B. Takacs, J. Stocker, V. Miggiano, C. Stähli, M. Rubinstein, W. P. Levy, R. Hershberg, and S. Pestka, *Proc. Natl. Acad. Sci. U.S.A.* **78**, 1848 (1981).

[2] S. Rubinstein, P. C. Familletti, and S. Pestka, *J. Virol.* **37**, 755 (1981).

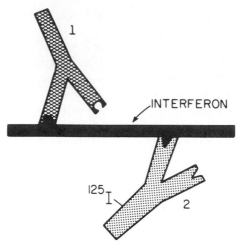

Fig. 1. Schematic diagram of sandwich radioimmunoassay with two monoclonal antibodies. The monoclonal antibodies bind to different epitopes of the interferon molecule.

Principle of the Assay

Two monoclonal antibodies are selected which fulfill two conditions: (1) they each recognize a different epitope (antigenic determinant) of human leukocyte interferon, and (2) they bind to interferon simultaneously without mutual interference. One antibody (antibody-1) is immobilized on a solid surface (e.g., plastic microtiter well or plastic bead), the second antibody (antibody-2) is labeled, e.g., with ^{125}I or with a covalently conjugated enzyme (Fig. 1). Interferon solution and labeled antibody are incubated simultaneously or in sequence with the immobilized antibody-1. Binding of labeled antibody-2 is dependent on and proportional to interferon bound by antibody-1. After incubation and washing, interferon concentration is determined by measuring the amount of labeled antibody-2 bound. A standard curve is established with known amounts of interferon.

Selection of Suitable Pair of Monoclonal Antibodies

We have characterized 12 monoclonal antibodies to human leukocyte interferon with regard to their heavy and light chain isotypes, interferon-neutralizing activity, and binding to different purified leukocyte interferon species.[1] From this group, two monoclonal antibodies, LI-1 and LI-9, which apparently produce optimal complementation, were chosen for assay of human leukocyte interferons.

Solutions and Materials

Monoclonal antibodies LI-1 and LI-9 were purified as described.[1] Twenty micrograms of LI-1 was labeled with ^{125}I by the chloramine-T method[3] to a specific activity of about 20 $\mu Ci/\mu g$. Free iodine was separated from labeled antibody by gel filtration on Sephadex G-25 in the presence of PBS and 1% bovine serum albumin (BSA).

Coating of PVC-Microtiter Plate

Each well of a polyvinylchloride (PVC)-microtiter plate (Dynatech Laboratories, Alexandria, Virginia) was coated with 50 μl (or 100 μl) of LI-9 (15 $\mu g/ml$ in PBS) at room temperature overnight (≥ 0.5 hr) or at 4° for days or weeks in a humidified chamber. Before use, the solution of LI-9 was removed and the wells filled with 3% BSA for at least 15 min to block all protein-binding sites. The wells were then rinsed four times with PBS.

Interferon and Interferon Titrations

Unfractionated and purified human leukocyte interferon species were prepared as described.[4,5] Antiviral interferon assays were performed by a cytopathic effect inhibition assay with bovine MDBK cells as described.[2] The laboratory standard of unfractionated human leukocyte interferon was titrated against the reference standard of human leukocyte interferon (G-023-901-527) supplied by the Antiviral Substances Program of the National Institute of Allergy and Infectious Diseases, National Institutes of Health, Bethesda, Maryland.

Assay Procedure

The LI-9-coated and BSA-saturated PVC-microtiter plates are prepared as described above. Fifty microliters (or 100 μl) of interferon solution in PBS containing 1% BSA are added to each well, incubated for 30 min at room temperature, and then rinsed once with PBS. Then, 50 μl (or 100 μl) of [^{125}I]LI-1 (about 150,000 cpm, 4 ng) in PBS containing 1% BSA and 150 μg of human IgG per ml are placed into each well and incubated for 2 hr at room temperature followed by four rinses with PBS. The individual wells are then counted in a gamma scintillation spectrometer. The sensitivity of this assay may be increased by prolongation of the incubation time.

[3] F. C. Greenwood and W. M. Hunter, *Biochem. J.* **89**, 114 (1963).

[4] D. S. Hobbs, J. Moschera, W. P. Levy, and S. Pestka, this series, Vol. 78 [68].

[5] M. Rubinstein, W. P. Levy, J. A. Moschera, C.-Y. Lai, R. D. Hershberg, R. T. Bartlett, and S. Pestka, *Arch. Biochem. Biophys.*, in press.

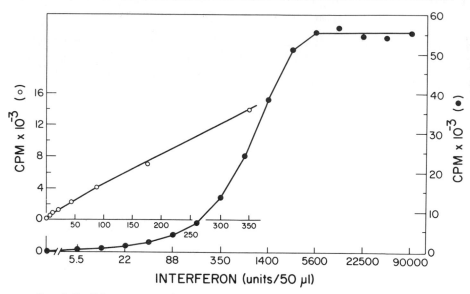

Fig. 2. Radioimmunoassay for human leukocyte interferon. The data are plotted with linear and logarithmic scales on the abscissa. The assay was performed as described in the text. Crude human leukocyte interferon (specific activity approximately 1×10^6 units/mg) was used in the assay.

The results of the assay are shown in Fig. 2. The counts of [^{125}I]LI-1 bound were essentially linear up to 350 units/50 μl, a range where estimates were most accurate. Nevertheless, interferon levels up to 2800 units/50 μl (56,000 units/ml) could be estimated with reasonable precision. Above 56,000 units/ml, maximal [^{125}I]LI-1 was bound.

The assay can be performed by incubating interferon and the second antibody simultaneously rather than sequentially. In this case, excess interferon reduces the amount of [^{125}I]LI-1 bound (Fig. 3).

The interferon assay can be made more sensitive by using monoclonal antibody LI-1 with a higher specific activity of labeling (about 160,000 cpm/ng). A crude mixture of leukocyte interferons can be measured with this assay to a concentration as low as 5 units/ml (Fig. 4). The absolute amounts of interferon determined per 0.10-ml assay are as little as 0.5 unit (1 pg, based on a specific activity of about 5×10^8 units/mg interferon). The assay was performed with 40% human plasma or serum without any change in the results.

Since a pair of monoclonal antibodies was utilized in the assay, only those interferons which were recognized by each of the monoclonal antibodies would be detected in the assay. Twelve purified human leukocyte

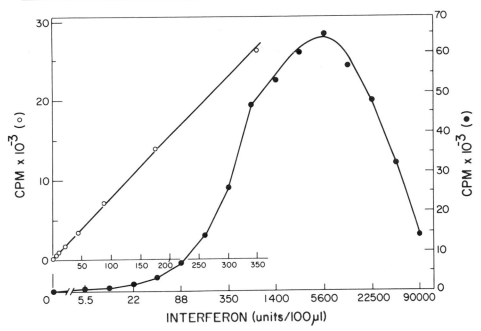

FIG. 3. Radioimmunoassay for human leukocyte interferon. Each well of a PVC-microtiter plate was coated with 100 μl of LI-9 (15 μg/ml in PBS) as described in the text. About 140,000 cpm of [^{125}I]LI-1 (4 ng) in 0.1 ml of PBS containing 1% bovine serum albumin and 150 μg/ml of human IgG were placed into each well. Appropriately diluted interferon solution in a volume of 1 to 20 μl was added to each well followed by brief mixing. After incubation of interferon and [^{125}I]LI-1 together for 3 hr at room temperature, the solution was removed, the wells rinsed, and [^{125}I]LI-1 bound determined as described in the text. Crude human leukocyte interferon (specific activity approximately 1 × 10^8 units/mg) was used in the assay. The data are plotted with linear and logarithmic scales on the abscissa.

interferon species from chronic myelogenous leukemia cells[5] were tested in the radioimmunoassay. The results showed that different leukocyte interferon species were detected with different sensitivities.[1] Some of the interferon species, for example, IFL-γ$_3$ and IFLrD, which appears to be equivalent to the natural IFL-γ$_3$, were not detected. Since monoclonal antibody LI-1 does not recognize IFL-γ$_3$,[1] this result was expected. To assay for leukocyte interferon species which are not detected by this assay, another pair of monoclonal antibodies must be chosen. Specifically, since LI-1 does not recognize IFL-γ$_3$, another monoclonal antibody must be used in place of LI-1 in the assay. The assay can be used to detect purified or unfractionated leukocyte interferons. As anticipated, human fibroblast and immune interferons gave no positive response.

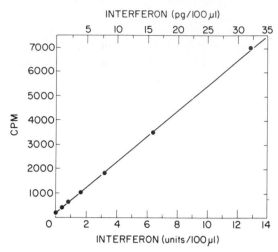

FIG. 4. Sensitive radioimmunoassay for human leukocyte interferon. The assay was performed as described in the legend to Fig. 3, except that 250,000 cpm of [^{125}I]LI-1 (1.5 ng) was present in the 0.10-ml volume with the interferon dilutions and incubation was performed at room temperature for 16 hr.

Concluding Comments

Monoclonal antibodies provide a novel opportunity to select two or more antibodies directed against different epitopes of an antigen. For a sandwich immunoassay, these antibodies should have the following properties: (1) lack of mutual interference with antigen binding; (2) optimal characteristics for solid phase function, i.e., antigen collection; (3) suitability for revealing antigen, i.e., stability to labeling with radioisotope, enzyme, fluorescent dye, etc.; (4) high antigen affinity and specificity of both antibodies. Our use and proper selection of two suitable monoclonal antibodies for measuring an antigen in a sandwich immunoassay provides a novel and universal strategy for designing and developing this type of assay for any antigen that has at least two independent epitopes. For some antigens, this approach may be the only possible solution; for other large substances such as carcinoembryonic antigen it should improve the specificity, sensitivity, and long-term reproducibility of already existing sandwich immunoassays.

This interferon radioimmunoassay has already proven extremely useful for monitoring interferon during purification. It has proven to be suitable for surveillance of interferon levels in patients for pharmacokinetic studies with interferon, and might be useful for differential diagnosis of various infectious and other diseases. By covalently linking horseradish peroxidase to monoclonal antibody LI-1, the assay also has been modified into a convenient and highly sensitive enzyme immunoassay for interferon.

Section IX

Cloning of the Interferon Genes and Complementary DNAs

[77] Cloning of the Human Interferons

By SIDNEY PESTKA

In the past several years, recombinant DNA technology has begun to elucidate the structure and regulation of eukaryotic genes. Although many achievements have been made and much more needs to be accomplished, the isolation and identification of recombinants containing specific eukaryotic DNA sequences provide initial probes to isolate full-length cDNA sequences as well as genomic clones containing the structural genes.

Up to 1980, most cDNAs that had been cloned represented mRNAs that consisted of greater than 1% of the total mRNA population. The construction and identification of recombinants containing the human interferon sequences extended the technology to mRNAs that represented less than 1% of the cellular mRNA content. Thus, the general procedures utilized could be employed to isolate almost any cDNA recombinant, whatever its abundance. Some of these procedures are outlined in the chapters of this section.

The isolation and identification of recombinants containing sequences for human leukocyte interferon were accomplished by Nagata et al.[1] and by Maeda et al.[2] Sequences containing human fibroblast interferon were isolated by several groups[2-7] that used different approaches. The recombinants were utilized to express the human interferons in *Escherichia coli*.[1,7-9] The recombinants containing human leukocyte and fibroblast

[1] S. Nagata, H. Taira, A. Hall, L. Johnsrud, M. Streuli, J. Ecsodi, W. Boll, K. Cantell, and C. Weissmann, *Nature (London)* **284**, 316 (1980).

[2] S. Maeda, R. McCandliss, M. Gross, A. Sloma, P. C. Familletti, J. M. Tabor, M. Evinger, W. P. Levy, and S. Pestka, *Proc. Natl. Acad. Sci. U.S.A.* **77**, 7010 (1980).

[3] T. Taniguchi, M. Sakai, Y. Fujii-Kuriyama, M. Muramatsu, S. Kobayashi, and T. Sudo, *Proc. Jpn. Acad. B* **55**, 464 (1979).

[4] R. Derynck, J. Content, E. DeClercq, G. Volckaert, J. Tavernier, R. Devos, and W. Fiers, *Nature (London)* **285**, 542 (1980).

[5] M. Houghton, A. G. Stewart, S. M. Doel, J. S. Emtage, M. A. W. Eaton, J. C. Smith, T. P. Patel, H. M. Lewis, A. G. Porter, J. R. Birch, T. Cartwright, and N. H. Carey, *Nucleic Acids Res.* **8**, 1913 (1980).

[6] T. Taniguchi, S. Ohno, Y. Fujii-Kuriyama, and M. Muramatsu, *Gene* **10**, 11 (1980).

[7] D. V. Goeddel, H. M. Shepard, E. Yelverton, D. Leung, R. Crea, A. Sloma, and S. Pestka, *Nucleic Acids Res.* **8**, 4057 (1980).

[8] D. V. Goeddel, E. Yelverton, A. Ullrich, H. L. Heyneker, G. Miozzari, W. Holmes, P. H. Seeburg, T. Dull, L. May, N. Stebbing, R. Crea, S. Maeda, R. McCandliss, A. Sloma, J. M. Tabor, M. Gross, P. C. Familletti, and S. Pestka, *Nature (London)* **287**, 411 (1980).

[9] T. Taniguchi, L. Guarente, T. M. Roberts, D. Kimelman, J. Douhan, III, and M. Ptashne, *Proc. Natl. Acad. Sci. U.S.A.* **77**, 5230 (1980).

interferon sequences were used as hybridization probes to isolate the corresponding human genes. In this way, a family of human genes was isolated corresponding to human leukocyte interferon.[10-14] In contrast, only a single gene was isolated corresponding to the predominant human fibroblast interferon.[11,12,15] The isolation of a family of human leukocyte interferon genes was consistent with the prior isolation of multiple species of human leukocyte interferon from normal leukocytes[11,16] and leukocytes from patients with chronic myelogenous leukemia.[11,17-20] In addition, multiple leukocyte interferon species were isolated from various leukocyte cell lines.[21,22] It was surprising, however, to find that none of the human leukocyte interferon genes contained any intervening sequences.[10-12,14] Furthermore, the human fibroblast interferon gene also did not contain any intervening sequences.[11,12,15]

The various human leukocyte interferons, both natural[23-25] and recombinant,[25-27] exhibit differences in their biological activities although

[10] M. Streuli, S. Nagata, and C. Weissmann, *Science* **209**, 1343 (1980).
[11] S. Pestka, S. Maeda, D. S. Hobbs, W. P. Levy, R. McCandliss, S. Stein, J. A. Moschera, and T. Staehelin, *in* "Cellular Responses to Molecular Modulators" (W. A. Scott, R. Werner, and J. Schultz, eds.), Miami Winter Symposia, Vol. 18. Academic Press, New York, 1981 (in press).
[12] S. Maeda, R. McCandliss, T.-R. Chiang, L. Costello, W. P. Levy, N. T. Chang, and S. Pestka, *in* "Developmental Biology Using Purified Genes" (D. Brown, and C. F. Fox, eds.), ICN-UCLA Symposia on Molecular and Cellular Biology, Vol. XXIII. Academic Press, New York, 1981 (in press).
[13] D. V. Goeddel, D. W. Leung, T. J. Dull, M. Gross, R. M. Lawn, R. McCandliss, P. H. Seeburg, A. Ullrich, E. Yelverton, and P. W. Gray, *Nature (London)* **290**, 20 (1981).
[14] R. M. Lawn, J. Adelman, T. J. Dull, M. Gross, D. Goeddel, and A. Ullrich, *Science* **212**, 1159 (1981).
[15] G. Gross, W. Mayr, F. Grosveld, H. M. Dahl, R. A. Flavell, and J. Collins, *Hoppe Seylers Z. Physiol. Chem.* **362**, 205 (1981).
[16] M. Rubinstein, S. Rubinstein, P. C. Familletti, R. S. Miller, A. A. Waldman, and S. Pestka, *Proc. Natl. Acad. Sci. U.S.A.* **76**, 640 (1979).
[17] W. P. Levy, J. Shivley, M. Rubinstein, U. Del Valle, and S. Pestka, *Proc. Natl. Acad. Sci. U.S.A.* **77**, 5102 (1980).
[18] M. Rubinstein, S. Rubinstein, P. C. Familletti, L. D. Brink, R. D. Hershberg, J. Gutterman, J. Hester, and S. Pestka, *in* "Peptides: Structure and Biological Function" (E. Gross and J. Meienhofer, eds.), p. 99. Pierce Chemical Company, Rockford, Illinois, 1979.
[19] S. Pestka, M. Evinger, R. McCandliss, A. Sloma, and M. Rubinstein, *in* "Polypeptide Hormones" (R. F. Beers, Jr. and E. G. Bassett, eds.), p. 33. Raven, New York, 1980.
[20] W. P. Levy, M. Rubinstein, J. Shively, U. Del Valle, C.-Y. Lai, J. Moschera, L. Brink, L. Gerber, S. Stein, and S. Pestka, *Proc. Natl. Acad. Sci. U.S.A.* (in press).
[21] G. Allen and K. H. Fantes, *Nature (London)* **287**, 408 (1980).
[22] D. S. Hobbs, J. Moschera, W. P. Levy, and S. Pestka, this series, Vol. 78 [68].
[23] M. Evinger, M. Rubinstein, and S. Pestka, *Arch. Biochem. Biophys.* (in press).
[24] N. Kaplan, personal communication.
[25] R. B. Herberman, J. R. Ortaldo, M. Rubinstein, and S. Pestka, *J. Clin. Immunol.* **1**, 149 (1981).

the corresponding native and recombinant species show identical activities. Accordingly, it appears that differences in their primary sequences lead to differences in their interaction with a variety of cells. In addition to the large number of natural interferons that can be isolated, it is possible to prepare hybrid and modified interferons by recombination, reconstruction, and alteration of the natural recombinants. Thus, it may be possible to tailor interferons for specific biological activities.

Several cloned human interferons have been expressed[1,7-9] and we have purified several of them. Recombinant leukocyte interferon A (IFLrA) was purified to homogeneity.[28,29] On January 15, 1981, clinical trials with IFLrA were begun by Hoffmann-La Roche. Crystals of IFLrA were obtained with the intention of beginning X-ray crystallography to determine its tertiary structure.[30,31]

It is evident that interferon research has taken several giant steps forward. The availability of large amounts of pure interferon for both basic and clinical research will stimulate the advancement of knowledge about interferon much further. The availability of the cDNA and genomic recombinants will lead to understanding how interferon synthesis is regulated by viruses and double-stranded RNA. Recombinants containing the IFLrA sequence have been expressed in an $E.\ coli$ DNA-dependent cell-free extract.[32] Such studies with the genomic clones in a eukaryotic cell-free system would be most rewarding. As with most productive areas of research, there is now very much more to do than there was before.

[26] M. Evinger, S. Maeda, and S. Pestka, $J.\ Biol.\ Chem.$ **256**, 2113 (1981).
[27] R. B. Herberman, J. R. Ortaldo, A. Mantovani, D. S. Hobbs, H.-F. Kung, and S. Pestka, $Nature\ (London)$ (in press).
[28] T. Staehelin, D. S. Hobbs, H.-F. Kung, and S. Pestka, $J.\ Biol.\ Chem.$ (in press).
[29] T. Staehelin, D. S. Hobbs, H.-F. Kung, and S. Pestka, this series, Vol. 78 [72].
[30] D. L. Miller, H.-F. Kung, T. Staehelin, and S. Pestka, this volume [1].
[31] D. L. Miller, H.-F. Kung, T. Staehelin, and S. Pestka, in preparation.
[32] H. Weissbach, D. V. Goeddel, R. McCandliss, S. Maeda, P. C. Familletti, B. Redfield, T. Staehelin, and S. Pestka, $Arch.\ Biochem.\ Biophys.$ (in press).

[78] Preparation of cDNA, ds cDNA, and Tailed ds cDNA from Interferon mRNA Templates

By RUSSELL MCCANDLISS, ALAN SLOMA, and HIDEO TAKESHIMA

In order to isolate recombinants containing specific sequences corresponding to messenger RNA (mRNA), it is necessary to make complementary DNA (cDNA) copies of the mRNA, to make the cDNA copy double-stranded, and to modify the double-stranded cDNA (ds cDNA) so that it can be incorporated into a cloning vehicle. Many cDNAs, including those for insulin,[1,2] growth hormone,[3] globin,[4] ovalbumin,[5] dihydrofolate reductase,[6] and interferons,[7–12] have been cloned by this approach. General techniques for these reactions have been presented in Vol. 68 of this series. In this chapter, specific details of the preparation of interferon cDNA for cloning[7] are described.

Synthesis of cDNA

Messenger RNA can be copied into a cDNA molecule with reverse transcriptase.[13] Interferon mRNA has been shown to be polyadenylated,[14–16] which permits oligo(dT) to be used as a primer for cDNA synthesis. Complementary DNA is synthesized by the following method.

Stock Solutions and Materials

Tris · HCl, 0.5 M, pH 8.3
KCl, 1.4 M
MgCl$_2$, 0.25 M

[1] A. Ullrich, J. Shine, J. Chirgwin, R. Pictet, E. Tischer, W. J. Rutter, and H. M. Goodman, Science 196, 1313 (1977).
[2] L. Villa-Komaroff, A. Efstratiadis, S. Broome, P. Lomedico, R. Tizard, S. P. Naber, W. L. Chick, and W. Gilbert, Proc. Natl. Acad. Sci. U.S.A. 75, 3727 (1978).
[3] P. H. Seeburg, J. Shine, J. A. Martial, J. D. Baxter, and H. M. Goodman, Nature (London) 270, 486 (1977).
[4] F. Rougeon, P. Kourilsky, and B. Mach, Nucleic Acids Res. 2, 2365 (1975).
[5] P. Humphries, M. Cochet, A. Krust, P. Gerlinger, P. Kourilsky, and P. Chambon, Nucleic Acids Res. 4, 2389 (1977).
[6] A. C. Y. Chang, J. H. Nunberg, R. J. Kaufman, H. A. Erlich, R. T. Schimke, and S. N. Cohen, Nature (London) 275, 617 (1978).
[7] S. Maeda, R. McCandliss, M. Gross, A. Sloma, P. C. Familletti, J. M. Tabor, M. Evinger, W. P. Levy, and S. Pestka, Proc. Natl. Acad. Sci. U.S.A. 77, 7010 (1980).
[8] S. Nagata, H. Taira, A. Hall, L. Johnsrud, M. Streuli, J. Ecsödi, W. Boll, K. Cantell, and C. Weissmann, Nature (London) 284, 316 (1980).
[9] T. Taniguchi, M. Sakai, Y. Fujii-Kuriyama, M. Muramatsu, S. Kobayashi, and T. Sudo, Proc. Jpn. Acad. Ser. B 55, 464 (1979).
[10] R. Derynck, J. Content, E. De Clercq, G. Volckaert, J. Tavernier, R. Devos, and W. Fiers, Nature (London) 285, 542 (1980).
[11] M. Houghton, A. G. Stewart, S. M. Doel, J. S. Emtage, M. A. W. Eaton, J. C. Smith, T. P. Patel, H. M. Lewis, A. G. Porter, J. R. Birch, T. Cartwright, and N. H. Carey, Nucleic Acids Res. 8, 1913 (1980).
[12] D. V. Goeddel, H. M. Shepard, E. Yelverton, D. Leung, R. Crea, A. Sloma, and S. Pestka, Nucleic Acids Res. 8, 4057 (1980).
[13] D. L. Kacian, and J. C. Myers, Proc. Natl. Acad. Sci. U.S.A. 73, 2191 (1976).
[14] F. H. Reynolds, Jr., and P. M. Pitha, Biochem. Biophys. Res. Commun. 59, 1023 (1974).
[15] S. Pestka, J. McInnes, E. A. Havell, and J. Vilček, Proc. Natl. Acad. Sci. U.S.A. 72, 3898 (1975).
[16] R. McCandliss, A. Sloma, and S. Pestka, this volume [8].

dATP, 0.05 M, pH 7.0
dGTP, 0.05 M, pH 7.0
dCTP, 0.05 M, pH 7.0
TTP, 0.05 M, pH 7.0
[α-^{32}P]dCTP, 400 Ci/mmol, 1 mCi/ml (Amersham)
Dithiothreitol (DTT), 0.01 M
Oligo(dT)$_{12-18}$, 250 μg/ml (Collaborative Research)
Actinomycin D, 500 μg/ml (Calbiochem)
Disodium ethylenediaminetetraacetate (EDTA), 0.2 M, pH 8.0
Avian myeloblastosis virus (AMV) reverse transcriptase, approxi-
     mately 10,000 units/ml (obtained from Research Resources Branch,
     Viral Oncology Program, National Cancer Institute)

All buffers and salt solutions are autoclaved. The other solutions are prepared with sterile glass-distilled water and stored in sterile containers. All stock solutions are stored frozen.

Procedure. As a template for cDNA synthesis, 12 S interferon mRNA isolated as previously described[16] is used. In order to follow synthesis, either [^{3}H]dCTP or [α-^{32}P]dCTP may be used. For preparation of recombinants, the cDNA must be located on polyacrylamide gels; therefore, it is more convenient to use ^{32}P as the radioactive label. The radioactive compound is dried by lyophilization, by blowing off the solvent with a stream of nitrogen, or by drying in a Savant Speed Vac Concentrator. For each microgram of mRNA, 5 μCi of [α-^{32}P]dCTP at a specific activity of 400 Ci/mmol are used. The dried material is dissolved in a 2$\times$ reaction mixture consisting of 0.1 M Tris · HCl, pH 8.3, 140 mM KCl, 20 mM MgCl$_2$, 1 mM dATP, 1 mM dCTP, 1 mM dGTP, 1 mM TTP, and 0.4 mM DTT. This solution is kept on ice. To this solution are added mRNA (50 μg/ml, final concentration), oligo(dT)$_{12-18}$ (25 μg/ml), actinomycin D (40 μg/ml), AMV reverse transcriptase (800 units/ml), and enough water to dilute the 2$\times$ mix to 1$\times$. After 5 min on ice, the reaction mixture is incubated at 46° for 10 min. After the incubation, EDTA is added to a final concentration of 25 mM. The solution is extracted one time with an equal volume of phenol : chloroform (1/1; v/v), and the aqueous phase is chromatographed on a column of Sephadex G-100 (0.7 $\times$ 20 cm) equilibrated with 10 mM Tris · HCl, pH 8.0, 1 mM EDTA, 0.1 M NaCl. The cDNA in the excluded volume is precipitated by addition of 0.1 volume of 2.4 M sodium acetate, pH 5, and 2.5 volumes of ethanol. To remove the mRNA template, the cDNA is sedimented by centrifugation, dissolved in 0.3 ml of 0.1 M NaOH, and incubated at 70° for 20 min. The solution is neutralized with 1.0 M HCl and precipitated with ethanol as described above. The yield of cDNA is 10–20% of the mRNA used. The conditions described are optimized for full-length cDNA synthesis, but not for maximal incorporation, which would be important in the synthesis of a cDNA probe.

Synthesis of ds cDNA

Synthesis of ds cDNA from cDNA is performed with the use of DNA polymerase I (Klenow fragment), which lacks the $5' \rightarrow 3'$ exonuclease activity.[17] No additional primer is needed because of the 3' loop on most cDNA molecules made with AMV reverse transcriptase.[18] In order to make ds cDNA, the 3' loop must be cleaved by *Aspergillus oryzae* S1 nuclease.[19] To select for full-length copies of the mRNA, the ds cDNA is sized on a polyacrylamide gel, and long fragments are located, excised, and electroeluted from the gel.

Stock Solutions and Materials

Potassium phosphate, 0.5 M, pH 7.4
MgCl$_2$, 0.25 M
DTT, 0.1 M
dATP, 0.05 M, pH 7.0
dGTP, 0.05 M, pH 7.0
dCTP, 0.05 M, pH 7.0
TTP, 0.05 M, pH 7.0
[8-^{3}H]dCTP, 22 Ci/mmol, 1 mCi/ml (Amersham)
Escherichia coli DNA polymerase I (Klenow fragment), approximately 1000 units/ml (Boehringer-Mannheim)
5× S1 nuclease buffer: 0.167 M sodium acetate, pH 4.5; 5 mM ZnCl$_2$

Procedure. For cloning, it is not necessary to use a radioactive label in the second strand if the first strand is labeled. However, to follow second-strand synthesis, it is convenient to use [^{3}H]dCTP. The labeled compound is dried as described above. Ten microcuries of [^{3}H]dCTP (specific activity, 22 Ci/mmol) are used for each microgram of mRNA used for cDNA synthesis. The dried [^{3}H]dCTP is dissolved in a 2× reaction mixture consisting of 0.2 M potassium phosphate, pH 7.4, 20 mM MgCl$_2$, 2 mM DTT, 0.4 mM each of dATP, dGTP, dCTP, and TTP. This mixture is kept on ice, cDNA in water is added, *E. coli* DNA polymerase I (Klenow fragment) is added to 100 units/ml, and water is added to dilute the reaction mixture to 1×. The solution is incubated a minimum of 2 hr at 15°. It is convenient to allow this reaction to incubate overnight. After the incubation, EDTA is added to 25 mM, the solution is extracted once with an equal volume of phenol : chloroform (1/1; v/v), and the aqueous phase is chromatographed on a 0.7 × 20 cm column of Sephadex G-100 equilibrated with 10 mM Tris · HCl, pH 8.0, 1 mM EDTA, and 0.1 M NaCl. The DNA in the excluded fractions is precipitated with ethanol as described

[17] H. Klenow, K. Overgaard-Hansen, and S. A. Patkar, *Eur. J. Biochem.* **22**, 371 (1971).
[18] A. Efstratiadis, F. C. Kafatos, A. M. Maxam, and T. Maniatis, *Cell* **7**, 279 (1976).
[19] V. M. Vogt, *Eur. J. Biochem.* **33**, 192 (1973).

above. The yield of dsDNA is 50–100% of the amount of cDNA used as template.

At this point, the ds cDNA contains a hairpin loop. The single-stranded loop is removed by digestion with *Aspergillus oryzae* S1 nuclease, prepared by the method of Vogt.[19] The ds cDNA is dissolved in water, and 0.25 volume of 5× S1 buffer is added. An appropriate amount of S1 nuclease is added, and the solution is incubated 20 min at 37°. The amount of enzyme to be added must be determined empirically for each enzyme preparation, since the activity varies from one preparation to another. This is done by measuring the decrease in trichloroacetic acid-precipitable counts from the ds cDNA. Usually, 50–75% of the ds cDNA is resistant to S1 nuclease. However, care must be exercised in order to avoid over-digestion due to low levels of contaminating nucleases in the S1 nuclease preparation. The S1-digested ds cDNA is extracted once with phenol : chloroform, and the aqueous phase is precipitated with ethanol as described above.

It is desirable to have full-length copies of the gene. Since interferon mRNA is about 12 S on sucrose gradients,[16] ds cDNA having about 900 base pairs should contain the entire sequence. Molecules of this size can be isolated by electrophoresis on an 8% polyacrylamide gel run in Tris–borate–EDTA (TBE) (0.09 M Tris, 0.09 M borate, 0.025 M EDTA, pH 8.3) as described by Maniatis *et al.*[20] Plasmid pBR322 digested with *Taq* I restriction endonuclease provides molecular weight markers of convenient size.[21] After electrophoresis, the gel is stained with 0.5 μg of ethidium bromide per milliliter to visualize the markers. The wet gel is then exposed to Kodak X-Omat R film for a short period of time (dependent upon the number of counts applied to the gel) to locate the labeled ds cDNA. The region of the gel containing DNA molecules between 500 and 1300 base pairs long is cut out, and the DNA is eluted electrophoretically. The gel slice is placed in a dialysis bag containing 0.3–0.6 ml of 0.1× TBE, and the bag is placed in an electrophoresis chamber similar to that described by Smith.[22] Electrophoresis is performed for 1.5 hr at 100 V. The solution from inside the dialysis bag is extracted once with phenol : chloroform (1/1; v/v), and the ds cDNA is precipitated with ethanol. Recovery can be estimated by monitoring Cerenkov radiation of the gel slice and the solutions.

Homopolymer Tailing of ds cDNA

A convenient means of inserting ds cDNA into a cloning vector is to add single-stranded homopolymer tails to the 3′-OH termini of the ds

[20] T. Maniatis, A. Jeffrey, and H. van deSande, *Biochemistry* 14, 3787 (1975).
[21] J. G. Sutcliffe, *Nucleic Acids Res.* 5, 2721 (1978).
[22] H. O. Smith, this series, Vol. 65, p. 371.

YIELDS OF PRODUCTS DURING SYNTHESIS OF
dC-TAILED ds cDNA SYNTHESIS

Step	Amount (μg)	Percent yield
mRNA	13	—
cDNA	1.5	11.5
Sized ds cDNA	0.24	1.18 (15)[a]
dC-ds cDNA	0.23	1.8 (96)

[a] Numbers in parentheses represent the stepwise yields.

cDNA and complementary homopolymer tails to the 3'-OH termini of the vector, and to anneal the two tailed DNA molecules. Methods of homopolymer addition have been described in detail by Nelson and Brutlag.[23]

Because cloning into the plasmid pBR322 at the *Pst* I site by G-C tailing regenerates the *Pst* I recognition sequence at each end of the inserted DNA sequence,[24] it is convenient to use this procedure.

Stock Solutions and Materials

Potassium cacodylate, 1.4 M; 0.3 M Tris; pH 7.6 (pH becomes 7.2 when diluted 1 : 10)
$CoCl_2$, 15 mM
DTT, 10 mM
dCTP, 4 mM, pH 7.0

Procedure. Sizing of the ds cDNA allows an accurate estimation of the concentration of 3' termini. Addition of dCMP residues to the 3' ends may be followed by incorporation of [³H]dCTP into acid-precipitable material. Twenty-five microcuries of [³H]dCTP are used for each microgram of mRNA used for the original cDNA synthesis. The radioactive compound is dried and redissolved in the reaction mixture. Double-stranded cDNA is dissolved in appropriate amounts of stock solutions to give final concentrations as follows: 0.14 M potassium cacodylate, 0.03 M Tris, pH 7.2; 1 mM DTT; 0.1 mM [³H]dCTP (1 Ci/mmol); 1.5 mM $CoCl_2$; and $2 \times 10^{-8} M$ 3' termini. The solution without $CoCl_2$ is warmed to 37°; the $CoCl_2$ is then added. Terminal deoxynucleotidyltransferase purified by the method of Chang and Bollum[25] is added to a final concentration of 100 units/ml. The

[23] T. Nelson and D. Brutlag, this series, Vol. 68, p. 41.
[24] F. Bolivar, R. L. Rodriguez, P. J. Greene, M. C. Batlach, H. L. Heyneker, H. W. Boyer, J. H. Crosa, and S. Falkow, *Gene* **2**, 95 (1977).
[25] L. M. S. Chang and F. J. Bollum, *J. Biol. Chem.* **246**, 909 (1971).

reaction is allowed to proceed at 37° for 5 min. At this time, a sample is taken to measure incorporation of [³H]dCMP into acid-precipitable material. If tails of sufficient length have not been added, the reaction can be continued by placing the solution at 37° again for the desired length of time. We have obtained the best transformation efficiencies with homopolymer tails of about 10–20 dCMP residues.[26] When tails of the desired length have been generated, EDTA is added (10 mM, final concentration). The solution is extracted once with phenol : chloroform (1/1; v/v), and the aqueous phase is precipitated with ethanol. The dC-tailed ds cDNA at this point is ready for insertion into a dG-tailed pBR322 vector as described by Maeda et al.[26] Results of a representative preparation are given in the table.

[26] S. Maeda, this volume [80].

[79] Preparation of Plasmids and Tailing with Oligodeoxyguanylic Acid

By SHUICHIRO MAEDA

The plasmid pBR322[1] was used for the cloning of complementary DNA (cDNA) copies of mRNA from interferon-producing human cells. Deoxyguanylic acid (dG) residues are added to the PstI restriction termini with terminal deoxynucleotidyltransferase.[2] The tailed plasmid DNA is annealed with cDNA which has been tailed with oligodeoxycytidylic acid (dC) residues.[3] Because the homopolymer-joining method reconstructs the PstI sites at each end of the inserted cDNA,[1] the inserted cDNA and flanking oligo(dG) · oligo(dC) sequences can be excised from the recombinants with PstI. Since the PstI site lies in the β-lactamase gene of the plasmid, the cDNA may be expressed as a fusion protein[4] or, in some cases, as an independent protein.[5]

[1] F. Bolivar, R. L. Rodriguez, P. J. Green, M. C. Betlach, H. L. Heyneker, H. W. Boyer, and S. Falkow, Gene 2, 95 (1977).
[2] R. Roychoudhury and R. Wu, this series, Vol. 65, p. 43.
[3] R. McCandliss, A. Sloma, and H. Takeshima, this volume [78].
[4] L. Villa-Komaroff, A. Efstratiadis, S. Broome, P. Lomedico, R. Tizard, S. P. Naber, W. L. Chick, and W. Gilbert, Proc. Natl. Acad. Sci. U.S.A. 75, 3727 (1978).
[5] A. C. Y. Chang, J. H. Nunberg, R. J. Kaufman, H. A. Erlich, R. T. Schimke, and S. N. Cohen, Nature (London) 275, 617 (1978).

Preparation of the Plasmid pBR322

The plasmid DNA is prepared essentially as described by Clewell and Helinski.[6] Covalently closed circular plasmid DNA is further purified by CsCl density gradient centrifugation.[7]

Bacterial Growth and Plasmid Amplification

Bacterial Strain and Medium. *Escherichia coli* K-12 derivative RR1[8] (F$^-$, *pro, leu, thi, lac* Y, *str*^{r}, $r_k^-m_k^-$, endoI$^-$) carrying the plasmid pBR322 is used for preparation of the plasmid DNA. M-9 medium is used for culturing *E. coli* RR1-harboring pBR322. Each liter of M-9 medium contains Na_2HPO_4 (6.0 g), KH_2PO_4 (3.0 g), NaCl (0.5 g), NH_4Cl (1.0 g), 1 ml of 1 M $MgSO_4$, 1 ml of 0.1 M $CaCl_2$, 25 ml of 20% (w/v) casamino acids, 8 ml of 50% (w/v) glucose, and 0.4 ml of 0.5% (w/v) thiamine · HCl.

Bacterial Growth. One hundred milliliters of M-9 medium, containing 25 μg of sodium ampicillin per milliliter, are inoculated with 3 ml of overnight culture of *E. coli* RR1/pBR322 and are incubated with shaking at 37° until the culture reaches mid-log phase (about 0.4 A_{590}). The 100-ml culture is then added to 9.9 liters of M-9 medium containing 25 μg of ampicillin per milliliter. The 10-liter culture is incubated until it reaches an optical density of 0.35 at 590 nm, at which time chloramphenicol is added to a final concentration of 250 μg/ml to amplify the plasmid DNA. The culture is incubated at 37° for an additional 12 hr. Cells are then harvested by centrifugation for isolation of the plasmid DNA. If not used immediately, the cell pellets can be stored frozen at $-70°$ or below.

Isolation of Plasmid DNA

1. The cells from a 10 liter culture (about 7 g) are resuspended in 150 ml of a solution containing 10 mM Tris · HCl, pH 6.8, and 1 mM Na_2EDTA.
2. The cells are pelleted by centrifugation for 20 min at 4000 rpm at 4° in a Sorvall GSA rotor.
3. The pellet is resuspended in 100 ml of cold 50 mM Tris · HCl, pH 6.8.
4. Then 4 ml of cold 0.25 M Na_2EDTA and 20 ml of a solution of egg white lysozyme (20 mg/ml in cold H_2O; freshly prepared; EC 3.2.1.17) are added. The suspension is kept on ice for 30 min with occasional gentle mixing.

[6] D. B. Clewell and D. R. Helinski, *Biochemistry* **9**, 4428 (1970).
[7] R. Radloff, W. Bauer, and J. Vinograd, *Proc. Natl. Acad. Sci. U.S.A.* **57**, 1514 (1967).
[8] R. L. Rodriguez, R. Tait, J. Shine, F. Bolivar, H. Heyneker, M. Betlach, and H. W. Boyer, *Miami Winter Symp.* **13**, 73 (1977).

5. The cells are lysed by the addition of 302 ml of cold lysis solution [2 ml of 2% (w/v) bentonite, 100 ml of 4% (v/v) Brij 58, 100 ml of 0.4% (w/v) sodium deoxycholate, 50 ml of 0.25 M Na$_2$EDTA, and 50 ml of 0.3 M Tris · HCl, pH 6.8].

6. The cell debris is removed by centrifugation for 1 hr at 30,000 rpm at 4° in a Spinco 42.1 rotor.

7. Sufficient NaCl is added to the supernatant to bring the final concentration to 1 M.

8. To deproteinize the solution, an equal volume of phenol : chloroform : isoamyl alcohol (50 : 50 : 2; v/v/v) is added to the solution and vigorously shaken.

9. To separate the phases, the mixture is centrifuged for 10 min at 4000 rpm at 4° in a Sorvall GSA rotor.

10. The aqueous phase is reextracted once with the phenol : chloroform : isoamyl alcohol mixture. The interphase is extracted with 200 ml of a solution containing 20 mM Tris · HCl, pH 6.8, 1 M NaCl, and 5 mM Na$_2$EDTA.

11. The DNA is precipitated from the combined aqueous extracts by addition of two volumes of 95% ethanol. After sitting overnight at $-20°$, the DNA precipitate is sedimented by centrifugation for 30 min at 9000 rpm at 0° in a Sorvall GSA rotor.

12. The pellet is resuspended in 240 ml of a solution containing 10 mM Tris · HCl, pH 8, and 5 mM Na$_2$EDTA.

CsCl-Buoyant Density Gradient Centrifugation

1. To each 14.8 ml of the DNA solution, 17.8 g of solid CsCl and 4.9 ml of propidium iodide solution (0.5 mg/ml in H$_2$O) are added.

2. The solution is centrifuged in a Beckman 60 Ti rotor for 42 hr at 36,000 rpm at 10°.

3. The lower fluorescent band, visualized with a long-wave UV lamp, contains the plasmid DNA and is removed with a Pasteur pipette.

4. The solution containing this DNA band is extracted several times with an equal volume of CsCl- and H$_2$O-saturated n-butanol to remove propidium iodide.

5. The solution of plasmid DNA is then dialyzed against a solution of 10 mM Tris · HCl, pH 8, and 1 mM Na$_2$EDTA.

6. The DNA is precipitated by addition of 0.1 volume of 3 M sodium acetate, pH 5.5, and 2.5 volumes of 95% ethanol as described above.

Yield of the plasmid DNA is approximately 1.0 to 1.5 mg from 1 liter of culture.

Adding Homopolymer Oligodeoxyguanylic Acid Tails to Plasmid pBR322

Digestion of the Plasmid with PstI Endonuclease

The reaction mixture in a total volume of 0.50 ml contains:
pBR322 DNA, 0.5 mg/ml
NaCl, 50 mM
Tris · HCl, pH 7.4, 6 mM
MgCl$_2$, 6 mM
2-Mercaptoethanol, 6 mM
Bovine serum albumin, 100 μg/ml
*Pst*I enzyme (New England Biolabs), 10 units/ml

The reaction is performed at 37° and is monitored by 0.7% agarose gel electrophoresis.[9] After the reaction is completed, 0.1 volume of 0.25 M Na$_2$EDTA is added. The reaction is extracted with phenol : chloroform : isoamyl alcohol, and the DNA is precipitated with ethanol as above (Step 6).

Addition of Homopolymer Oligodeoxyguanylic Acid Tails

Approximately 15–30 deoxyguanylic acid residues are added to each 3′ end of the *Pst*I-cleaved pBR322 DNA with terminal deoxynucleotidyltransferase in the presence of 1.5 mM Co^{2+}.[2] The reaction mixture (1.5 ml) contains:
*Pst*I-digested pBR322 DNA, concentration of 3′ ends = 2 × 10^{-8} M
Cacodylic acid, 140 mM
Tris(hydroxymethyl)aminomethane base, 30 mM
KOH, 110 mM
Dithiothreitol, 0.1 mM
CoCl$_2$, 1.5 mM
[^{3}H]dGTP (3 Ci/mmol, Amersham), 0.1 mM
Terminal deoxynucleotidyltransferase, 700 units/ml

The pH of the reaction is 7.6. The reaction is performed at 37°. Terminal deoxynucleotidyltransferase was purified by modifications of the method of Chang and Bollum.[10] In a control experiment, circular pBR322 DNA is added to the above transferase reaction instead of the linear plasmid DNA.

The reaction is monitored by taking samples of 5 μl from the reactions for acid precipitation. The number of deoxyguanylic acid residues added to each 3′ end of the linear DNA is determined by subtracting the number of residues added to the circular plasmid DNA from the number added to

[9] P. A. Sharp, B. Sugden, and J. Sambrook, *Biochemistry* **12**, 3055 (1973).
[10] L. M. S. Chang and F. J. Bollum, *J. Biol. Chem.* **246**, 909 (1971).

the linear plasmid DNA. After 15–30 residues have been added to each 3′ end, 0.1 volume of 0.25 M Na$_2$EDTA and carrier yeast tRNA (final concentration, 50 μg/ml) are added. The reaction is extracted with phenol : chloroform : isoamyl alcohol, and the DNA is precipitated with ethanol as above. Incubation for 5–10 min usually adds 15–30 deoxyguanylic acid residues.

Purification of Oligo(dG)-Tailed Plasmid DNA by 0.7% Agarose Gel Electrophoresis

The plasmid DNA containing homopolymer tails of oligo(dG) is purified by electrophoresis on a 0.7% agarose gel in order to remove small amounts of undigested circular plasmid DNA. The linear plasmid DNA is recovered from the gel by the KI-hydroxyapatite column chromatography method as described by Smith.[11] The eluted DNA is extracted with n-butanol to remove ethidium bromide, dialyzed against a solution of 10 mM Tris · HCl, pH 8, and 1 mM Na$_2$EDTA, and then precipitated with ethanol as above. Recovery with this purification step is 50–80%.

[11] H. O. Smith, this series, Vol. 65, p. 371.

[80] Annealing of Oligo(dC)-Tailed ds cDNA with Oligo(dG)-Tailed Plasmid and Transformation of *Escherichia coli* χ1776 with Recombinant DNA

By Shuichiro Maeda

For preparing recombinants, DNA fragments can be joined to plasmid or virus vectors in a number of ways. Vectors linearized with restriction enzymes can be annealed to DNA fragments terminating in homologous sequences.[1,2] Alternatively, vectors and DNA fragments can be terminated in the complementary oligodeoxynucleotides oligo(dA) and oligo(dT) or oligo(dG) and oligo(dC).[3,4] With the use of the plasmid pBR322 linearized with the restriction enzyme *Pst*I, the addition of oligo(dG) restores the *Pst*I site.[5] This conveniently permits the excision of the inserted DNA

[1] V. Sgaramella, *Proc. Natl. Acad. Sci. U.S.A.* **69**, 3389 (1972).

[2] J. E. Mertz and R. W. Davis, *Proc. Natl. Acad. Sci. U.S.A.* **69**, 3370 (1972).

[3] D. A. Jackson, R. H. Symons, and P. Berg, *Proc. Natl. Acad. Sci. U.S.A.* **69**, 2904 (1972).

[4] F. Rugeon, P. Kourilsky, and B. Mach, *Nucleic Acids Res.* **2**, 2365 (1975).

[5] F. Bolivar, R. L. Rodriguez, P. J. Greene, M. C. Betlach, H. L. Heyneker, H. W. Boyer, and S. Falkow, *Gene* **2**, 95 (1977).

sequence from the recombinants. The *Pst*I site is within the β-lactamase gene, so that these pBR322 recombinants with inserts are penicillin-sensitive, but retain their tetracycline resistance.

Solutions and Media

χ-Broth: Each liter of broth contains 10 g of Bacto-tryptone, 5 g of Bacto-yeast extract, 10 g of NaCl, diaminopimelic acid (DAP, 100 μg/ml), thymidine (20 μg/ml), *d*-biotin (1 μg/ml), and glucose (0.08% w/v).

χ-Agar: χ-broth supplemented with tetracycline (20 μg/ml)

Solution I: 5 mM Tris · HCl, pH 7.4, 5 mM MgCl₂, 0.1 M NaCl

Solution II: 70 mM MgCl₂, 40 mM sodium acetate, pH 5.6, 30 mM CaCl₂.

Solution III: 80 mM sodium acetate, pH 5.6, 0.6 M NaCl

Annealing of Tailed Plasmids and DNA Fragments

The oligo(dC)-tailed ds cDNA[6] is annealed with the oligo(dG)-tailed pBR322 DNA in a reaction volume of 0.90 ml containing 10 mM Tris · HCl, pH 7.5, 0.25 mM Na₂EDTA, 0.1 M NaCl, 468 ng of oligo(dG)-tailed plasmid DNA, and an equal molar amount (about 110 ng) of oligo(dC)-tailed ds cDNA. The DNA concentration used should be optimal for the formation of circular monomer hybrid molecules.[7] The reaction is heated to 70°C for 10 min, then gradually cooled to 37° and kept at this temperature overnight. The mixture is then allowed to cool slowly to room temperature and used for transformation.

Transformation of Escherichia coli χ1776

Escherichia coli K-12 derivative χ1776 (F⁻, *ton*A53, *dap*08, *min*A1, *sup*E42, Δ40[*gal-uvr*B], λ⁻, *min*B2, *rfb*-2, *nal*A25, *oms*-2, *thy*A57, *met*C65, *oms*-1, Δ29[*bio*H-*asd*], *cyc*B2, *cyc*A1, *hsd*R2) was obtained from Curtiss.[8] For transformation, the following procedure is used.

1. A single colony of χ1776 is inoculated into 5 ml of χ-broth supplemented with nalidixic acid (Nx, 50 μg/ml) and incubated at 37° with shaking overnight.

2. The overnight culture is diluted 1 : 100 into 300 ml of broth (without nalidixic acid) and incubated at 37° with shaking until the culture reaches an optical density of 0.3 at A_{550}.

---

[6] R. McCandliss, A. Sloma, and H. Takeshima, this volume [78].

[7] A. Dugaiczyk, H. W. Boyer, and H. M. Goodman, *J. Mol. Biol.* **96**, 171 (1975).

[8] R. Curtiss, III, M. Inoue, D. Pereira, J. C. Hsu, L. Alexander and L. Rock, *Miami Winter Symp.* **13**, 99 (1977).

3. The cells are sedimented at 4° by centrifugation for 5 min at 5000 rpm in a Sorvall GSA rotor.
4. The cell pellet is gently suspended in 150 ml of cold solution I and kept on ice for 5 min.
5. The cells are then sedimented as in step 3 above.
6. The cell pellet is resuspended in 30 ml of cold solution II and permitted to stand for 20 min on ice.
7. The cells are then pelleted as in step 3 above.
8. The cell pellet is gently resuspended in 10 ml of the same cold solution II.
9. Then 0.90 ml of cold solution III is added to the 0.90 ml of annealed DNA sample prepared above and kept on ice for 10 min or longer. (When plasmid DNA vector is used instead of a recombinant plasmid DNA, the DNA is resuspended in 40 mM sodium acetate, pH 5.6, 0.3 M NaCl.)
10. The competent cells (9 ml) are added to the 1.8 ml of DNA with a chilled pipette. This mixture is kept on ice for 60 min with gentle shaking every 10 min. It is important to keep all solutions cold and to use a cold pipette for the transfer of solutions.
11. After 60 min, the mixture is incubated at 37° for 2 min and then cooled to room temperature.
12. Portions of 75 μl are immediately plated onto freshly prepared 150-mm petri dishes containing χ-agar.
13. The plates are incubated for 2 days at 37°.

From 100 ng of cDNA, between 10,000 and 20,000 tetracycline-resistant recombinant transformants are obtained. The transformation efficiency of χ1776 by intact plasmid pBR322 by the above procedure yields approximately 1 to 3 × 10^6 transformants per microgram of pBR322 DNA.

[81] Screening of Colonies by RNA–DNA Hybridization with mRNA from Induced and Uninduced Cells

By SHUICHIRO MAEDA, MITCHELL GROSS, and SIDNEY PESTKA

Interferons are inducible proteins.[1] In most cases, there is no detectable level of synthesis of interferon by uninduced cells. The intracellular quantity of translatable interferon mRNA can be measured by translation

[1] S. Baron and F. Dianzani, eds., *Tex. Rep. Biol. Med.* **35**, 1 (1977).

CLONE SCREENING: HYBRIDIZATION

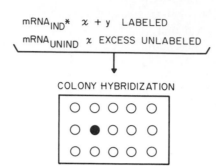

FIG. 1. Schematic outline of hybridization procedure. In the presence of excess unlabeled mRNA from uninduced cells, [^{32}P]mRNA from induced cells binds preferentially to sequences specific for the induced cells. The induced-specific sequences include interferon sequences as well as others that are induced concomitantly with interferon.

in *Xenopus laevis* oocytes.[2–4] In uninduced cells, no translatable interferon-specific mRNA is detected, but after induction with a variety of inducers interferon mRNA is found.[3,4] On the basis of the above observations, the following strategy for the screening for interferon-specific cDNA was considered.

Partially purified interferon mRNA from induced cells was used for synthesis of cDNA for transformation of *Escherichia coli* χ1776.[5–7] This same mRNA is hydrolyzed,[8] and the fragments are end-labeled with [γ-^{32}P]ATP.[9] Unlabeled mRNA from uninduced cells is mixed with the labeled, induced mRNA at a ratio of 200:1. This mixture is then used as a probe for screening transformants by colony hybridization.[10] Hybridization of the ^{32}P-labeled mRNA should occur preferentially to cDNA sequences present only in the induced cells[11] (Fig. 1).

[2] J. B. Gurdon, C. D. Lane, H. R. Woodland, and G. Marbaix, *Nature (London)* **233**, 177 (1971).

[3] R. L. Cavalieri, E. A. Havell, J. Vilček, and S. Pestka, *Proc. Natl. Acad. Sci. U.S.A.* **74**, 3287 (1977).

[4] A. Sloma, R. McCandliss, and S. Pestka, this volume [10].

[5] R. McCandliss, A. Sloma, and S. Pestka, this volume [82].

[6] R. McCandliss, A. Sloma, and H. Takeshima, this volume [78].

[7] S. Maeda, this volume [79].

[8] R. M. Bock, this series, Vol. 12A, p. 218.

[9] J. R. Lillehaug, R. K. Kleppe, and K. Kleppe, *Biochemistry* **15**, 1858 (1976).

[10] M. Grunstein and D. S. Hogness, *Proc. Natl. Acad. Sci. U.S.A.* **72**, 3961 (1975).

[11] A. Okuyama, J. McInnes, M. Green, and S. Pestka, *Arch. Biochem. Biophys.* **188**, 98 (1978).

Alkaline Hydrolysis and Terminal Labeling of Purified mRNA from
   Induced Cells

Purified mRNA from interferon-producing cells is hydrolyzed to fragments of about 100 bases in 1 N KOH[8] for terminal labeling. Free 5'-
hydroxyl termini of the fragments are labeled with T4-polynucleotide
kinase and [γ-^{32}P]ATP.[9] The procedures for the hydrolysis and labeling are
given below.

1. Two micrograms of partially purified interferon mRNA are hydrolyzed in 120 μl of 1 N KOH at 27° for 40 sec.[8]
2. Then 120 μl of 1 N HCl and 15 μl of 2 M Tris · HCl, pH 7.9, are
   added to neutralize the reaction mixture.
3. The hydrolyzed mRNA is precipitated with 2.5 volumes of 95%
   ethanol. After remaining overnight at $-20°$, the mRNA precipitate
   is sedimented by centrifugation.
4. The pellet is resuspended in 14 μl of H_2O.
5. Terminal labeling is performed at 37° for 1 hr in a 20-μl reaction
   mixture containing:
   Alkaline-hydrolyzed mRNA, 0.1 mg/ml
   Tris · HCl, pH 8.9, 50 mM
   MgCl$_2$, 10 mM
   2-Mercaptoethanol, 10 mM
   [γ-^{32}P]ATP (2300 Ci/mmol, Amersham), 25 μM
   T4-polynucleotide kinase (New England Biolabs), 200 units/ml
6. The end-labeled mRNA is separated from unincorporated [γ-
   ^{32}P]ATP by chromatography on a Sephadex G-50 (fine) column
   (Pharmacia).

The specific activity of the labeled mRNA obtained by this method is
3–5 × 10^7 cpm/μg.

Colony Hybridization

Colony hybridization is carried out by a modification of the procedure
described by Grunstein and Hogness.[10]

Preparation of Nitrocellulose Filters and Colonies for Hybridization

1. Nitrocellulose filters (82 mm, Schleicher & Schuell, BA 85/22) are
   boiled in three changes of H_2O to remove detergent.
2. The washed filters are sandwiched between two sheets of Whatman
   No. 1 filter paper, wrapped in aluminum foil, and autoclaved for 5
   min followed by fast exhaust of the autoclave.

3. The sterile filter is placed on a 100-mm χ-agar[12] plate without trapping air bubbles.
4. About 120 transformants are streaked onto the filter and onto two reference plates with toothpicks.
5. The plates are incubated at 37° until distinct colonies appear.
6. The filter is peeled off, laid on a χ-agar plate containing chloramphenicol (12.5 μg/ml), and incubated at 37° for 18 hr to amplify plasmid DNA in the transformants.[13] The two reference plates are incubated for 12–18 hr longer for the cells to grow well and are stored at 4°.

Cell Lysis, DNA Denaturation, and Fixation

The filter from above is then prepared for hybridization as follows:

1. Colonies on the filter are lysed on two sheets of Whatman 3 MM paper saturated with 0.5 N NaOH. (The nitrocellulose filter, with colony side up, is placed on Whatman 3 MM paper flooded with 0.5 N NaOH.)
2. After 7 min, the filter is blotted briefly on a paper towel.
3. The above steps 1 and 2 are repeated twice with 1 M Tris · HCl, pH 7.4, once with 2 × SSC (1 × SSC: 0.15 M NaCl, 0.015 M trisodium citrate), and twice with 90% ethanol.
4. The filter is dried in air and then baked at 80° in a vacuum oven for 2–3 hr.

Hybridization

The dry filter is probed for induced-specific cDNA sequences by colony hybridization performed essentially as described by Wahl *et al.*[14]

Solutions

Pretreatment buffer: 50% recrystalized formamide, 5 × SSC, 5 × Denhardt's solution[15] (1 × Denhardt's solution: 0.02% (w/v) each of bovine serum albumin, polyvinyl pyrrolidine, and Ficoll 40,000), and 50 mM sodium phosphate, pH 6.5.

Hybridization buffer: 50% recrystallized formamide, 5 × SSC, 1 × Denhardt's solution, 20 mM sodium phosphate, pH 6.5, 10% sodium dextran sulfate 500 (Pharmacia), and 6 μg each of alkaline-hydrolyzed mRNA from uninduced human leukocytes and poly(A).

---

[12] S. Maeda, this volume [80].
[13] D. Hanahan and M. Meselson, *Gene* **10**, 63 (1980).
[14] G. M. Wahl, M. Stern, and G. R. Stark, *Proc. Natl. Acad. Sci. U.S.A.* **76**, 3683 (1979).
[15] D. Denhardt, *Biochem. Biophys. Res. Commun.* **23**, 641 (1966).

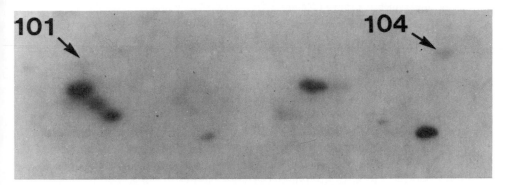

Fig. 2. Autoradiogram of colony hybridization procedure. The colonies could be classified into three groups by intensity of autoradiograph: intense, less intense (from barely visible to moderately intense), and no intensity (not visible). Colonies 101 and 104 were shown subsequently[18] to contain human fibroblast and leukocyte interferon sequences, respectively.

Procedure

1. The filter is placed in a polyethylene bag (Dazey Seal-a-Meal) containing 5 ml of pretreatment buffer, and incubated at 42° for 2 hr.
2. The pretreatment buffer is extruded from the open bag with the use of a glass rod.
3. Then, 2 ml of hybridization buffer containing 30 ng of the [^{32}P]mRNA probe (partially purified labeled mRNA from induced cells, prepared as described in the section on nitrocellulose filters and colonies for hybridization, above, with specific activity of 3 to 5 × 10^7 cpm/μg) is added to the bag.
4. The bag is sealed without trapping air bubbles and incubated at 42° for 16–20 hr.
5. After hybridization, the filter is washed three times with 100-ml portions of 2× SSC containing 0.1% sodium dodecyl sulfate for 5 min at room temperature, and then twice with 100-ml portions of 0.1× SSC containing 0.1% sodium dodecyl sulfate for 15 min each at 40–50°.
6. The washed filter is dried in air, placed on a sheet of Whatman 3 MM paper, covered with a plastic sheet, and exposed to Kodak XR2 X-ray film with a DuPont Cronex intensifying screen at −80° for 18–20 hr.
7. After development of the autoradiogram to localize all colonies, the filter is treated with RNase A (60 μg/ml) and RNase T1 (33 units/ml) in 25 ml of 2× SSC at 37° for 45 min with gentle shaking. Background levels of radioactivity on the autoradiogram are minimized with the use of RNase.

8. The RNase-treated filter is exposed to X-ray film as in step 6, and the autoradiogram is developed to localize colonies containing cDNA sequences specific for induced cells.

By this screening procedure, two classes of positive clones were detected (Fig. 2). One class, comprising 2–3% of the total, hybridized very strongly to the probe and appeared as intense signals on the autoradiogram. The second class, which consisted of 10% of the total, were significantly less intense than those of the first class and varied in intensity over a wide range. The positive colonies were examined further for the presence of interferon-specific cDNA sequences by an mRNA hybridization translation assay.[16,17] Intense colonies (60) were shown to have no interferon-specific sequences, but two interferon cDNA clones were identified among 59 of the less intense positive colonies.[18]

[16] J. C. Alwine, D. J. Kemp, B. A. Parker, J. Reiser, J. Renart, G. R. Stark, and G. M. Wahl, this series, Vol. 68, p. 220.

[17] R. McCandliss, A. Sloma, and S. Pestka, this volume [82].

[18] S. Maeda, R. McCandliss, M. Gross, A. Sloma, P. C. Familletti, J. M. Tabor, M. Evinger, W. P. Levy, and S. Pestka, Proc. Natl. Acad. Sci. U.S.A. 77, 7010 (1980).

[82] Use of DNA Bound to Filters for Selection of Interferon-Specific Nucleic Acid Sequences

By RUSSELL McCANDLISS, ALAN SLOMA, and SIDNEY PESTKA

The identification of cloned genes is dependent upon a specific selection method. For the selection of a cloned interferon cDNA sequence,[1] the most sensitive and specific method depends upon hybridization of interferon mRNA to plasmid DNA covalently bound to diazobenzyloxymethyl (DBM) paper followed by elution and translation of the mRNA (Fig. 1). This selection method is based upon the procedures developed and described by Alwine et al.[2] and has been used to select plasmids carrying segments of the Xenopus laevis vitellogenin gene.[3] Modifications of this procedure have also been used to identify plasmids containing part of the gene for human leukocyte interferon.[1,4]

[1] S. Maeda, R. McCandliss, M. Gross, A. Sloma, P. C. Familletti, J. M. Tabor, M. Evinger, W. P. Levy, and S. Pestka, Proc. Natl. Acad. Sci. U.S.A. 77, 7010 (1980).

[2] J. C. Alwine, D. J. Kemp, B. A. Parker, J. Reiser, J. Renart, G. R. Stark, and G. M. Wahl, this series, Vol. 68.

[3] D. F. Smith, P. F. Searle, and J. G. Williams, Nucleic Acids Res. 6, 487 (1979).

[4] S. Nagata, H. Taira, A. Hall, L. Johnsrud, M. Streuli, J. Ecsödi, W. Boll, K. Cantell, and C. Weissmann, Nature (London) 284, 316 (1980).

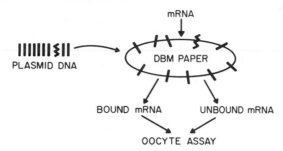

FIG. 1. Schematic illustration of screening of recombinants with DBM paper. Plasmid DNA is covalently linked to DBM paper as described in the text. Plasmid DNA isolated from single colonies or from several pooled colonies can be bound to the paper. To detect a recombinant containing an interferon sequence (zigzag plasmid DNA in the illustration), mRNA prepared from cells synthesizing interferon is hybridized to the DNA on the filter as described in the text. If a recombinant plasmid bound to the filter contains a sequence homologous to the interferon mRNA, bound mRNA yields active interferon after translation in *Xenopus laevis* oocytes (see the text and table).

Preparation of DNA Filters

Plasmid DNA to be analyzed is isolated by the cleared lysate procedure of Clewell and Helinski[5] and purified by centrifugation in a CsCl equilibrium density gradient. For each clone, 5–10 μg of plasmid DNA are linearized by digestion with *Hin*dIII restriction endonuclease. Plasmid DNA is denatured by heating at 70° in 0.1 M NaOH for 20 min. The solution is neutralized with 1 M HCl, and the DNA is precipitated with ethanol. Prior to reaction with DBM paper, the DNA is dissolved in 10 μl of 0.2 M sodium acetate, pH 4.0, and heated for 5 min at 95°. Forty microliters of dimethyl sulfoxide (DMSO) (Eastman, Spectra grade) are added, and the solution is placed in an ice bath.

DBM paper is prepared and diazotized exactly as described by Alwine *et al.*[2] Briefly, Whatman 540 paper is cut into a 15 × 15-cm square and placed in a square glass baking dish. The dish is floated in a 60° water bath in a fume hood, and a solution containing 0.52 g of 1-[(*m*-nitrobenzyloxy)-methyl]pyridimium chloride (British Drug House) and 0.16 g of sodium acetatetrihydrate in 6.4 ml of water is placed over the paper. The solution is rubbed on the paper so that it is evenly distributed, and the paper is allowed to dry. The paper is dried further for 10 min in a 60° oven and then heated to 130–135° for 30–40 min. The nitrobenzyloxymethyl (NBM) paper is washed with water for 20 min and acetone for 20 min, and then it is air-dried. NBM paper prepared in this manner is stored at 4°. Prior to reduction and diazotization, the NBM paper is cut into circles about 1 cm in

[5] D. B. Clewell and D. R. Helinski, *Biochemistry* **9**, 4428 (1970).

diameter and the papers are numbered with a pencil. Circles of this size fit into 8-ml plastic scintillation vials, in which the hybridization is performed. To convert the NBM paper to aminobenzyloxymethyl (ABM) paper, the filters are incubated at 60° for 30 min with a freshly prepared 20% solution of sodium dithionite (0.4 ml/cm²) in water. The filters are then washed at room temperature extensively with water, briefly with 30% acetic acid, and again extensively with water. For diazotization, the filters are blotted on a paper towel, placed in ice-cold 1.2 M HCl (0.3 ml/cm²), and 0.027 ml of a freshly prepared 1% solution of NaNO₂ in water is added per milliliter of 1.2 M HCl. The solution is kept ice-cold for at least 30 min. During this time, the filters turn light yellow. After diazotization, the DBM paper circles are washed with ice-cold water and ice-cold 0.2 M sodium acetate, pH 4, and are blotted on cold paper towels. The denatured DNA in DMSO is placed on the filter papers on a plastic petri dish, and the reaction is allowed to proceed overnight at room temperature. During this time, the light yellow filters turn a very deep orange color. After the reaction, the filters are washed extensively with water. Each filter is then placed in the bottom of a plastic scintillation vial. To remove any noncovalently bound DNA and to protect against RNase activity, the filters and vials are washed with 0.4 M NaOH for 30 min at 37°, before washing with a large volume of water. The color of the filter is an indication of the pH change, with a deep red-orange color in alkali and deep orange in neutral solution. The filters are washed with a solution containing 10 mM Tris · HCl, pH 8, and 1 mM EDTA. They are stored in this buffer at 4°.

Hybridization

The hybridization buffer contains 0.02 M piperazine-N,N'-bis(2-ethanesulfonic acid) (PIPES), pH 6.8, 0.2% sodium dodecyl sulfate (SDS), 1 mM EDTA, 0.9 M NaCl, 1 mg of yeast tRNA per milliliter, and 50% formamide (recrystallized according to Casey and Davidson[6]). The filters are washed once with 0.5 ml of hybridization buffer, and the buffer is removed by aspiration with a Pasteur pipette. The mRNA to be hybridized to the filter is dissolved in 40 μl of hybridization buffer, which is added to the filter. Approximately 1 μg of 12 S mRNA[7] is used for each filter. The vials are capped or sealed tightly with Parafilm and are placed at 48° for 6 hr.

After the hybridization, the unbound RNA is removed and the bound mRNA is eluted as follows. The hybridization solution is removed from the filters, and they are washed for 15 min at 37° with 1 ml of 1× SSC (0.15

---

[6] J. Casey and N. Davidson, *Nucleic Acids Res.* **4**, 1539 (1977).
[7] R. McCandliss, A. Sloma, and S. Pestka, this volume [8].

YIELD OF INTERFERON mRNA BOUND TO DBM FILTERS[a]

Expt. No.	Plasmid bound to filter	Interferon titer
1	Pool K16	128
	pBR322	32
	Pool 17	<4
2	Pool K16	16
	pBR322	<2
	Pool K18	<2
	pβG1	<2
3	p101	48
	p104	48
	p106	<2
	pBR322	<2

[a] Pools K16, K17, and K18 were plasmids prepared from ten induced-specific transformants as described by Maeda *et al.*[1]; p101, p104, and p106 were plasmids prepared from individual colonies from pool K16; interferon titers were determined by injection of eluted mRNA into *Xenopus laevis* oocytes[8] and assay of the translation products on AG-1732 cells with vesicular stomatitis virus as the challenge virus.[10] Units are expressed relative to the reference standard for human leukocyte interferon (G-023-901-527) provided by the Antiviral Substances Program of the National Institute of Allergy and Infectious Diseases, National Institutes of Health. For each assay, a titer of 2000–4000 would have been expected if 100% of the mRNA was bound to the filters and was recovered.

M sodium citrate, 0.15 M NaCl), 0.2% SDS, 1 mM EDTA, and 50% formamide. The wash buffer is removed with a Pasteur pipette, and the washing is repeated four times. After the last wash, the buffer is aspirated with a Pasteur pipette so that the filter is just moist. The bound mRNA is eluted by washing the filter at 70° for 2 min with 150 μl of 0.02 M PIPES, pH 6.8, 2 mM EDTA, 95% formamide. The eluate is removed, and the elution is repeated once. The two eluates (0.3 ml) are combined, 20 μg of yeast tRNA are added as carrier, and the RNA is precipitated with ethanol by addition of 30 μl of 2.4 M sodium acetate, pH 5.5, and 0.75 ml of 95% ethanol. After standing overnight at −20°, the RNA is collected by centrifugation, washed once with 70% ethanol, air-dried and dissolved in 5 μl of water. The mRNA is assayed by injection into *Xenopus laevis* oocytes.[8]

[8] A. Sloma, R. McCandliss, and S. Pestka, this volume [10].

The yield of mRNA (see the table) has varied from 1 to 10% of the input mRNA. The results were found to be reproducible. Negative controls performed with either pBR322 or pβG1[9] plasmid DNA yielded low base levels of interferon mRNA bound.[10] In most of the experiments we have performed, the amount of interferon mRNA bound to the control filters was below our detection level, but occasionally there was a very low level of interferon mRNA bound. Accordingly, we found it necessary to include the above negative controls to identify plasmids containing interferon sequences.[1]

[9] T. Maniatis, S. G. Kee, A. Efstratiadis, and F. C. Kafatos, *Cell* **8**, 163 (1976).
[10] P. C. Familletti, S. Rubinstein, and S. Pestka, this series, Vol. 78 [56].

[83] Detection of Expressed Polypeptides by Direct Immunoassay of Colonies

By DAVID J. KEMP and ALAN F. COWMAN

Bacterial colonies that express antigenic polypeptides can be detected by a variety of *in situ* immunological methods.[1-6] Such "colony immunoassays" provide an approach to screening clone banks for defined genes, which is useful when purified nucleic acid probes are not available. Colony immunoassays are also potentially useful for detecting strains that express optimal levels of a defined polypeptide. Antigenic polypeptides that are applicable include "fused" polypeptides, created by cloning the desired DNA segment within genes present on plasmids,[7-9] protein fragments internally initiated from cloned sequences,[6] and intact proteins ini-

[1] A. Skalka and L. Shapiro, *Gene* **1**, 65 (1976).
[2] B. Sanzey, O. Mercereau, T. Ternynck, and P. Kourilsky, *Proc. Natl. Acad. Sci. U.S.A.* **73**, 3394 (1976).
[3] H. A. Erlich, S. N. Cohen, and H. O. McDevitt, this series, Vol. 68, p. 443.
[4] S. Broome and W. Gilbert, *Proc. Natl. Acad. Sci. U.S.A.* **75**, 2746 (1978).
[5] L. Clarke, R. Hitzeman, and J. Carbon, this series, Vol. 68, p. 436.
[6] D. J. Kemp and A. F. Cowman, *Proc. Natl. Acad. Sci. U.S.A.* **78**, 4520 (1981).
[7] K. Itakura, T. Hirose, R. Crea, A. D. Riggs, H. L. Heynecker, F. Bolivar, and H. W. Boyer, *Science* **198**, 1056 (1977).
[8] L. Villa-Komaroff, A. Efstratiadis, S. Broome, P. Lomedico, R. Tizard, S. P. Naber, W. L. Chick, and W. Gilbert, *Proc. Natl. Acad. Sci. U.S.A.* **75**, 3727 (1978).
[9] T. H. Fraser and B. J. Bruce, *Proc. Natl. Acad. Sci. U.S.A.* **75**, 5936 (1978).

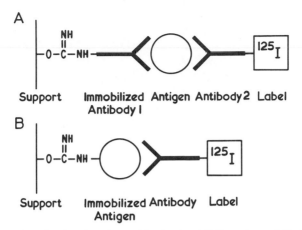

Fig. 1. Detection of antigenic polypeptides *in situ*. (A) Procedure of Clarke, Hitzeman, and Carbon,[5] wherein antibody 1 is bound to the support. (B) The present method wherein the antigenic polypeptide is covalently bound to the support.

tiated from synthetic sequences.[10] In most colony immunoassays,[3-5] two antibody molecules must bind to the antigenic polypeptide (Fig. 1A). We have developed a direct colony immunoassay[6] that requires binding of only one antibody molecule per antigenic polypeptide molecule. This method might therefore be expected to detect a greater proportion of fused polypeptides and small polypeptide fragments than other methods. The immunoassay is suitable in principle for use with monoclonal antibodies, which recognize only one antigenic determinant per molecule. The principle of the direct colony immunoassay is shown in Fig. 1B. Proteins from colonies lysed *in situ* are bound covalently to CNBr paper. Antigenic polypeptides are detected by autoradiography after reaction with ^{125}I-labeled antibodies or, alternatively, after reaction with unlabeled antibodies followed by a second reaction with ^{125}I-labeled protein A from *Staphylococcus aureus*. Protein A binds to the C_H regions of many (but not all) antibodies, including rabbit, mouse, and human IgG,[see 11,12] and hence only one radioiodinated preparation is required for use with a variety of antisera. A schematic diagram outlining our direct colony immunoassay is shown in Fig. 2.

[10] D. V. Goeddel, E. Yelverton, A. Ullrich, H. L. Heynecker, G. Miozzari, W. Holmes, P. H. Seeburg, T. Dull, L. May, N. Stebbing, R. Crea, S. Maeda, R. McCandliss, A. Sloma, J. M. Tabor, M. Gross, P. C. Familletti, and S. Pestka, *Nature (London)* **287**, 411 (1980).
[11] J. Goudswaard, J. A. van der Donk, A. Noordzij, R. H. van Dam, and J. F. Vaerman, *Scand. J. Immunol.* **8**, 21 (1978).
[12] P. L. Ey, S. J. Prouse, and C. R. Jenkin, *Immunochemistry* **15**, 429 (1978).

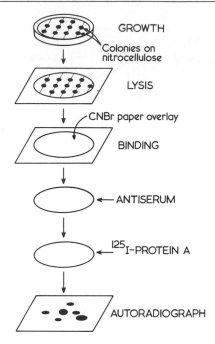

FIG. 2. Schematic diagram of the colony immunoassay.

Reagents and Solutions

Reagents

Whatman 540 filter paper (W & R. Balston, Maidstone, England)
CNBr-Sepharose 4B (Pharmacia, Uppsala, Sweden)
CNBr (Eastman Organic Chemicals, Rochester, New York)
Acetonitrile (Waters Associates, Milford, Massachusetts)
Bovine albumin (Fraction V, Armour Pharmaceutical Co., Eastbourne, England)
Protein A (pharmacia, Uppsala, Sweden)
Chloramine-T (BDH Chemicals Ltd., Poole, England)
$Na^{125}I$, carrier free (Radiochemical Centre, Amersham, England)
$Na_2S_2O_5$ (BDH Chemicals Ltd., Poole, England)
Nitrocellulose (type HA; Millipore Corp., Bedford, Massachusetts)
Lysozyme (Boehringer Mannheim GMbH, West Germany)
DNase I, Grade I. (Boehringer Mannheim GMbH, West Germany)
BioGel P6 (Bio-Rad Laboratories, Richmond, California)

Solutions

CNBr, dissolved in acetonitrile (1 g/ml). This is most safely prepared by adding the appropriate volume of acetonitrile (with a safety pipette) to a fresh bottle of CNBr (in a fume hood) and is stored sealed at $-20°$.

K_3PO_4, 0.5 M

$NaHCO_3$, 0.1 M

Wash buffer: 50 mM Tris · HCl, pH 8.0, 0.5 M NaCl, 0.1% Triton X-100, 1% w/v BSA

Binding buffer: 0.1 M $NaHCO_3$, 0.1% Triton X-100

Lysis buffer: 0.1 M $NaHCO_3$, 1% Triton X-100, 2 mg of lysozyme per milliliter. The lysozyme is added immediately before use.

Phosphate-buffered saline (PBS): 8.7 g of NaCl, 2.85 g of Na_2HPO_4 · 2 H_2O, 0.625 g of NaH_2PO_4 · 2 H_2O

Glycine HCl, 0.2 M, pH 2.8

L agar plates: 10 g of Tryptone, 5 g of yeast extract, 12 g of agar, 10 g of NaCl per liter of H_2O, adjusted to pH 7.0. For growth of pBR322, the plates contain tetracycline (10 μg/ml); and for pTrpED5-1 they contain ampicillin (30 μg/ml).

DNase I solution: 2 μg of DNase I per milliliter in 10 mM Tris · HCl, pH 7.4, 10 mM $MgCl_2$, 100 mM NaCl, 100 μg of gelatin per milliliter. The DNase is stored at 1 mg/ml in the same solution plus 50% v/v glycerol at $-20°$ and diluted immediately before use.

Preparation of Materials

Preparation of CNBr Paper. The procedure is a modification of that of Clarke, Hitzeman, and Carbon.[5] Twenty-four Whatman 540 filter paper circles (80 mm in diameter) in a 2-liter plastic beaker are washed briefly with water. The water is decanted, and 0.5 M K_3PO_4 (500 ml) is added, followed by 12 ml of CNBr solution (in a fume hood, with a safety pipette). The beaker is vigorously swirled for 8 min by hand to keep the filters suspended. Immediately after decanting the solution, the filters are washed rapidly by swirling the beaker for a few seconds with, successively, about 500-ml portions of 0.1 M $NaHCO_3$ (twice), 50% acetone/H_2O, and finally acetone (twice). The filters are then spread out on blotting paper to dry for 5 min; drying is completed in a vacuum desiccator at room temperature. They are stored at 2–4° over desiccant and are stable for several months.

Pretreatment of Antisera. We treat all sera by absorption with bacterial extract bound in CNBr Sepharose to remove antibacterial antibodies that may bind to colonies.[3] Preparation of the absorbant is identical to the procedure of Ehrlich, Cohen, and McDevitt[3] except that a 10-fold higher concentration of extract is used.

Before use in the immunoassay, serum (20 μl) in 400 μl of wash buffer is added to about 0.1 ml of packed Sepharose-bacterial extract in a 1.5-ml Eppendorf centrifuge tube. The mixture is rocked at room temperature for 20 min and centrifuged at 12,800 g for 5–10 sec; the supernatant is diluted further as required with wash buffer.

Affinity Purification of Antisera. Procedures for affinity purification have been discussed in Vol. 34 of this series. For purification of anti-μ serum, IgM is coupled to CNBr Sepharose according to the method of Axén *et al.*[13] The anti-μ antiserum (5 ml) is filtered through a column containing 5 ml of IgM-Sepharose. The column is washed with PBS (50 ml), and bound material is eluted with 0.2 M glycine · HCl, pH 2.8. The major protein-containing fractions are pooled, neutralized with 1 M NaOH as quickly as possible, dialyzed against PBS, and stored at $-20°$ in small aliquots.

Preparation of ^{125}*I-Labeled Protein A and* ^{125}*I-Labeled Antibodies.* Protein A is iodinated by the chloramine-T procedure of Hunter and Greenwood.[14] The reaction mixture contains 25 μl of protein A (1 mg/ml in PBS), 5 μl of PBS, 15 μl of carrier-free Na^{125}I (1.14 mCi), and 5 μl of chloramine-T (1 mg/ml in H$_2$O). After 10 min at 0°, the reaction is terminated with 5 μl of Na$_2$S$_2$O$_5$ (2.4 mg/ml in H$_2$O) followed by carrier (20 μl 0.1 M KI and 20 μl of 5% BSA). Unincorporated label is removed by fractionation on BioGel P6 equilibrated in PBS. The specific activity is 5 to 10 $\times$ 10^7 cpm/μg. Antibodies (12 μg) are iodinated (after affinity purification) by the same procedure except that 1 μl of chloramine-T and 0.15 mCi of Na^{125}I are used. The specific activity is ~10^7 cpm/μg.

Growth of Colony Replicas on Nitrocellulose Filters. Colonies are grown overnight at 37° on a nitrocellulose filter placed on an L-agar plate containing the appropriate antibiotic. The nitrocellulose filter is then placed colony-side up on a dry sterile filter paper. Another nitrocellulose filter is wetted on an L-agar plate, placed on the first filter, and overlaid with a second sterile filter paper. After rubbing the filter paper firmly all over, it is discarded and the replica is placed on another L plate. The first filter can be used immediately for the immunoassay, or the replica can be used after growth overnight. At least 30 replicas can be made from the one filter. Freshly grown colonies are always used for the immunoassay.

Procedure for the Direct Colony Immunoassay

Immobilization of Proteins from Escherichia coli Colonies by Binding to CNBr Paper. All steps are performed at room temperature. Colonies on a

---

[13] R. Axén, J. Porath, and S. Ernbäck, *Nature* (*London*) **214**, 1302 (1967).
[14] W. M. Hunter and F. C. Greenwood, *Nature* (*London*) **194**, 495 (1962).

nitrocellulose filter are lysed by placing the filter colony-side up on blotting paper saturated with lysis buffer for 20 min in an atmosphere of $CHCl_3$ vapor. During this step the blotting paper is supported on a sheet of glass standing (on glass rods) about 2 cm above the bottom of a large stainless steel dish with a tight-fitting lid. Chloroform (50 ml) is poured into the bottom of the dish at the start of the 20-min incubation. The nitrocellulose filter is placed on blotting paper saturated with binding buffer for an additional 20 min in the $CHCl_3$ atmosphere and then on damp blotting paper (i.e., paper soaked in binding buffer, blotted, and placed on a sheet of aluminum foil). A CNBr-paper filter is wetted with binding buffer, blotted, and placed on the lysed colonies. A sheet of dry blotting paper is placed on the CNBr paper, rubbed firmly all over for a few seconds, and discarded. Several filter "sandwiches" are wrapped in aluminum foil to minimize evaporation and left at least 4 hr or overnight to allow covalent binding to proteins.

Detection of Antigens in Colony Lysates Bound to CNBr Paper. Before reaction with antiserum, the CNBr papers are treated with glycine (5% w/v) in wash buffer (minus BSA) for 16 hr, then with DNase I solution for 1 hr, and finally with wash buffer for at least 30 min. The colonies appear gelatinous after the glycine treatment and are not visible after the DNAse step. For each step (and all subsequent washes) we use 5–10 ml of solution per CNBr paper, in a 1-liter beaker on a rotary shaker.

For the protein A procedure, the CNBr papers are now blotted and placed in petri dishes containing 1 ml of diluted antiserum (generally 1 : 500, see below) in wash buffer. They are inverted several times to ensure even distribution of the antiserum and then incubated for 1 hr. Excess antiserum is removed by washing each CNBr paper as above at least five times with 20 ml of wash buffer for a total time of at least 1 hr. The CNBr papers are blotted and treated with 1 ml of ^{125}I protein A (10^6 cpm/ml, in wash buffer) for 1 hr, then washed as above, blotted, dried at room temperature, wrapped in Saran wrap, and autoradiographed overnight at −80° with preflashed film (Kodak RP/S X-Omat) and an intensifying screen (Cronex Lightning-plus) as described by Laskey and Mills.[15]

If ^{125}I-labeled antibodies are to be used instead of protein A, the pretreated CNBr papers are incubated with ^{125}I-labeled antibodies (10^6 cpm/ml) in wash buffer as above, then washed, blotted, dried, and autoradiographed as above.

Results

We tested the colony immunoassay with two vectors and two different test inserts. First, we inserted a cloned immunoglobulin μ cDNA fragment

[15] R. A. Laskey and A. D. Mills, *FEBS Lett.* **82**, 314 (1977).

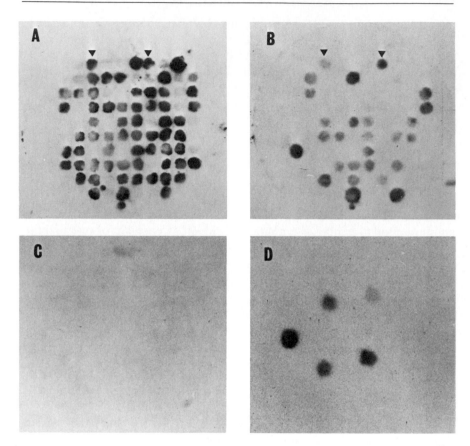

FIG. 3. Detection of ptrpED-μ and pBR-γ_2 clones by the colony immunoassay.   (A) The ptrpED-μ colonies were scored with a [^{32}P]μDNA fragment[6] by the procedure of M. Grunstein and D. S. Hogness [*Proc. Natl. Acad. Sci. U.S.A.* **72**, 3961 (1975)] to reveal clones that contain μ cDNA.   (B) A replica of the colonies shown in A were scored by the colony immunoassay with 1 : 500 dilution of rabbit anti-μ serum followed by ^{125}I-labeled protein A (10^6 cpm/ml, 34.4 μCi/μg). Arrows show the alignment of the two arrays.   (C) A replica was scored by the colony immunoassay as for (B), but with normal rabbit serum (1 : 500).   (D) Sixteen pBR322 colonies containing mouse γ_{2A} cDNA inserts were scored with ^{125}I-labeled sheep anti-mouse γ_{2A} antibodies (10^6 cpm/ml, 8.0 μCi/μg). The antibodies were purified by affinity chromatography on IgG$_2$-Sepharose prior to iodination. Autoradiographs were exposed for 10 hr (A), 14 hr (B and C), and 70 hr (D), respectively.

by the dG : dC tailing procedure into the *Hin*dIII site within the trpD gene of plasmid ptrpED5-1[16] and also into the *Pst*I site of plasmid pBR322.[17] In each case, the cloning method should give rise to inserts in both orienta-

[16] R. A. Hallewell and S. Emtage, *Gene* **9**, 27 (1980).
[17] F. Bolivar, R. L. Rodriguez, P. J. Green, M. C. Betlach, H. L. Heyneker, H. W. Boyer, and S. Falkow, *Gene* **2**, 95 (1977).

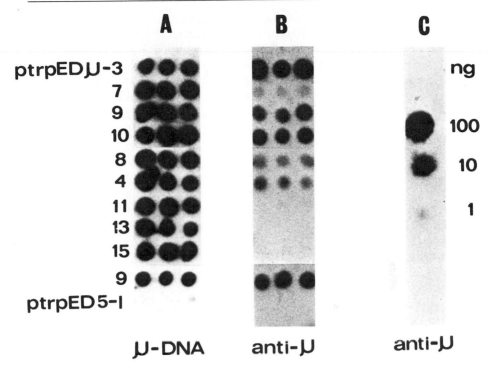

Fig. 4. Reproducibility of the colony immunoassay. Single colonies from representative positive (ptrpEDμ3-10) and negative (ptrpEDμ11-15) clones were picked in triplicate. (A) A replica was scored by the Grunstein–Hogness procedure with [^{32}P]μDNA. (B) A replica was scored by the colony immunoassay, with affinity-purified anti-μ antibodies (5.4 μg/ml) followed by ^{125}I-labeled protein A (10^6 cpm/ml, 34.4 μCi/μg). (C) Aliquots (1 μl) containing 1, 10, and 100 ng of IgM were spotted onto CNBr paper dampened with binding buffer. After 4 hr at room temperature, the CNBr paper was treated as for the colony immunoassay in Fig. 3B except that the DNase step was omitted. Autoradiographs were for 12 hr (A), 24 hr (B), and 24 hr (C).

tions and in all coding frames, and hence one in six of the clones should encode fused polypeptides. We scored the colonies with anti-μ serum and ^{125}I-labeled protein A (Fig. 3). About 50% of the ptrpED-μ (Fig. 3B) and pBR-μ (not shown) clones were positive.[6] Clones containing irrelevant inserts were negative (not shown), and controls with normal or irrelevant sera were negative (Fig. 3C). Individual colonies obtained after plating out several different ptrpED-μ clones gave strikingly reproducible signals (Fig. 4B). The unexpectedly high frequency of expression probably results from internal initiation within the cDNA segment[6] and will be extremely useful if it occurs commonly with other cloned genes. As a second test system we used mouse γ_{2a} cDNA cloned in the *Pst* site of pBR322. The colonies were scored with affinity-purified ^{125}I-labeled sheep anti-γ_2 antibodies because sheep antibodies bind protein A poorly.[11] Of the 32 independent colonies tested, 8 were clearly positive (Fig. 3D). One of these contained an insert that was only ~260 bases long.

Background Problems and Limitations. Filter background problems[6] were due to incomplete blocking of active sites. The 16 hr glycine treatment described here permits use of much more concentrated antiserum (1:10, cf. 1:500[6]). The appropriate dilution is determined by spotting antigen onto CNBr paper and scoring the "spots" with several dilutions of the serum (Fig. 4C). Calibration experiments suggested that most positive colonies contain less than 10 ng of antigenic polypeptide. Hence the antiserum must be potent enough to detect small quantities of antigen when diluted sufficiently to overcome background. Affinity purification of the antibodies greatly improved the signal-to-noise ratio, as expected. One batch of anti-μ serum gave strong false-positive signals with mucoid colonies present in one set of clones containing irrelevant inserts. Presumably, the mucoid colonies reacted with anti-carbohydrate antibodies, commonly present in antisera. It is important to appreciate that exhaustive controls for specificity of antisera are necessary in any serological technique. Although we have not yet tested monoclonal antibodies, they should overcome problems of sensitivity and specificity. However, because they bind to only one antigenic determinant, many monoclonal antibodies will not bind to short polypeptide fragments or to fused proteins or protein subunits that differ in conformation from the native protein at the relevant site.

Acknowledgments

We thank J. Goding, A. Wilson, and I. Walker for antisera. This work was supported by grants from the National Institutes of Health and the National Health and Medical Research Council (Canberra).

Author Index

Numbers in parentheses are reference numbers and indicate that an author's work is referred to although the name is not cited in the text.

Subject Index

A

Acetone, purification for amino acid analysis, 24

Acetonitrile, reverse-phase high-performance liquid chromatography, 18

Acetylcholinesterase, unaffected by interferon-treated cells, 343

N-Acetylglucosaminyl lipids, decreased in interferon-treated cells, 304

Actinomycin D
  complementary DNA synthesis, 603
  fibroblast interferon induction, 89, 90
  inhibition of host RNA synthesis, 319, 323
  mouse interferon induction, 93
  potentiation of interferon, 432
  preparation of mengovirus RNA, 266
  radioactive messenger RNA synthesis, 254
  retrovirus complementary DNA preparation, 316
  Rous sarcoma virus complementary DNA preparation, 316
  translation of heterologous messenger RNA, 100, 101, 110
  viral RNA preparation, 457

Adenosine diphosphate-agarose
  (2'-5')-oligoadenylate synthetase assay, 228–233
  (2'-5')-oligoadenylate synthetase binding, 229, 230

Adenosine diphosphate-ribose, substrate for (2'-5')-oligoadenylate synthetase, 138

Adenosine diphosphate-Sepharose, see Adenosine diphosphate-agarose

Adenosine monophosphate, assay of 2'-phosphodiesterase, 160, 161

Adenosine triphosphate
  adenylate cyclase assay, 164–168
  assay of (2'-5')-oligoadenylate synthetase, 150

cyclic adenosine monophosphate synthesis, 164–166

(2'-5')-oligoadenylate synthesis, 138, 139, 150–153, 160, 161, 184–197, 244, 273

phosphorylation of proteins, 170, 173

radioactive, purification, 186

substrate for (2'-5')-oligoadenylate synthetase, 138, 184–197, 244, 273

S-Adenosylmethionine, in S-adenosyl-L-methionine decarboxylase assay, 347, 348

S-Adenosylmethionine decarboxylase
  assay, 347–349
  induction, 345
  inhibition by interferon, 343, 344

Adenovirus
  DNA, isolation, 313
  preparation, 313

Adenylate cyclase, 162, 163
  activities, 167
  assay, 164–168
  binding subunit, 167
  catalytic subunit, 167
  receptor, 167

African fish oocyte, see Tilapiamosaiabica peters oocyte

African green monkey kidney cell, see Vero cell

Agarose gel electrophoresis, 115–118
  for plasmid DNA, 610, 611
  RNA, 428, 429

Alkaline phosphatase
  assay of (2'-5')-oligoadenylate synthetase, 151, 157, 158
  of 2'-phosphodiesterase, 161

Alkylphenyl-silica, high-performance liquid chromatography, 14

P

Pactamycin
  inhibitor of initiation of protein
    synthesis, 289
  of protein synthesis, 271, 289
Parainfluenza virus, preparation, 317
Paramyxovirus, preparation, 317
Particle, intramembraneous, changes in
  interferon-treated cells, 459
Parvovirus
  DNA, preparation, 313–315
  isolation, 313–315
  preparation, 313–315
Penicillium chrysogenum phage dsRNA,
  in (2′-5′)-oligoadenylate synthesis,
  188
Peptide
  amino acid analysis, 20–25, 25–27
  chain elongation, inhibition in
    interferon-treated cells, 145
  chain initiation, inhibition in
    interferon-treated cells, 142, 147
  high-performance liquid
    chromatography, 7–20
  maps, by high-performance liquid
    chromatography, 16–20
  microsequence analysis, 27–31
  picomole-level detection, 7–16
  reverse-phase chromatography, 14–20
  sequence analysis, 27–48
Percoll, density gradients for preparation
  of lymphocytes, 478, 482–484
Phenol
  in adenovirus DNA preparation, 313
  in fibroblast interferon messenger
    RNA preparation, 106
  in hamster interferon messenger
    RNA preparation, 106
  in influenza viral RNA preparation,
    318
  in mengovirus RNA preparation, 268,
    269
  in messenger RNA preparation, 52,
    53, 129, 308–310
  in mouse messenger RNA
    preparation, 93–96
  in plasmid preparation, 609
  in vesicular stomatitis virus RNA
    preparation, 316
  in viral RNA preparation, 428

Phenol-sodium dodecyl sulfate,
  Bunyavirus RNA preparation, 318
Phenylthiohydantoin amino acids,
  identification by HPLC, 29–31,
  41–48
Phosphatase
  binding to poly(I)·poly(C) paper, 197
  in phosphoprotein phosphatase assay,
    180–184
Phosphocellulose, in (2′-5′)-
  oligoadenylate synthetase
  purification, 158, 159
Phosphodiesterase
  degrading (2′-5′)-oligoadenylate, 139,
    147
  interferon induction, 149–161
    assay, 149–161
2′-Phosphodiesterase
  assay, 158–161
  in cells, 160
  reaction, 158–161
  requirements, 158–161
  unit, 161
Phosphoprotein phosphatase, inhibition
  by poly(I)·poly(C), 179–184
Phosphorylation
  assay, 173–174
  in interferon-treated cells, 168–178
  protein P_1, 141, 169, 174, 293
  protein, *see* Protein kinase
  ribosome-associated proteins,
    168–178, 301, 302
  small subunit of initiation factor eIF-
    2, 142, 169, 285
O-Phosphoserine, 176
Photomicrography, of fibroblasts,
  404–408
Phytohemagglutinin, in cytotoxicity
  assay, 490, 491
Phytohemagglutinin P
  immune interferon induction, 437
  in preparation of interferon action
    inhibitor, 441, 442
Pipette, for microinjection of oocytes, 70,
  126
Placental conditioned medium, for
  colony-stimulating activity, 522
Plaque-forming cell
  assay, 513–515
  inhibition by interferon, 506–517